# Thermodynamics

# Thermodynamics

**S. C. Gupta**
*Professor*
*Department of Mechanical Engineering*
*Engineering College*
*Kota, Rajasthan*
*India*

PEARSON

ISBN 978-81-317-1795-0

**First Impression, 2009**

Published by Dorling Kindersley (India) Pvt. Ltd., licensees of Pearson Education in South Asia.

Head Office: 7th Floor, Knowledge Boulevard, A-8(A), Sector-62, Noida – 201309, U.P, India.
Registered Office: 11 Community Centre, Panchsheel Park, New Delhi 110 017, India.

Laser typeset by Le Studio Graphique.

Digitally printed in India at Saurabh Printers Pvt. Ltd in the year 2014.

*To the next generation of engineers,*
*who through their skill and knowledge will leave an imprint*
*on the world and forever change it*

*To my family members*

# Preface

Thermodynamics relates the thermal interactions between a system and its surroundings with the changes in the system and predicts whether such interactions can take place or not.

This text is designed for the first course on thermodynamics offered to undergraduate students of mechanical engineering. The book first discusses the basic concepts of the subject and then supports the concepts with applications and solved numerical problems. This approach will help the students in developing an analytical mind. An engineer with an analytical mind and approach would be able to face any problems encountered in the field of work.

The examination pattern followed in most of the universities, shifts the interest of the students towards solving problems. Hence, the students do not give attention to the theoretical concepts. This is evident from competitive examinations where majority (around 90%) of the students are not able to answer the theoretical questions. Most often, they do not attempt to solve problems. Students expect the book to give all the steps so that they may read the derivations as well as the solutions of the numerical problems. I have been guided by these while presenting the material and hope that by using this book, the students will develop skills for deriving various relations and equations. A number of problems have been solved in this book with this objective.

Two approaches have been followed in the thermodynamic analysis—macroscopic and microscopic. This book covers both. Chapters 1 to 8 cover the macroscopic analysis and Chapter 9, the last chapter, deals with the development of the microscopic approach, which includes both kinetic theory and statistical thermodynamics. Chapter 1 covers some basic definitions and the concept about thermodynamic work, heat, system equilibrium, temperature, zeroth law and units and dimensions. Chapter 2 is devoted to conservation of energy (first law) and Chapter 3 deals with the second law of thermodynamics. Both the first and second law of thermodynamics along with the behaviour of a pure substance are discussed in Chapter 4. Chapter 5 is an extension of Chapter 4 dealing with mixture of non-reactive gases, gases and vapours, and solutions of two pure substances, called binary mixture. It includes applications of conservation of mass and energy to various psychrometric processes. Chapter 6 explores the determination of final composition when the constituents of the mixture react with each other, the amount of interaction and the loss of energy in the process.

I have been teaching thermodynamics for decades and have read a very large number of books. I feel indebted to all the authors who shaped my thoughts on the subject and whose work has formed the basis for this book. I also owe thanks to my students who gave feedback and inspired me to write the book.

I thank my parents, wife, children and their family for their whole-hearted supported in completing the project. I am also thankful to my colleagues, Mr. Praveen Bhandari for helping me in preparing the illustrations and Mr. Shiv Lal for selecting the problems and arranging books for reference.

Any constructive criticism for the improvement of the book will be appreciated. For any query or difficulty, you are most welcome to write to me at dr_scg@hotmail.com.

**S.C. Gupta**

# Contents

# CHAPTER ONE

# Basic Concepts



## 1.1 INTRODUCTION

With globalisation, the quality consciousness for consumer products has increased all over the world. Also, the choice for the same product is different for different people. The consumer is not ready to compromise with their liking. The new scenario has changed production techniques drastically. This put together has increased automation in the industry, resulting in more consumption of electricity for operating the machines. It is produced in power plants housed outside the cities from the available sources of energy in nature; we know energy cannot be created; it can only be changed from one form to another. Therefore, our hopes hinge only on nature, which has large amount of fossil and nuclear fuels, flowing water, wind power, solar energy, tidal wave energy, geothermal energy, etc.

Fossil and nuclear fuels release energy as heat during chemical or nuclear reactions. This heat is utilised in a steam turbine system to produce steam in boilers at high pressure and temperature. Passing steam through a converging passage, called nozzle, produces a high-speed steam jet. This passage is convergent, and the mass conservation principal demands that velocity should increase along the flow direction. Bernoulli's principal says that pressure would then decrease. So, a nozzle is used whenever a high-speed jet is required to be produced, with the pressure at its inlet being more than that at its outlet, that is, the pressure decreases and velocity increases in a nozzle. In a gas turbine system, we compress atmospheric air to increase pressure, burn some fossil fuel, normally natural gas, into it to increase temperature and then pass the gases through a nozzle to produce high velocity gas jet. We may also harvest the vast energy of flowing water by constructing a dam across the river to store water, converting that kinetic energy of flowing water into potential energy. Water is then transported through pipes or tunnels to nozzles (placed at a much lower level compared to the level of the free surface of water in the dam) to produce a high velocity water jet. Such a system is most suitable in hilly regions as rainfall is more, dams can easily be constructed on rivers and the nozzle can be kept at a much lower level than that of the storage tank.

So, we know now how to produce a high velocity jet of fluids like gas, steam or water from the naturally available sources of energy. These jets have linear kinetic energy and may be used to produce force by impinging them on plane or curved surfaces. The jet strikes the moving plate (surface) and changes its direction while moving over it. We know from Newton's laws of motion that because of inertia, a mass does not change its direction of motion, and to do so, a force should act on it. This implies that the plate exerts the force, and in return, receives reaction, which is equal in magnitude but in opposite direction, as per Newton's third law of motion. So, there is action (force on the plate) and movement of the plate (displacement). That is, mechanical power is produced in this process. This procedure of producing power has limitations, as it is not practically possible to move the plate like this any distance we wish, and then to also ensure that the jet would follow it. So, one has to think of developing a system where, while one plate goes out of the jet path, another comes in that position so that the show goes on.

Moreover, we have not yet produced electricity. We generate electricity by rotating a rotor in a magnetic field, and therefore, require a source of rotational and not linear power. We may think of curved plates fitted over the periphery of a drum and the fluid jet flowing through the space between two adjacent plates. The fluid changing its direction, while moving over the surface, generates a torque, which rotates the drum and generates rotational power. Such a machine is known as a turbine. There are other thermal systems also, like internal combustion engines, windmill, solar mill, etc., that produce rotational motion. The machine producing power is known as *prime mover.*

Technology has also been developed for harnessing energy from other sources of energy like wind, solar, ocean, or geothermal. A prime mover is one example of a thermal system. Power plants are recognised as thermal, gas or nuclear if the fuel used as source of energy is coal, gas or nuclear, respectively. A power plant based on use of water is known as hydel (power) plant.

Refrigeration and air conditioning is another area of importance, which falls under the category of thermal systems. Their importance can be understood from the fact that it would not have been possible to preserve perishable food and medicines for a long period of time, had these not been developed. The other thermal system of importance in industry is a compressor or a pump. We use fans in our day-to-day life and we know the hardship caused when electricity is not supplied during summer. We know the importance of pumps, had they not been there, how could water be supplied to our houses? We fill air in our scooter or car using air at high pressure. We know that this high pressure is developed by compressors. There are many such systems. We shall learn about them in detail as we proceed. This also brings out an important fact, that a thermal system does not mean a system in which there has to be heat transfer. We get this notion sometimes, because of the word thermal, which we in our subconscious mind attach to heat transfer. In fact, work transfer is also very much part of any *thermal system.*

In fact, all these systems have something in common—there is transfer of mass, heat and/or work between the device and its surroundings; here and/or means that both or one of these exchanges. We would be interested to know what are system and surrounding, how much exchange of mass and energy is taking place between them, how the system properties are changing, and what is the final state of the system and the surrounding after completion of a process. We may also be interested to find out what processes should be used for achieving a desired goal, and to know whether it will be possible or not?[1] All these questions are answered by studying the analytical procedures developed in a course on Thermodynamics. Thermodynamics, is a subject in which the effect of energy transfer to or from the device is studied when it is in thermodynamic equilibrium with the surroundings (we shall discuss this later). In this sense, the word thermodynamic is a misnomer, as thermostatic would have been a better choice to represent the static condition of the system. Whatever the arguments may be, the name Thermodynamics has been adopted for this science, which encompasses a very large area of application, since energy, mass (or both) transfer is bound to take place in any system subjected to a process.

In this chapter, we shall study the definition and concept of various terms used, thermodynamic equilibrium, reversible and partial irreversible processes, quasi-static process, zeroth law and the concept of temperature. The philosophy of temperature measurement and dimensions is also discussed.

### 1.1.1 Macroscopic and Microscopic Analysis

Two methods for analysis of system have been used. In one approach, the system is considered at its molecular level as in the kinetic theory of gases. That is, the system is assumed to be made up of molecules in incessant random motion with wide range of velocities. This kind of analytic approach is called the *microscopic approach,* and has been useful to predict pressure, temperature and internal energy of gases contained in a container in terms of the density of molecules (number of molecules per unit volume) and their

---

1. We know heat transfers from higher temperature to lower one, and if we think of a process in which it is other way, then it will not be possible.

velocity distribution, expressed by the root mean square velocity. These relations were obtained using the statistical tools for finding out the average values of the square of the velocity. An important result is obtained for mean free path, which is average of the free path, defined as distance travelled by a molecule between two successive collisions. This distance will vary, and would depend on proximity of the other molecules around it. The mean free path $\lambda$ is given by the following equation

$$\lambda = 1/(\sigma n)$$

where $\sigma$ is the collision cross-sectional area and is equal to $\pi d^2$, $d$ is diameter of the molecule and $n$ is the number of molecules in unit volume. In statistical mechanics, which is also part of microscopic analysis, we even go to the level of electrons and the radiations produced by their jump from one energy level to another, which has given very useful predictions for the energy radiated from a hot body and for its spectrum.

However, the analysis is very time consuming and in most of the applications, it is not required to go to the details of the individual molecule and then taking out the average. This is possible when the size of the system is much larger than the mean free path of the molecules. In such situations, what happens to the individual molecule is of no concern, and the only interest is to know the interaction between the system and its surroundings and the overall effect left on the system (as well as the surroundings). Such an approach is called the *macroscopic approach*. Thermodynamics deals with both kinds of approaches. However, in this book, we shall concentrate more on the macroscopic approach. It should be borne in mind, that we may follow any approach, but the results should be the same, as the results do not depend on the method of analysis used. So, a method which gives the correct result with the least labour input, will be best suited.

We mentioned earlier that the macroscopic approach is applicable only when the smallest size of the system is large enough compared to the mean free path. Under such circumstances, the matter may be considered as continuous. This is called the continuum concept, that is, the matter is uniformly and continuously distributed and not made up of pebble-like molecules followed with space.

We would like to study the behaviour of a system as its size or volume is reduced. Let $\delta V$ be the volume occupied by some mass $\delta m$ of the system at any time. We know that their ratio, $\delta m/\delta V$, is called the density, $\rho$. Figure 1.1 shows variation of $\rho$ as volume decreases. It is observed that when volume is $\delta V'$ or less, then density changes sharply; but the density approaches asymptotically to a value which is used in day-to-day life as density when volume of the system increases. This happens because the matter can no longer be considered as continuous when volume of the system has reached $\delta V'$ and the presence of molecules affects the results. This means that we may ignore molecules when volume is more than a certain volume, which is related to the mean free path. That is why the concept of macroscopic approach is applicable only when the smallest size of the system is more than the mean free path, and is thus closely linked with the continuum concept. This needs redefining of the pressure concept, which is based on the limit area tends to zero, as we study in Physics. If $\delta A'$ is the smallest area up to which the continuum concept holds true, then pressure is defined as

$$p = \lim \delta A \to \delta A'(F/\delta A)$$

where $F$ is the force acting on surface area $\delta A$. Note the difference lies in taking the limit. Here it is not zero, as then the continuum concept is not applicable.

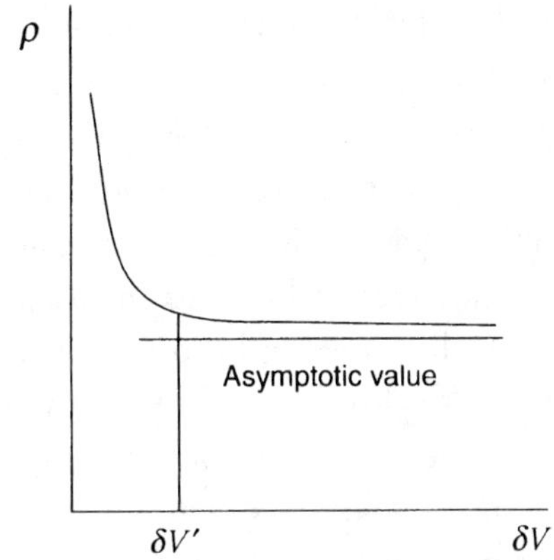


**Fig. 1.1** Variation of density with volume

## 1.2 TERMS USED IN ANALYSIS

As our discussions would be more general and not restricted to a particular thermal system, we would need to define a few terms, as done in the subsequent paragraphs.

### 1.2.1 System Boundary and Surroundings

A system is defined as a quantity of matter or a region in space chosen for study.[2] For example, when a fluid jet moves through a space formed by a curved surface, its direction can be changed only if some external force acts on it, as discussed above. How much force acts or who exerts this force on this jet, or whether there is any effect on the agent exerting the force on the jet are a few questions that have to be answered. That is, in this case, the flow of fluid jet flowing through the space formed by the curved surface is to be analysed. Therefore, we choose this space as a system whose boundaries are open to flow. Consider another example; say we are interested in widening a particular section of a road in a congested area of a city for which we have to count the rate of traffic flow. That is,

2. It need not always be a physical entity and may even be a thought like God. This depends on what is to be studied.

how many vehicles are crossing the road in an hour at different hours of the day? Here, a section of the road is of our interest. Then this portion of the road is a system. In both these examples, we notice that there is flow--fluid particle in the first case and vehicles in the second case. But we are not worried about which fluid particle or vehicle. Flow of mass across the system is what we are interested in. This is needed to demarcate the system from the rest of the world or universe and is done by drawing a boundary, called system boundary, around the system (Fig. 1.2). As a matter of fact, this boundary may be real or imaginary, rigid or flexible, stationary or moving; it depends upon the type of system.

Everything lying outside the system boundary that affects the system is called the surrounding. To understand the surrounding, consider a person (system) travelling in a train and a bus moving on the road running parallel to the railway track. This bus is not its surrounding as there is no interaction between the train and the bus. However, if the man is travelling in a car, then the same bus becomes his surroundings, as he has to save the car from any collision with the bus. Therefore, only that part of the universe, which has a relation with the system, but is not part of it, is its surrounding. No specific name is being given to the rest of the universe. A system interacts with its surroundings in the form of mass transfer, energy transfer or both. We shall study the various forms of interactions later.

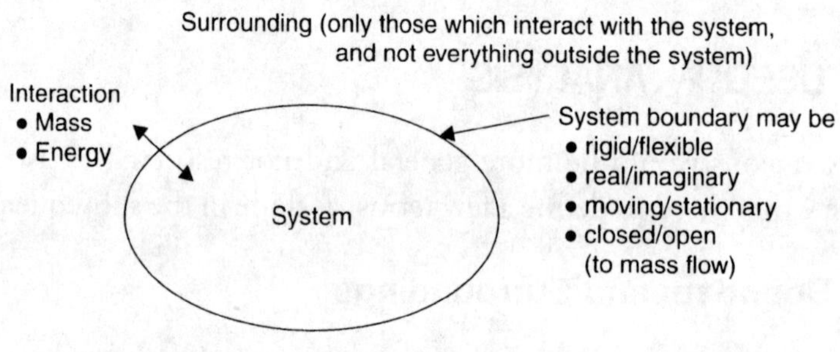


**Fig. 1.2** System, system boundary, and the surrounding

A system may be thought of like a black box in which we are not interested in knowing what is inside. Our interest is limited to the input to and output from the black box, that is, the type of interaction, and to analyse the performance of the black box. For example, let us consider turbines. There are hydraulic turbines, steam and other vapour turbines, and gas turbines. We would not go into the details of what type of turbine it is, as our interest is only to find out how much power is developed by it. We need such details only when a detailed study of the system is required. These various turbines can be represented by a rectangle or a circle or any other shape in which mass enters from one end and leaves from another, and in this process, work and/or heat interaction occurs with the surrounding.

We shall learn in Chapter 2 that flowing mass has energy in the form of kinetic, potential, internal and flow energies, and a change in one or many of them are responsible for turbine or compressor work. That is all we need to know. It does not matter whether the system is a compressor or a turbine and which type. In other way, a system approach leads to a macroscopic analysis.

### 1.2.2 Types of Systems

We have discussed systems in which there is mass flow across its boundaries; such systems are named Open System. Steam, gas or water turbines, centrifugal and axial flow compressors, heat exchangers and nozzles are examples of open thermal systems. A turbine is a device which converts kinetic, pressure and internal energy of the fluid at its entrance into mechanical work, and in this way, the energy of the fluid is decreased. On the contrary, pressure and velocity of the fluid increase in a compressor and pump to which work is supplied by an electric motor. A nozzle is a converging passage used to convert a pressure head into a kinetic head. A heat exchanger is another thermal system of much importance, as it is used to cool one fluid and heat another by passing one fluid through a tube and another over the tube. We shall discuss these components in detail later.

Nevertheless, there are also systems in which there is no mass flow across their boundaries, such as water contained in a bucket, gas confined within the cylinder and piston, car battery, etc. Such a system is called a Closed System. The name has been chosen, as there is no mass flow across the system boundary, or, the *system boundary is closed to mass flow*. However, if the bucket is placed below a running tap, then it is an open system. Similarly, a pressure cooker is a closed system till the steam does not come out. Let us consider another example of a not-connected-through train.[3] When the train is standing on a station passengers can board in or get down from a coach, and the coach is an open system. However, in a running train, it is not possible for the passengers to get in or get down, and the same coach is a closed system. Thus, the same system may be closed at one time and open at another time, and how to take it depends upon what we want to find out. In a nutshell, while classifying a system to be open or closed, one has to see conditions for that situation and observe for any mass flow across the system boundary.

While talking about a system, we develop a concept that a system will have some mass in it, or it may be crossing its boundary. But a system need not have any mass and may contain only energy as in electromagnetic field in vacuum, radiation field in a furnace, etc. Such systems are also in use. However, we shall not discuss these.

---

3. In these trains, passengers cannot move from one compartment to another within the train itself.

### 1.2.3 Description of a System

Everything in the universe, physical or abstract, is given some tag to identify it. For example, Ram is the name given to God by the Hindus, and Rahim to another God by Muslims; the moment we use either of these words, we recognise the religion, as these names carry their identity. Or, take a pen. It can be described by mentioning its make, trademark, and colour of ink, and so on. In other words, a few attributes are assigned to the system to describe it. We choose these parameters to describe the system, which can be easily described, measured with suitable instruments, or calculated directly using relations. Thus, for a gas in a container, pressure and temperature are the preferred parameters, since these can be measured by a pressure gauge and a thermometer. Volume may also be used for a perfect gas, as it can be determined by using algebraic relations. Besides pressure, temperature and volume, we also use many other parameters like entropy, internal energy, enthalpy, etc. to describe a system. These are determined by integrating differential equations discussed in the fourth chapter. These parameters are called the property of the system. Thus, a system is described by a set of its properties.

**State.** Consider a gas contained in a cylinder fitted with a piston, and that the piston is moving. In that case, the properties will go on changing with the changing position of the piston. However, the set of all the properties of this system will have a unique value for one position of the piston, and this set will change with different positions of the piston. So, we cannot specify the set of properties, unless we specify the position of the piston as well. We use the term 'state' to indicate the particular position of the piston in this case. In general, properties are defined only for a given state of the system, whereas a state is defined by a set of properties of the system, and this set has a unique value for that state only. Thus, state and property cannot be defined without each other.

It may be mentioned that change in a property of the system depends only on the initial and final state of the system, and in no way on how this change is brought about. Such functions are called point functions. Point functions have mathematical importance, since the differential operator, which we have studied in differential calculus, operates on such functions only. An equation $dz = Mdx + Ndy$, where $M$ and $N$ are functions of independent variables $x$ and $y$, can be integrated only because $x$, $y$ and $z$ are point functions; if not, integration cannot be done. Recall that before integrating any differential equation, you first check whether or not the condition $(\partial M/\partial y)_x = (\partial N/\partial x)_y$ is satisfied. Integration is permissible only if the condition is satisfied. This means that if we develop differential equations in various properties, these can be integrated to give the desired change in the dependent property, knowing the change in independent properties. Later we shall discuss about the use of different letters, namely, $d$, $\delta$, $\partial$ and $\Delta$ which are used in different contexts, and do not mean the same thing.

### 1.2.4 Equilibrium of a System

Again, consider the above example. When the piston is moving, then pressure or temperature, if measured at different points within the system, will be very likely different. But once the piston comes to rest, then after a few seconds, pressure, etc. shall be the same throughout the system, as redistribution of mass and energy within the system would end. Further, the piston moves outward because the system pressure is more than the surrounding pressure; the more the pressure difference, the more is the speed with which the piston moves, and more is the variation in pressure within the system. This means that the piston should move so slowly that pressure everywhere within the system remains the same. It is possible only if the pressure difference between the system and the surrounding is infinitesimal—had there been no pressure difference, no movement of the piston would have been possible. Such a condition is referred to as *mechanical equilibrium* of the system. Under this condition, system pressure is equal to surrounding pressure, and is the same within the system at its every point. Notice that pressure, like electric potential and temperature, is a potential, which is responsible for mechanical equilibrium. Similarly, a system is said to be in thermal equilibrium when the temperature of the system equals that of the surrounding, in chemical equilibrium when chemical potential of the system and the surrounding is the same and so on so forth. That is, for each potential, some equilibrium is associated, which requires its equality within the system and with the surrounding. When a system is in equilibrium for all possible potentials that are responsible for interaction between the system and its surrounding, then we say that the system is in *thermodynamic equilibrium.* That is, for a system in thermodynamic equilibrium, pressure, temperature, chemical potential, electrical potential, and so on will not only be same throughout the system, but shall have the same value as that of its surrounding. If even any one of them, say pressure, is not the same throughout the system or equal to that of the surrounding or both, the system is not in mechanical equilibrium and hence thermodynamic equilibrium .

Here, an example of a liquid contained in a test tube is worth quoting. Pressure at the free surface is atmospheric, and as we move down, pressure increases. That is, pressure within the system is not the same, though it is in thermodynamic equilibrium. We have learnt in school that this increase in pressure is due to the presence of earth's gravity field around the system. Had there been no such field, then pressure everywhere within the tube would be the same. Therefore, while determining the equilibrium of a system, it is equally important to consider the absence of external fields. We use the term *simple system* to refer to such a condition.

### 1.2.5 Intensive and Extensive Property

Have you ever heard that 1 kg of iron has a certain density? No, because density of any matter does not depend on its mass or size. The kinetic theory of gases tells us how to calculate pressure exerted by the gas on the walls of its container, and temperature of the gas in terms of motion of the molecules of the gas, and also that both do not depend on mass of the gas (Eqs. 1.1 and 1.2). In fact, all the thermodynamic potentials, like density, pressure, temperature, and electric potential, do not depend on size, extent or mass of the system. A system property that does not depend on system mass is called an *intensive property*. On the contrary, volume, energy, and entropy of the system depend on the extent or mass of the system, and are termed as *extensive property*. There is an easy way to find out the dependence of a given property on the mass of the system. Imagine a simple system to be in thermodynamic equilibrium and divided into a number of sub-systems by drawing lines across it so that the size of sub-systems formed is different from each other. Now, travel from one sub-system to another, and see if there is any change in the value of that property. If not, it is intensive, else it is extensive.

It has been the practice to express an extensive property for 1 kg of the system and the property is called *specific property* like specific volume, specific entropy, and is represented, by convention, using lower case letters $v$ and $s$, respectively.

### 1.2.6 Intrinsic and Extrinsic Property

While travelling in a train if someone asks a question "what is your speed", your answer to the question will be zero with respect to the train, and with respect to the ground the speed of the train. We get different answers because speed needs a reference frame to specify it. Similarly, there are many other properties which need a reference frame that can be chosen as per our convenience to specify their values. While specifying velocity or position of the system, we normally choose the reference frame attached to the ground, but for space study, the reference may be the sun or some other planet. We choose a reference point to specify enthalpy, entropy or internal energy of a pure substance, such that their values are never negative for the entire range of pressure and temperature. In such case, the property tables of different pure substances may have different reference datum, and has to be taken care of while analysing systems consisting of mixtures of such substances, as in combustion system.

A property such as entropy, enthalpy, velocity or elevation, whose value depends on the reference point, is called *extrinsic property*. On the other hand, pressure, volume, temperature or internal energy, of a system arises from its particular characteristic of mass within the system and is called *intrinsic property*.

### 1.2.7 Dependent and Independent Property

Thermal systems can be described using intensive properties such as pressure ($p$), specific volume ($v$), temperature ($T$), specific internal energy ($u$), specific enthalpy ($h$), and specific entropy ($s$). The question is, do we need all these properties to define the system? The answer to this question is a big No. We have studied equation of state for a gas in school education. We know that if any two intensive properties, out of the list given above, are known, then using algebraic or differential equations, we can determine all other properties of the gas. Pressure, temperature and specific volume are convenient properties, and therefore, any two of them are generally used to describe the gas system. These properties can be varied independently of each other to determine the effect of variation in them on the system and are called *independent property* and the rest are called the *dependent property*. So, a gas system has two independent properties, which normally are taken as $p$ and $T$, though $p$ and $v$, or $v$ and $T$ are also taken in the analysis. But for a real gas obeying Van der Waals equation, it may not be a right choice to choose $v$ and $T$ as independent variables, since this equation is implicit in $v$. The study of a system helps in choosing independent variables. A system consisting of gas is a specific case of a more general study made by Gibbs.

Gibbs considered a system made up of $\alpha$ number of distinct chemical species, each species present in all the phases numbering $\beta$ and in thermodynamic equilibrium with each other. It is assumed that all the species are present in all the phases. For a given phase of such a system, we need pressure, temperature and the mole fraction $(\alpha - 1)$ of $\alpha$ species in each phase to specify its state. Thus, each phase requires $(\alpha + 1)$ variables and $(\alpha + 1)\beta$ for system to specify these completely. However, thermodynamic equilibrium of the system requires pressure and temperature to be the same for all the phases yielding $2(\beta - 1)$ equations. Similarly, chemical equilibrium has to be the same in all the phases yielding $(\beta - 1)$ equations. This will hold true for all the species. Thus, chemical equilibrium consideration yields $\alpha(\beta - 1)$ equations. Thus, equilibrium consideration results in $\alpha(\beta - 1) + 2(\beta - 1) = (\alpha + 2)(\beta - 1)$ equations. As a result, the number of independent intensive properties,[4] $f$, needed to specify the system is $f = (\alpha+1)\beta - (\alpha+2)(\beta-1) = \alpha-\beta+2$. This rule is known as *Gibbs phase rule*. For example, wet steam, which is a mixture of liquid and gas, has one chemical component ($H_2O$) and two phases (liquid and gas). Gibbs phase rule says it shall require one intensive property for describing the system. Moreover, the ratio of masses of liquid and vapour in wet steam will give its composition. We shall discuss more about it in Chapter 4.

### 1.2.8 Interactions and Processes

We considered mass interaction between the system and the surrounding on the basis of which a system was classified as open or closed. The other interaction of importance is energy interaction, which takes place through a carrier of energy like heat or work. Here,

4. In mathematics, we learn that if there are $n$ equations in $m$ unknowns ($n < m$), then $(m - n)$ is the degree of freedom to choose that many number of variables arbitrarily and determine the rest in terms of these chosen variables. In the cited example, $f$ is the degree of freedom of the system.

we lay emphasis on the fact that heat and work are carriers of energy and not energy. Energy is a property and is possessed by the system, whereas heat and work are boundary phenomena. Change in energy of the system between its two states is the same, irrespective of how this change in state is brought out, but the amount of work and heat transfer will depend on how the change is made. These differences should be understood. In the next section, we shall discuss work interaction. It is sufficient to mention that heat transfer is an interaction between the system and the surrounding due to temperature difference between them and takes place by conduction, convection or radiation. A system is called *isolated* if there is no kind of its interaction with the surrounding, and *adiabatic* when there is no heat transfer between the system and the surrounding.

**Process.** Interaction between the system and the surrounding brings a change in the state of the system, and we say the system is undergoing or subjected to a process. In a process, one or more property of the system may remain constant. This is indicated by adding a prefix 'iso' before the property, like iso-temperature, iso-pressure, iso-entropy, and iso-volume. However, in practice, these processes are instead called isothermal, isobaric, isentropic and isometric processes, respectively. Adiabatic process is another process of interest and is defined as one in which there is no heat transfer between the system and the surrounding. In isentropic process too, there is no heat interaction, but it differs from an adiabatic process because it is a reversible process, whereas an adiabatic process is not.

**Reversible, Irreversible, and Quasi-static process.** In our discussions throughout the book, we shall assume that the system is in thermodynamic equilibrium during the process, so that its properties are well defined throughout its path and we may represent it on a two-dimensional plot by a solid line. *Such a process is a reversible process, because if the process is reversed from the final state to the initial state of the system following the same path, then no effect is left on the system and the surrounding at the conclusion of the reverse process.* That is, if first a process 12 is executed and then 21, then it will appear as if nothing has happened. The importance of a reversible process is demonstrated by taking an example of a lift in which, say 10 persons are standing on the platform, which is driven by gas pressure acting below it. Let $p$ be pressure of the gas when the lift is not moving and $A$ be platform area so that the platform is in equilibrium due to gas force equal to weight 10 $W$ of persons, where $W$ is the average weight of the persons in the lift. Let one of the passengers remove his tie and throw it out of the lift causing decrease in downward force. This results in movement of the lift, say by distance $dz$. Work done by gas is 10 W$dz$. However, if, say, 8 or all persons get down, then work done by the gas will be 2 W$dz$ or zero, respectively, for the same movement of platform. That is, maximum work is done by the gas on the surroundings only when the system (gas) was in thermodynamic equilibrium with the surrounding (persons in lift). The first process is classified as reversible and rest, irreversible processes.

However, it should be realised that the system is static in the state of thermodynamic equilibrium, and only under these conditions, minimum number of properties will be required to define the state of the system. Therefore, we assume that the system executes a process in such a manner that the potentials are the same throughout the system during the process. Such a process is called *quasi-reversible process* or *quasi-static process,* and occurs only when the potential difference between the system and the surrounding is infinitesimal. Moreover, the change in system state is the same for the same process. That is, the process is single-valued and monotonic. This means that it is not possible to have different processes between the same two states.

Further, when a system is subjected to a series of processes one after another, so that at the conclusion of the last process, it restores its initial state, then the system is said to have executed a *thermodynamic cycle.* There may be any number of processes in a cycle. If all the processes are reversible, the cycle is known as a *reversible cycle.* Carnot, Otto, Diesel, Sterling, Ericsson cycles are examples of reversible cycles. Unfortunately, students somehow develop a wrong concept that the Carnot cycle is the only reversible cycle. This may be the result of the fact that they are introduced to this cycle at the first opportunity, whereas the others are introduced at a much later stage. Even after reading the other cycles, students have another wrong notion that the Otto, Diesel cycles, etc., can be implemented in practice, but the Carnot cycle cannot, and that is another reason for such misconception. However, the fact is that the other power cycles are also reversible cycles and we shall take care of these points later in Chapter 8 while analysing power cycles.

These cycles are made up of four processes. Is it necessary to have four processes in a reversible cycle? Let us examine it. If there is only one process then system state is not restored at the end of process, being single-valued and monotonic. There cannot be two processes in a cycle either because then we know that no effect is left on the surrounding and the system, as discussed before. In a power or refrigeration cycle, there have to be two isothermals, which do not intersect. Therefore, it will not be possible to have a single process, which would yield a three-process cycle. That is why a cycle is made up of four processes. Mechanical considerations in case of a reciprocating system would demand an even number of processes in a cycle for its implementation.

We discussed a quasi-static process with reference to a closed system. However, the situation in an open system is different in the sense that there is significant change in the state from the inlet to the outlet. Therefore, the question that arises is whether there is thermodynamic equilibrium or not in such a system. Let us measure a few potentials at points at an infinitesimal distance from the system boundary on both sides, at the inlet and outlet sections. We shall notice that difference in any of the potentials will be infinitesimal at all such sections. Therefore, open systems also undergo quasi-static processes. We are not bothered how the potential changes from the inlet to the outlet, as the analysis considers only the end states. A flow is called a steady flow if the property at a given point does not change with time. Note that as we move from the inlet to the outlet section, the property changes, but also that the point of observation has shifted. The catch in the steady flow definition is the same point of observation.

### 1.2.9 Work Transfer

Let us consider a force $F$ acting on a mass, $m$, which is a system. The mass will start moving slowly with some acceleration given by Newton's second law of motion. If $d$ is displacement in time $t$, then the scalar product $F.d$ of the two vectors is the amount of work done by the force $F$ on the mass. The amount of work done on the system increases its kinetic energy by amount $Fd$. Now, consider the case of an alternator, an electric device that generates electric current. It is driven by a prime mover. The amount of current flowing through the stator of the generator produces a torque opposing the motion, and the motion would occur only when the applied torque is more than this resisting, or what we call, load torque. If the applied torque of the prime mover is more than the load torque, the speed increases, and becomes constant when the two are equal. This is a steady state. Under this situation, power supplied by the prime mover is the actual amount of work done on the generator, and if excess work is supplied, then it is used up in increasing speed only. Therefore, while determining power transmitted to a machine, we consider the resisting force, which is equal to the applied force in equilibrium.

However, the mechanical definition of work is not enough to cover other cases of work transfer, like flow of electricity through a wire. As a matter of fact, we associate electricity more with heat, as we use it for heating purpose in heaters, ovens, and microwaves, where heat is generated. On the other hand, electricity is also used for turning motors and we express that work is done. This difference in opinion is caused due to perception of our senses, and can be circumvented by extending the definition of work, as given by mechanics, beyond the application of force and displacement.

Thermodynamics, while encompassing definition of mechanics, defines work as an interaction with the surrounding, if its *sole effect* on the surrounding *could be reduced to raising of a weight* and is equal to the product of the weight of the mass raised and the distance moved by it. The phrase 'could be reduced' is a key point in the definition. Figure 1.3 shows a system 'battery' with two surroundings, that is, a heater in Fig. 1.3(a) and an electric motor in Fig. 1.3(b). In this example, a heater or a motor is only the surrounding, and in both the cases, the interaction between the system and the surrounding is flow of current through wires connecting the battery with them. Thus, the arrangement in Fig. 1.3(a) could be reduced to that in Fig. 1.3(b), in which the effect of interaction between the battery and its surrounding has been reduced to raising of a weight, and therefore, electricity flow across the system boundary is an example of work and not heat transfer. Another key word in the definition of work is sole effect, which means that raising of weight is the only effect on the surrounding and nothing else. For example, consider a hot body whose only interaction with the surrounding can be to produce steam and run a turbine to raise a weight. But in this process, some heat is also rejected to the surrounding when steam after expansion in the turbine is left into the atmosphere. Thus, work is produced by the system, but that is not the only or sole effect on the surrounding, and therefore, this interaction does not qualify as the definition of work; it is heat transfer, since interaction is due to temperature difference.

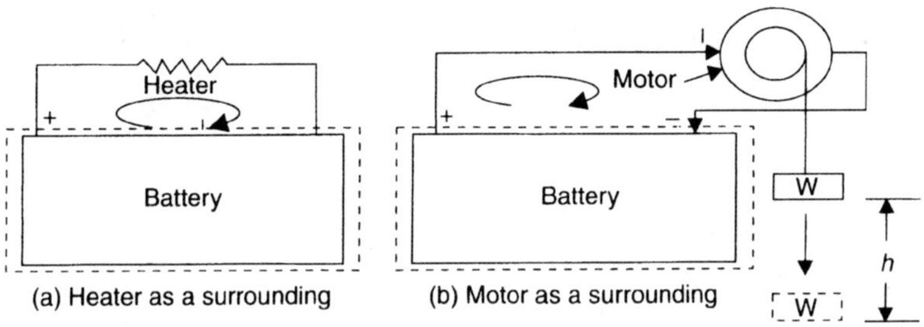


**Fig. 1.3** A system with two surroundings

### *1.2.9.1 Sign Convention*

Sign convention is needed for both heat and work transfer to indicate whether it is from or to the system. Many conventions were used in the past, but now, work transfer from the system to the surrounding and heat transfer from the surrounding to the system are considered positive; the other direction naturally is negative. This may be remembered easily since work is what we want to get from the system. Heat is required to be supplied to the system in order to get work, and therefore, this direction of heat transfer is positive.

### *1.2.9.2 Boundary, Displacement, or pdv Work*

When a system is subjected to a process, then its boundary may move. This movement of the boundary occurs against the surrounding pressure, and therefore, there is work transfer between the two. This work transfer, due to boundary displacement, is called *boundary* or *displacement work*. Therefore, displacement work is due to external forces acting over the system surface, and can be evaluated by dividing the surface into finite number of surfaces and summing over the work exchange at each surface in the limit that the number of surfaces tends to infinity. This way we ensure that pressure on each surface element $dS$ may be taken constant. Consider the system shown by full lines in Fig. 1.4 before the process, and by the dotted line after it. The system is subjected to a reversible process, which means the system pressure during the process was equal to the surrounding pressure $p$ acting over the surface area $dS$ of the boundary, producing infinitesimal displacement $dl$. Therefore, work done by the system on the surrounding through its surface $dS$ is = $(pdS)\,dl = pdv$, where $dv = dSdl$ is change in volume of the system at its surface $dS$. Net work done by the system can be obtained by visualising the whole surface to be made up of surfaces like $dS$ and adding work so obtained. Thus, boundary work is given by the $\int pdv$ over the system surface, and for this reason, it is also named *pdv* work.

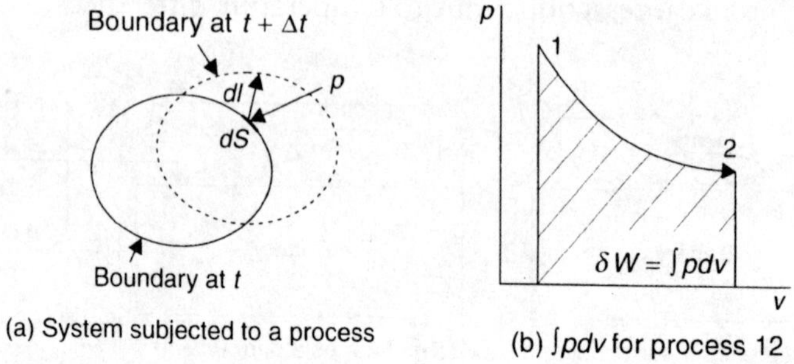


**Fig. 1.4** *pdv* work

Evaluation of the boundary work requires knowledge of variation of $p$ with $v$, that is, the path of the process, and hence work is a path function. We use the symbol $\delta$ to express differential amount for path functions like work and heat transfer during a process. In contrast, the symbol $d$ is used for exact differential of properties that are point functions. Thus,

$$\delta w = \int_1^2 pdv$$

where integration is from state 1 to 2, as shown in Fig. 1.4(b). This is the area formed by the curve with the $v$ axis and two verticals passing through state points 1 and 2 on the $pv$ diagram.

A cycle, as discussed before, is normally made up of four reversible processes executed one after another, so that at the end of these processes, the system restores the initial state; say these processes are 12, 23, 34 and 41, taken in the direction as shown on $pv$ diagram in Fig. 1.5. It can be seen that the area under the curve 12 formed by two verticals through points 1 and 2, the curve 12 and the $v$ axis, is the work transferred from the system to the surrounding, as work is given by $\int pdv$, and it is positive. Thus, work is done on the surrounding during processes 12 and 23, and by the surrounding on the system during processes 34 and 41. Therefore, the area of the cycle 1234, when executed in the clockwise direction, is the net work output received by the surrounding during the execution of the cycle. That is, the cycle develops work, and therefore, it is a power cycle. Such a system executing a thermodynamic cycle, is named as cyclic device, and the one that produces power, an engine.

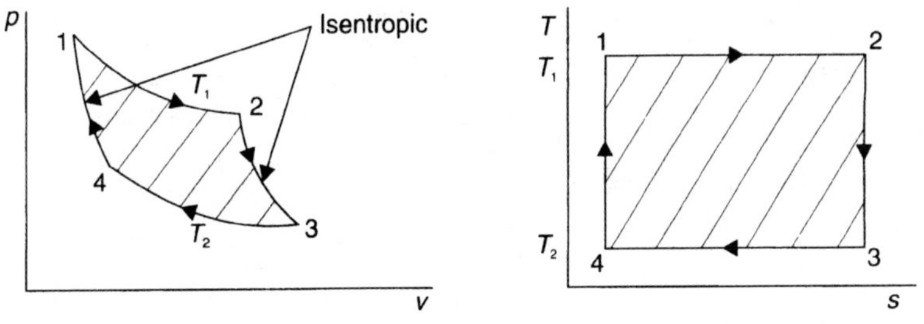


**Fig. 1.5** Carnot cycle on $p$–$v$ and $T$–$s$ diagrams

Since the cycles are reversible, it is also possible that the event of processes may be executed in the anti-clockwise direction, that is, in the order 14, 43, 32 and 21. In this case, working fluid system (i) does work on surroundings during 14 and 43 and (ii) receives work in process 32 and 21; sum effect is to receive work (area of the cycle) from the surrounding. System receives heat during 43 at low temperature $T_2$ and rejects heat to sink at higher temperature $T_1$. Thus, the same cycle, when executed in anti-clockwise direction, produces a cooling effect by supplying work to the system. Such a cycle is known as refrigeration cycle. The cyclic device is called to as a refrigerator.

Another point to be kept in mind is that air is always present around all the thermal systems, though it may not be part of the surrounding. For example, consider an internal combustion engine. It is a reciprocating device consisting of a piston and cylinder arrangement in which the piston reciprocates. Linear motion of the piston is changed to rotary motion by connecting the piston to the crankshaft with the help of a connecting rod and crank, as shown in Fig. 1.6. This mechanism is known as the crank mechanism. Components of this mechanism are shown in Fig. 1.6. The gas inside the cylinder is the system and the piston and crank mechanism are its surrounding. However, air is also present around the piston. When the gas exerts pressure on the piston, then the air surrounding the piston also opposes motion of the piston and some useful work is lost in this process, which is beyond our control. We should take care of this while determining work exchange between the system and the surrounding, since air is not forming the surrounding. Lastly, some work is lost in overcoming friction at different surfaces of contact. Thus, the work $W$ produced during expansion of gas in the piston is equal to the sum of the work done on the atmosphere, overcoming friction and transmitted to the other machine through the crank mechanism. We shall later study how to take these into account.

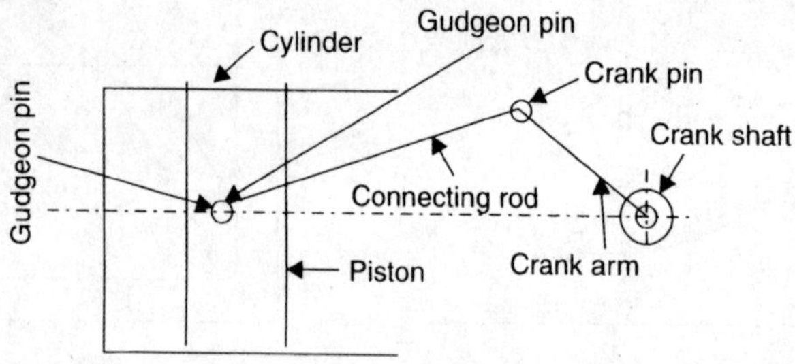


**Fig. 1.6** The crank mechanism

### *1.2.9.3 Other Forms of Work*

In physics, various forms of work and how to determine their magnitude is taught. We would summarise those relations here for completeness. Displacement work was shown to be *pdv*, which is a product of force and linear displacement. This can be extended for other forms of energy by taking appropriate force and corresponding generalised displacement. For example, the generalised force and displacement are torque and angular displacement in shaft work, voltage (electrical potential) and electric charge in electrical work, magnetic field strength and magnetic dipole moment in magnetic work, electric field intensity and polarisation of the medium in electrical polarisation work, and so on. If we represent a generalised force by $Y$ and generalised displacement by $X$, then work transfer is given by

$$W = \int YdX$$

The unit of the work exchange is watt, $W$. However, it is too small a unit and in engineering, kW is used; conversion factor is 1 KW = $10^3$ W.

1. Electrical, magnetic, and electrical polarisation work

   If $I$ amp current flows through an electrical element such as armature of the electric motor because of a potential $E$ across the ends of the element, then electric charge $dq = Idt$ flows through the element in time $dt$. Energy transferred as work in time $t$ is given by

   $$W = \int_0^t EIdt$$

   where $I$ is the time rate of electrical charge and is measured in ampere, the SI unit of current strength.

2. Shaft work

   When a machine is connected to an electric motor, then work is transferred to it through the shaft coupling the two. If the shaft rotates at $N$ revolutions per minute (abbreviated rpm) while transmitting torque $T$, then

$$\text{Power transmitted} = T\omega$$
$$= 2\pi NT/60$$

where $\omega$ represents angular speed of rotation, rad/s.

3. Stirring or paddle work

   A stirrer churns fluid inside the container. It requires some work to be done on its shaft. Work done in stirring is transferred to the fluid. To calculate the amount of work transfer, we may assume a pulley attached at the end of the stirrer with a rope around it, at whose free end a weight $W$ is attached (Fig. 1.7). If $h$ is the height through which the weight descends, then the decrease in potential energy of the weight is equal to the amount of work transfer. Thus, we may write the equation

$$\text{Work transfer} = Wh$$

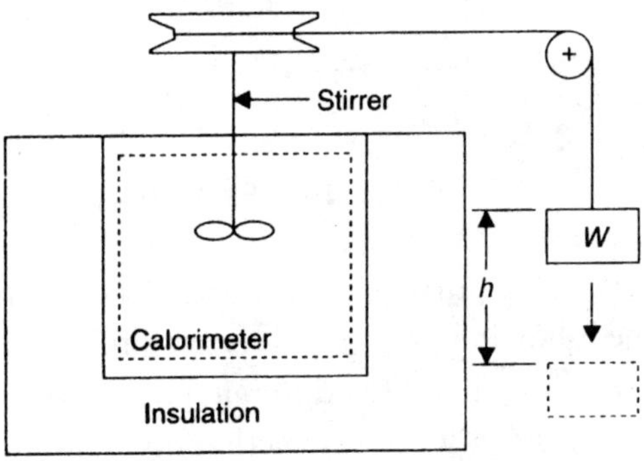


**Fig. 1.7** Paddle work

4. Spring work

   A spring exerts force when it is stretched or compressed; the force exerted by it increases with more and more displacement. If $x$ is change in length of the spring, then force exerted by it is $kx$, where $k$ is the spring constant. Work done in stretching it further by distance $dx$ is then given by $kxdx$, and the total work done in changing its stretch from $x_1$ to $x_2$ is obtained by integrating this expression for work. It is

$$W = \tfrac{1}{2}\, k\left(x_2^2 - x_1^2\right)$$

5. Flow work

   This occurs in an open system, and represents the energy imparted to the flow mass by a pump, blower or a fan to make it flow across the system boundary. This is a kind of displacement work, and may be obtained by assuming an imaginary piston (Fig. 1.8) pushing the fluid element through a distance $dl$ in the direction of flow causing this mass = $\rho A dl$ to enter into the open system, shown by the dotted boundary. Let $p$ be pressure on the piston surface during this movement, then work done over the fluid by piston to push 1 kg of fluid is

$$pA\,dl/(\rho A\,dl) = p/\rho = pv$$

where, $A$ is the cross-sectional area of flow, and $v$ is the specific volume $(= 1/\rho)$ of the fluid at the entrance condition of the open system. That is, work $pv$ is done by the pumping agency on the fluid to make it flow and is stored in the fluid. It is called *flow work* or *flow energy*, because the work has ultimately been stored as energy.

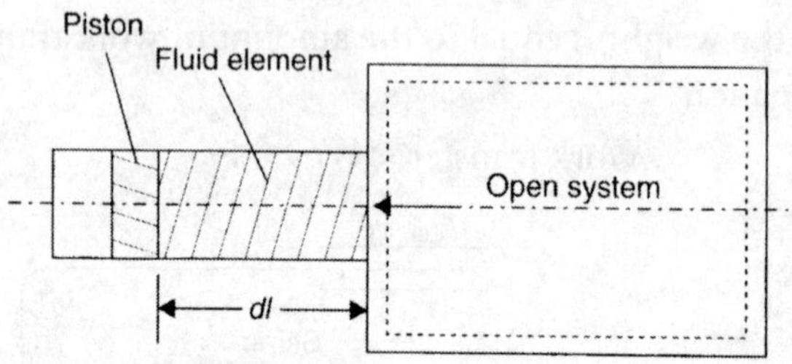


**Fig. 1.8** Flow work

We shall now demonstrate these concepts and the importance of system definition through the following examples. It is to emphasise here that a system may have different work interactions at different points on its surface at the same time. If $W_i$ is the amount of work interaction associated with each of these, where $i$ represents different interactions, then the net work transfer is, $W = \sum W_i$, summed over all $i$, while taking into consideration their algebraic sign.

**Example 1.1** A train consists of an engine coupled with a number of coaches. The engine pulls the coaches and the train moves. If we consider the engine as a system, then the coaches are its surroundings. Work is transferred to the surroundings, and so is positive by sign convention. On the contrary, if coaches are taken as the system then the engine is the surrounding and work transfer is to the system, and hence negative. If, however, the system consists of both engine and coaches, then there is no boundary phenomenon. Whatever is happening is within the system, and hence work transfer is zero. So, our answer depends very much upon the choice of system. Note that there is no physical change in any of the situations described above.

**Example 1.2** Figure 1.9(a) shows an electric oven in which electricity flows through the heating element. The oven is on and the wires are live. It is insulated for human safety and heat loss reduction. Heat produced in the heater is passed on to the oven space to cook the food placed therein. We have shown in the figure two system boundaries, namely, *a* encompassing only the oven space, and another *b*, which includes heating elements as well. The system inside the boundaries is a closed system in both the cases, as there is no mass flow across the boundary. The only interaction is energy transfer. For system *a*, energy transfer is due to temperature difference between the heating element and the oven space, and hence it is heat transfer; by sign convention, it is positive. However, for system *b* interaction between the system and the surrounding is through a live wire carrying electric current, and we know that this is work transfer from the surrounding to the system and hence negative. Physically, the phenomenon is the same, but our answers differ for the two systems.

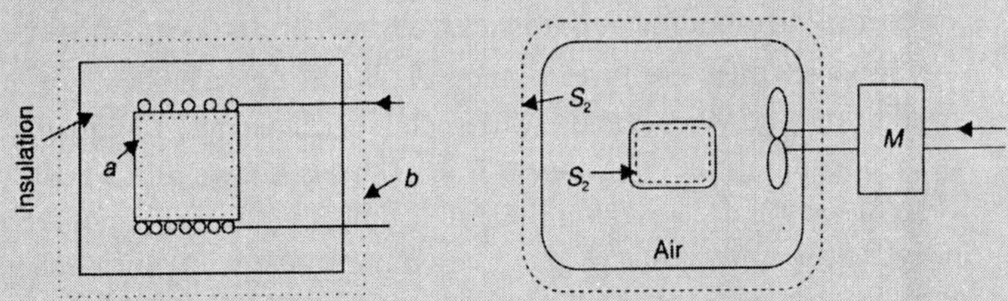


(a) Electric oven

(b) Closed duct fitted with a fan

**Fig. 1.9** Work/heat transfer in different closed systems

Figure 1.9(b) shows another arrangement in which an electric motor is placed outside the duct fitted with a fan that circulates air inside it when the motor is on. This is a typical system, as there are two unconnected boundaries $S_1$ and $S_2$ used for defining the system, which is air in the duct. This highlights an important fact that more than one boundary may be used to define a system, and these need not be connected with each other. In this case, energy transfer is through the shaft connecting the motor to the fan, and therefore, work is transferred to the system when the motor is running. The moment electricity is switched off, the shaft is still rotating because of its inertia, and work continues to transfer to the system till the fan rotation stops completely. Now, consider that the motor is also placed inside the duct and is forming part of the system. When power is on, there is work transfer again to the system, but it is now electric work and not shaft work. Moreover, the moment power is switched off, interaction stops and there is no more work transfer, though the motor shaft is still rotating.

The above examples clearly bring to focus the importance of defining a system. While making analysis of any thermal system, the first step therefore is to clearly define the system and then to see whether it is closed or open.

## 1.3 ZEROTH LAW

A system is said to be in thermal equilibrium, as discussed before, when temperature throughout the system is the same, and is equal to that of its surrounding. We may take the surrounding as another system in contact with the given system, and then both the system and the surrounding may be taken as isolated systems. Or, vice versa is also possible—that is, one of the two systems in contact may be taken as the system and another its surrounding; it is immaterial which one is taken as the system.

Suppose, there are three systems $A$, $B$ and $C$, such that $A$ and $B$ are in thermal equilibrium and so are $B$ and $C$ with each other. Then question is, will $A$ and $C$ also be in thermal equilibrium? The answer is yes. This is known as the zeroth law, because this observation was declared after the establishment of the first law of thermodynamics. The *zeroth law says that if two systems A and B are in thermal equilibrium and so are systems B and C, then A and C would also be in thermal equilibrium.* It is not an unexpected answer, since equivalence of temperature between the two systems is the crux of thermal equilibrium.

This can be established by considering variation of the variables pressure and volume of the systems in thermal equilibrium, which change in both (systems) till the equilibrium is established between them. We may write a functional relation among the variables of the two system $A$ and $B$ or $B$ and $C$ (in thermal equilibrium) in the following form:

$$p_B = f_{AB}(p_A, v_A, v_B) \text{ or}$$
$$p_B = f_{BC}(p_C, v_C, v_B) \tag{1.1}$$

Here, subscripts $A$ and $B$ refer to the corresponding systems and $AB$ to the systems in thermal contact. This relation says that pressure in system $B$, when systems $A$ and $B$ are in contact, will depend on volumes of $A$ and $B$ and pressure of $A$. Similarly, we may write for pressure in system $B$, when systems $B$ and $C$ are in thermal equilibrium. Equation 1.1 implies:

$$f_{AB}(p_A, v_A, v_B) = f_{BC}(p_C, v_C, v_B) \tag{1.2}$$

When we consider equilibrium of systems $A$ and $C$ and write such functional relation as in Eq. 1.1, then the functional relation will not contain the term $v_B$. This suggests that the functional relation should be of the following form

$$f_{AB} = \phi(v_B)\theta_1(p_A, v_A) + \eta(v_B)$$

$$f_{BC} = \phi(v_B)\theta_3(p_C, v_C) + \eta(v_B)$$

since equating the two functions to satisfy Eq. 1.2, we get

$$\theta_1(p_A, v_A) = \theta_3(p_C, v_C)$$

which does not contain the term $v_B$. Also, substituting for function in Eq. 1.1, we get

$$p_B = \phi(v_B)\theta_3(p_C, v_C) + \eta(v_B)$$

$$\rightarrow \theta_3(p_C, v_C)$$

$$= p_B/\phi(v_B) - \eta(v_B)/\phi(v_B)$$

$$= \theta_2(p_B, v_B),$$

since the right hand side is a function of $p_B$ and $v_B$. This implies that it is possible to find out a function of the type $\theta(p, v)$ for all the systems in thermal equilibrium, such that it is the same for all these systems. Thus, this function $\theta$ represents some property of the system, which remains the same in all the systems in thermal equilibrium. The property is given the name temperature.

Thus, the zeroth law implies that all the systems in thermal equilibrium will have the same temperature, irrespective of their size. That is, temperature is an intensive property of the system. We get the same result from the kinetic theory of gases, where we study motion of the molecules and determine the average energy of the molecules. The average kinetic energy of the molecules, which are in random motion, is

$$\tfrac{1}{2}\, m\, C_{rms}^2 = 3/2\ kT \text{ or}$$

$$T = 1/3\ (m/k)\ C_{rms}^2 \tag{1.3}$$

where, $m$, $k$ and $C_{rms}$ are mass of the molecule, Boltzman constant having the value $1.38 \times 10^{-23}$ J/K and root mean square velocity of the molecules, respectively.

### 1.3.1 Temperature Measurement

Accurate measurement of temperature is of importance both in laboratories and industries. An instrument used for measurement of temperature is called the thermometer, which works on the principle of zeroth law. For this reason, the thermometer is brought in contact with the system, whose temperature is to be measured, for enough time, so that both come in thermal equilibrium. Let us assume that we are using a gas thermometer and that the system is large enough, so that contact of the thermometer with it does not affect the system properties significantly while attaining thermal equilibrium. As discussed in Section 1.3, both pressure and volume of the working substance used in a thermometer would change. However, we may fix one of these variables and measure change in the other variable. For example, volume is fixed in a constant volume gas thermometer and pressure is measured. Such property used for measurement of temperature is called *thermometric property*. Pressure would depend on the temperature of the system; more is the temperature more is the pressure. Thus, by calibrating the thermometer, it is possible to find out the actual temperature of the system. Calibration is a process in which thermometer pressure is recorded at known temperatures. These values are then plotted on a graph, which would look like Fig. 1.10 for a constant volume ideal gas thermometer. Here, we see a number of straight lines because for a perfect gas, we know that $p = \rho RT$, which gives $p \propto T$, where the constant of proportionality depends on the product of density and gas contants, and hence is different for different gases. Once the graph is available, then we read the actual value of the thermometric property from the thermometer when it has reached in thermal equilibrium with the system, and from graph, the temperature is read. Normally, a scale is attached to the thermometer for convenience and the temperature is read directly from it, as we do in a mercury thermometer. Calibrating the thermometer time to time gives more accurate results, because the performance of the thermometer changes with time. In such a case, the graph is to be used. The measurement using a constant volume thermometer is tedious and the thermometer is bulky. It gives accurate results that can be reproduced again and again. For these reasons, the gas thermometer is used in laboratory for accurate temperature measurement.

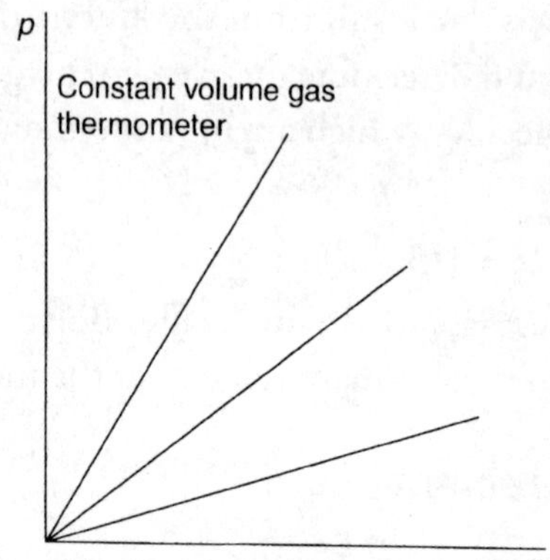


**Fig. 1.10** Calibration process

We may take advantage of the linearity of the curve by just measuring pressure at two known standard temperatures, and then divide the range linearly into equal number of divisions. Normally, ice and the boiling temperature of pure water (free from salt contamination) at sea level pressure is used for the purpose, as these two temperatures are known accurately.[5] This way, a scale is developed, which is very handy in measurement, just like the length scale, and is attached to the thermometer itself. Such a scale is called empirical scale, as the fixed points have been assumed arbitrarily and does not imply that it means the same temperature as used in thermodynamic relations. The old temperature scales, namely Celsius[6] and Fahrenheit, use two fixed-point temperatures, as just discussed. However, it is also possible to use one fixed-point temperature, since a straight line between the dependent variable $y$ and independent variable $x$ can be written as

$$y = ax \text{ or } y = ax + b$$

where $a$ and $b$ are constants to be determined. Lord Kelvin developed a scale in which he used triple point temperature of water as the fixed point, which is at 273.16 K. Example 1.3. explains the procedure of developing the scale and relating it to other temperature scales. The graph shown in Fig. 1.10 is a simplified model, as actually, the behaviour of gases at high pressure deviates from a linear relation. In such a case, we may choose a polynomial of degree $n$ or even a logarithmic scale, as was used by Lord Kelvin initially and dropped later because temperature measured on this scale did not satisfy the thermodynamic relations in use at that time. However, the linear temperature scale developed by him satisfied all the relations and matched very well with those given by a gas thermometer.

In the above discussions, we took pressure variation as a measure of temperature. In fact, we may take any other property as a thermometric property. Various thermodynamic properties used in thermometers are (i) increase in length, such as in mercury or other liquid thermometers; (ii) change in electric resistance, as in a platinum resistance thermometer used normally in laboratories; (iii) change in Seeback voltage as in thermocouples, which are very popular in industrial temperature measurement; (iv) change in volume at constant pressure, as in a gas thermometer; and (v) change in frequency of radiations, as in a radiation thermometer. Variation of these thermometric properties with temperature is linear only in a certain range of temperature, beyond which it is non-linear. The linear scale used in thermometers therefore limits their use for this temperature

5. These are not the only temperatures that are used for calibration. Other temperatures such as boiling point of sulphur, carbon tetrachloride, etc. and melting point of lead are also used.
6. Celsius fixed ice and steam temperatures at 0 and 100°C, respectively, and Fahrenheit at 32 and 212°F. They divided the range linearly in 100 and 180 divisions. These scales are known as Celsius and Fahrenheit temperature scales, which are in use even today. Prior to this scale, Celsius fixed ice and steam temperatures at 100 and 0, but this scale was dropped by him later. In fact, all these numbers assigned to different temperature levels were arbitrary and many more scales were used. But with time, only those scales survived, which could yield consistent results when used with thermodynamic relations. Ultimately, this led to the thermodynamic temperature scale, given by Lord Kelvin and discussed in Chapter 3. (Temperature on this scale is called Kelvin and ice point on this scale corresponds to 273.15 K.)

range only. This temperature range is different for different thermometers. This explains why different thermometers are used in practice in different ranges. Taking $y = t_3 = 273.16$ K at the triple point of water where $x$ is $x_3$, we get $y = 273.16\, x/x_3$

Table 1.1 lists some of the thermometers using different thermometric substances, the property $x$ and $y$ for that thermometer, relation used for the scale, thermometric fluid used and the range of application.

**Table 1.1** Types of thermometers based on different thermometric properties used

| Type of thermometer | x | Relation | Thermometric substance | Range (°C) | Resolution |
|---|---|---|---|---|---|
| Liquid-in-glass | $L$ | $t = 273.16\ L/L_3$ | Mercury, alcohol, organic liquids, etc. | –180 to 50 | 0.5–6.0 |
| Resistance | Bridge balance | $t = 273.16\ R/R_3$ | Pt, thermistor, constantan, carbon, etc. | –273 to 1000 | 0.001–10 |
| Thermocouple | $V$ | $t = 273.16\ V/V_3$ | Copper, constantan, iron, etc. | –200 to 500 | |
| Gas | $p$ | $t = 273.16\ L/L_3$ | Helium, hydrogen, nitrogen, etc. | | |
| Radiation | $J$ | | | > 1000 | |

*Note*: $J$ is the intensity of radiation falling on the thermometer, which are compared with those of a standard temperature source. Radiation thermometers are used for measurement of high temperatures only, while a gas thermometer is used only in laboratories for accurate measurement.

However, we should not forget that variation of thermodynamic property with temperature for any thermometer, is close to, but not exactly linear in the range for which these are used, and depend very much on the material used. This means that if we use two thermometers of the same type, say resistance thermometer, made from different material for measuring the same temperature level, their readings will not be exactly the same, except at the temperature at which these were calibrated. Reproducibility is another important parameter in measurement. It means that if we measure a variable $n$ times, then we must get the same result. But because of various kinds of errors involved in measurement, it seldom happens. Thus, accurate measurement of temperature is a very challenging job.

We have discussed a constant volume gas thermometer before. Instead of volume, we may keep pressure, or for that matter, even $pv$ constant. The reason was that we wanted to develop a thermometer, which gives readings independent of the mass and property of the thermometric substance used in it. Historically, the symbol $\theta$ has been used to represent temperature on the gas thermometer. A thermometer is made of a bulb B (Fig. 1.11) having a capillary attached at its upper end. *U* tube manometer is connected to the other capillary end to measure pressure in the bulb *B* for which the bottle is raised, so that the capillary end is just touching the level of the manometric fluid used. The mass of the gas filled in the bulb is so small that pressure in the bulb is close to zero, since then, any gas would behave like a perfect gas. This means that thermometers having different gases filled in the bulb will give the same reading, irrespective of the temperature level being measured. The following equation defines the gas thermometer scale, obtained using the relation of the type $\theta = a(pv)$

$$\theta = \lim.(p_b \to 0)\left[273.16 \times \left\{(pv)_b / (pv)_{tp}\right\}\right]$$

where, subscripts *b* and *tp* refer to the state inside the bulb and the triple point of water, respectively.

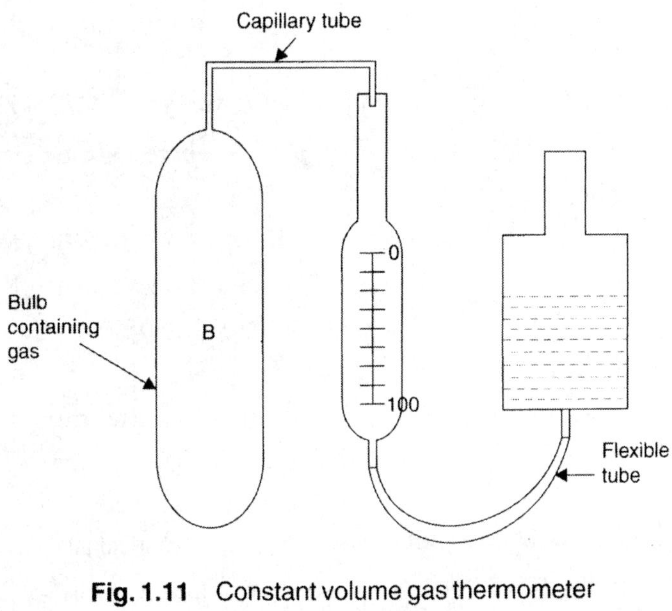


**Fig. 1.11** Constant volume gas thermometer

**Example 1.3** It is proposed to develop a temperature scale with values 4 and 16 assigned to ice and steam points, respectively. The pressure of an ideal gas at constant volume is to be used as the thermometric property. (a) Find a linear relationship between the pressure and the temperature (represented as *S*) on the new scale. (b) What is the Kelvin absolute zero on this scale? (c) Find the relation between *S* and *K*, where *K* is used for temperature measured on the thermodynamic temperature scale. (d) If the pressure at the steam point were 10 atm, what is the pressure at 0 *S*, and what is the temperature when pressure is 1 atm?

***Solution*** (a) We shall assume linear relation of the type $t = ap + b$, since the thermodynamic property is pressure (given). Using two fixed points, we get $a$ and $b$. This results in the following equations for the new scale

$$t = (16 - 4)(p - p_i)/(p_s - p_i) + 4 \quad \text{(A)}$$

Here, $t$ is measured in *S*. Similarly, we may obtain the Celsius scale, which is

$$t_c = (100 - 0)\ (p - p_i)/(p_s - p_i)$$

where, $t_c$ is the temperature on the Celsius scale in °C. From these two equations, we get

$$(p - p_i)/(p_s - p_i) = (t - 4)/\ 12 = t_c/100 \quad \text{(B)}$$

This is the relation for converting temperature on the two scales.

(b) At absolute zero, $t_c = -273$°C. Substituting it in relation *B*, we get $t = -28.8\ S$

(c) Writing $t_c = T - 273$ in relation *B*, we get $T = 273 + 25\ (t - 4)/3$

(d) For a perfect gas, $p \propto T$ or $p = kT$, where k is the constant of proportionality.

Given: $p = 10$ atm at $T = 373$ K, we get $k = 10/373$. Therefore,

$$\begin{aligned} p &= (10/273) \times [273 + 25\ (t - 4)/3] \\ &= 7.32\ (0.879 + 0.0305\ t) \quad \text{(C)} \end{aligned}$$

When $t = 0\ S$, $p = 7.32 \times 0.879 = 6.43$ atm

Also, when $p = 1$ atm, then from Eq. C, after simplification,

$$t = -24.3\ S$$

**Example 1.4** Show that the ice point temperature $Ti$ on an ideal gas temperature scale is given by $Ti = 100/(r_s - 1)$, where $t_s = \lim(p_b \to 0)(p_s/p_i)$ at constant volume, $p_s$ and $p_i$ are the pressure in the bulb at steam and ice point, respectively. If a fractional error of 1% is made in the measurement of $r_s$, then show that the error in $Ti$ will be approximately 3.73%, and that the fractional error in $T$ is $dT/T = dr/r + 3.73\ dr_s/\ r_s$, where $T = r\ Ti$.

***Solution*** We assume the relation between temperature and pressure in the constant volume bulb of the thermometer of the type $T = ap$. Let $p_s$ and $p_i$ be the pressure measured by it at steam and ice point temperatures, respectively. Then

$$Ts = ap_s,\ Ti = ap_i$$

$$\rightarrow\ a = (Ts - Ti)/(p_s - p_i) \text{ and } T = 100\, p/(p_s - p_i)$$

since $(Ts - Ti)$ is 100 on the Celsius scale. It may be noted that the absolute scale and the Celsius scale differ by a constant of $273.15 \approx 273$, and hence there is no difference in the two. Substituting $p = p_i$, we get

$$Ti = 100/[(p_s/p_i) - 1] = 100/(r_s - 1) \qquad \text{Q.E.D.}$$

Taking its logarithmic differentiation, we get

$$dTi/Ti = -d\,r_s/(r_s - 1)$$

$$\rightarrow \text{ percentage error in } Ti = (dTi/Ti) \times 100$$

$$= -[r_s/(r_s - 1)] \times [(dr_s/r_s) \times 100]$$

But $\quad r_s = Ts/Ti = (Ti + 100)/Ti$

$$= 1 + 100/Ti,\ Ti = 273.15 \text{ K and}$$

$$[(dr_s/r_s) \times 100] = 1 \text{ (given)}$$

$$\therefore \text{ percentage error in } Ti = -3.73\%$$

Temperature $T$ is $rTi$. This gives

$$dT/T = dr/r - d\,r_s/(r_s - 1) = dr/r + 3.76\, d\,r/r \qquad \text{Q.E.D.}$$

## 1.4 UNITS AND DIMENSIONS

A careful study of the various properties reveals that all these properties are expressible in terms of the few quantities called fundamental dimensions. The property itself is called secondary or derived dimension. For example, the dimension velocity is the rate of change of displacement. Here, displacement is the length dimension and time is another dimension. The dimension pressure is the ratio of force and area; the latter is the product of two lengths, and the former may be taken as another dimension (or we shall see that it maybe expressed in terms of some other dimensions). Thus, in the said examples, length and time are taken as fundamental dimensions and the velocity or pressure as the secondary dimensions. However, it is still our choice to take force as the fundamental or secondary dimension. That is, we decide some dimensions as fundamental dimensions, such that all the other dimensions may be expressed in terms of these chosen fundamental dimensions from their definition and using some mathematical relation. Experience has taught us that mass, length and time can be taken as these dimensions, and are known as *fundamental dimensions*. These are written as $M$, $L$ and $T$ respectively, and their size has been fixed after long discussions over time by the International Scientific Society, which has the onus to set standards. These are measured in kg, metre and s, respectively. These are known as S.I. units, though other units, notably FPS (the name comes from the units pound, ft and s used for mass, length and time), CGS (cm, g and s) are also in use. In these units, kg, g, m, cm and s are the abbreviations for kilogram, gram, metre, centimetre and second, respectively, adopted by the International Society of Weights and Measures.

Any dimension is expressed in terms of adopted fundamental dimensions using mathematical relations like volume = (length)$^3$. It is a simple relation but maybe complex algebraic or even differential relation. Today, we have three fundamental dimensions in Mechanics. Depending upon the area of application, some other variables have also been fixed as fundamental dimensions. For example, current strength I in a circuit has been taken as a fundamental dimension in Electric Engineering or temperature $\theta$ in Thermodynamics. However, it is not necessary that there have to be only three fundamental dimensions. In fact, force had also been used as a fundamental dimension in the MKS system of dimensions and units in Mechanics. The nomenclature MLT or MLT$\theta$, or FMLT is used to indicate the fundamental dimension used.

In a four-dimensional system FMLT, force is also taken as a fundamental dimension, in addition to mass, length and time. This system has been used by engineers for a long time, and is known as the MKS system, in which mass, length and time were measured in metre, kg and s, respectively. In this system, force is defined as that which would act on a mass of 1 kg placed in the earth's field, where $g = 9.80665$ m/s$^2$, and was called 1 kg; the symbol $kg_f$ has been used to differentiate it from 1 kg mass. Force $F$ acting on a mass $m$ placed in the earth's gravity $g$ is defined by Newton's second law of Mechanics, which is written in mathematical form as follows

$$F \propto mg \quad or \quad F = (1/g_c)\, mg \tag{1.4}$$

where, $(1/g_c)$ is a constant of proportionality. It can be seen that it has dimensions [$F^{-1}MLT^{-2}$], since $g$ has dimensions [$LT^{-2}$], the dimensions of acceleration. It may be viewed as a conversion factor between the units of force and mass. Equation 1.4 then demands that the numerical value of $g_c$ should be that of $g$ to satisfy it, and that this value will depend on the system of unit used. This system was a bit confusing, as the same unit kg was used as unit for mass and force. However, in the SI system of units, which is now adopted universally as the system of units to be used everywhere in the world, force is not taken as a fundamental dimension. The unit of force is Newton (N), and is defined as that force acting on 1 kg mass that would accelerate it by 1 m/s$^2$. Equation 1.4 is automatically satisfied if we choose $g_c$ as 1, having no dimension. Thus, the MLT system is more coherent and is known as a coherent or consistent system, whereas the FMLT system is a non-coherent type.

It is also possible to have less number of fundamental dimensions. We shall not go into more details. A few problems solved at the end of book dwell more on this aspect.

Another interesting point with dimensions is that an equation is dimensionally homogeneous, meaning that all the terms have the same dimensions. This fact has been used in many useful ways. For example, we may check whether the relation is correct, at least dimensionally. If not, we may check for the error. Also, if we divide one term of the equation by another, then the resultant will be a dimensionless quantity. This fact has been vastly used in decreasing the number of unknowns in similitude study and model development of the machine, so that we are sure about its performance after its manufacture. The simulation technique has proved to be a boon. Nowadays, computer simulation is replacing the conventional laboratory testing and modelling techniques. But it is quite a complex process, and laboratory testing has its own advantages.

A word related to dimension, but not the same, is unit. We may understand it from unit of length metre, which is assigned to length of a bar, kept in a museum in France between two fixed points on it at a fixed temperature. So, kg and second have been fixed. These have some standard reference preserved at France. That is, all the fundamental dimensions have been defined at some standard conditions and given some name to them. These are called the SI units. However, in practice, other units are also used to meet the requirement, and these differ from the SI unit by a factor called the conversion factor. For measurement of small distances we use cm or mm or even some other units. The idea is to keep the number associated with the size of the dimension reasonable. We would prefer 1 km instead of 1000 m for length and so on.

Each dimension has a unit, which depends on the system of unit used. Different system of units such as FPS, CGS, MKS and SI have been in use for a long time and are still in use. However, SI system of units is most commonly used. Table 1.2 gives units of fundamental dimensions in these system of units.

**Table 1.2** Units for fundamental dimensions in different system of units

| Dimension | FPS | CGS | MKS | SI |
|---|---|---|---|---|
| Length | ft | cm | m | m |
| Mass | pound | g | kg | kg |
| Time | s | s | s | s |
| Force | – | – | $kg_f$ | – |
| Current | amp | amp | amp | amp |

Units for other quantities are obtained from the dimensional relation of these quantities with the fundamental dimension. We have learnt some of such units in other courses and shall learn as we proceed. Besides these units, we also use secondary units for practical reasons. For example, length can be measured in mm, cm, m, km, and so on, though the unit for it in SI system is m, and all the relations relating different dimensions need these units only. Therefore, whenever we make a substitution in any relation, we have to ensure that we are using the appropriate unit, and not the unit in which the value is given. For example, pressure has to be in N/m$^2$ in any equation, whereas it is measured in bar, kPa or Mpa. So, when substituting pressure, say in equation of state or any other such equation, then its value in N/m$^2$ is to be obtained using a conversion factor. Consider equation of state for an ideal gas, $PV = mRT$ which you have read in school. Here, $p$ is in N/m$^2$, $V$ in m$^3$, $m$ in kg, $R$ in J/kg K and $T$ in K. But pressure normally is given in bar

(1 bar = $10^5$ N/m$^2$) and $R$ is given in kJ/kg k (1 kJ = 1000 J). After substitution using the conversion factor, we are left with a factor of 100 on LHS, which is a conversion factor for the set of units used above for this equation; this has been used frequently in solutions.

Regarding temperature, we may use the relations

$$K = °C + 273 \quad C/5 = (F - 32)/9$$

Here, $K$, $C$ and $F$ refer to values in Kelvin, Celsius and Fahrenheit scales, respectively.

## SUMMARY

For thermal analysis of any system, the first step is to define clearly the system and whether it is a closed, open or isolated system. We established through examples that the answer depends on the choice of the system. Once the type of system is known, then we can decide whether to use non-flow, steady-flow, or general-flow energy equation (discussed in Chapter 2).

A system may interact with its surrounding through mass and energy interactions. In thermodynamics, work transfer $\delta w$ and heat transfer $\delta q$ are important interactions.

$\delta w$ and $\delta q$ are forms of the energy interactions of the surrounding with the system, and are taken as positive when work is done by the system and heat is supplied to the system. Both are boundary phenomena and the amount of energy transfer depends upon the process. That is, both heat and work transfer, unlike property, are path functions. That is, an infinitesimal amount of heat transfer cannot be represented as ~~dq~~; it is written as $\delta q$. Similarly, a finite change in a process 12 is written as $w_{12}$ or $q_{12}$, and not as $\Delta w$ or $\Delta q$. Whereas heat transfer takes place due to temperature difference between the system and the surrounding, work transfer is equal to the product of some kind of generalised force and generalised displacement associated with it. The generalised force and displacement are: torque and angular displacement in shaft work, voltage (electrical potential) and electric charge in electrical work, magnetic field strength and magnetic dipole moment in magnetic work, electric field intensity and polarisation of the medium in electrical polarisation work, pressure and volume in displacement work, etc. This way, various energy interactions could be brought under the umbrella of work transfer (as the force and displacement is not seen by our eyes in many of these work forms), while retaining the mechanics concept of work.

A system is specified by a set of variables that identifies the system. These variables have unique value for a given system when it is in thermodynamic equilibrium with the surrounding; the set of variables define state of the system. The system has different values of variables in its different states though one or more such variables maybe same for many states. These variables used to define system are called property which are point functions meaning that change in their value does not depend on the manner system state is changed.

We consider the system to be subjected to a quasi-static process, as we can then safely assume that property within the system is the same everywhere, and the system state is then well defined. Also, in such a situation, surroundings potentials will also be equal to that of the system. Thus, we consider the system to be almost in thermal equilibrium when subjected to a quasi-static process, and can be shown by firm lines on any two or three-dimensional diagram.

A quasi-static process is considered a reversible process if executed in reverse direction, and leaves no effect on the system and the surrounding. This is possible when (a) potential difference between the system and the surrounding causing interaction is infinitesimal, and (b) there is complete absence of dissipating forces such as friction, eddy formation, magnetic or electrical hysteresis, etc. in the system. These are also referred to as external and internal irreversibility, respectively. When a system is subjected to a reversible process, maximum work is produced. If a closed system is subjected successively to different reversible processes, so that at the end of the last process it restores the initial state, then the system is said to have executed a thermodynamic cycle. Similarly, the fluid may move from one unit to another in succession, and return back to the first unit. Thus the fluid has undergone a cyclic change. The whole cycle consisting of different units can be taken as a closed system.

# EXERCISES

## Review Questions

1.1 What is the concept of continuum? Define pressure and volume in this light.

1.2 Under what circumstances does the concept of continuum become invalid?

1.3 What is thermodynamic equilibrium? Discuss its various aspects. Explain the significance of a quasi-static process.

1.4 How do you find out whether a given quantity is a property or not? Are mass, heat and work transfer properties of a system? Make a list of some of the properties used in thermodynamic analysis and categorise them as intensive or extensive and intrinsic or extrinsic.

1.5 A can of soft drink at room temperature is put into a refrigerator so that it will cool. Would you model the can of soft drink as a closed system or as an open system? Explain.

1.6 Explain the concept of thermodynamic work and discuss its consistency with the definition of work read in physics course.

1.7 Can we define lowering of a weight as negative work? Give reasons.

1.8 Does a work transfer between a system and surrounding always mean change in volume of the system? Justify your answer with reason and examples.

1.9 Do temperature and heat transfer represent the same thing? Explain with examples.

1.10 Does increase in temperature of a system always require heat transfer? Justify with examples.

1.11 What is a temperature scale? How is it established? What specifications are necessary for a temperature scale?

1.12 Are we justified in using a linear scale for temperature measurement as is commonly used? When will we choose a linear equation with one or two unknowns?

1.13 What are the units used for mass, length and time in FPS, CGS, MKS and SI system of units?

1.14 What is the difference in $g$, the acceleration due to gravity, and the constant $g_c$, used in non-coherent system of units?

1.15 What does MFLT stand for in dimensions? Is it of a coherent type?

1.16 One bar is equivalent to $10^6$ dyne/cm$^2$. Find the conversion factor between bar and psi, where psi is the unit of pressure in FPS system and dyne in CGS system.

1.17 Classify each of the following systems into an open or closed system:

(a) air compressor (b) carburettor (c) pressure cooker (d) radiator of an automobile (e) tea kettle

1.18 What do you understand by intensive and extensive properties? Classify the following properties with their units in the metric system:

(a) specific weight (b) volume (c) specific volume (d) density (e) molecular weight (f) pressure (g) temperature (h) velocity

1.19 Roughly one-third of the energy liberated in a car is rejected to the air by the radiator through the circulating water. Should the radiator be analysed as a closed system or as an open system? Explain.

1.20 State and prove the Gibbs phase rule. How is it used?

1.21 What is a property? What are the methods of classifying properties? Give examples.

1.22 What is a quasi-static process? What are the common quasi-static processes? Explain how to execute a quasi-static expansion from pressure $p_1$ to pressure $p_2$.

1.23 A gas is compressed in a piston cylinder arrangement resulting in increase in temperature. Is it work or heat transfer? Explain.

1.24 A room heater heats air. Is this a heat or work interaction, if the system is (a) a heater, (b) air in a room, (c) heater and room air, and (d) the entire room, including the heater?

1.25 The air in a room is heated by solar radiation coming in through the window. Is it heat or work transfer?

1.26 A candle heats an insulated room. Is it a heat or work interaction? Take the entire room, including the candle, as the system.

1.27 A potato, initially at room temperature, is being baked in an oven that is maintained at 200°C. Is there any heat or work transfer?

1.28 Consider the process of heating water on top of an electric range. What are the forms of energy transformation involved during this heating process?

1.29 Consider two identical rooms, one with a refrigerator and the other without it. If all the windows and rooms are closed, will the room containing the refrigerator be cooler or warmer than the other one?

1.30 What are the different mechanisms for transferring energy to a control volume?

1.31 An experimentalist claims to raise the temperature of water to 150°C by transferring heat to it from steam at 120°C. Examine his claim.

1.32 Show how the ideal gas temperature scale is developed. State its connection with the thermodynamic scale.

1.33 What are primary fixed-points? What are their values in Fahrenheit and Celsius scales?

1.34 Compare and contrast the characteristic of control mass and control volume.

1.35 Name three types each of reversible and irreversible processes, and state the criterion in terms of the second law of thermodynamics to recognise an irreversible process. A system is taken from state 1 to 2 by (a) reversible, and (b) irreversible process. Is change in entropy of the system same or different in the two processes?

## Problems for Practice

1.1 An insulated copper vessel contains a quantity of oil at a temperature of 15°C. A steel bar of mass 0.1 kg at a temperature of 800°C is suddenly dropped into the oil. When the temperatures have equalised, the temperature of the oil is 16.5°C. Determine the magnitude and sign of heat transfer for each of the following systems: (i) the vessel, (ii) the steel bar, and (iii) the oil.

1.2 A 1 kg mass is weighed on a beam balance at a location where $g = 9$ m/s². What is the reading on the balance? If the same mass is weighed on a spring scale originally calibrated in a region where $g = 9.81$ m/s², what is the reading on the spring scale?

1.3 The temperature $t$ on a thermometric scale is defined in terms of a property by the relation, $t = a\ln p + b$, where $a$ and $b$ are constants. The temperature at ice point and the steam point are assigned the numbers 3 and 212, respectively. Experiment gives values of $p$ as 1.86 and 6.81 at these two temperatures. Evaluate the temperature corresponding to $p = 2.5$ on the thermometer.

1.4 A thermometer is calibrated at 0 degree mark with boiling tetrachloride and at 100 degree mark with melting lead. When used to measure temperature of a system, the device indicates 62 degrees. Convert this reading to the equivalent reading on the centigrade scale. Carbon tetrachloride boils at 76.75°C under mercury pressure of 760 mm and lead melts at 327.4°C.

1.5 A Reamur thermometer scale is graduated so that it reads ice point and steam point as 0 and 80°, respectively. If this scale is used in conjunction with a gas thermometer operated at constant volume, determine (a) the absolute zero of this scale, and (b) the absolute Reamur temperature at the ice and steam point.

1.6 The equation $Rt = R_0(1 + \alpha_0 t)$ is applicable to a resistance thermometer, where $R_t$ and $R_0$ are the resistance at temperature 0 and $t$°C, and $\alpha_0$ is the thermal coefficient of resistance at 0°C. The thermometer is calibrated by immersion in boiling water (100°C) and boiling sulphur (445°C). The instrument indicates resistance at these two temperatures to be 14.7 and 29.2 ohms, respectively. Determine temperature of a fluid for which the corresponding thermometer resistance is 23.5 ohms.

## Multiple-choice Questions

1.1 Which statement is correct?

(a) An isolated system not influenced by surroundings is called the universe.
(b) The system and surrounding combined constitute the universe, whether these interact with each other or not.
(c) A system which interacts with the surrounding is part of the universe.
(d) The system and surrounding put together form the universe only if there is interaction between them.

1.2 A centrifugal fan forms a/an

(a) closed system. (b) open system. (c) isolated system. (d) none of the above,

1.3 For constant volume process on a gas, the value of index $n$ for the path followed on $p - v$ coordinates is equal to

(a) 1. (b) zero. (c) $\infty$. (d) 1.4.

1.4 A gas expands following the law $pV^n = C$ from initial pressure $P_1$ and volume $V_1$ to final volume 0.5 $V_1$. The work obtained is maximum when $n$ is

(a) 1.4. (b) 1. (c) zero. (d) $\infty$.

1.5 Air expands from initial pressure $P_1$ and volume $V_1$ to final volume $0.5V_1$ following the law $P\,V^n = C$. Here

(a) greater the value of $n$, greater the work obtained.

(b) smaller the value of $n$, smaller the work obtained.

(c) for $n = 0$, the work obtained is the greatest.

(d) for $n = 1.4$, the work obtained is the greatest.

1.6 Which of the following can be used as properties of a system?

(a) $\int(dT/T - v\,dp/T)$. (b) $\int(dT/T - p\,dv/T)$. (c) $\int pdv$. (d) $\int vdp$.

1.7 Zeroth law of thermodynamics defines

(a) internal energy. (b) enthalpy. (c) temperature. (d) pressure.

1.8 1 mm of $H_2O$ is equal to

(a) $100 \times 10^{-6}$ bar. (b) 0.001 kgf/cm$^2$. (c) 9.80665 N/m$^2$. (d) 0.077 mmHg.

1.9 –40°C is equal to

(a) –40°F. (b) 230 K. (c) 400 R. (d) –72 °F.

CHAPTER TWO

# Laws of Thermodynamics—I



## 2.1 INTRODUCTION

We learnt about system and processes in Chapter 1. The questions that come to our mind are (i) whether or not the system can execute the given process, (ii) if yes, then what type and how much interaction is taking place between the system and the surrounding, and lastly, (iii) what effects are left on the system and the surrounding after completion of the process. The questions are answered by thermal analysis of the system, which is based on the second and the first law of thermodynamics, and knowledge of the behaviour of working substances used in the system. Both of these laws are based on our observations for millennia, and the only proof is that these have not been proved wrong so far. Thus, a complete analysis of a thermal system is carried out in steps using the second law of thermodynamics to ascertain the existence of a process, and the first law to find out the amount of interaction of the system with the surrounding and the final state of the system after the completion of a process. This analysis is called a *thermal analysis.*

Section 2.2 deals with the first law for a closed and an open system, which are subjected to a process or a cyclic change and in the process, a new property energy of the system is introduced.

## 2.2 FIRST LAW OF THERMODYNAMICS

This law is nothing but the law of conservation of energy. Joule gave a mathematical foundation to it by carrying out an experiment on water (system) contained in an insulated calorimeter. Water was agitated by a stirrer (connected to a pulley, as shown in Fig. 2.1) by lowering weight $W$ through a known distance $h$. In this process, amount of work is equal to decrease in potential energy of lowered weight = $W \times h$ is transferred to the system. He measured the increase in water temperature to calculate the amount of heat to be removed from water in order to bring it back to the original temperature, which can be implemented by removing the insulation over the calorimeter. We notice here that the system under consideration is a closed system, which is subjected to a cycle consisting of only two processes; in the first, there is work transfer and in the second, heat transfer only. He observed from his calculations that there is proportionality between the amount of heat and work transfer in the cycle. That is,

$$\oint \delta q \propto \oint \delta w \quad \text{or} \quad \oint \delta w = J \oint \delta q \tag{2.1}$$

where, $J$ is a constant of proportionality and is known as Mechanical Equivalent of Heat, because this equation, titled Joule's law, sets a relation between units used for measurement of work and heat. In the SI system of units, both heat and work are measured in Joule ($J$), and therefore, $J$ is a pure constant having numerical value of unity. In practice, however, kJ is used as the unit, which is equal to $10^3$ J, and therefore, care has to be taken while making substitutions. For example, potential energy is given by $mgh$, where $m$ is the mass in kg, $g$ is the acceleration due to gravity in m/s$^2$, and $h$ is the elevation in $m$. With these units for $m$, $g$ and $h$, unit for potential energy is N-m or $J$, and not kJ. Similarly, we may see that kinetic energy will have units of $J$.

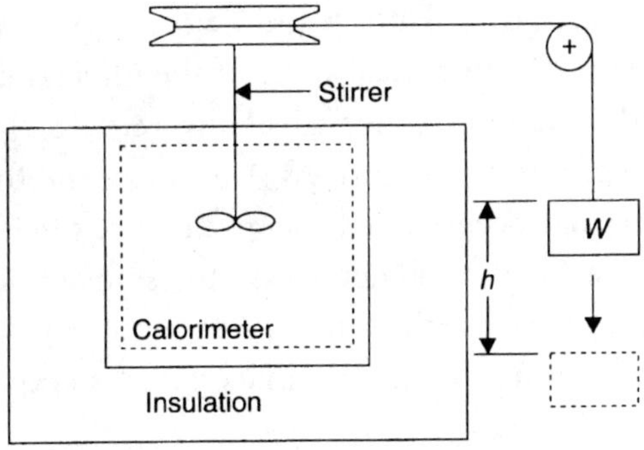


**Fig. 2.1** Joule's experimental set-up

## 2.3 NON-FLOW ENERGY EQUATION

Application of Joule's law to a closed system subjected to a process gives a very interesting result. Points 1 and 2 represent state of a closed system in Fig. 2.2, on say, p–v diagram, before and after the system has been subjected to a process A. We may restore the system state from 2 to 1 by subjecting it to arbitrary processes *B* or *C*. Applying Joule's law to cycles 1A2B1 or 1A2C1 for 1 kg of system, we get

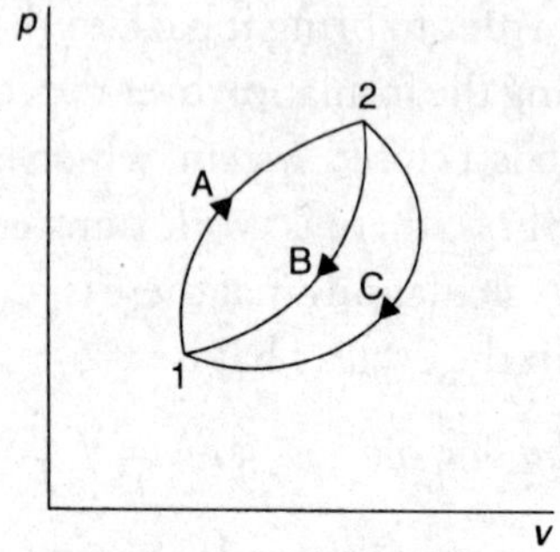


**Fig. 2.2** Different processes

$$\int_{1A}^{2}\delta q + \int_{2B}^{1}\delta q = \int_{1A}^{2}\delta w + \int_{2B}^{1}\delta w \quad \text{and} \quad \int_{1A}^{2}\delta q + \int_{2C}^{1}\delta q = \int_{1A}^{2}\delta w + \int_{2C}^{1}\delta w \qquad (2.2a)$$

$$\text{or} \qquad \int_{2B}^{1}(\delta q - \delta w) = \int_{2C}^{1}(\delta q - \delta w) \qquad (2.2b)$$

after subtracting the second equation from first in Eq. 2.2(a) and rearranging the terms in the manner given in Eq. 2.2(b).[1] Paths B and C are arbitrary, and therefore, we conclude from this equation that the numerical value of the integral depends only on the initial and final states 1 and 2 and not on the path. Thus, $(\delta q - \delta w)$ represents change in some property of the system during the process A between the final and initial state of the system. This property has been named *energy,* because the change is due to transfer of heat and work, which are carriers of energy as discussed before. The unit used for energy is Joule, written as *J*, and we use the symbol *E* to represent it. Since the amount of heat and work transfer to bring a given change in its state depends on the mass of the system,

1. We could have arranged the terms as $\delta w - \delta q$ as well in the integral of Eq. 2.3.(b). Following the sign conventions for heat and work transfer (given in Chapter 1), this means that the energy of the system will increase if the system delivers heat or work to the surrounding; that is not true as per our perception, because in both the cases, the energy of the system will decrease.

energy is also a function of mass, and is therefore an extensive property; $e$ then represents specific energy, by adopted convention. We may now write $de$ for $(\delta q - \delta w)$, or $dE$ for $(\delta Q - \delta W)$. Thus,

$$\delta Q - \delta W = dE \quad \text{or} \quad \delta q - \delta w = de \tag{2.3}$$

gives change in energy of the closed system for given heat and work interactions, and is known as the non-flow energy equation, written in abbreviated form as nfee. We observe that though both $\delta q$ and $\delta w$ are path functions, their difference is a point function.

A system possesses energy because of its position with respect to its surroundings, like kinetic or potential energy, or due to motion of molecules (from which the substance is made) within the system. Energy associated with molecules[2] in the system is called internal energy, represented by the symbol $U$ or $u$ for specific internal energy and is measured in kJ (not $J$, like kinetic or potential energy) for most of the substances; Joule, however, is the SI unit. Normally, the magnitude of change in internal energy is much large compared with those in kinetic and potential energies for most of the thermal systems, and hence $dE \cong dU$, and we may write nfee as

$$\delta Q - \delta W = dU \tag{2.4}$$

In Eq. 2.3, $\delta Q$ and $\delta W$ are the heat and work transfer, respectively, during an infinitesimal part of the process 12. That is, the process 12 is divided into a number of processes from 1 to 2. The question is whether the system is undergoing reversible or an irreversible process, and in what way it shall affect the equation. If the process 12 is reversible, then $dE$ will represent the change in internal energy only, and it takes the form of Eq. 2.4. However, if the process is irreversible, then; (i) energy of segments of the system will be different, which is not an easy job to determine; and (ii) the energy will not merely be internal energy, but there may be other forms of energy too, such as kinetic and potential energies. Thus, in that case, Eq. 2.4 will not be applicable. However, if we consider the initial and final state of the system subjected to the process 12, then the system will only have internal energy, which is well defined, as the system is then in equilibrium. For the process, Eq. 2.4 is written in its integral form as

$$Q - W = \Delta U \tag{2.4a}$$

**Example 2.1** When a system is taken from state $a$ to state $b$, as shown in Fig. 2.3, along the path $acb$, 20 kJ of heat flows into the system and the system does

2. A system posseses kinetic energy because of translation and rotation of its molecules, and potential energy due to intermolecular force between them, which is maximum in solids and minimum in gases. Molecule posseses chemical energy associated with atomic bonds in a molecule, and nuclear energy associated with bonds within the nucleus of the atom. The sum of all such forms of energy in a system is labelled as the internal energy of the system.

7.5 kJ of work. Find the heat transfer to the system along path *adb*, if the work done is 2.5 kJ. The system is returned back from *b* to *a* along the curved path, the work done on the system is 5 kJ. Does the system absorb or liberate heat, and how much? Also, find heat and work transfer in the processes *ac* and *cb*, taking $U_a = 0$ and $U_c = 10$ kJ.

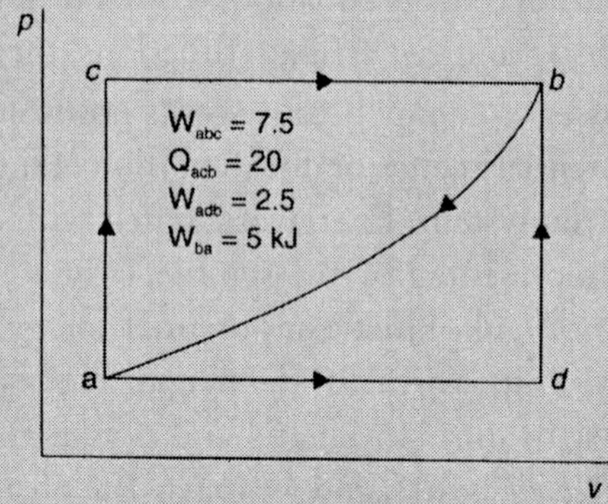


**Fig. 2.3** Different processes

***Solution*** The processes are shown in Fig. 2.3. As it is not mentioned that there is mass flow across the system boundary, we shall assume it closed system. The amount of heat and work transfer is given in different processes, analysis requires considering each process one by one.

*Path acb* Given: $Q = 20$ kJ and $W = 7.5$ kJ. Using nfee,

$$dU = U_b - U_a = Q - W = 12.5 \text{ kJ}$$

*Path adb* Given: $W = 2.5$ kJ. Using nfee, $U_b - U_a = Q - W$, for this path. We get $Q = 12.5 + 2.5 = 15$ kJ. Since this is positive, heat is supplied to the system.

*Path ba* given: $W = -5$ kJ (negative because work is done on the system) and change in internal energy $U_a - U_b = -12.5$ kJ. Applying nfee to this process, we get $Q = W + (U_a - U_b)$

$$= -5 - 12.5 = -17.5 \text{ kJ}$$

The negative sign indicates that heat is rejected by the system in the process.

*Process ac* given: $W = 0$ (constant volume process), $U_a = 0$ and

$U_c = 10$ kJ. Using nfee, we get

$Q - 0 = 10$, or $Q = 10$ kJ is supplied to the system in this process.

*Process cb* We determined the amount of heat transfer in the process *acb* and ac to be 15 kJ and 10 kJ, respectively. We may write heat transfer in process *acb* = heat transfer in process *ac* + heat transfer in process *cb*. This gives heat transfer in process *cb* = 5 kJ. Similarly, work transfer in process *acb* = work transfer in process *ac* + work transfer in process *cb*. This gives work transfer in process *cb* = 2.5 kJ.

**Example 2.2** A mass of 1 kg of air contained in a closed system at 1 bar, 300 K is stirred with a constant torque of 1 Nm at a speed of 100 rpm till the volume doubles. The initial and final temperatures were found to be the same. If 10 kJ of heat is absorbed during the experiment, calculate its duration.

***Solution*** Given: $\tau$ = 1 Nm, $\omega = 2\pi \times 100/60$ = 10.472 rad/s,

$v_2/v_1 = 2$, $m$ = 1 kg of air and $q$ = 10 kJ.

In this problem, work is being done by the system as its volume is changing, whereas work is supplied to it by the stirrer. Expansion decreases the temperature, whereas stirring will increase it. The system is a closed system, and hence nfee gives $Q - (W + Ws) = dU = 0$, where *Ws* is the stirrer work and *W* the boundary work; *dU* is zero since temperature is constant and air is a perfect gas. However, *W* can be determined from the relation, $W = p_1 v_1 \ln v_2/v_1$ (see Eq. 4.29). From perfect gas equation, $V_1 = \mathrm{mRT}_1/p_1 = 0.8061$ m$^3$. Therefore, $W = 1 \times 0.8061 \times \ln 2 \times 100$ kJ = 55.88 kJ. Substituting different values in nfee, we get

*Ws* = 10 − 55.88 = −45.88 kJ. But from mechanics, we know that work done in rotating a shaft in time *t* is equal to $\tau\,\omega\,t$, where $\tau$ and $\omega$ (both given) are torque in Nm and angular speed of rotation of shaft in rad/s. Equating it to *Ws* and simplifying, we get, *t* = 73 min.

**Example 2.3** A cylinder and piston assembly contains 0.5 g (g stands for gram, the unit of mass, and not acceleration due to gravity) of gas, as shown in Fig. 2.4. Initially, the piston face is at $x = 0$ and the spring exerts no force on the piston. As a result of the heat transfer, the gas expands, raising the piston until it hits the stops. At this point the piston location is $x = 0.05$ m, and the heat transfer stops. The force exerted by the spring on the piston as the gas expands varies linearly with *x* according to the equation $F = kx$, where *k* is the spring constant = 10000N/m. Take mass of piston = 10 kg, area of piston = 0.0078 m$^2$, atmospheric pressure = 1.0 bar and specific internal energies at initial and final states as 214 and 337 kJ/kg, respectively. Neglect piston and cylinder wall friction. Determine (a) the initial pressure of the gas, (b) the work done by the gas on the piston, and (c) the heat transfer.

***Solution*** Given: closed system consisting of gas confined in piston cylinder having $m = 0.5$ g.

$k = 10000$N/m, $Mp= 10$ kg, $Ap = 0.0078$ m$^2$

$ui = 214$ kJ/kg, $u_f = 337$ kJ/kg

$x = 0.05$ m, $p_a = 1$ bar

(a) Initial pressure should be enough to oppose the downward force on it due to atmospheric pressure and the weight of the piston, since the piston is in equilibrium and static. This gives,

$p_i = p_a + Mp/Ap = 1 \text{ (bar)} + [\{10 \times 9.81/0.0078\}(\text{N/m}^2) \times 10^{-5}] \text{ (bar)}$
$= 1.12577$ bar

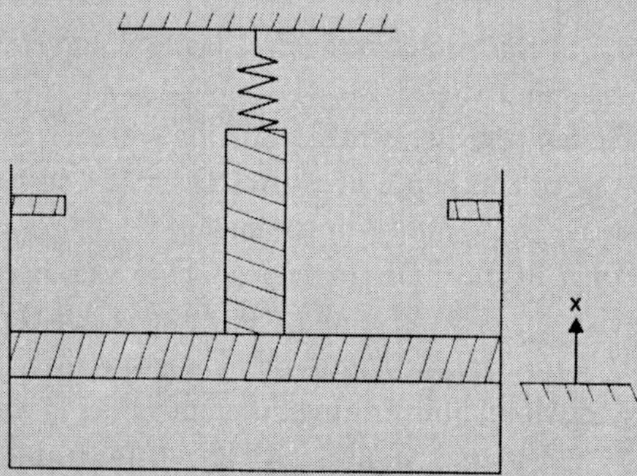

**Fig. 2.4** Piston loaded with spring

Units in parentheses are the units of the quantity up to that stage. For example, N/m$^2$ is the unit of the quantity to its left and similarly bar. Note that the conversion factor $10^{-5}$ is used to convert N/m$^2$ to bar. Student should not only keep in mind the dimensional homogeneity of any equation, but also the homogeneity of the units. This happens because the SI units recommended for any quantity is different from the actually used units for it and needs conversion from one to another. It is my experience that even the best student commits this mistake and the average number of students making such silly mistakes is alarmingly high. It is therefore, advisable to be cautious about it, so that it becomes integral part of your life.

(b) For a small displacement of the piston $dx$, work done is $p\ Ap\ dx$, where $p\ Ap$ is gas force at any position of the piston, which is equal to

$p_a Ap + Mp g + kx$; the last term represents spring force. Substituting it is the expression for displacement work and integrating from $x = 0$ to $x = 0.05$ m, we get

$$W = \int_0^{.05} (p_a\ Ap + Mp\ g + kx)\ dx = 56.4\ \text{J}$$

(c) Using Eq. 2.4, we get

$Q = W + (U_f - U_i)$

$= 56.4$ (J) + [{0.5/1000 (kg) × (337 – 214) (kJ/kg)} (kJ) × $10^3$] (J) = 118 J

**Example 2.4** A fully insulated electric battery discharges at constant volume. It is thermally insulated. During a one hour test, it is found that a current of 20 amp at 10 volts flows, while the temperature increases from 20°C to 35°C. The battery is used to heat the heating element, which heats water. The temperature of the heating element and water is raised by 20°C. The heating element has heat capacity equivalent of ¼ kg of water. Determine the change in internal energy, work done and heat transfer when, (a) only the battery is treated as a system, (b) only the heating element is considered as a system, (c) water and the container are considered as a system, and (d) heating element, water, container and battery are considered as a system.

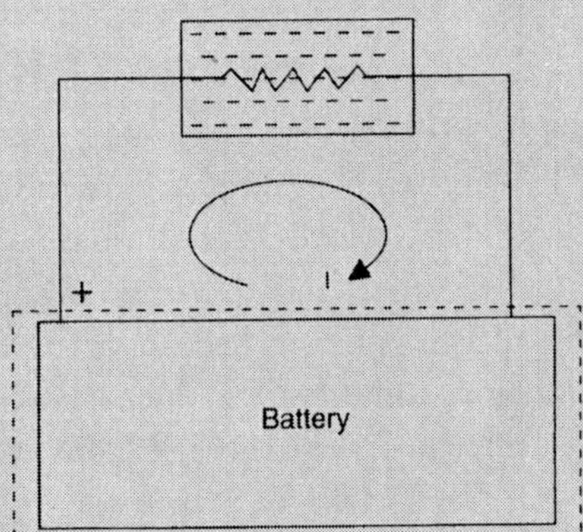


**Fig. 2.5** Heating of water

***Solution*** Figure 2.5 shows the arrangement in which current flows from the battery to the heating element. The battery is thermally insulated, and hence heat transfer from it is zero. The amount of energy flowing with the current is given by $\int EIdt = EI$ per unit time, since both $E$ and $I$ are constant in the given problem. Substituting values of $E$ and $I$, we get the amount of energy conveyed through the current per second = 20 × 10 = 200 W. We would now consider the different systems.

(a) **Battery is the system.** Work transfer is only due to flow of current, since the volume of the battery does not change in the process, and hence boundary work is zero. The battery supplies current, and hence the work transfer is positive. Thus, $W$ = 200 W. Moreover, $Q$ = 0. Using Eq. 2.4, we get $dU$ = –200 W. It is worth noting that the problem mentions rise in temperature of the battery content, giving a feeling as if internal energy has increased, but it is not so. Actually, there is decrease in internal energy. Increase in temperature is due to a chemical reaction, in which the chemical energy of the contents decreases to produce current and an increase in temperature. However, the process decreases the internal energy.

(b) **Heating element is the system.** In this case, the battery is the surrounding which supplies work 200 W to the element, and hence is negative. As a result, its temperature increases by 20°C in an hour. That is, the internal energy of the element is increased by

$dU$ = thermal capacity of the element × rise in temperature in one second

$= \frac{1}{4} \times 4.2 \times (20/3600) \times 1000$ J/s = 5.82 W

Using Eq. 2.4, we get $Q = dU + W$ = 5.82 – 200 = –194.18 W. The negative sign shows that heat is given off by the element. This heat heats the water.

(c) **Water and its container are the system.** This does not include the element, which is the surrounding, and the type of interaction between the system and the surrounding is heat transfer that is equal to +194.18 W; +ve because it is going to the system. The container is rigid, and therefore, there is no boundary work. So, work transfer in this case is zero. Increase in internal energy of water and the container is $Q$ = 194.18 W.

(d) **Battery, heating element, water and the container are the system.** The system is an isolated system and hence $W = Q = dU$ = 0; whatever is happening physically is within the system.

In Chapter 1, we discussed that one of the important mode of work transfer is boundary work given by $\int pdv$. This is a simple integration and requires knowledge of the relation between $p$ and $v$, that depends on the type of working substance. This integration has been done for ideal gas in Section 4.5 of Chapter 4. You may see it first, or since this equation is taught at class 12, you may also do it, as we will need some of the results while applying the energy equations to the systems, closed or open. In Chapter 3, the method to calculate change in entropy of the closed and open system during a reversible process is explained, and also that heat transfer in the system is related to the change in entropy. Therefore, heat transfer can also be calculated using this path. However, it is seldom used, and Eq. 2.4 is normally used to calculate it. We shall take a few more examples to illustrate the procedure, which is to determine (i) $dU$, (ii) $\delta W$, and finally (iii) $\delta Q$. Results from Section 4.5 are reproduced below

$$W = \frac{p_1 v_1 - p_2 v_2}{n-1} = \frac{R(T_1 - T_2)}{n-1}, \quad \text{if } n \neq 1 \text{ and}$$
$$= p_1 v_1 \, ln(v_2/v_1) = RT_1 \, ln(v_2/v_1) \text{ if } n = 1 \qquad (4.29)$$

Here, $n$ is the index of the process represented by the process equation $pv^n$ = constant and has value of 0, 1, $\gamma$ and $\infty$ respectively, for constant pressure, constant temperature, isentropic and constant volume process.

**Example 2.5** Air at 30°C and 1 bar is compressed reversibly and polytropically from 5 m$^3$ to 1 m$^3$. Calculate the final temperature, pressure and also the work done and heat transferred if the index of compression is 1, 1.4 and 0.9, respectively. Assume $c_p$ = 1.005 and $c_v$ = 0.7178 kJ/kg K.

***Solution*** Given: $T_1 = 273 + 30 = 303$ K,

$p_1$ = 1 bar, $V_1$ = 5 and $V_2$ = 1m$^3$

$c_p$ = 1.005 and $c_v$ = 0.7178 kJ/kg K

$\rightarrow \gamma = c_p/c_v = 1.4$

$R = c_p - c_v = 0.2872$ kJ/kg K

Also $m$ = mass of the gas = $m = p_1 V_1 / RT_1$

$= 1 \times 10^5 \times 5/(0.2872 \times 10^3 \times 303) = 5.734$ kg

(a) For $n = 1$

$p_1 V_1 = p_2 V_2$

$\rightarrow 1 \times 10^5 \times 5 = p_2 \times 1$

or $p_2 = 5$ bar

Also, $n = 1$ is an isothermal process and therefore $T_2 = T_1 = 303$ K and $dU = 0$.

From Eq. 4.29

$W = p_1 \, V_1 \, ln(V_2/V_1)$

$$= 1 \times 10^5 \times 5 \times \ln(1/5)/1000 = -804.7 \text{ kJ}$$

work is done on system.

From Eq. 4.29, $\delta Q = \delta W$

$$= -804.7 \text{ kJ (heat is rejected)}$$

(b) For $n = 1.4$, the process is isentropic and therefore

$$\delta Q = 0 \rightarrow \delta W = dU = m\,c_v\,(T_2 - T_1)$$

$$T_2 = T_1\,(V_1/V_2)^{\gamma-1} = 303\,(5)^{0.4} = 577 \text{ K}$$

and $$p_2 = p_1\,(V_1/V_2)^{\gamma} = 1 \times 5^{1.4} = 9.5 \text{ bar}$$

$$\therefore\; dU = 5.734 \times 0.7178 \times (577 - 303) = -740 \text{ kJ}$$

(c) For $n = 0.9$, we have the process equation

$$p_1 V_1^{\;0.9} = p_2 V_2^{\;0.9} \rightarrow p_2 = 4.26 \text{ bar}$$

$$T_2 = T_1 (V_1/V_2)^{n-1} = 258 \text{ K}$$

$$\delta W = (p_1 V_1 - p_2 V_2)/(n-1)$$

$$= 10^5\,(1 \times 5 - 4.26 \times 1)/\{(0.9 - 1) \times 10^3\}$$

$$= -740 \text{ kJ} \quad \text{(work is done)}$$

$$dU = m\,c_v\,(T_2 - T_1)$$

$$= 5.734 \times 0.7178\,(258 - 303) = -185.21 \text{ kJ}$$

$$\delta Q = -185.21 + (-740) = -925 \text{ kJ (heat is rejected)}$$

## 2.4 ENERGY EQUATION FOR OPEN SYSTEM

Energy transfer between an open system and the surrounding, besides heat and work transfer, is also through flowing mass, as different amount of energy will be carried with mass entering or leaving the system. We mentioned before that any mass in motion would have kinetic energy, potential energy and internal energy. However, flowing mass has one more form of energy, called flow energy, which is equal to $pv$, where $p$ and $v$ are the pressure and specific volume at the inlet or outlet section, (see Section 1. 9.3). Thus, each kg of fluid shall have energy, $e = u + pv + ke + pe$. Mollier realised the importance of the term $u + pv$ in the analysis of a power cycle, and combined these terms to form another property called *specific enthalpy,* symbolised as $h$, only for convenience. Nevertheless, it should always be kept in mind that the $pv$ term exists only for flow system, or during phase change, where energy is required to overcome the forces of attraction among the molecules. Therefore, use of $h$ should be made carefully. For example, consider air in a room at temperature $T$ and pressure $p$ having specific volume $v$ and internal energy $u$. Then, $u + pv$ is not specific enthalpy of air, as it is not flowing mass.

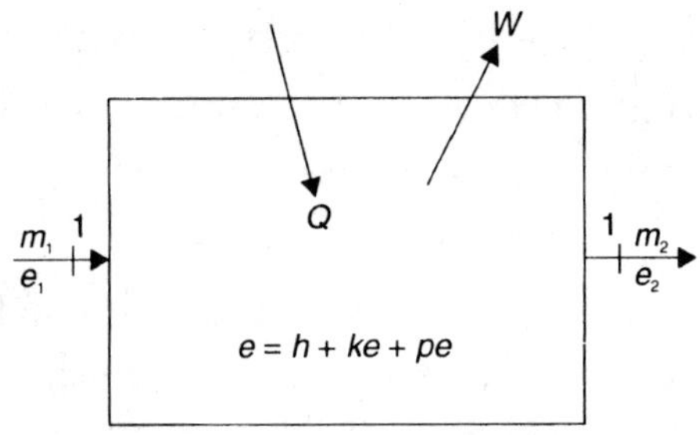


**Fig. 2.6** Open system

Now, we can give mathematical shape to the energy equation for an open system, considering energy flow from or to the system and change in its energy $dE_\sigma$ during this period. Figure 2.6 shows an open system.[3] For simplification, we assume that fluid enters the open system at one section, and leaves at another section, being shown as Sections 1 and 2 in the figure. However, the process can be extended for systems where mass flows across the boundary at many points. Let $Q$ and $W$ be the heat and work transfer to the system in unit time, and $m$ the mass flow rate. (Normally symbols $\dot{W}$, $\dot{Q}$, etc. are in use to indicate time rate; a. over the variable has been dropped in this book.) Using the law of energy conservation, we can write the following equation,

energy in per unit time = energy out per unit time + change in energy of the system in unit time

But energy in = energy flow rate with incoming mass + heat transfer rate

$= m_1e_1 + Q$ and

energy out = energy flow rate with outgoing mass + work transfer rate = $m_2e_2 + W$.

The energy equation for an open system thus becomes

$$m_1 e_1 + Q = m_2 e_2 + W + dE_\sigma$$

or
$$Q - W = m_2 e_2 - m_1 e_1 + dE_\sigma \tag{2.5}$$

This is the energy equation for a flow system, and is called the flow energy equation. In most of the thermal systems, the flow is a steady flow.[4] In such systems, the net change in mass and energy of the system will be zero, that is,

3. The term control volume has also been used for an open system, following the Eulerian approach in Fluid Mechanics. A control volume is a space fixed in the flow field, and in this approach, we do not bother which fluid particle is entering or leaving.

4. Steady flow means that at any given point in the flow field, properties of the fluid do not change with time. Consider a flow through a diverging pipe. We know that as we move in the direction of flow, pressure increases and velocity decreases. But if we fix ourselves at any point in this flow region, then we shall find that pressure and velocity do not change with time, and this flow is steady.

$$dE_\sigma = 0 \text{ and } m_1 = m_2 = m \text{ (say)}$$

Using this knowledge, the flow equation reduces to

$$(Q - W)/m = e_2 - e_1 = de$$

or $$q - w = d(h + ke + pe) \cong dh \tag{2.6}$$

using the small case letter $q$ and $w$ respectively for $Q$ or $W$ for unit mass flow rate, and neglecting change in kinetic and potential energies, which is true for most of the thermal systems. This equation is known as the steady flow energy equation, abbreviated as sfee. We shall discuss its application later.

It is interesting to determine the expression for boundary work transfer for an open system as well. We know that there is change in flow energy as the fluid moves from one section to another. Writing for $h = u + pv$ in sfee, we get

$$\delta q - \delta w = dh = d(u + pv) = du + pdv + vdp \tag{2.7}$$

We may treat the fluid element entering into the open system as a system, and follow it as it moves from the entrance to the exit. It is a closed system. In that case, the boundary work transfer $pdv$ occurs between the system and the surrounding (assuming there is only boundary work transfer) during the process, nfee gives $du = \delta q - pdv$, or $\delta q = du + pdv$. Substituting it in Eq. 2.7, the boundary work for an open system comes out to – ($vdp$) after simplification.

Another interesting feature of sfee is observed if we assume that there is no heat and work transfer from or to the system, the fluid is incompressible and there is no change in temperature (and hence internal energy) of the fluid from the entrance to the exit. In this case, sfee gives

$$d(u + pv + c^2/2 + gz) = 0 \text{ or } dp/\rho + c\,dc + g\,dz = 0 \tag{2.8}$$

which is Bernoulli's equation. Thus, Bernoulli's equation is a special form of sfee. Here, $\rho = 1/v$ is the density of fluid at a given section or point.

## 2.5 APPLICATION OF SFEE TO THERMAL SYSTEMS

Commonly used thermal systems in practice are boilers, condensers, evaporators, heat exchangers, compressors, turbines, power plants, refrigerators, and air-conditioners. These are the names given to devices used for a specific purpose. For example, a *boiler* is a system that changes liquid to vapour by application of heat. Water enters the boiler at some pressure and temperature, and part of it is changed into steam at that pressure. Steam so produced may be used for heating a solution in a process industry, or producing power by expanding it in a turbine or engine. Since there is flow across boundaries, a boiler can be analysed as an open system. It is represented as a circle in Fig. 2.7(a), with water entering at Section 1 and steam leaving it at Section 2. Further, the flow conditions will not change with time and hence it is a steady flow, and therefore mass flow rate at

Sections 1 and 2 will be equal. Therefore, the steady flow, energy equation is the right tool to be used for analysis. This equation gives $\delta q = de$,[5] where $e = (u + pv) + ke + pe$. So, energy of the fluid increases in this process. Normally, change in potential and kinetic energy is negligible and there is an increase in internal and flow energy ($pv$ term) or enthalpy of steam.

In absorption refrigeration, a generator is used to separate ammonia from its solution in water by application of heat. This generator is nothing but a boiler in which liquid is transformed to vapour; a different name is given, as its purpose is different from that of a boiler. Here too, energy of the working substance is increased by addition of heat. In a refrigeration system, we use an *evaporator* in which the dryness fraction of vapour is improved by extraction of heat from the space where cooling is required. On the other hand, energy of the fluid is reduced by taking out heat in a *condenser,* which is primarily used to condense vapour into liquid. In a *heat exchanger*, shown in Fig. 2.7(e), there is, as the name suggests, exchange of heat between two flowing fluids, so that energy of one increases by the amount it decreases for another fluid. In fact, both the evaporator and condenser are heat exchangers. We represent an evaporator or condenser as shown in Fig. 2.7(d).

In a *compressor,* Fig. 2.7(c), the energy of the fluid is increased by pumping work (and not heat) into the fluid. The result is an increase in internal energy, flow energy or both. If the fluid is an incompressible liquid like water, there is insignificant change in internal energy (normally neglected) and only flow energy increases, whereas in case of a compressible fluid, like air and vapour, both increase. When we say flow energy is increasing, it includes an increase in pressure also. Because of increase in pressure, the term compressor has been used for such a system. It is the property of gases and vapours that their volume decreases as pressure increases, and that is why we represent a compressor as a converging section. Moreover, we use the term pump for a compressor, when it is used for raising the pressure of a liquid.

A *turbine* is a device that, contrary to a compressor, reduces total energy *e* of the fluid by producing work. It is shown in Fig. 2.7(b). Fluids like water, steam, various refrigerants and gases like air, $CO_2$, and helium are used as working substances. Water is stored by

5. Note that the symbol $\delta$ and not *d*, has been used with *q*. This is our way of representing that heat and work transfer are path and not point functions. Use of *d* in *de* then means that *e* is a point function. Another mistake students commonly make is that they use $\partial$ or $\Delta$ in place of $\delta$. This is wrong because the symbol $\partial$ represents a partial derivative, and $\Delta$ the total change. If someone calls Ram as Rahim, Rahim will not know that he is being called. Names and symbols have their own importance, and care should be taken while using them. Another mistake a student commonly makes is to call a fluid to be steady; it is not the fluid, but the flow that is steady. Care should be taken of these points while communicating with another person, so that one is properly understood.

constructing a dam over a running water stream to increase its potential energy, whereas in other fluids, their pressure and temperature are increased. These fluids, after expansion in a device called nozzle, are changed to a jet having kinetic energy, as explained in Chapter 1. This jet of fluid enters the turbine and leaves it with almost 10% of the kinetic energy at the inlet. So, a turbine is primarily a device that converts translation kinetic energy into rotational kinetic energy of the turbine wheel (work) by making the fluid flow over curved vanes fixed to the wheel. During this process, enthalpy also decreases in case of a compressible fluid.

What we observe is that all these devices change the energy of the fluid which flows through them, and the flow remains steady over the entire period of operation. Therefore, all these systems can be analysed using the steady flow energy equation. Thus, a thermal system is just like a house. A tenant arranges all his belongings in it, optimising it with his needs, so that it looks attractive and purposeful. If his need is different, he will go to another house that matches his requirements. The same is true here also. If power is required, we use a turbine, and if high-pressure air is required, then we use a compressor. Like different houses, these thermal systems too will have different structures to meet our needs. Even a turbine for water will differ in many constructional features from a steam turbine, and a steam turbine will differ from a gas turbine, since the type of fluid used with them is different.

We design all these devices so that they have maximum efficiency (or rather, a good word is performance) to reduce running cost, minimum size to reduce the first cost and problems associated with their transportation from the place of manufacture to the place of use, and to have maximum life for giving us better return for our investment over a long period. That is, the overall economy of the system is the main criterion in its design.

In fact, all these devices are individual units, and are combined to form a family that delivers goods to the society. For example, if a boiler, turbine, condenser and a pump are placed in that order, then the whole system is a closed system, whose purpose is to deliver mechanical power that is then converted into electrical energy by connecting an electrical generator to the turbine shaft. Such a system is called a power plant. Input to it is the heat released from fossil or nuclear fuels and the output is work; heat is also rejected (requirement of the second law of thermodynamics, discussed later). However, all individual components are open systems and flow through them is steady. Heat or work transfer in each of these units is evaluated applying the steady flow energy equation to individual units. Once this has been done, then Joule's law is applied to the whole cycle to determine the performance of the cycle.

Similarly, when a compressor, a condenser, an expansion valve and an evaporator are put together in that order, it results in a refrigeration system, whose function is to produce temperatures lower than the atmospheric temperature. It too works on a cycle. It is natural for a system to operate on a cycle, so that we get desired results continuously. It is like our life cycle, water cycle, etc. that we observe in our life.

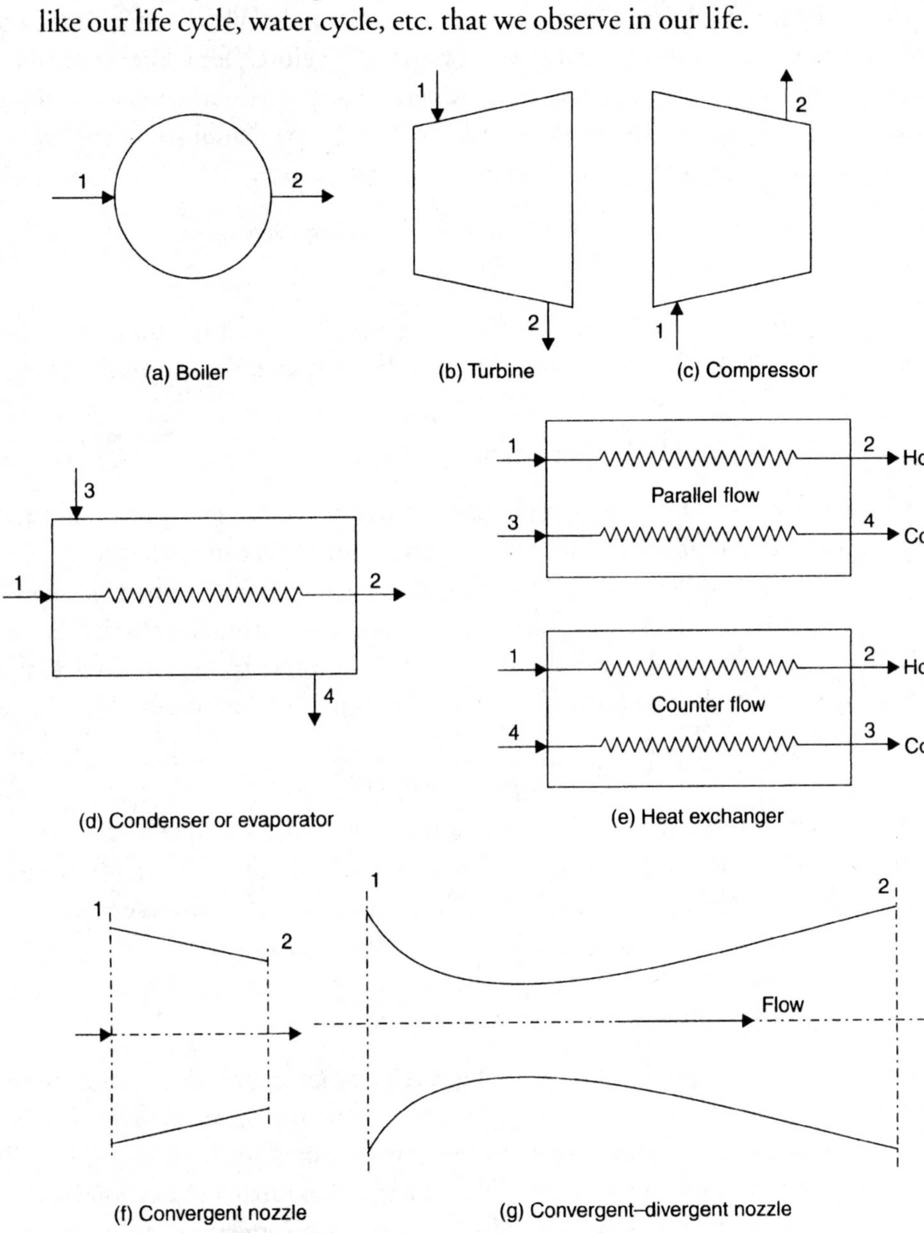


**Fig. 2.7** Representation of different thermal systems

Figure 2.7 shows a representation of the above mentioned systems. We note that a turbine is shown such that the area of flow is increasing, since due to expansion in the turbine, pressure decreases, and hence volume flow rate increases. The opposite is true for a compressor. A coil in a condenser, evaporator, or heat exchanger indicates that the two fluids flowing through the unit do not come in contact physically. The direction of flow is indicated by arrows. We may use any notation for inlet and outlet; 1 and 2 are commonly used. All these systems are open and flow is steady. Therefore, sfee relates work and heat interaction to it with change in property. Property transported at the inlet and outlet are $e_1$ and $e_2$ respectively, per kg mass flow rate. We shall now demonstrate application of sfee one by one to different thermal systems described above.

**Boiler.** Input is heat $\delta q$ and no work transfer. Therefore, sfee gives

$$\delta q = h_2 - h_1$$

*Compressor or Turbine.* We assume isentropic compression or expansion, respectively. Therefore, $\delta q = 0$. Both of these units interact through shaft work, $\delta w$ sfee gives

$$-\delta w = h_2 - h_1$$

$\delta w$ is positive for a turbine and negative for a compressor.

**Condenser, evaporator, or heat exchanger.** Since there are two fluids, we use the symbols $A$ and $B$ to represent them. Here, $A$ is a hot fluid entering the unit at Section 1 and leaving at 2, whereas $B$ is a cold fluid with notations 3 and 4 for inlet and outlet sections, respectively. We may consider one fluid confined within the physical boundary of the device as one system, then naturally, the other becomes its surrounding, as there is interaction between the two only. This interaction is in the form of heat. Simple energy balance gives

heat given by hot fluid = heat received by cold fluid.

When there is no change in phase, amount of heat transfer is equal to $mc\Delta T$, where $\Delta T$ is change in fluid temperature from the inlet to the outlet, $m$ is mass flow rate and $c$ the specific heat. However, when there is change in phase, then this is equal to $m\ \Delta h$, where $\Delta h$ is determined using appropriate tables, discussed in Chapter 4. Normally, in a heat exchanger, both fluids are single phase. But in both condenser and evaporator, one fluid is liquid and another wet vapour.

**Nozzle.** It is a device used for producing high velocity jet at its exit. This is achieved by expanding fluid through it. By Bernoulli's principle, as pressure decreases, velocity increases. Such a nozzle is called a subsonic nozzle, and is shown as a convergent passage (Fig. 2.7(f)). However, when fluid velocity reaches sonic velocity, then further expansion is achieved by a diverging section. This nozzle is called convergent-divergent or supersonic nozzle, and is shown in Fig. 2.7(g). Since the purpose of the nozzle is to produce a high velocity jet, change in kinetic energies from inlet to outlet cannot be neglected, and therefore, the corresponding term in sfee is to be retained. Moreover, expansion is assumed adiabatic.

Since there is no work exchange between the system and the surrounding, $\delta w = 0$. Many students have raised the question that in a closed system, we determine boundary (or displacement) work transfer by evaluating $\int pdv$, then why do we not use $-\int vdp$ for determining boundary work transfer in a nozzle, which is an open system. We must understand that in power cycles, reversible expansion or compression of the working substance will occur only and only when movement of the piston is opposed by the surrounding, otherwise, free expansion would occur and there will be no work transfer. That is, there is physical work transfer between the system and the surrounding. This is not so in a nozzle. In fact, $\int vdp$ gives change in enthalpy from the inlet to the outlet, which can be easily seen from *TdS* equations (Chapter 4). Therefore, sfee gives

$$0 - 0 = (h_2 - h_1) + (C_2^2 - C_1^2)/2000$$

Knowing $(h_2 - h_1)$ and $C_1$, we can determine $C_2$. A nozzle has been discussed in detail in Chapter 7.

**Example 2.6** In a gas turbine installation, the gases enter the turbine at the rate of 5 kg/s with a velocity of 50 m/s and enthalpy of 400 kJ/kg, and leave with 150 m/s and enthalpy of 400 kJ/kg. The loss of heat from the gases to the surroundings is 25 kJ/kg. Assume $R = 0.285$ and $c_p = 1.004$ kJ/kg K and inlet conditions to be at 1 bar and 27°C. Determine the work output and the diameter of the inlet pipe.

***Solution*** Given: $p_1 = 1$ bar, $T_1 = 300$ K, $C_1 = 5$ m/s, $C_2 = 150$ m/s, $h_1 = 900$ kJ/kg, $h_2 = 400$ kJ/kg, $m = 5$ kg/s, $Q = -25$ kJ/kg.

Using sfee Eq. 2.5, we get

$$-25 \times 5 - W = 5 \times [\{400 + 150^2/2000\} - \{900 + 50^2/2000\}]$$

$$\rightarrow W = 2325 \text{ kW}$$

Volume of the gases entering into the turbine/s

$$V = mRT_1/p_1$$

$$= 5 \times (0.285 \times 1000) \times 300/(1 \times 10^5)$$

$$= 4.275 \text{ m}^3/\text{s}$$

$$= \pi/4\, d^2\, C_1 = \pi/4\, d^2 \times 50 \rightarrow d = 0.33 \text{m}$$

**Example 2.7** The air speed of a turbojet engine is 270 m/s. Ambient air temperature is −15°C. Gas temperature at the outlet of the nozzle is 600°C. Corresponding enthalpy values for air and gas are respectively 260 and 912 kJ/kg. Fuel air ratio is 0.019. Chemical energy of the fuel is 44.5 MJ/kg. Owing to incomplete combustion, 5% of the chemical energy is not released in the reaction. Heat loss from the engine is 21 kJ/kg of air. Calculate the velocity of the exhaust jet.

***Solution*** Here, the system is moving and ambient air, the surrounding, is stationary. We may change the coordinate system to bring the system to rest by fixing it to the jet itself. In that case, air is moving in the direction opposite to the direction of the jet flight. The system is shown in Fig. 2.8. Also, we know

$C_1 = 270$, $h_1 = 260$, $h_2 = 912$, $C.V. = 44500$ kJ/kg, 5% fuel remains unburnt $\rightarrow$ $Qa = .95 \times 44500 = 42275$ kJ/kg, $Q = -21$ kJ/kga and $f = F/A$ ratio $= 0.019$, $W = 0$. Assume $m_a = 1$ kg.

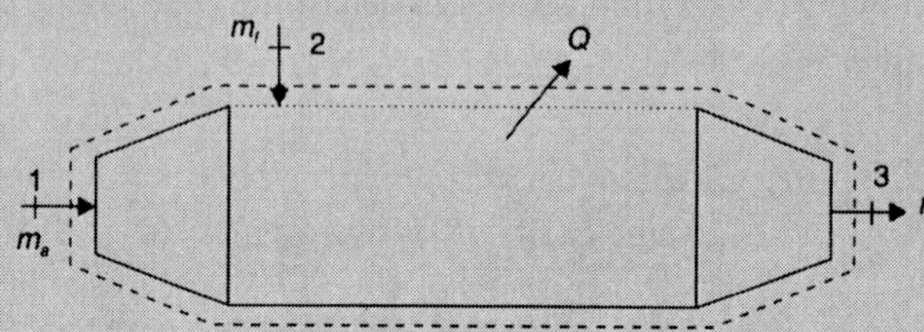


**Fig. 2.8** Turbojet engine

Mass and energy conservation gives

$$m_g = m_a + m_f \text{ and } (Q + m_f \times C.V.) - W$$
$$= m_g\,(h_2 + C_2^2/2000) - m_a\,(h_1 + C_1^2/2000)$$

or $$(-21 + 0.019 \times 42275)$$
$$= (1 + 0.019)(912 + C_2^2/2000) - 1 \times (260 + 270^2/2000)$$
$$\rightarrow C_2 = 560 \text{ m/s. (w.r.t jet)}$$

## 2.6 ANALYSIS OF AN OPEN SYSTEM AS A CLOSED SYSTEM

Though not normally required, yet, it is possible to change an open system into a closed system. We used this approach while determining work transfer in an open system in Section 2.2.2. Example 2.4 a shows the procedure for analysing an open system as a closed system for a reciprocating engine inhaling charge from outside during downward movement of a piston. The amount of air to be inhaled is also included with the reciprocating engine as a system, as then it will be a closed system. Example 2.8 is another example of analysing open system as closed system.

**Example 2.8** A bullet, $m = 2$ g, flying horizontally at 1 km/s strikes a fixed wooden block ($M = 5$ kg, $c = 0.14$ kJ/kg k), and is embedded in it. Assume that there is no heat lost from the block, and that the block does not change its volume. What will be the change in the temperature of the block?

***Solution*** If we consider the block as a system, then it is an open system, since the bullet strikes it and penetrates into it. But we may consider the bullet and the block as one system, then it is a closed system, and the bullet-hitting block process occurs within the system. This is a redistribution of energy within the system, and not an interaction. Moreover, there is no heat transfer (given). Work transfer is also zero. So, nfee gives, $\Delta E = 0$.

Consider the situation when the bullet is just about to strike the block as the initial state, and after bullet has hit the block and come to rest as the final state. Initially, energy $E$ consists of the kinetic energy of the bullet $\frac{1}{2} \times (2/100) \times 1000^2$ J = 1 kJ plus the internal energy of the bullet and the block. Finally, the energy of the system is only the internal energy of the two. Since specific heat of the bullet material is not given, we shall neglect change in internal energy of the bullet. This means that change in internal energy of the block $McdT$ should be equal to the initial kinetic energy of the bullet (before hitting the block). Equating the two and taking care of proper units (kinetic energy is in $J$, but energy of the block is in kJ), or we get, $McdT = 1$ change in temperature of the block = 1.43 K.

**Example 2.9** Figure 2.9 shows a cylinder piston arrangement (vertical) for an internal combustion engine, in which a piston of area 40 cm$^2$ moves downwards 8 cm drawing 250 cm$^3$ of fresh air from the atmosphere. Pressure inside the cylinder during the suction process may be assumed to be constant at 0.8 bar when atmospheric pressure is 1 bar. Calculate the work required to pull the piston in this process.

***Solution*** When the piston moves downwards, then air enters into the cylinder, since the pressure outside is 1 bar and inside 0.8 bar. Assume the piston to be at its top dead centre (TDC)[6] position. If the clearance space inside the cylinder formed by the piston crown in the TDC position, cylinder head and its walls is taken as a system, then it is an open system, as air flows across the system boundary during downward motion of the

---

6. Piston reciprocates in cylinder. Therefore, at two extreme positions of its travel, its velocity has to be zero, as it is changing the direction of movement. These two ends are known as dead centre. In a vertical engine, these are named as top dead centre (TDC) and bottom dead centre (BDC) when the piston is at the top position and BDC for the bottom position.

piston. To change the open system to a closed system, we assume a balloon containing 250 cm$^3$ of air attached to the suction valve of the engine. The system $S_1$ consists of air inside the balloon and the clearance space (assumed zero for simplification). Let $S_1$ take the shape $S_2$ formed by the space above the piston up to the cylinder head. Thus, during the process, volume of $S_2$ increases from 0 to the volume swept by the piston (piston area × stroke length) and $S_1$ decreases to zero. Thus, the net work required is the algebraic sum of the work transfer on two boundaries. Noting that pressure $p$ remains constant for both system boundaries $S_1$ and $S_2$ during the process, it can be taken out of the integral sign in ($pdv$); the integral is simply equal to the product of pressure and change in volume of the system. It is positive if the volume increases, and negative if it decreases. Thus,

$$\text{Net work done} = (pdv)_{S_1} + (pdv)_{S_2}$$
$$= [0.8 \times 10^5] \times [40 \times 8 \times (10^{-2})^3] - 1 \times 10^5 \times 250 \times 10^{-6} \approx -0.6\ \text{J}$$

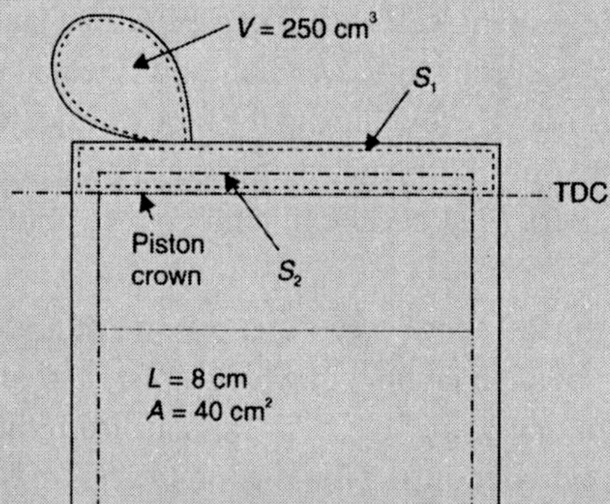


**Fig. 2.9** System analysis

A minus sign shows work is done in this process by the mechanism. That is, work is required to be done on the piston to pull it down.

In the same way, we may consider an exhaust operation during which burnt gases are pushed out, representing the volume of the gases going out by another boundary, say $S_3$, attached to the exhaust valve. Thus we can consider the three boundaries, namely $S_1$, $S_2$, and $S_3$ taken together that form a closed system; volume of the space within the individual boundaries will change during different processes.

## 2.7 TRANSIENT PROCESS

Analysis of a transient process is a little involved, as it requires determination of the change in energy of the control volume representing the open system. The evaluation process depends on the actual problem, and there is only one general rule, that the change in internal energy is equal to the difference of final and initial energy. Here, initial and final refer to the commencement and end of the process, respectively. Sometimes, the problem can be simplified by encompassing the system and the surrounding to make an isolated system, and then applying the conservation of mass and energy principle to the resulting system. We shall demonstrate the process through examples. In some cases, the mass and the state of the fluid entering or leaving a control volume may be known, as in Example 2.10. In such cases, we can convert the control volume into a closed system as discussed there.

**Example 2.10** A tank is connected through an infinite source of steam at 7 bar and 250°C. The valve is opened and steam is allowed to flow into the tank until the pressure just reaches 7 bar. Calculate the condition of steam in the tank when: (a) the tank is empty completely, (b) A non-heat conducting piston is placed in the tank containing air at 1.4 bar and 40°C, as shown in Fig. 2.10, and a cooling coil placed inside the tank to maintain the temperature of air at 40°C.

***Solution*** As steam flows to the tank, it is filled, and as a result, the mass and the energy content of the tank increase with time. Therefore, the filling is a transient process. Fortunately, steam is coming from an infinite source, which means that withdrawal of any amount of steam from the source would not change the condition of steam coming from the source. However, this does not ensure steady flow requirements, though the state of the steam at the entrance to the tank does not change with time. We must remember that sfee can be used only and only when the steady conditions exist in entirety of the application, including the region falling within the control volume. Take tank volume 0.1 m$^3$ in both the cases.

Given: (a) Let $i$ and $f$ refer to the time of opening and closing of the valve, and 1 refer to the entrance section of the tank.

$m_i = 0 \quad Q = 0$ (insulated tank) $p_1 = 7$ bar $t_1 = 250$°C

$h_1 = 2953.6$ (from steam table)

$m_2 = e_2 = 0$ (no outflow) $C_1 = 0$ (assumed to be negligible)

Equation 2.5 and mass conservation gives

$$m_f = m_1 \text{ and } -m_1h_1 + (m_fu_f - 0) = 0 \rightarrow u_f = h_1 = 2953.6$$

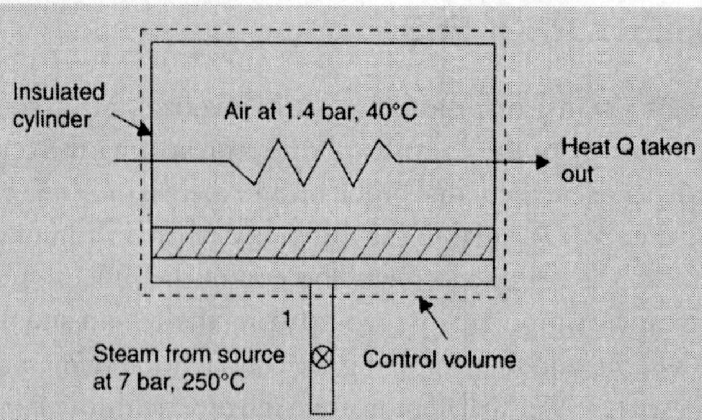


**Fig. 2.10** Transient process

Using linear interpolation on temperature, superheated steam table gives:

Temperature in tank just when the valve is closed = 394.4°C.

(b) $p_{ai}$ = 1.4 bar $T_1$ = 313 K $V_{ai}$ = 0.1 m$^3$

$$\rightarrow m_a = p_{ai}V_{ai}/RT_1 = 0.153 \text{ kg}$$

As steam enters, the piston moves in and the air is compressed isothermally. Work done in piston movement, $W = m_a RT_1 \ln(p_2/p_1) = 9.53$ kJ $= Q$, since the process is isothermal; it is rejected through the cooling coil. Moreover, during filling of the tank, there is no change in internal energy of the air, as it is maintained at the same temperature. This implies that change in energy of the control volume will be due to the steam that fills the space below the piston. Again, as there is no mass outflow, the mass of the steam collected in the tank = $m_1$, the mass of the steam that enters through the Section 1 during the filling period. Using Eq. 2.5 simplifying, we get

$$m_1 h_1 = Q + m_1 u_f \rightarrow u_f = 2953.6 - 9.53/m_1 \qquad \text{(A)}$$

Consideration of the movement of the piston gives volume of the steam in the tank. Final air volume = $p_{ai} V_{ai}/p_f$ = 0.02 m$^3$. Therefore, volume of the steam in the tank = 0.1 – 0.02 = 0.08 m$^3$ = $m_1 v_f$. Substituting for $m_1$ in Eq. A, we get

$$u_f = 2953.6 - 9.53 \times v_f/0.08 = 2953.6 - 119\, v_f \qquad \text{(B)}$$

This equation can be solved by hit and trial method for which we assume some steam temperature, read the values required in Eq. B from the steam table, and verify that Eq. B is satisfied; if not, the procedure is repeated. This process yields that the steam temperature in the tank is approximately equal to 365°C.

**Example 2.11** A tank of volume 0.5 m$^3$ is filled with air at a pressure of 30 bar and temperature of 50°C. The air is allowed to discharge through a turbine into the atmosphere until the pressure in the tank has fallen to the atmospheric pressure of 1 bar. Find the maximum work that could be delivered by the turbine.

***Solution*** In this problem, fluid supply is from the source to the atmosphere, where the state does not change. However, as air is discharged from the tank, the tank pressure would decrease due to expansion of the remaining air, which could be assumed isentropic. Therefore, the state of the air would be changing at the exit of the tank.

Given: $T_i = 323$ K, $p_i = 30$ bar, $p_f = 1$ bar, $V = 0.5$ m$^3$

If subscripts $i$ and $f$ represent the conditions before and after emptying of the tank, then the final air temperature in the tank is

$$T_f = T_i \left(p_a / p_i\right)^{(\gamma-1)/\gamma} = 121 \text{ K}$$

As usual, we use subscript 2 for the outlet section. Then, $m_2 = m_i - m_f$, since there is no inflow. Using Eq. 2.5, we get

$$W = m_2 h_2 + (m_f u_f - m_i u_i)$$

$$(m_i - m_f)\, c_p T_2 + c_v (m_f T_f - m_i T_i) \qquad \text{(A)}$$

where, $W$ is the amount of work produced by the turbine and $h = c_p T$ and $u = c_v T$ for air. Equation of state gives $m_i = 15.88$ and $m_f = 1.41$ kg. Taking $c_p = 1.005$ and $c_v = 0.73$ kJ/kg K and substituting for different variables in Eq. A, we get, after simplification, $W = 1594$ kJ.

**Example 2.12** Air is contained in a tank $A$, which is connected to tank $B$ through a valve. Initially, when the valve is closed, then tank A contains 5 kg air at 6 bar and 400 K, whereas B contains 2 kg air at 2 bar and 300 K. The valve is kept opened till air flow stops. Determine the final pressure and temperature of air in both the tanks.

***Solution*** In this problem, it is possible to combine both the tanks into one system. In that case, there is no process occurring, except that there is redistribution of mass and energy within the system because of pressure difference. (Is there any distribution of energy because of temperature difference? Comment.) We use subscripts 1 and 2 to represent the condition in tanks before commencement and after stopping of flow, and $a$ and $b$ for tanks $A$ and $B$, respectively. Due to flow of fluid, isentropic expansion would occur in tank $A$ and compression in tank $B$.

Given: $m_{a1} = 5$ kg, $p_{a1} = 6$ bar, $T_{a1} = 400$ K

$\rightarrow V_{a1} = 5 \times 0.287 \times 400/(6 \times 100) = 0.9567$

$m_{b1} = 2$ kg, $p_{b1} = 2$ bar, $T_{b1} = 300$ K

$\rightarrow V_{b1} = 2 \times 0.287 \times 300/(2 \times 100) = 0.8610$

$\gamma = 1.4 \rightarrow (\gamma - 1)/\gamma = 0.2857$

applying the principle of conservation of mass and energy, and using the relation, $u = c_v T$, for air, we get

$$m_{a2} + m_{b2} = 7 \quad \text{(A)}$$

and

$$m_{a2}T_{a2} + m_{b2}T_{b2} = m_{a1}T_{a1} + m_{b1}T_{b1}$$

that is $5 \times 400 + 2 \times 300 = 2600$ (B)

Using perfect gas equation, we can change temperature in terms of pressure. Let $p$ be the final pressure in both the tanks when flow stops, then from Eq. B, we get

$$\text{LHS} = (p \times 100/0.287)\,(V_{a1} + V_{b1}) = 633.35\,p = 2600$$

$$\rightarrow p = 4.105 \text{ bar}$$

There are four unknowns namely $m_{a_2}$, $m_{b_2}$, $T_{a_2}$ and $T_{b_2}$ and two equations, and therefore, two more equations need to be developed. These come from the expansion and compression process in tanks *A* and *B*. This gives

$$T_{a2} = (4.105/6)^{0.2857} \times 400 = 359 \text{ K and}$$

$$T_{b2} = (4.105/2)^{0.2857} \times 300 = 368.4 \text{ K}$$

Now, Eqs. A and B can be solved to give mass in tanks A and B, respectively.

**Discussion.** In this problem, flow is due to pressure difference and not temperature difference, and hence at the end of the process, temperature in tank *B* may be less, equal or more than that in tank *A*. In this example, it is more.

**Example 2.13** A large vessel contains steam at pressure of 20 bar and a temperature of 350°C. This large vessel is connected to a steam turbine through a valve, followed by a small, initially evacuated, tank with a volume of 0.8 m$^3$. During an emergency power requirement, the valve is opened and the tank fills with steam until the pressure is 20 bar. The temperature of the tank is then 400°C. Assume that the filling process takes place adiabatically and the changes in potential and kinetic energies are negligible. By drawing the control volume, calculate the amount of work developed by the turbine.

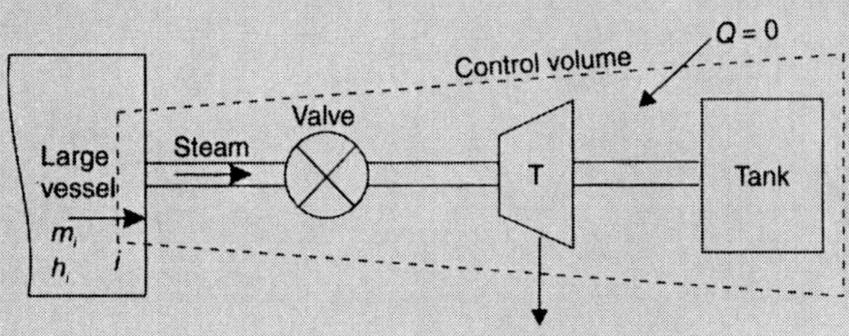


Fig. 2.11

***Solution*** Figure 2.11 shows the schematic arrangement of the problem. Initially the tank is empty. When the valve is opened, steam flows through the turbine and is collected in the tank, whereby mass and energy of the contents of the tank increase. As such, the system is a flow system, during which work is produced. Since pressure in the tank is increasing, the back pressure on the turbine is also increasing, and therefore, the power output will go on decreasing with time. The whole flow process is unsteady. We may consider the control volume, as shown in the figure, with steam entering into it at section *i*. If $m_i$ is the mass of the steam entering in unit time with $h_i$ enthalpy, and heat transfer is assumed to be zero in the process, then rate of change of mass and energy of the control volume is

$$dm_{cv}/dt = m_i \quad \text{and} \quad dU_{cv}/dt = -\delta W + m_i h_i$$

Here, $\delta W$ is the amount of work done by the turbine at that instant. Writing for $m_i$ in the rate of change of energy equation and integrating the equation, we get

$$dU_{cv}/dt = -\delta W + h_i \, dm_{cv}/dt$$

and $$\Delta U_{cv} = -W + h_i \, \Delta m_{cv} \tag{A}$$

where, $\Delta$ represents the total change in the quantity of the tank during the whole process. Let subscripts 1 and 2 represent the initial and final conditions in the tank, respectively. Then

$\Delta U_{cv} = m_2 u_2 - m_1 u_1 = m_2 u_2$, since the tank is empty initially, and therefore $m_1 = 0$.

Since condition in the large steam vessel does not change with withdrawal of any amount of steam from it $h_i$ remains constant. From steam table, we read $h_i = 3138.6$ kJ/kg. Also, at the end of the process, we are given $p_2 = 20$ bar, $t_2 = 400$°C. From steam table, we read

$$u_2 = 2947.5 \text{ kJ/kg and } v_2 = 0.1511 \text{ m}^3\text{/kg}$$

Since volume of tank is 0.8 m$^3$ (given)

$$\rightarrow m_2\, v_2 = 0.8 \text{ or } m_2 = \Delta m_{cv} = 5.294 \text{ kg}$$

$$\Delta U_{cv} = m_2 u_2 - m_1\, u_1 = m_2 u_2 = 5.294 \times 2947.5$$

Substituting these values in Eq. A and simplifying, we get,

$$W = 1011.7 \text{ kJ}$$

## SUMMARY

For thermal analysis of any system, the first step is to define clearly the system—whether it is a closed, open or an isolated one. We established through examples in Chapter 1 that the answer depends on the choice of the system. Once the type of system is known, then we can decide whether to use the non-flow, steady-flow, or the general energy equation. Fortunately, in most of the situations, flow in an open system is steady, and therefore, sfee is frequently used. Students are advised to master the application of nfee and sfee, as these are frequently used in analysis of thermal systems. Non-flow and steady-flow energy equations are written as $\delta q - \delta w = du$ and $\delta q - \delta w = dh$, respectively. Here, *du* and *dh* are change in internal energy per kg mass of a closed system and enthalpy for 1 kg/s flow rate of mass in an open system, and are equal to final minus initial value and not vice versa. It is assumed that change in external energies such as kinetic or potential is negligible, which in some cases may not be true. Therefore, care has to be taken when to neglect these while analysing a given system.

There are many steps involved in a systematic analysis, and in bringing a clarity in the solution. These are:

1. The first step is to state the problem. The data should be recorded on paper such that a glimpse over it gives what is given and what not. Then a schematic diagram is drawn showing given data at appropriate places and finally show different processes on the relevant 2-D diagram(s). Also, the assumptions, if required to be made, are to be mentioned. It is said, a question well understood is half done.
2. The second step is identifying the system and its type. If it is an open system, then whether the flow is steady or not, and take a decision on the energy equation used.
3. Look for the working substance and decide whether to use the equation of state or charts. For a gas, many equations of state, discussed in Chapter 4, are in use and a decision is to be made for the appropriate one. If the working substance is

a two-phase system, then charts are to be used to know the property and change in it during a process.

4. Look for type of work interactions, to determine net work transfer given by $\sum Wi$, where $Wi$ represents work interaction for mode $i$. Boundary work is obtained from integration $pdv$ for a closed system and $dvdp$ for an open system. Integration will depend on the relation between $p$ and $v$ for a given process. Usually, such a relation is of the form $pv^n$ = constant, where $n$ is 0, 1, $\gamma$ and $\infty$ for isobaric, hyperbolic, isentropic and isometric process, respectively. This equation is known as the process equation. For a perfect gas, hyperbolic and isothermal processes are the same.
5. We may also write the equation for change in internal energy for gases, or use the chart for determining change in internal energy. This is discussed in Chapter 4.

This way, we may determine the final state of the system and the amount of work and heat interaction. The second law analysis now can be done by determining change in entropy, availability and exergy for each process. Finally, exergetic efficiency and exergy balance is obtained; these are discussed in detail in Chapter 3.

# EXERCISES

## Review Questions

2.1 Why is the first law of thermodynamics needed?

2.2 What basic question does the first law answer?

2.3 What are the characteristics of energy, which follow directly from the first law?

2.4 What is the difference between internal energy and other forms of energy? List the forms of energy that contribute to the internal energy of a system.

2.5 Consider the process of heating water on top of an electric range. What are the forms of energy transformation involved during this heating process?

2.6 Consider two identical rooms, one with a refrigerator and the other without it. If all the windows and rooms are closed, will the room containing the running refrigerator be cooler or warmer than the other one?

2.8 When we wind a spring in a toy or stretch a rubber band, what happens in terms of work, energy and heat transfer? What happens when the spring is later released?

2.9 A constant mass goes through a process, whereby 100 W of heat transfer comes in and 100 W of work leaves. Is there any change in the state of the mass?

2.10 Water is heated from 1 bar 20°C to 10 bar 200°C. In one case, pressure is raised at $T = C$, then $T$ is raised at $p = C$. In the second case, the opposite order is followed. Does that make a difference for heat and work transfer?

2.11 A mass of 3 kg of nitrogen gas at 2000 K, cools at constant volume giving off 500 W. What is dT/dt?

2.12 A mass flow rate into a control volume is given by $PA \times C$ where $c$ is velocity component normal to $A$. Why?

2.13 A temperature difference causes heat transfer. Does a similar concept apply to mass flow rate?

2.14 How does a nozzle generate kinetic energy? Are we justified in taking work transfer zero in sfee?

2.15 What is the difference between nozzle flow free expansion and a throttle process?

2.16 If saturated steam is throttled, will it become superheated? Justify your answer.

2.17 Two streams of air enter a constant volume chamber at different pressure but at the same temperature. How would you determine the final pressure?

2.18 If you compress air in a compressor, its temperature goes up. Why?

2.19 An empty cylinder is filled with air from an air-line having air at 20°C and 1 bar pressure. Assuming no heat transfer, will the air temperature in the cylinder be more, equal or less than 20°C after the filling process? Why?

## Problems for Practice

2.1 A closed steel bottle contains ammonia at –20°C, $x = 0.2$ and the volume is 0.05 m$^3$. It has a safety valve that opens at a pressure of 14 bar. By accident, the bottle is heated until the safety valve opens. Find the temperature and heat transfer when the valve first opens.

2.2 A vertical cylinder fitted with a piston contains 5 kg of R 22 at 10°C. Heat is transferred to the system, causing the piston to rise until it reaches a set of stops, at which point the volume has doubled. Additional heat is transferred until the temperature inside reaches 50°C, at which the pressure inside the cylinder is 13 bar. (a) What is the quality at the initial state? (b) Calculate the heat transfer for the overall process.

2.3 A perfectly insulated vessel containing 0.05 m$^3$ of $H_2$ at 25°C and 5 bar is stirred at constant pressure till the temperature reaches 50°C. Determine (a) the heat transferred, (b) the internal energy, (c) the net work done, and (d) the stirring work done. Assume $H_2$ to be an ideal gas with $M = 2$ and $\gamma = 1.4$.

2.4 Hydrogen is contained in a piston-cylinder device at 1.01 bar and 0.125 $m^3$. At this state, a linear spring ($f \propto x$) with a spring constant of 219 kN/m is touching the piston, but exerts no force on it. The cross-sectional area of the piston is 0.279 $m^2$. Heat is transferred to the hydrogen, causing it to expand until its volume doubles. Determine (a) the final pressure, (b) the total work done by hydrogen, (c) the fraction of this work done against the spring. Also, show the process on a pv diagram.

2.5 3 kg of air at 2.5 bar and 77°C is compressed polytropically to 7.5 bar, $n = 1.2$. It is then cooled isothermally to its original state. Find the net work and heat transferred. [*Ans.* 485 kJ, –72.6 kJ) (IES 1992).

2.6 A rigid insulated tank is separated into two rooms by a stiff plate. Room A of 0.5 $m^3$ contains air at 2.5 bar and 300 K, and room B of 1 $m^3$ has air at 1.5 bar and 1000 K. The plate is removed and the air comes to uniform state. Find the final pressure and temperature.

2.7 A piston cylinder assembly has 1 kg of propane gas at 7 bar and 40°C. The piston cross-sectional area is 0.5 $m^2$, and the total external force restraining the piston is directly proportional to the cylinder volume squared. Heat is transferred to the propane until its temperature reaches 700°C. Determine the final pressure inside the cylinder, the work done by the propane and the heat transfer during the process.

2.8 The heaters in a spacecraft suddenly fail. Heat is lost by radiation at the rate of 100 kJ/h and the electronic instruments generate heat at the rate of 75 kJ/h. Initially, the air is at 1 bar and 25°C with a volume of 10 $m^3$. How long will it take to reach an air temperature of – 20°C?

2.9 In a jet engine, a flow of air at 1000 K, 2 bar and 40 m/s enters a nozzle, where the air exits at 500 m/s, 0.9 bar. What is the exit temperature, assuming no heat loss?

2.10 A sluice gate holds water 5 m high. A 1 cm diameter hole at the bottom of the gate allows water at 20°C to come out. Neglect any changes in the internal energy. Find the mass flow rate.

2.11 A diffuser has air entering at 1 bar and 300 K with a velocity of 200 m/s. The inlet cross-sectional area of the diffuser is 100 $mm^2$. At the exit, the area is 860 $mm^2$ and the exit velocity is 20 m/s. Determine the exit pressure and temperature of the air.

2.12 Helium is throttled from 12 bar and 20°C to a pressure of 1 bar. The diameter of the exit pipe such that the inlet and exit velocities are equal. Find the exit temperature of the helium and the ratio of pipe diameters.

2.13 Methane at 30 bar, 300 K is throttled through a valve to 1 bar. Calculate the exit temperature assuming no changes in the kinetic energy and ideal gas behaviour.

2.14 A blower handles 1 kg/s of air at 20°C and consumes a power of 15 kW. The inlet and outlet velocities of air are 100 and 150 m/s. Find the exit air temperature, assuming adiabatic conditions. [*Ans.* 28.38°C].

2.15 At the inlet of a nozzle, the enthalpy of fluid is 3000 kJ/kg and the velocity is 60 m/s. At the discharge end, the enthalpy is 2762 kJ/kg. The nozzle is horizontal and there is negligible heat loss from it. (a) Find the velocity at the exit from the nozzle. (b) If the inlet area is 0.1 $m^2$ and the specific volume at the inlet is 0.187 $m^3$/kg, find the mass flow rate. (c) If the specific volume at the exit is 0.498 $m^3$/kg, find the exit area of the nozzle. [*Ans.* 692.5 m/s, 32.08 kg/s, 0.3 $m^2$].

2.16 Consider a water pump that receives liquid water at 15°C and 1 bar and delivers it to a same diameter pipe having a nozzle with the exit diameter of 1 cm from which water is discharged to the atmosphere at 1 bar. Neglect the kinetic energy in the pipes and assume constant internal energy for water. Find the exit velocity and the mass flow rate if the pump draws 1 kW of power. [*Ans.* 29.4 m/s, 2.31 kg/s].

2.17 A household fan of diameter 0.75 m takes air in at 0.98 bar, 22°C and delivers it at 1.05 bar, 23°C with a velocity of 1.5 m/s. What are the mass flow rate, the inlet velocity and the outgoing flow rate?

2.18 A gas flows steadily through a rotary compressor. The gas enters the compressor at a temperature of 16°C, pressure of 1 bar and an enthalpy of 391.2 kJ/kg. The gas leaves the compressor at a temperature of 245°C, pressure of 6 bar and an enthalpy of 534.5 kJ/kg. There is no heat transfer to or from the gas as it flows through the compressor. (a) Evaluate the external work done per unit mass of gas, assuming the gas velocities at the entry and exit to be negligible. (b) Evaluate the external work done per unit mass of gas when the gas velocity at the entry is 80 m/s, and at the exit 160 m/s. [*Ans.* 143.3, 152.9 kJ/kg].

2.19 An air compressor takes in air at 1 bar and 17°C, and delivers it at 10 bar and 600 K to a constant pressure cooler, from which the air exits at 300 K. Find the specific heat transfer in the cooler.

2.20 An air turbine forms part of an aircraft refrigeration plant. Air at a pressure of 2.95 bar and a temperature of 58°C flows steadily into the turbine with a velocity of 45 m/s. The air leaves the turbine at a pressure of 1.15 bar, a temperature of 2°C and a velocity of 150 m/s. Neglecting changes in elevation, determine the magnitude and sign of the heat transfer per unit mass of air flowing. [*Ans.* +7.96 kJ/kg].

2.21 The air conditioner in a house or car has a cooler that brings atmospheric air from 30 to 10°C with both states at 1.01 bar. If the flow rate is 0.5 kg/s, find the rate of heat transfer.

2.22 Carbon dioxide enters a heater at 3 bar and 15°C, and exits at 2.75 bar. Changes in kinetic and potential energies are negligible and flow is steady. Calculate the required heat transfer per kg of $CO_2$ flowing through the heater.

2.23 A condenser brings 1 kg/s water flow at 10 kPa from 300°C to a saturated liquid at 10 kPa. The cooling is done by lake water at 20°C that returns to the lake at 30°C. For an insulated condenser, find the flow rate of cooling water.

2.24 An automotive radiator had glycerine enter at 95°C and return at 55°C. Air flows in at 20°C and leaves at 25°C. If the radiator should transfer 25 kW, what is the mass flow rate of the glycerine and what is the volume flow rate of air in at 1 bar?

2.25 Two air streams are combined to a single stream. One steam is 1 m$^3$/s at 20°C and the other is 2 m$^3$/s at 200°C, both at 1 bar pressure. They mix without any heat transfer to produce an exit flow at 1 bar. Find the exit temperature and volume flow rate, neglecting changes in energies.

2.26 Two steady flows of air enter a control volume. One is 0.025 kg/s of flow at 3.5 bar, 250°C (state 1), and the other enters at 4.5 bar, 15°C (state 2). A single flow exits at 1 bar, –40°C (state 3). The control volume rejects 1 kW of heat to the surroundings and produces 4 kW of power output. Determine the mass flow rate at state 2. Neglect the kinetic energies.

2.27 A closed cylinder contains a free frictionless piston. Initially, there is 0.03 m$^3$ of oxygen at 4 bar, 50°C on one side of the piston, and 0.06 m$^3$ of methane at 4 bar, –10°C on the other side. The cylinder wall and the piston may be regarded as perfect thermal insulators, but the oxygen may be heated electrically. Heating takes place so that the volume of oxygen is doubled. Find the final state, work done by the piston and the heat transfer to oxygen. Treat both gases as perfect gas for which $c_p$ and $R$ are $O_2$ 0.92 kJ/kg K, 0.25 kJ/kg K and $CH_4$ 2.01 kJ/kg K, 0.52 kJ/kg K. [*Ans.* $O_2$ side: $p_2$ = 10.2 bar, $T_2$ = 335 K, heat transfer = 147.56 kJ].

2.28 Air is contained in an insulated cylinder as shown in Fig. 2.12. The pressure, temperature and volume of air are 150 kPa, 25°C and 0.015 m$^3$. The area of cross-section of the piston is 0.05 m$^2$. The spring force varies linearly with distance with spring stiffness of 40 kN/m. The valve is now opened and air from the line at 800 kPa, 25°C flows into the cylinder until the pressure inside reaches 800 kPa. Determine the temperature inside the cylinder at the instant. Assume $c_p$ = 1.005 and $c_v$ = 0.71 kJ/kg K. [*Ans.* 352.6 K].

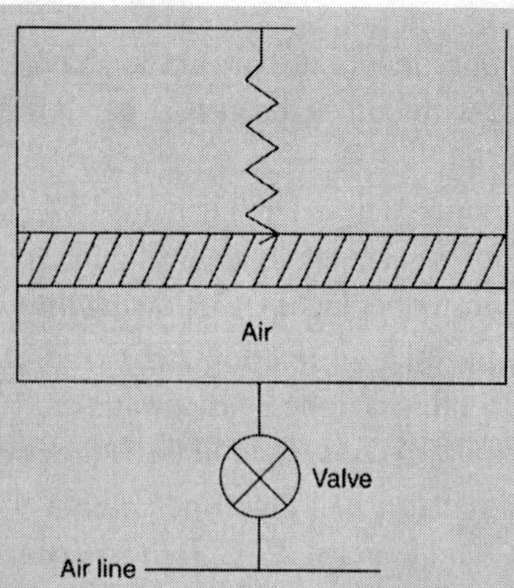


**Fig. 2.12** Piston loaded with spring

2.29 A simple carburettor has to supply 5 kg/min of air. The air is at 1.013 bar and 27°C. Calculate the throat area of the choke for air velocity of 90 m/s. Take velocity coefficient of 0.8. Assume isentropic flow. [*Ans.* 325 mm](IES 1992).

2.30 A compressed air bottle of volume 0.15 $m^3$ contains air at 40 bar and 27°C. It is used to drive a turbine, which exhausts to the atmosphere at 1 bar. If the pressure in the bottle is allowed to fall to 2 bar, determine the amount of work that could be delivered by the turbine. [*Ans.* 585 kJ.] (IES 98).

2.31 A rigid tank A of volume 0.6 $m^3$ contains 3 kg of water at 120°C, and rigid tank B is 0.4 $m^3$ with water at 6 bar and 200°C. They are connected to a piston cylinder assembly through valves. The pressure in the cylinder should be 8 bar to float the piston. Now, the valves are slowly opened and heat is transferred so the water reaches a uniform state at 250°C with the valves open. Find the final volume and pressure, and the work and heat transfer in the process.

## Multiple-choice Questions

2.1 During a cycle of processes, the heat transfers are the following: 120 kJ, –16 kJ, –48 kJ and + 12 kJ. The net work for the cycle is

(a) 60,000 N-m. (b) 68,000 N-m. (c) 120,000 N-m. (d) 44,000 N-m.

2.2 A system undergoes a cycle of processes *AB, BC, CD* and *DA.* The internal energy changes for various processes in kJ are

| | AB | BC | CD | DA |
|---|---|---|---|---|
| (a) | 32.5 | –18.5 | –30 | 16 |
| (b) | 18.5 | –32.5 | –30 | 16 |

(c) 30 −18.5 −32.5 16

(d) 16 −18.5 −32.5 +30

2.3 Internal energy of ideal gas is a function of

(a) temperature and volume. (b) pressure and volume.

(c) pressure and temperature. (d) temperature alone.

2.4 How is the internal energy change of an ideal gas expressed as?

(a) $du = C_v \, dT$ (b) $du = C_p \, dT$

(c) $du = (C_p - C_v) \, dT$ (d) $du = C_v \, dT + (\partial u/\partial v)_T \, dv$

2.5 A closed system undergoes a process 1–2 for which the value of $W_{12}$ and $Q_{12}$ are 50 kJ and + 20 kJ, respectively. If the system is returned to state 1 and the $Q_{21}$ is −10 kJ, the work $W_{21}$ is

(a) −80 kJ. (b) +40 kJ. (c) −20 kJ. (d) −40 kJ.

2.6 A closed system goes from state 1 to state 2 in a process for which $Q = 0$ and $W = 100$ kJ. Then, the system is returned to state 1 in a second process for which $W = -80$ kJ. The heat transfer for the second process will be

(a) 20 kJ. (b) zero. (c) −20 kJ. (d) indeterminate.

CHAPTER THREE

# Laws of Thermodynamics—II



## 3.1 SECOND LAW OF THERMODYNAMICS

The first law of thermodynamics helps in finding out change in energy of a system for a given amount of heat and work exchange in a process. However, it fails to answer questions like why heat flows from high temperature to low temperature, why work cannot be produced by a system undergoing a cycle change if it exchanges heat with a single thermal reservoir, why an apple falls to the ground after leaving the branch of a tree, or why water flows from high altitude to low altitude and so on. That is, energy analysis cannot answer phenomena taking place in one direction, which is a common characteristic of processes occurring in nature. Naturally, there must be some other law of nature to answer such questions. The answer came in the form of the second law of thermodynamics. This law is a collection of our experiences with nature regarding a particular direction in which a process occurs. These experiences are not limited to a few, but encompass the whole universe. Therefore, we have as many versions for this law as are our experiences. However, we shall restrict ourselves to the following four statements, which are of importance for analysis of thermodynamic systems.

These statements not only gave answer to many of our questions related to the directionality of a process, but also gave many important concepts because of which we were able to make improvements and use the available energy of the fuel in a better way. Clausius gave a very important inequality principle, which led to the concept of entropy, the law of increase in entropy for the universe, and lastly, the concept of exergy. These concepts gave a mathematical base to study a given process and to make fruitful inferences for the betterment of the process. The other important outcome was that it was possible to develop a scale for measurement of temperature. This chapter deals with all these aspects of the system, closed or open.

### 3.1.1 Kelvin–Planck Statement

This statement is useful in respect of systems undergoing a thermodynamic cycle to produce net work. Such a system is recognised as an engine. *Kelvin and Planck stated that it is impossible to construct an engine whose sole effect is to extract heat from a thermal reservoir and completely convert it into work.* However, work can be produced by an engine if the engine exchanges heat with at least two thermal reservoirs at different temperatures—extracting heat from the higher temperature reservoir and rejecting to the lower temperature reservoir. These thermal reservoirs may be finite or infinite; an infinite thermal reservoir means a reservoir of very large thermal capacity, so that extraction of any finite amount of heat from it does not change its temperature. The atmosphere, oceans, rivers, ponds, two-phase systems and industrial furnaces are examples where heat addition or removal does not affect their temperature. A thermal reservoir need not be large. Any physical body whose thermal energy capacity is large relative to the heat supplied or received by it may be used an as infinite thermal reservoir. A reservoir that supplies heat is called the source and the one receiving it, the sink.[1]

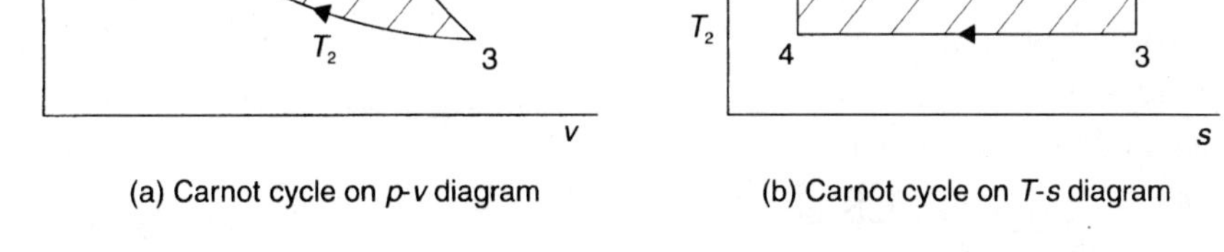


(a) Carnot cycle on $p$-$v$ diagram (b) Carnot cycle on $T$-$s$ diagram

**Fig. 3.1** Carnot cyle on $p$-$v$ and $T$-$s$ diagrams

1. Since the temperature of the source and sink is different, students usually take the reservoir at high temperature as the source and the one at low temperature as the sink, which is not true. The confusion probably arises as we are introduced with the concept of thermal reservoir with reference to an engine, and we live with that notion even in refrigeration, where the source is at low temperature.

Carnot developed a power cycle. He assumed an ideal cylinder with a perfect insulated wall and piston (no heat transfer across these) and its head made from a metal which is a perfect heat conductor so that heat transfer is possible across it even with infinitesimal temperature difference. The cylinder contains air, which is a perfect gas. Air confined within boundaries formed by the cylinder and piston is the system for analysis.

Figure 3.1 shows the Carnot cycle on *p-v* and *T-s* diagrams. The piston moves out very slowly from position 1 to 2 with the cylinder head at that moment in contact with the source at temperature $T_1$. Thus, heat $Q_1$ is transferred to air at constant temperature $T_1$ during process 12. At point 2, the source is removed, a perfect insulator is brought in contact with the cylinder head, and the piston is moved with infinite speed, so that the time taken for completion of process 23 is infinitesimally small for allowing any heat transfer, and the process may be considered isentropic. The system does work on the surroundings in both processes 12 and 23 given by the area under these curves with the *v*-axis on the *p–v* diagram. Now, the sink at temperature $T_2$ ($T_2 < T_1$) comes in contact with the cylinder head, and the piston moves inside slowly up to point 4; heat $Q_2$ is rejected by the system during this isothermal compression process 34 at temperature $T_2$. Finally, the isentropic process 41 completes the cycle. We may notice that this cycle is hypothetical only, as the situations demanded for realisation of the cycle cannot be implemented. Moreover, friction will always be present in the system, which makes any process irreversible. Nevertheless, we study this cycle as it has maximum efficiency. We notice that the cycle has been executed in the clockwise direction, and the area of the cycle is the net work output of the cycle, as explained in Section 1.2.9.2. Such a cycle is therefore, known as a power cycle. We observe that there is no mass flow across the system boundary during these processes, and therefore, it may be analysed using nfee.

Heat $Q_1$ is added to the system in the process 12, and $Q_2$ rejected in 34. By applying Joule's law to the Carnot cycle, work output of the cycle is $Q_1 - Q_2$. For the cycle, we define a performance parameter named *efficiency*, as a fraction of the input heat converted into work. Therefore, engine efficiency is $1 - Q_2/Q_1$. This expression suggests that efficiency can be made 100% by reducing the heat rejected to the sink to zero. However, the second law tells us that this is not possible.

We shall show in Chapter 7 that the efficiency of the Carnot cycle is equal to $1 - T_2/T_1$. Thus, efficiency of a reversible engine will be more if the source temperature is high and/or the sink temperature is low. This then means heat at high temperature has high quality, as more work is obtained. Therefore, both quantity and quality of heat matters for production of work. Nonetheless, temperature of the sink is not much in our control, as the atmosphere or water in lakes, ponds, etc. is normally taken as the sink. Equating two expressions for efficiency obtained above, we get

$$Q_2/Q_1 = T_2/T_1 \quad \text{or,} \quad Q_1/T_1 = Q_2/T_2 \tag{3.1}$$

Performance of any system, in general, can be expressed in terms of desired output and the required input. For example, for a power cycle, work is the desired output and heat supplied the input. Same way these are: for a refrigeration cycle the cooling produced in the evaporator and external work required for the refrigeration cycle, or that for a heat pump heat rejected from condenser for heating space and work done on the heat pump cycle.

### 3.1.2 Clausius Statement

It is our experience that heat flows out of the system only if the temperature of the system is more than that of the surrounding. But once the system acquires the surrounding temperature, then its temperature cannot be decreased further. The question that arises is—how can we achieve a temperature lower than the surrounding temperature? Clausius answered this question stating that *It is impossible to construct a cyclic device which produces no effect other than transfer of heat from a low temperature heat source to a high temperature heat sink.* Or, in other words, heat cannot flow by itself from low temperature to high temperature. Cooling, however, can be produced by using a Carnot refrigeration cycle, which extracts heat from the system (to be cooled) at temperature $T_2$, and rejects it to the sink at higher temperature $T_1$. Performance of such a cycle is measured by the term coefficient of performance (*COP*), defined as the ratio of heat extracted from cooled space to the amount of work done on the cycle. We will show later that COP for the Carnot cycle is $T_2/(T_1 - T_2)$. Such a cyclic device is known as a refrigerator.

### 3.1.3 Principle of Increase of Entropy

We shall define entropy later in Section 3.1.4. The principal of increase of entropy states that entropy of the universe increases in all processes except a reversible process for which there is no change. Section 3.1.5 deals more about it.

### 3.1.4 Principle of Degradation of Energy

We learnt that a given amount of work can be converted into equivalent amount of heat, but the reverse is not true. The amount of work that may be obtained from a given amount of heat depends on the temperature of the source and the sink, as well as the amount of irreversibility present in the cycle. If temperature of the source is high, then more work is obtained because efficiency of the cycle is more. In other words, no work can be obtained from a vast source of energy like atmospheric air or water in a lake, since then there is only one thermal reservoir with which an engine can exchange heat. So, we use the phrase *poor or good quality of heat.* This led to the development of exergy analysis of the system, which is very important, and any project on a thermal system is to be evaluated on this basis, too. Only the first law or energy analysis is not enough.

We get maximum work from the system when its state is changed reversibly to standard sink temperature taken at 27°C; it is called *availability* of the system. Dissipating agents such as friction, eddies, etc. are present in the system, resulting in less work than decrease in availability. This fact is expressed by the *principle of degradation of energy*, which states that when a system executes a process, then availability of the system decreases more than the work obtained during the process. That is, availability of a system is lost in all real processes.

**Example 3.1** An inventor claims that he has developed an engine that takes 2500 kJ of heat at a temperature of 500 K, rejects 1000 kJ heat at a temperature of 300 K and delivers 1285 kJ of work. Would you advise your client to put this engine to use?

***Solution*** We have learnt that maximum efficiency of a reversible engine will be equal to 1 – 300/500 = 0.4. Therefore, such an engine can deliver work = efficiency × heat intake = 0.4 × 2500 = 1000 kJ, whereas the claim is for 1285 kJ, which is not possible. Hence, the claim of the inventor cannot be accepted. The advice therefore would be not to put this engine to use.

**Example 3.2** A Carnot cycle operates between two heat sources at temperatures $T_1$ and $T_2$ ($T_1 > T_2$). However, the temperature of the working fluid is $\theta_1$ and $\theta_2$, where $\theta_1 < T_1$ and $\theta_2 > T_2$. If $T_1 = (\theta_1 + KQ_1)$ and $T_2 = (\theta_2 - KQ_2)$ where, $Q_1$ and $Q_2$ are the amount of heat extracted from the source and rejected to the sink, respectively, and $K$ is a positive constant, find out the efficiency of the cycle.

***Solution*** This is an example illustrating the effect of finite temperature difference between the temperature of the source or sink and the working fluid, and also to realise whether to use $\theta_1$ and $\theta_2$ or $T_1$ and $T_2$ in Eq. 3.1. Recall that the temperatures $T_1$ and $T_2$ in Eq. 3.1 are temperatures of the working fluid, and hence that of the thermal reservoir, because of the infinitesimal temperature difference between the working fluid and the reservoir. Equation 3.1 gives

$$\frac{Q_2}{Q_1} = \frac{\theta_2}{\theta_1} = \frac{T_2 + KQ_2}{T_1 - KQ_1}$$

$$\Rightarrow Q_2\,(T_1 - KQ_1) = Q_1\,(T_2 + KQ_2)$$

or $$\frac{Q_2}{Q_1} = \frac{T_2}{T_1 - 2KQ_1} \quad \text{and} \quad \eta = \frac{Q_2}{Q_1} = 1 - \frac{T_2}{T_1 - 2KQ_1}$$

We observe that the denominator has decreased due to finite temperature difference between the sink or source and the working fluid. This decreases the efficiency of the Carnot cycle. *So a finite temperature difference is responsible for decrease in work output and efficiency of the engine and to introduce irreversibility in the process.* This may be generalised for any other thermodynamic potential. The other cause of increase in irreversibility in any process is the presence of energy dissipating agents in a system like friction, turbulence, and eddies, which change high grade of energy such as kinetic energy to low grade internal energy. Thus, irreversibility is caused by finite potential difference and presence of dissipating agents. These are known as external and internal irreversibility, respectively.

**Example 3.3** A heat pump working on a reversed Carnot cycle takes in energy from a reservoir at 3°C and delivers it to another reservoir at 77°C. The heat pump gets power from a reversible engine, taking heat from the reservoir at 1077°C and rejecting to the reservoir at 77°C. For 100 kJ/s of energy supplied to the reservoir at 77°C, estimate the energy taken from the reservoir at 1077°C.

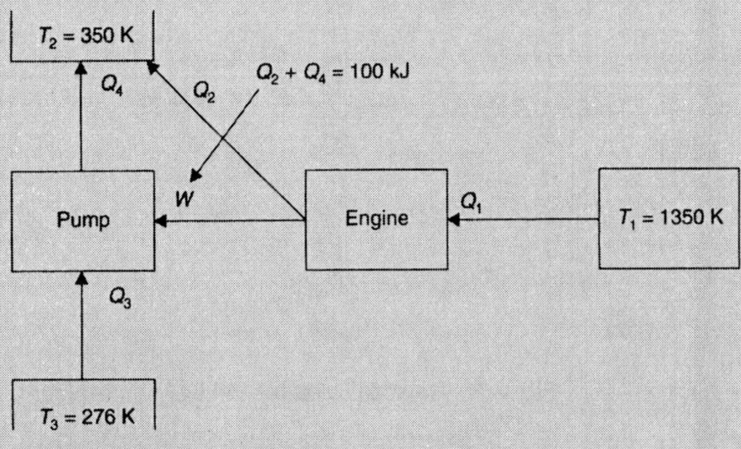


**Fig. 3.2** Heat pump run by a Carnot engine

***Solution*** Given: $T_1 = 1077 + 273 = 1350$ K, $T_2 = 77 + 273 = 350$ K, $T_3 = 276$ K, and $Q_2 + Q_4 = 100$ kJ.

The arrangement of heat flow, amount and direction of energy flow, and the relative position of the heat pump and engine are shown in Fig. 3.2. From the definition of engine efficiency and COP of heat pump, we write

$$\eta = \text{Engine output/input to engine} = (Q_1 - Q_2)/Q_1 = (T_1 - T_2)/T_1$$

$$Q_1/Q_2 = T_1/T_2 \text{ and } Q_1 - Q_2 = Q_1\,(T_1 - T_2)/T_1 \qquad \text{(A)}$$

$$\text{COP} = \text{Heat ejected to sink/work done} = Q_4/(Q_4 - Q_3)$$

$$= T_4/(T_4 - T_3)$$

$$\rightarrow Q_4 - Q_3 = Q_4\,(T_4 - T_3)/T_4 \qquad \text{(B)}$$

But work for running the pump is given by the engine. Therefore, $Q_1 - Q_2 = Q_4 - Q_3$. Equating the expression for these from Eq. *A* and *B*, substituting and simplifying, we get $Q_4 = 3.5\ Q_1$. Also,

$$Q_2 = T_2/T_1 \times Q1 = 350/1350\ Q_1 = 0.259\ Q_1$$

We are given $Q_1 + Q_2 = 100$. Substituting for $Q_2$ in terms of $Q_1$ gives after simplification, $Q_1 = 26.60$ kJ

### 3.1.5 Equivalence of the Second Law of Thermodynamics

Though there are many ways in which the second law has been stated, basically, they are one and the same. This means that if a system violates any of the statements, it will violate other statements as well. We shall demonstrate this for a device violating, say, the Kelvin–Planck statement.

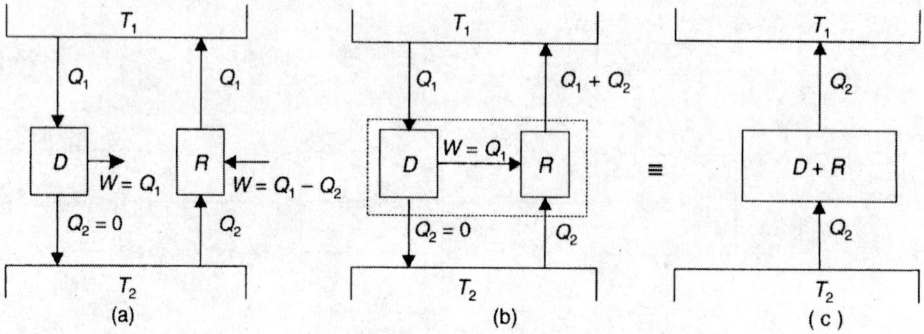


**Fig. 3.3** Device D violates the Kelvin statement

Figure 3.3 (a) shows an engine *D*, which violates the Kelvin–Planck statement. That is, it draws heat $Q_1$ from thermal reservoir at temperature $T_1$, rejects *no heat* to the thermal reservoir at temperature $T_2$ ($T_1 > T_2$) and completely converts $Q_1$ into work. *R* is a refrigerator, which works between same thermal reservoirs extracting heat $Q_2$ at $T_2$, rejecting heat $Q_1$ at $T_1$ and requiring work equal to $Q_1 - Q_2$ for completion of this cycle. We may couple *D* with *R* and run the two together (Fig. 3.3 b). We notice that R rejects heat $Q_1 + Q_2$ to the thermal reservoir at $T_1$ and takes heat $Q_2$ at $T_2$, whereas *D* takes heat $Q_1$ from the thermal reservoir at $T_1$ and rejects no heat during completion of the cycle by the combined system. Work produced by *D* is supplied to *R*. *D* and *R* together may be taken as a system, the situation is shown in Fig. 3.3c. Since work transfer from $D$ to $R$ is an internal phenomenon of the system, the work exchange with the surrounding is zero. Thus, there is heat transfer $Q_2$ from the thermal reservoir at $T_2$ to that at $T_1$ without requiring any work. So, we have demonstrated that heat is transferred by itself from low to high temperature, which is in violation of the Clausius statement. Therefore, existence of device $D$ violating the Kelvin–Planck statement ensures the existence of another device which shall violate the Clausius statement.

Similarly, we may show that if a device that violates the Clausius statement, then it is possible to combine it with an engine $E$ (not violating the Kelvin–Planck statement) to produce another system which shall violate the Kelvin–Planck statement. We leave it to the reader to verify it in the same way as we did above, keeping in mind that the combination will result in a situation where heat is extracted from a single thermal reservoir and completely converted into work.

## 3.2 SECOND LAW OF THERMODYNAMICS—COROLLARIES

We shall discuss only the important aspects of corollaries to see the power of this law.

### 3.2.1 Efficiency of Cycle

Amongst all the engines operating between a given source and sink, a reversible engine has maximum efficiency, which is equal to the Carnot efficiency and does not depend on the working fluid used. This is known as the Carnot principle. Further, all the reversible engines working in the same temperature limits have the same efficiency. This corollary has importance because Carnot efficiency is the maximum efficiency that may be achieved, and any exercise to achieve more than this value is futile. Similarly, coefficient of performance of the Carnot refrigeration cycle is maximum for given operating temperature limits.

**Proof.** Let $X$ and $R$ be irreversible and reversible engines operating between same temperature limits $T_1$ and $T_2$, as shown in Fig. 3.4, and that both are extracting the same amount of heat $Q$ from the reservoir at $T_1$. *We assume that X is more efficient.* Then, work produced $W_X$ by the irreversible engine is more than that produced by the reversible engine, $W_R$. Fig. 3.4 (a) shows the situation. Now, imagine both $X$ and $R$ are coupled, while $R$ is running as a refrigerator, taking the required work from $X$. Since $W_X > W_R$, the combined system will produce $W_X - W_R$ amount of work. Figure 3.4 (c) summarises the work and heat interactions. It is seen that the combined device is violating the Kelvin–Planck statement, and hence this is not possible. So *R is more efficient.*

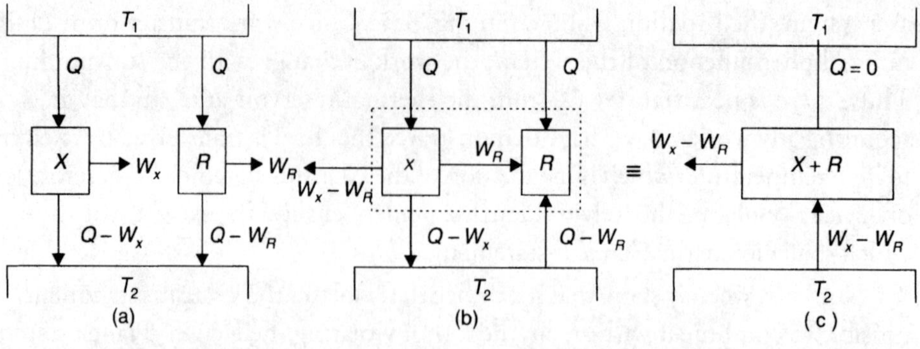


**Fig. 3.4** Efficiency of a reversible engine is maximum

Now, assume that $X$ is also a reversible engine. Then, the above theorem implies $\eta_x \le \eta_R$ and $\eta_R \le \eta_x$. This is possible only when efficiency of both the reversible engines is the same. Thus, all reversible engines operating between two thermal reservoirs at $T_1$ and $T_2$ have the same efficiency and it is maximum. In the same way, we may prove that the efficiency of a reversible engine does not depend on the working substance of the engine.

### 3.2.2 Absolute Temperature Scale

Temperature is an important parameter in the design of thermal systems and needs to be measured accurately for which a scale is needed, like the one used for measurement of length, mass or time. In fact, temperature has also been taken as a fundamental quantity, and therefore, its units are to be defined like any other fundamental quantity. In the past, many temperature scales such as Celsius, Fahrenheit, etc., were in use. In these scales, melting and boiling points of water were fixed, and then the range was linearly divided into a certain number of divisions. For example, Celsius assigned 0 and 100 value to these two temperatures and divided the range in 100 equal divisions to give the Celsius temperature scale. Fahrenheit, on other hand, assigned values 32 and 212 respectively, and divided the range in 180 parts. However; these scales did not have any thermodynamic base.

Lord Kelvin used the Carnot engine and Eq. 3.1 to define a Thermodynamic Temperature Scale. From Eq. 3.1, zero on this scale is that temperature of a sink when a Carnot engine, operating between it and a source at any temperature $T_1$, will convert completely a given amount of heat taken from the source into work, with no heat rejection to the sink. He decided the size of the scale by choosing the triple point[2] of water, and

2. Water can exist in three single phases (solid (ice), liquid (water) or gas (steam)), two phases (solid–liquid, liquid–gas (wet steam) or solid–gas), or three phases. The state when all the three phases of water are present simultaneously is called the triple point; it occurs only for one combination of pressure and temperature, which for water is 0.006028 atm and 0.01°C.

dividing the range from 0 to this temperature linearly. The unit of temperature is Kelvin; written as K. Thus, magnitude of a Kelvin is 1/273.16 of the temperature interval between absolute zero and the triple point temperature of water. This scale has an inherent characteristic that when a Carnot engine operates anywhere on the scale between different thermal reservoirs such that their temperature difference is the same produces equal work. That is, if two Carnot engines are operating between 100 and 200°C or 450 and 550°C, their output is the same. Figure 3.5 shows a number of reversible engines working between different temperature limits, such that they produce the same work. Engine $E_1$ operates between $T_1$ and $T_2$, $E_2$ operates between $T_2$ and $T_3$, and so on. Heat rejected from one engine is supplied to the next engine. The figure also shows the amount of heat transaction and the temperature at which it is occurring.

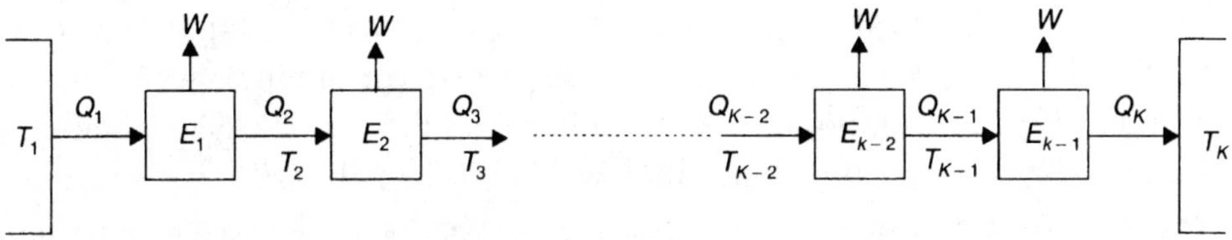


**Fig. 3.5** Reversible engines product equal work

From Eq. 3.1 and the fact that the engines are producing the same work $W$, we may write

$$Q_1/T_1 = Q_2/T_2 = Q_3/T_3 = \dots = Q_K/T_K$$

$$Q_1(T_1 - T_2)/T_1 = Q_2(T_2 - T_3)/T_2 = Q_3(T_3 - T_4)/T_3$$

$$= \dots = Q_{k-1}(T_{k-1} - T_{k-2})/T_{k-1} = W$$

Substituting from first equation into the second and simplifying, we get

$$T_1 - T_2 = T_2 - T_3 = T_3 - T_4 \dots = T_K - T_{K-1}$$

This proves the statement. This temperature scale is known as the absolute temperature scale.

However, the absolute temperature scale could not be implemented in practice, since the Carnot engine itself is an abstract concept. Therefore, the International Practical Temperature Scale has been adopted for industrial and scientific measurement of temperature, for which an error table giving deviation between the actual temperature measured by the standard thermometer prescribed for that temperature range and the absolute temperature is available for correction. That way, we can measure absolute temperature with reasonable accuracy. Further, temperatures on absolute and Celsius temperature scales are related by the relation $T = 273.15 + t$, where $T$ and $t$ are temperatures on absolute and Celsius scales. In practice, 273 and not 273.15 is used.

### 3.2.3 Entropy Concept—Clausius Inequality Principle

The first law of thermodynamics gives the concept of energy. Similarly, the second law predicts the presence of another property of the system. This could be possible because of the Clausius inequality principle, which states that *whenever a system executes a cycle,* $\int \delta Q/T$ *around the cycle is less than zero for a real cycle and zero for a reversible cycle.* This can be easily demonstrated by assuming that a closed system is executing a process in which heat $\delta Q$ goes in and work $\delta W$ is done by the system. In order that the heat transfer may be reversible, we assume that the system interacts with the surrounding at temperature $T_R$ through a Carnot engine. Let $T$ be the temperature of the system (Fig. 3.6a). (If $T$ is greater than the surrounding temperature, then heat will flow from the system to the surrounding.) The engine extracts heat $\delta Q_R$ from the surroundings and transfers $\delta Q$ to the system, such that $\delta Q_R/T_R = \delta Q/T$ (Eq. 3.1) and produces work $\delta W_R = \delta Q_R - \delta Q$. In the mean time, the system produces work $\delta W$. For the combined system comprising system and engine, the first law gives combined work output, $\delta W_C = \delta W_R + \delta W$. Using nfee, we get $\delta W_C = \delta Q_R - dE_c$, where $dE_C$ is change in energy of the system. Now, assume the system executes a cycle and the engine undergoes an integral number of cycles. The net output of the combined system has to be less than zero by the Kelvin–Planck statement of the second law of thermodynamics, since the combined system is extracting heat from a single source. Then,

$$\oint \delta W_C = \oint \delta Q_R = T_R \oint \delta Q/T \le 0 \quad \Rightarrow \quad \oint \delta Q/T \le 0$$

However, it is possible that work is done on the system and heat is transferred from it to the thermal reservoir. *Therefore, the combined work output of the system and engine can never be positive.* This proves the Clausius inequality principle.

Now, consider reversible processes A, B and C between two state points 1 and 2 or 2 and 1, as shown in Fig. 3.6 (b). Applying the Clausius principle to two cycles 1A2B1 and 1A2C1, we get

$$\int_{1A}^{2} \delta Q/T + \int_{2B}^{1} \delta Q/T = 0 \quad \text{and} \quad \int_{1A}^{2} \delta Q/T + \int_{2C}^{1} \delta Q/T = 0$$

or

$$\int_{2B}^{1} \delta Q/T = \int_{2C}^{1} \delta Q/T = S_1 - S_2 \tag{3.2}$$

Note that both the cycles are reversible. We observe that change in $\delta Q/T$ in a reversible process depends only on the end states and not on the path; hence it represents change in some property of the system. This property is named *entropy*, represented by the symbol $S$ having SI units J/kg K. However, it is measured in kJ/kg K. We may write

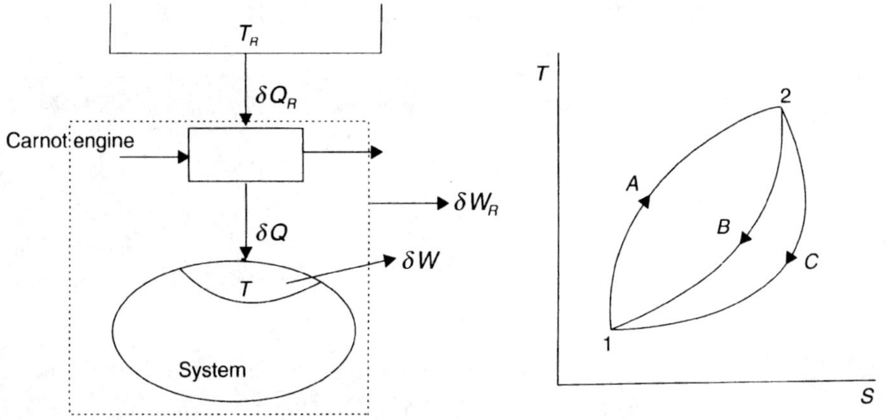


(a) System executing a process (b) Processes to which system is subjected

**Fig. 3.6** Clausius inequality and the entropy concept

$$dS = (\delta Q/T)_{rev} \tag{3.3}$$

Now, imagine that the process B in Fig 3.6 (b) is irreversible. Then, for the irreversible cycle 1A2B1 and reversible cycle 1A2C1, we can write the Clausius equation as

$$\int_{1A}^{2} \delta Q/T + \int_{2B}^{1} \delta Q/T < 0 \quad \text{and} \quad \int_{1A}^{2} \delta Q/T + \int_{2C}^{1} \delta Q/T = 0$$

or
$$\int_{2B}^{1} \delta Q/T < \int_{2C}^{1} \delta Q/T = S_1 - S_2 \tag{3.4}$$

Comparing these two equations, we observe that the integral of $\delta Q/T$ from 2 to 1 along path B is less. But entropy being a point function, change in it is the same along B or C. This means that change in entropy is more than that obtained by evaluating $\int \delta Q/T$, that is, *there is production of entropy in process B due to some other cause and this cause is irreversibility of the process.* This extra entropy is entropy production, which led to the formulation of the second law as principle of increase of entropy or degradation of energy, as mentioned in the beginning of the second law.

Next, for the irreversible cycle 1A2B1, we may show that the entropy of the universe increases at the conclusion of the cycle in Table 3.1.

$$\therefore \ (\Delta S)_{universe} = (\Delta S)_{system} + (\Delta S)_{surrounding} > 0 \tag{3.5}$$

Equation 3.5 is the universal law of increase of entropy, which states that *entropy of the universe increases in an irreversible process and cycle.*

**Table 3.1** Increase in entropy for an irreversible cycle 1A2B1

| Process | $(\Delta S)_{system}$ | $(\Delta S)_{surrounding}$ | $(\Delta S)_{universe}$ | Remarks |
|---|---|---|---|---|
| 1-A-2 | $S_2 - S_1$ | $S_1 - S_2$ | 0 | Reversible process |
| 2-B-1 | $S_1 - S_2$ | $< (S_1 - S_2)$ | $>0$ | Irreversible process |
| Cycle(row 1 + row 2) | 0 | $> 0$ | $> 0$ | Irreversible cycle |

**Example 3.4** A cylinder contains a mixture of water and vapour at 100°C and rejects 300 kJ heat to the surrounding at 27°C. Find out the change in entropy of the universe.

***Solution*** In the given question, water and its vapour is the system, and the cylinder plus outside air is the surrounding. We assume that the quantity of the mixture is sufficient that two phases co-exist even after rejection of 300 kJ of heat. Since the cylinder is in contact with two phases of water, its temperature will also not change and would remain at 100°C. However, the amount of heat received from the system is transmitted to air, and there is no change in its state, and hence in its entropy. In the process, heat is rejected at constant temperature of 273 + 100 = 373 K, and received by the atmosphere at constant temperature 273 + 27 = 300 K. Therefore, total change in entropy

$$\Delta S = -300/373 + 300/300 = 0.1957 \text{ kJ/K} > 0$$

Entropy of the universe has increased, since heat transfer is due to finite temperature difference being an irreversible process.

**Example 3.5** A cylinder contains 1 kg of air at 27°C and is heated by an electric heater placed inside it to a temperature of 57°C. Calculate change in entropy of the universe.

***Solution*** In this problem, if we take the contents of the cylinder including the heater as the system, then the interaction with the surrounding is through the electric current, and hence there is no change in entropy of the surrounding. However, temperature and pressure of air are increasing, resulting in increase in entropy of the system. This indicates that something is happening within the system that is causing entropy production. We may visualise it easily, as the current flowing through the heating element produces heat whose redistribution within the system is irreversible, and hence increase in entropy. The amount of entropy generation can be determined by using the entropy relations for a perfect gas, as we know the initial and final temperatures of air.

On the other hand, if only air contained in the cylinder is taken as the system, then the heating element is forming the surrounding. Energy is transferred to air because of temperature difference between the heating element and air, which is an irreversible process and causes entropy production.

Since heating occurs at constant volume in both the above situations, heat supplied to air for a small change in its temperature $dT$ at absolute temperature $T$ is $\delta q = c_v\, dT$ and change in entropy of air for this heat transfer is $ds = c_v\, dT/T$.

$$\therefore\ \Delta S = \int_1^2 c_v dT/T = c_v \ln(T_2/T_1) = 0.718 \ln(330/300) = 0.068 \text{ kJ/kg K}$$

**Example 3.6** A mass $m$ of water at $T_1$ is isobarically and adiabatically mixed with an equal mass of water at $T_2$. Find out the change in entropy of the universe and show that this is positive.

***Solution*** After mixing, both waters will have temperature, say $T_f$. We may find it out by considering an insulated container having a partition with water at $T_1$ on its one side, and at $T_2$ on other side, and its contents as the system. This way, the system is made a closed system. Let the partition be removed with negligible work being done. Due to removal of the partition, water flows from one side of the partition to another side, but it is within the system and hence no mass is crossing the system boundary. Mixing of two water lumps occurs within the system in which there is no work or heat transfer with the surrounding. Thus, change in internal energy of the system is zero, by nfee. That is, $dU = 0$. However, $dU$ for incompressible fluids such as water is equal to $(mcdT)$, where $c$ is the specific heat taken same for both the water lumps. Its integration from some reference temperature, $T_{ref}$ (at which $U = 0$) to the given temperature will give change in internal energy.

$$\therefore\ mc(T_1 - T_{ref}) + mc(T_2 - T_{ref}) = 2\,mc(T_f - T_{ref}).$$

or $$T_f = (T_1 + T_2)/2.$$

Further, the amount of heat, $\delta Q$, required to change the temperature of water by $dT$ is given by the equation, $\delta Q = mcdT$. Substituting it in the Eq. 3.3 for both water lumps, we get

$$\Delta S_1 = \int_{T_1}^{T_f} \frac{mcdT}{T} = mc\ln\frac{T_f}{T_1} \quad \text{and} \quad \Delta S_2 = \int_{T_2}^{T_f} \frac{mcdT}{T} = mc\ln\frac{T_f}{T_2}$$

or $$\Delta s = \Delta S_1 + \Delta S_2 = mc\ln\left(T_f^2/T_1T_2\right) > 0$$

$T_f$ is the arithmetic mean, whereas $\sqrt{(T_1T_2)}$ is the geometric mean of $T_1$ and $T_2$. We know that the arithmetic mean is greater than the geometric mean, and therefore augment of log function is more than 1. So, change in entropy is positive. Moreover, there is no change in entropy of the surrounding, as there is no interaction with the surrounding and change in entropy of a system is change in entropy of the universe.

**Example 3.7** Two identical bodies of mass $m$ and constant specific heat $c$ are at temperatures $T_1$ and $T_2$, respectively ($T_1 > T_2$). They are used as source and sink for a heat engine. Find out the maximum obtainable work.

***Solution*** Maximum work is obtained if heat transfer from the source to the sink takes place reversibly, which can be made by running a reversible engine between them. Since the source and the sink are finite thermal reservoirs, any amount of heat transfer from them will change their temperature. Thus, temperature of the source shall decrease, and that of the sink will increase, till both are at the same temperature $T_f$. Let $W$ be the work produced by the engine. If we consider the system to consist of the source, the sink and the engine, then it is a closed system, which gives work W to the surrounding, and there is no other interaction. From nfee, we get $W = -dU$. It may be emphasised here that heat transfer from the source to the engine, or from the engine to the sink is an internal phenomenon of the system, and hence is not an interaction; rather, it is redistribution of internal energy, and we note that internal energy of the system decreases in this process.

To calculate change in internal energy, we shall need the final temperature of the source and the sink. This is obtained by using the fact that work interaction with the surrounding does not change entropy of the surrounding, and that change in entropy of the system is also zero because of reversible heat transfer between the source and the engine, or the sink and the engine. We may find out change in entropy of the source and the sink in the same way as we did in Example 3.6 and write

$$\Delta s = \Delta S_1 + \Delta S_2 = mc\ln\left(T_f^2/T_1T_2\right) = 0 \quad \text{or} \quad T_f^2 = T_1T_2$$

$$\therefore\ W = -dU = U_i - U_f = mc(T_1 + T_2) - 2mc\,T_f = mc\left(\sqrt{T_1} - \sqrt{T_2}\right)^2$$

We note that the right hand side will always be positive, irrespective of whether $T_1$ or $T_2$ is greater. The only difference will be that mass at higher temperature shall be a source, and the other the sink.

**Example 3.8** A heat pump operates between two identical bodies. In the beginning, both the bodies are at the same temperature $T_1$, but operation of the heat pump cools down one of the bodies to temperature $T_2$. Determine the minimum work required for the pump. Assume specific heat of both bodies as $c$.

***Solution*** Minimum work will be done if heat is pumped from one body to another in a reversible manner, in which case, the total change in entropy of the two bodies would be zero. Let $T_f$ be the temperature of the other body when that of the cold body is $T_2$. Since masses of the two bodies are not given, we assume the same mass $m$ for these. Then, the change in entropy for both, using equation $dS = m\,c\,dT/T$, gives

$$mc\int_{T_1}^{T_f} dT/T + mc\int_{T_1}^{T_2} dT/T = 0 \ \rightarrow\ T_f = T_1^2/T_2$$

Consider both bodies and the heat pump as the system and assume the pump to execute an integral number of cycles, so that there is no change in its internal energy. We may write for the initial and final internal energy of the system, and use nfee to give the amount of work required.

$$U_i = 2mcT_1 \ \text{ and } \ U_f = mc\left(T_2 + T_f\right) = mc\left(T_2 + T_1^2/T_2\right)$$

$$\therefore\ W = U_f - U_i = mc\left(T_1^2/T_2 + T_2 - 2T_1\right) = mc(T_1 - T_2)^2/T_2$$

### 3.2.3.1 Entropy Change in an Open System

In Section 3.2.3, we determined change in entropy of a closed system due to heat transfer. However, in an open system, mass too crosses the system boundary through which there is convection of entropy. So, in this case, change in entropy of the system is given by the sum of both the effects. The entropy change due to heat transfer is to be obtained assuming it to be reversible. For the purpose, the system surface may be divided into a number of finite surfaces (as temperature may be different at different places on the surface, as we move from the entrance to the exit of the system), such that each such surface is at the same temperature, and heat is transferred from the surrounding to each of these surfaces through separate Carnot engines. Let $T$ be temperature of one such surface and heat transferred in unit time be $\delta Q$. Then time rate of change of entropy of the system due to heat transfer alone is the sum of change in entropy at all the surfaces = $\sum \delta Q/T$. Now, superimposing over it, the change due to entropy convection, the total time rate of change of entropy of the open system is

$$dS_\sigma = m_1 s_1 - m_2 s_2 + \sum \delta Q/T \tag{3.6}$$

for a steady flow, $dS_\sigma$ will be zero and in such a case,

$$\sum \delta Q/T = m_2 s_2 - m_1 s_1 \tag{3.7}$$

Equation 3.7 is true only when the system is undergoing a reversible process. If that is not so, then the various causes of irreversibility will generate entropy, which is not accounted for by this equation.

## 3.3 AVAILABILITY CONCEPT

The amount of work that may be obtained from a given system by subjecting it to a process, is limited by the surrounding temperature. Consider, for example, oceans or the air around us. Can we get any work from these? The answer is no, because we cannot get a thermal reservoir at a temperature lower than the temperature of these vast sources of internal energy. Hence, the engine cannot produce any work. Therefore, work is obtained from a given system only when its temperature is more than that of the atmosphere, oceans, river water or other such reservoirs, which are normally taken as the sink. Once the system acquires the sink's temperature, then no more work can be obtained from it by heat transfer. Some students suggest that a low temperature sink may be produced using the refrigeration cycle. Well, this can be done, but then the work done in producing the temperature of the sink will be more than the work that shall be obtained by running the engine with the surrounding as source. So, this possibility is ruled out.

However, work may still be obtained by reducing other potentials of the system to that of the sink. Once all the potentials of the system are reduced to that of the sink, then no more work can be obtained. This state of the sink is referred to as the dead state, because once the system attains this state, it is not possible to get any more work from it.

We know that when we rub our hands, heat is produced, thereby increasing temperature. But it will not be possible to recover this work again from the amount of heat so generated. Or, heat produced when the electric current flows through the resistance wire in a heater cannot be converted back to current. Thus, whereas work can be converted into heat completely, the vice versa is not true. Therefore, we accordingly grade different forms of energy into high-grade and low-grade. For example, kinetic, electric energy, etc., is high-grade energy, as these can be completely converted into work. On the contrary, energy from fossil and nuclear fuels cannot be completely changed to mechanical work, and is called low-grade energy.

We learnt earlier that finite potential difference between the system and the surrounding makes the process irreversible, and hence the capacity of maximum power production is reduced. Similarly, the presence of friction will reduce the amount of work actually obtained, as some of it will be used up in overcoming the friction. This may be generalised to all kinds of irreversibility, internal or external. In fact, the presence of dissipating agents in the system converts high-grade of energy into low-grade energy. Such dissipating agents and finite potential difference between the system and the sink cause irreversibility in the process, and results in decreasing capability of the system to deliver maximum work.

Maximum work is obtained from a system when its state is taken to a dead state by a reversible process; this maximum work is called *avan* of the system when the final state of the system is the dead state. However, Rant gave the name *exergy* to it in 1954. Another name, *assergy* is also found in literature. However, with time, the term 'exergy' has found use throughout the world. Determination of irreversibility in a process had proved to be very useful in improving the performance of the thermal system, utilisation of the available energy in a better way and reducing pollution. This is the availability concept. Loss of work in an irreversible process is termed *irreversibility*. That is, irreversibility is the difference of the exergy and the actual work obtained for the same change of state of the system, irrespective of whether the final system state is dead or any other state. On the other hand, the maximum useful work during the process is the difference in exergy of the system from the initial to the dead state.

### 3.3.1 Availability Function or Exergy of the System

While determining the availability function of a system, we have to keep in mind that surrounding air exerting pressure on the system boundary, may not be part of the surrounding of the system, as we discussed for internal combustion engine in Chapter 1. In such a case, the amount of work done by the expanding system boundary against the air pressure is lost, and therefore, has to be subtracted from the maximum work to get exergy. We shall now determine exergy of thermal reservoirs, closed system and open system.

#### *3.3.1.1 Thermal Reservoir*

It is a source of internal energy. Figures 3.7 (a) and (b) show variation of temperature of the reservoir with withdrawal of any amount of heat $\Sigma\,\delta Q$ from an infinite and finite thermal reservoir. We will find out the availability function of this amount of heat $\Sigma\,\delta Q$ for which we run the Carnot engine *n* times between the thermal reservoir extracting heat $\delta Q$ at temperature $T$ and rejecting heat $\delta Q_0$ at $T_0$ in each cycle; one cycle abdca is shown in the figure. Output of the engine in a cycle is $\delta Q - \delta Q_0 = \delta Q - T_0/T \times \delta Q$, usir

Eq. 3.1. Summing it over $n$ cycles and taking out the constant quantity from the summation sign, we get:

$$\text{Availability of heat from thermal reservior} = \Sigma(1 - T_0/T)\delta Q = \Sigma\delta Q - T_0\Sigma(\delta Q/T)$$

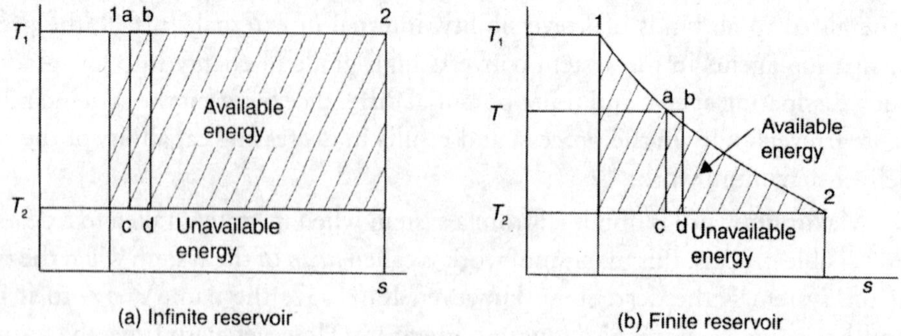


**Fig. 3.7** Availability of thermal reservoir

But $\Sigma(\delta Q/T)$ is the decrease in entropy $\Delta S$ of the reservoir as a result of heat withdrawal. For infinite and finite thermal reservoirs, this becomes $(\Sigma\delta Q)/T$ and $\Sigma(\delta Q/T)$, respectively. Note difference in two expressions: in the former, $T$ is outside the summation sign since $T$ remains constant for an infinite reservoir. If the number of cycles $n$ become infinite, then the sigma sign is replaced with the integral sign. Evaluation of the integral depends upon the type of the thermal reservoir. If it is made from a solid or a liquid, then $\delta Q = mc\, dT$. Thus, the integral can easily be determined. For a gaseous reservoir, this will be discussed in Chapter 4.

$T_0\Sigma(\delta Q/T)$ represents area below the heat rejection line on *T-s* diagram and represents the amount of unavailable energy. Thus, out of the total heat $\Sigma\delta Q$, $\Sigma\delta Q - T_0\Sigma(\delta Q/T)$ is available and $T_0\Sigma(\delta Q/T)$ is unavailable energy. Since heat is rejected at dead state temperature, the available energy is the exergy or availability function of the heat extracted from the thermal reservoir.

**Example 3.9** There are two thermal reservoirs $A$ and $B$, respectively at temperature 500 and 1500 K. Compare the availability of 1000 kJ heat drawn from these if the reservoirs are (a) infinite and (b) finite. Assume surrounding temperature of 300 K.

***Solution*** (a) Thermal reservoirs are infinite

Availability of heat from reservoir $A$

$$= 1000 - 300\,(1000/500) = 400 \text{ kJ}$$

Availability of heat from reservoir $B$

$$= 1000 - 300\,(1000/1500) = 800 \text{ kJ}$$

Note that availability for B is more than that of A, as its temperature is high.

(b) Thermal reservoirs are finite

The problem is incomplete as no information is given about the reservoir (since the procedure to find out change in entropy depends upon the type of working medium). Let us say that the reservoir consists of 10 kg mass of water. For water $c$ is 4.2 kJ/kg K. Since there is no mass transfer across its boundaries, it is a closed system. If $T$ is the temperature of water after heat transfer, then nfee gives

Case (a) $1000 = 10 \times 4.2\,(500 - T)$

$$\Rightarrow T = 476.2 \text{ K}$$

$$\int \delta Q/T = m\int_{500}^{T} c\,dT/T = mc\ln(T/500)$$

$$= -2.048 \text{ KJ/kg-K}$$

$$\text{Availability of heat}^{3} = 1000 - 300 \times 2.048 = 385.6 \text{ kJ}$$

Case (b) $1000 = 10 \times 4.2\,(1500 - T)$

$$\Rightarrow T = 1476.2$$

$$\int \delta Q/T = m\int_{1500}^{T} mc\,dT/T = mc\ln(T/500)$$

$$= -0.672$$

$$\text{Availability of heat}^{3} = 1000 - 388 \times 0.672 = 938.55 = 798.48 \text{ kJ}$$

It is observed that as temperature of the reservoir increases from 500 K to 1500 K, availability of 1000 kJ of heat increases too, from 385.6 kJ to 798.48 kJ.

### 3.3.1.2 Closed System

Figure 3.8 shows a closed system whose state is brought to a dead state by reversible processes. Heat transfer to the surroundings is made reversible by running the Carnot

---

3. Many students put a negative sign as well while making the substitution for $\int \delta Q/T$. It should be kept in mind that a negative sign in change in entropy is the result of decrease in entropy of the reservoir due to heat extraction from it, and hence this sign is already included in the equation for availability of heat.

engine between the system and the surrounding. Since the system is finite, its temperature will go on decreasing with successive withdrawal of heat. Let at some point of time, system temperature be $T$ when it is rejecting heat $\delta Q$ to the Carnot engine during an infinitesimal process. Let change in its volume, entropy and energy be $dV$, $dS$ and $dE$, respectively during the process and surrounding is receiving heat $\delta Q_0$ at temperature $T_0$. By Eq. 3.1

$$\delta Q/T = \delta Q_0/T_0 \quad \text{or} \quad \delta Q_0 = T_0\,\delta Q_0/T = T_0\,dS \tag{3.8}$$

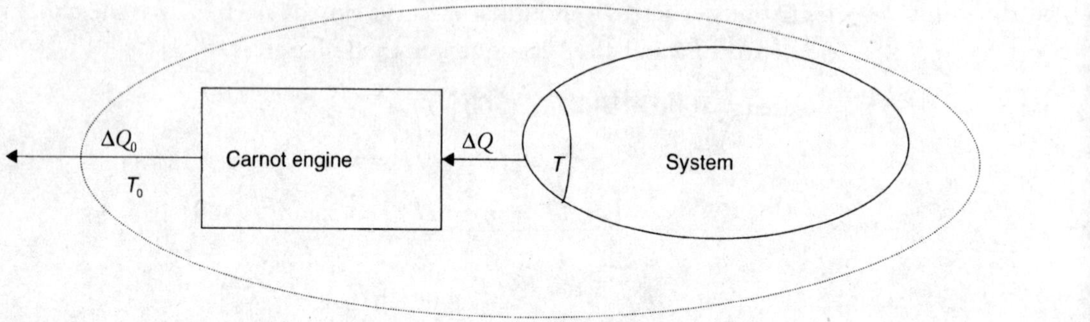


**Fig. 3.8** Availability of closed system

Let us assume that the Carnot engine has completed an integral number of cycles during this period (so that there is no change in energy of its working substance), and that both the system and the engine, being taken as new system, produce $\delta W$ amount of work. Then, nfee yields

$$\delta W = \delta Q_0 - dE = -\,dE + T_0\,dS$$

Recall our discussions in Section 1.2.9.1 about the extraneous surroundings such as air, which cannot be avoided and some useful work equal to $p_0\,dV$ is lost because of this, where $p_0$ stands for atmospheric pressure. Therefore, the system produces reversible work, $\delta W_{\text{rev}}$, given by

$$\delta W_{\text{rev}} = \delta W - p_0\,dV = -\,d\left(E + p_0 V - T_0 S\right) \tag{3.9}$$

since $T_0$ and $p_0$ are constant and can be taken inside the differentiation symbol. Integration of Eq. 3.9 from the initial to the dead state gives the availability of the closed system,

$$A_c = (U + p_0\,V - T_0\,S + KE + PE) - (U_0 + p_0\,V_0 - T_0\,S_0)$$

where, subscript '$_0$' refers to the dead state. The group $(U + p_0\,V - T_0\,S)$ is made from the properties of the system, and the pressure and temperature at the dead state, which remain constant. For this reason, this group may be taken as the system property and is named as exergy or *availability function*, $\phi$ of the closed system. So,

$$\phi = U + p_0\,V - T_0\,S \tag{3.10}$$

and $$A_c = (\phi - \phi_0) + KE + PE \cong (\phi - \phi_0)$$

neglecting kinetic and potential energies of the system. Further, we may choose $\phi_0 = 0$, since at the dead state, no work can be obtained, and so the availability function of the dead state is zero. Equation 3.9 may also be integrated from state 1 to state 2 to give (neglecting change in kinetic and potential energies) net reversible or useful work,

$$W_{rev} = \phi_1 - \phi_2 = \Delta A_c$$

the change in availability of the system during the process.

### 3.3.1.3 Open System

The procedure developed for a closed system is extended to the open system, as shown in Fig. 3.9, by adding mass flow across the system boundary. Recall our discussion in Section 3.2.3.1 of dividing the system into a number of surfaces, so that each surface may be assumed to be at the same temperature; Figure 3.9 shows heat transfer from one such surface. Using Eq. 3.6, rate of work produced by the Carnot engine is given by

$$\delta W_{Carnot} = \delta Q_0 - \delta Q = T_0 \Sigma \delta Q/T - \delta Q$$
$$= T_0 (dS_\sigma - m_1 s_1 + m_2 s_2) - \delta Q$$

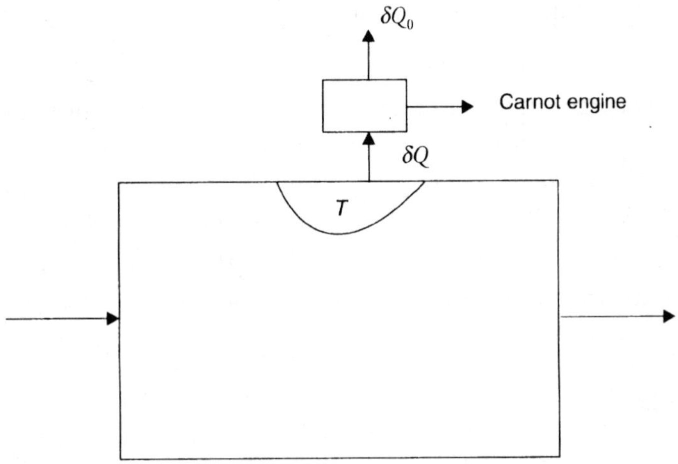


**Fig. 3.9** Availability of open system

where $\delta Q$ is the total rate of heat interaction between the system from its inlet to its exit and the surrounding. Reversible work produced during this process, using Eq. 2.5 and 3.6, is

$$\delta W_{rev} = \delta W + \delta W_{Carnot}$$
$$= \left[(m_1 e_1 - m_2 e_2) + \delta Q - dE_\sigma\right] + \left[T_0 (dS_\sigma - m_1 s_1 + m_2 s_2) - Q\right]$$
$$= m_1 (e_1 - T_0 s_1) - m_2 (e_2 - T_0 s_2) - d(E - T_0 S)_\sigma \quad (3.11)$$

For most of the systems, kinetic and potential energies can be neglected, and in that case, $e$ is equal to $h$ only. The group $(h - T_0 s)$ has been named as the Darrius function, also known as the Keenan function. So,

$$b = h - T_0 s \tag{3.12}$$

For a steady flow through an open system, we get from Eqs. 3.11 and 3.12

$$W_{rev} = b_1 - b_2 \tag{3.13}$$

where $b_1$ and $b_2$ are the Darrius function at the inlet and exit of the open system.

### 3.3.2 Exergy

Exergy of a system for its given state is defined as

$$\text{Exergy} = (h - h_0) - T_0 (s - s_0) = h - T_0 s \tag{3.14}$$

assuming exergy at the dead state to be zero. It can be seen that exergy is a property of the system for a given dead state, and can be presented in tabular or chart forms like other properties. Exergy is measured for all the substances from the same reference state, which is the dead state, and it is taken zero at the dead state. However, properties of pure substances are measured from different reference states, as mentioned earlier in Chapter 1. We may get rid of this problem for exergy by adding a constant $C$ to the right side of Eq. 3.14, so that all the substances will have zero exergy at the dead state. Therefore,

$$\text{Exergy} = h - T_0 s + C \tag{3.15}$$

$C$ can be easily obtained. For example, for air, reference is the absolute zero temperature, that is $h$ and $s$ are zero at 0 K. Enthalpy and entropy at 300 K (dead state temperature) can be read from the gas table and substituted in Eq. 3.15 with the left hand side equal to zero. This gives $C = 438.338$ kJ/kg. In the same manner, $C$ for steam, for which enthalpy and entropy are taken as zero at 0°C for water, is 5.02416 kJ/kg. In other words, Eq. 3.15 takes care of different reference states for working fluids, so that all of them have zero exergy at the dead state.

### 3.3.3 Irreversibility

Actual or useful work obtained from the system for a given change of its state is less than the reversible work obtainable for this change. This means that during the process in which the system delivers work, the decrease in its availability is more than the work obtained from it. This is due to the presence of irreversibility in the process. This fact is expressed by two terms, namely, irreversibility and effectiveness. The latter is discussed in the next section, while the former, is defined as the difference of the reversible and actual work. We have already obtained expressions for reversible and useful work before, and therefore, irreversibility can easily be determined in open and closed systems. We shall write this expression for an open system using Eqs. 3.11 and 2.5, respectively, for reversible and actual work. Thus, the irreversibility I, after simplification, is given by

$$I = m_1(-T_0 s_1) - m_2(-T_0 s_2) - d(T_0 s)_\sigma - Q$$
$$= T_0(m_2 s_2 - m_1 s_1 - dS_\sigma) - Q = T_0 \Sigma \delta Q/T - Q$$

using Eq. 3.6 for change in entropy of the fluid from its entrance to the exit of the system. However, $\Sigma \delta Q/T$ represents change in entropy of the system due to heat transfer from the surrounding and is positive, whereas

$$Q = \Sigma \delta Q = T_0 \Sigma \delta Q/T_0 = T_0\, dS_{\text{surrounding}}$$

the negative sign because entropy of the surrounding will decrease in this heat transfer.

$$\therefore \quad I = T_0\left(dS_\sigma + dS_{\text{surrounding}}\right) = T_0\, dS_{\text{universe}} \tag{3.16}$$

We note that irreversibility is related to entropy increase of the universe in the process and increases with it. That is, more is the process irreversible, more is the increase in entropy of the universe and more is the irreversibility, which is a measure of loss of work due to the presence of irreversibility in the system and the process.

### 3.3.4 Effectiveness

Irreversibility was defined as the difference of reversible and actual work. However, a fraction gives a better picture rather than the number itself. Effectiveness is, like efficiency in the first law analysis, a ratio. *It is defined the as the ratio of the actual work to reversible work.* That is, effectiveness

$$\varepsilon = \delta W_{\text{actual}} / \delta W_{\text{rev}} \tag{3.17}$$

We may put it in terms of gain or loss of availability, since reversible work is loss in availability of the system and actual work is gain in availability of the surrounding in a process when energy is crossing to the surroundings. Or, for such a case

$$\varepsilon = \left|\frac{\text{gain in availability of the surrouding}}{\text{loss in availability of the system}}\right| \tag{3.18}$$

However, in the process, when the system gains energy, then the definition of effectiveness becomes

$$\varepsilon = \left|\frac{\text{gain in availability of the system}}{\text{loss in availability of the surrouding}}\right| \tag{3.19}$$

The two vertical bars indicate that only the magnitude of the fraction is to be considered. We shall consider the case of a heat exchanger in which hot fluid flows through the tube and the cold fluid outside over the tube surface (it is also possible that cold fluid flows inside and the hot outside the surface) to demonstrate the importance of the term effectiveness.

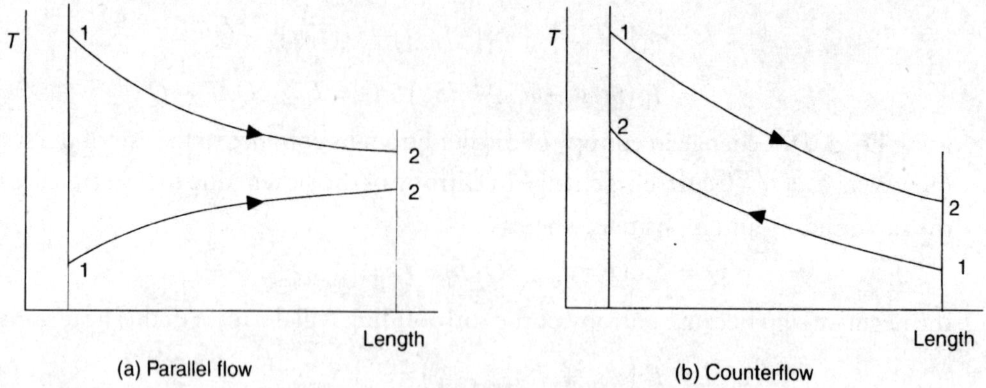


**Fig. 3.10** Temperature variation along the length of the heat exchanger

Figure 3.10 shows temperature distribution of hot and cold fluids along the length of the tube. In arrangement (a), the two streams are flowing in the same direction (parallel flow arrangement) and in (b) in the opposite direction (counter flow arrangement). We make the following observations.

1. Heat given by hot fluid is completely absorbed by the cold fluid. Thus, there appears no loss in energy.
2. In the parallel flow arrangement, temperature difference between hot and cold fluids at any section decreases along the direction of flow; whereas in the counter flow arrangement, the temperature difference between the two fluids remains almost the same throughout the length of the heat exchanger. Also, temperature difference at any section along the tube length in the parallel flow heat exchanger is more than that in the counter flow arrangement, except at its exit where it may be equal; it is assumed that some temperature difference is needed for heat transfer.
3. In the parallel flow arrangement, the lowest temperature of the hot fluid that may be attained is the temperature of the *cold fluid at the exit*. But in case of counter flow arrangement, by properly designing the heat exchanger, hot fluid can even attain the temperature of the *cold fluid at the inlet* to it. So, the latter arrangement helps in reducing the size of the heat exchanger and mass flow rate of the cooling media.

We may elaborate point 3 by taking an example of an air-to-air heat exchanger used in aircraft for air conditioning. Hot air from the discharge end of the compressor system is cooled by a cold air stream surrounding taken from the suction side of the compressor. (A compressor increases temperature and pressure.) Let subscripts $c$ and $h$ refer to cold and hot air streams, subscripts 1 and 2 to inlet and outlet conditions of the two streams, as shown in Fig. 3.10 and $m$ the mass flow rate. Since no heat is lost to the surrounding air, applying energy balance between the two streams, we can write

Heat given by hot fluid = heat received by cold fluid

or $$m_h (h_{h1} - h_{h2}) = m_c (h_{c2} - h_{c1}) = Q \text{ (say)}$$

$$\Rightarrow m_c = m_h c_h (T_{h1} - T_{h2})/c_c(T_{c2} - T_{c1}), \tag{3.20}$$

since $dh = cdT$ for air. Specific heat $c$ for hot and cold air streams may be taken as the same, as temperature variation is not that large to make substantial difference in their value. Lastly, the specific heat is taken at constant pressure since the flow through the tube or outside it may be assumed to be at constant pressure. Further, from Eqs. 3.12, 3.13 and 4.25

Loss of availability of the hot fluid $= m_h c_h (T_{h1} - T_{h2}) - T_0 m_h c_h \ln (T_{h1}/T_{h2})$
$= Q - T_0 m_h c_h \ln (T_{h1}/T_{h2})$

Whereas, gain in availability of cold fluid

$$= m_c c_c (T_{c2} - T_{c1}) - T_0 m_c c_c \ln (T_{c2}/T_{c1})$$
$$= Q - T_0 m_c c_c \ln (T_{c2}/T_{c1})$$

$$\therefore \text{ effectiveness} = [Q - T_0 m_c c_c \ln (T_{c2}/T_{c1})]/[Q - T_0 m_h c_h \ln (T_{h1}/T_{h2}] \tag{3.21}$$

using Eq. 3.18 for effectiveness. We assume that a temperature difference of at least 5°C is required for heat transfer to take place between two streams, and that the exchanger is long enough that the two streams acquire this temperature difference. This means that $T_{h2} - T_{c2}$ in case of a parallel flow heat exchanger, or $T_{h1} - T_{c2}$ in case of a counter flow arrangement is 5°C.

Let us now compare the performance of the parallel flow with the counter flow arrangement for which $m_h$, $t_{h1}$, $t_{h2}$ and $t_{c1}$ shall be given for the given application. This means $m_c$ and $t_{c2}$ will depend on the type of heat exchanger used, which can be determined using energy balance. Since $T_{c2}$ has to be less than $T_{h2}$ in the parallel flow case, and may be even higher than $T_{h2}$ in the counter flow case, mass flow rate of cooling air required is more in the parallel flow heat exchanger, as inferred from Eq. 3.20. We observe from Eq. 3.21, that $m_c$ appears in the numerator and decrease in its value therefore increases the numerator, whereas the effect of the log term in the numerator and denominator does not affect the result much, as it is a slow varying function compared to a linear function. So, the net effect of reduction in flow rate of the coolant is to increase the effectiveness of the counter flow heat exchanger. *Thus, a counter flow heat exchanger will have more effectiveness in addition to a smaller size and requiring less amount to cool the media compared to a parallel flow arrangement.*

The same results may be inferred from the fact that more temperature difference in parallel flow arrangement causes more loss in availability of the hot fluid, as discussed in the law of degradation of energy. In actual practice, the temperature difference in the counter flow arrangement is almost the same throughout the length of the heat exchanger, and it is of the order of that at the outlet of the parallel flow arrangement. The overall effect is to increase the effectiveness of the counter flow arrangement. Here, we took air as fluid for the simple reason that change in enthalpy can be written in terms of temperature change, and it is easy to find out change in entropy. Otherwise, the results hold true for any kind of fluid.

From these discussions, we conclude that a counter flow arrangement is better than the parallel flow one. This conclusion we fail to obtain from the first law analysis.

## 3.4 SECOND LAW ANALYSIS

We can analyse a process to find how much close or away it is from a reversible process. We may see the importance of this kind of analysis from the fact that all our hopes hinge around building a reversible process, so that the cycle formed from such processes is reversible, since such a cycle has maximum efficiency. However, the efficiency of a given cycle does not speak of quality of energy. Efficiency may even give a misleading picture, and quite often it does so. A glaring example is that of a boiler in a steam power plant. Combustion efficiency may be around 98%, and heat balance may result in boiler efficiency of around 90%. But we know that a very large difference exists between temperature of the gases and the working fluid in the furnace, and hence great loss of exergy. Therefore, in no way does the efficiency give a correct picture in this case. Another example that may be cited is the parallel and counter flow arrangements used in heat exchangers. From the point of view of the first law, there is no difference in the two, but we have discussed before that a counter flow arrangement is preferred, as it gives a smaller heat exchanger for given change in temperature of hot and cold fluids.

We may define efficiency based on exergy, like thermal efficiency, for a power cycle. This is called *exergetic efficiency* defined as ratio of work produced by the cycle to the exergy input to it. That is,

$$\text{exergetic efficiency} = \text{exergy output of the system/exergy input to the system} \times 100 \tag{3.22}$$

Such kind of analysis is known as *second law analysis*. Figure. 3.11 shows availability v/s crank angle for a four stroke petrol engine during compression and expansion processes to get a glimpse of the importance of the second law analysis. We observe that availability increases during the compression process, which is obvious, as work is being done on the mixture in compression stroke. Further, availability decreases in the expansion process because of (i) expansion work and (ii) irreversibility. We also observe that during the expansion process, irreversibility is comparable to the expansion work, and that it is partly due to finite temperature difference and partly due to the combustion process. Therefore, improvement in the combustion process is the need, and a lot of progress has been made in this direction. We shall demonstrate the procedure for the first law and second law analysis with the help of an example from a steam power plant.

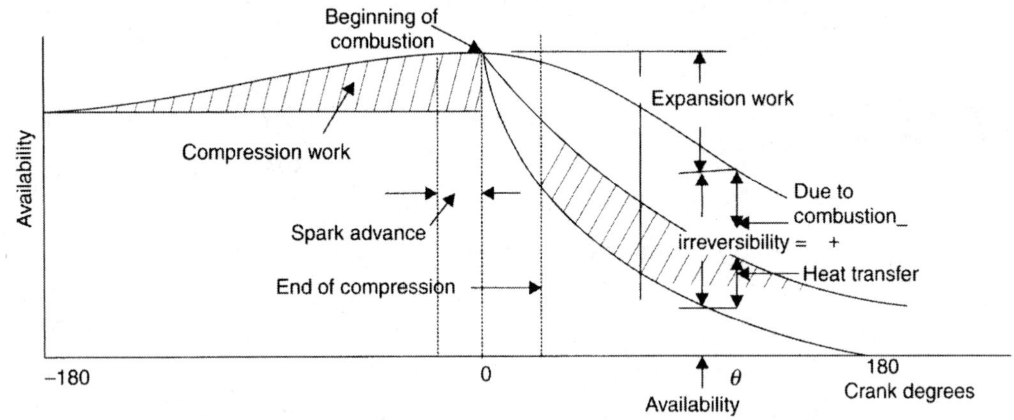


**Fig. 3.11** Availability of fuel air mixture and products of combustion

**Example 3.10** Calculate the availability of air in the following cases

(a) $p = 4$ bar, $T = 150$ K

(b) $p = 4$ bar, $T = 400$ K

(c) $p = p_0$, $T = T_0$

***Solution*** Given that ambient pressure and temperature are 1 bar and 27°C respectively. Neglect kinetic and potential energies in all cases. Assume $c_v = 0.716$ KJ/kg K and $c_p = 1.005$ KJ/kg K.

The availability of a closed system, when kinetic and potential energies are negligible, is given by

$$a_c = (u - u_0) + p_0 (v - v_0) - T_0 (s - s_0)$$

$$= c_v (T - T_0) + p_0 (v - v_0) - T_0 (s - s_0), \text{ for perfect gases}$$

$$= c_v (T - T_0) + R p_0 - (T/p - T_0/p_0) - T_0 (s - s_0)$$

where $s - s_0 = c_p \ln (T/T_0) - R \ln (p/p_0)$

and $R = c_p - c_v = 0.289$

(a) $(s - s_0) = c_p \ln (T/T_0) - R \ln (p/p_0) = -0.197$ KJ/kg K

Availability $= c_v (T - T_0) + R \times 1 (T/p - T_0/p_0) - T_0 (p - p_0)$

$= 145.70$ KJ/kg K

(b) Like (a), we get, $s - s_0 = -0.4$ KJ/kg K

$s - s_0 = c_p \ln (T/T_0) - R \ln (p/p_0) = 0.111$

$$\text{Availability} = c_v (T - T_0) + R(T/p - T_0/p_0) - T_0 (s - s_0)$$
$$= 346.68 \text{ KJ/kg}$$

(c) The system is in dead state and hence its availability is zero.

**Example 3.11** Calculate the availability of steam in steady flow for the following cases.

(a) $p = 200$ bar, $T = 500$°C; $C = 300$ m/s; $z = 20$ m

(b) $p = 0.007$ bar, $d.f = 0.9$, neglect KE and PE

(c) Saturated water at 100°C, neglect KE and PE.

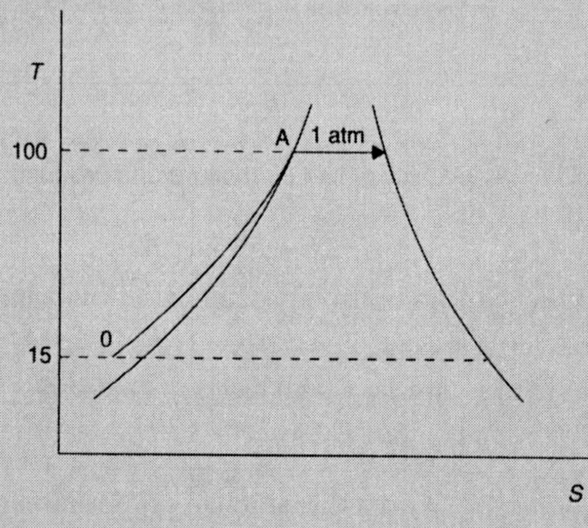


**Fig. 3.12**

***Solution*** Eq. 3.12 will be used to determine the availability of flowing steam. To find $s_0$, the entropy of fluid in dead state either we can use tables for supercooled water or proceed as follows. The dead state is represented on the *T-s* diagram by the point *O* (see figure) whereas the saturated liquid state corresponding to 1 atm pressure is shown by point *A*. From steam tables, the corresponding saturation temperature is 100°C. For constant pressure heating *OA* of the liquid, we can write

$$s_0 - s_A = c_p \ln\left(\frac{T_0}{T_A}\right) = c_p \ln\frac{288}{373} = -1.099 \text{KJ/kg K}$$

Also from steam tables, $s_A = 1.3063$ KJ/Kg K

$$\therefore s_0 = 2.4053 \text{ KJ/kg K}$$

At dead state, $h_0 = 62.97$ KJ/kg

(a) At 200 bar, 500°C, $h = 3245$ KJ/kg

$s = 6.1591$ KJ/kg K

$$\therefore \text{ availability} = (h - h_0) - T_0\,(s - s_0) + \frac{c^2}{2000} + \frac{20 \times 9.81}{1000}$$

$$= 2146.13 \text{ KJ}$$

(b) At 0.07 bar, 0.9 d.f

$h = 2328.81$

$s = 7.503$

$$\therefore \text{ availability} = h - h_0 - T_0\,(s - s_0) = 797 \text{ kJ}$$

(c) For saturated water at 100°C; $h = 419.12$ and $s = 1.306$

$\therefore$ availability = 28.47 KJ/kg

**Example 3.12** In a boiler plant steam is produced at 40 bar. The hot gases in the fire box are at a temperature of 1500°C and the ambient temperature is 35 °C. In the process of heat transfer from gases to boiler water, 420 kJ are transferred per second. What is the distribution of available and unavailable energy:

(a) as the energy leaves the hot system.

(b) as it enters the steam at saturation temperature corresponding to the pressure.

(c) Show that the relation $I = T_0(\Delta S)_{\text{universe}}$ holds good.

***Solution*** (a) The fire box can be treated as infinite thermal reservoir as its temperature does not change with the withdrawal of heat.

$\therefore$ available energy of heat transferred from fire box is

$$= \delta Q - T_0\,\Delta S$$

$$= 420 - 308\left(\frac{420}{1733}\right) = 346.1 \text{ kJ/s}$$

Unavailable energy = 420 – 346.1 = 73.9 kJ/s

(b) From steam tables, the saturation temperature corresponding to 40 bar/cm$^2$ is (249.2 + 273) = 522.2 K. As a result of heat transfer the temperature potential is decreased from 1773 K to 522.2 K. The available and unavailable energy of the heat received by steam is

$$\text{Available energy} = 420 - 308\left(\frac{420}{522.2}\right) = 171.67 \text{ kJ/s}$$

Unavailable energy = 248.33 kJ/s

(c) From above we see that increase in available energy is
248.33 –73.9 = 174.43 KJ/s, which is due to irreversible heat transfer. This, by definition, is the irreversibility.

Also $(\Delta S)_{steam}$ = increase in entropy of steam

$$= \frac{420}{522.2} = 0.803 \text{ kJ/s}$$

$(\Delta S)_{gas}$ = decreases in entropy of gas

$$= \frac{420}{1773} = -0.2361 \text{ kJ/s}$$

$\therefore\ (\Delta S)_{universe} = (\Delta S)_{steam} + (\Delta S)_{gas} = 0.5668 > 0$ and

$T_0\,(\Delta S)_{universe} = 174.43$ kJ/s = Irreversibility

**Example 3.13** A steam turbine is supplied with steam at 50 bar and temperature 450°C and exhausts at 0.05 bar, 0.9 dry. Calculate the turbine efficiency, effectiveness and loss of available energy. Assume the process to be adiabatic. Assume $T_0$ = 298 K.

***Solution*** From steam table

at 50 bar, 450°C, $h_1 = 3316$, $s = 6.825$

at 0.05 bar, 0.9 dry, $h_1 = 2313.74$, $s_1 = 7.633$

Applying *Sfee* to the expansion process in the turbine, with subscripts 1 and 2 respectively at inlet and exit of the turbine, we get after neglecting kinetic and potential energies

$$h_1 + \delta q = h_2 + \delta w \quad (\text{Assume } \delta q = 0)$$

or

$$\delta w = 1002 \text{ kJ/kg}$$

Also decreases in availability of steam

$$= b_2 - b_1 = (h_2 - h_1) - T_0\,(s_2 - s_1)$$

$$= -\ 1243.12 \text{ kJ/kg}$$

$$\therefore \text{ effectiveness} = \left|\frac{h_1 - h_2}{b_1 - b_2}\right| = \left|\frac{1002}{1243.12}\right| = 0.806$$

**Isentropic process.** In order to find turbine efficiency, we have to assume the expansion of steam from 50 bar to 0.05 bar as isentropic process. That is, the entropy of steam after expansion is 6.825. This shows that steam is wet at turbine exit. Let $x$ be the dryness fraction after expansion; then

$$6.825 = 0.473 + x \times 7.926$$

or $$x = 0.8$$

$$\therefore\ h_2' = 2072.6 \text{ kJ/kg}$$

$$\text{Isentropic enthalpy drop} = h_1 - h_2' = 1243.54 \text{ kJ}$$

and $$\text{turbine efficiency} = \frac{\text{actual enthalpy drop}}{\text{isentropic enthaply drop}} = 0.85$$

$$\text{Loss of available energy} = T_0(\Delta S)_{\text{universe}}$$

However, the process is adiabatic and as such $(\Delta S)_{\text{surr}} = 0$

$$\therefore\ (\Delta S)_{\text{universe}} = (\Delta S)_{\text{system}} = 7.633 - 6.825 = 0.808$$

or $$\text{loss of available energy} = 298 \times 0.808 = 240.81 \text{ KJ}$$

**Example 3.14** The following data applies to a thermal power plant working on a Rankine cycle. Temperature of the combustion gases 2000 K; inlet pressure of steam 50 bar at saturated condition; condenser pressure 0.1 bar; stack temperature 600 K; surroundings temperature 300 K; $c_p$ of combustion gases and air 1 kJ/kg K. Make a first and second law analysis of this plant. Given, heating value of fuel is 34000 kJ/kg.

***Solution*** *Availability of combustion products.*

Given: $T_g$ = temperature of combustion gases = 2000 K, steam pressure = 50 bar and condenser pressure = 0.1 bar, $T_0$ = 300 K, $c_p$ = 1 kJ/kg K, $c_V$ = 34000 kJ/kg.

**Assumption.** We assume (i) combustion air enters at the surrounding temperature, as it is not given in the question, (ii) $m$ kg of fuel gases are formed at temperature $T_g$ for 1 kg of fuel burnt in the boiler furnace, (iii) there is complete combustion of fuel at constant pressure, (iv) the fuel gases behave like perfect gas and (v) all the heat produced during burning of the fuel is used to raise temperature of the fuel gases. Energy balance gives

$$m\, c_p\, (T_g - T_0) = 34000$$

or $$m = 34000/1 \times (2000 - 300) = 20 \text{ kg}$$

The change in entropy $\Delta S$ of fuel gases during the process is given by Eq. 4.25.

$$\Delta S = m\,c_p \ln (T_0/T_g) = 20 \times 1 \times \ln 300/2000 = -37.94 \text{ kJ/kg K}$$

$\therefore$ availability of heat given off during the combustion process = availability of combustion products

$$= 34000 - T_0\,\Delta S = 22617.28 \text{ kJ/kg}$$

and loss in availability during combustion

= availability before combustion – availability after combustion

$$= 34000 - 22617.28 = 11382.72 = 11382.72/34000 \times 100$$
$$= 33.48\%$$

**Availability of stack gases.**

Gases leaving chimney (stack) are at a temperature of 600 K and carry

internal energy = $m\,c_p\,(600 - 300) = 6000$ kJ, which can be transformed to work by running a Carnot engine between it and the surrounding. Change in entropy during this reversible constant pressure process is

$$\Delta S = 20 \times 1 \times \ln 300/600 = -13.863 \text{ kJ/kg K} \quad \text{(using Eq. 4.25)}$$

$\therefore$ availability of stack gases = $6000 - 300 \times 13.863 = 1841.1$

Unavailable energy = $300 \times 13.863 = 4158.9$

$= 4158.9/34000 \times 100$

$= 12.43\%$ of that of fuel

$= 4158.9/6000 \times 100$

$= 69.32\ \%$ of heat available from stack gases

**Availability of heat supplied to cycle fluid (steam).**

The Rankine cycle is shown in Fig. 3.12. From the steam table, for saturated steam at 50 bar: $h_1 = 2797.7$, $s_1 = 6.0191$ and at 0.1 bar, $h_f = h_3 = 191.8$, $s_f = s_3 = 0.649$, $h_{fg} = 2392.8$, $s_{fg} = 7.501$.

Since expansion in turbine is isentropic, the process equation is $s_1 = s_2$. This gives $6.0191 = 0.649 + x_2 \times 7.501$, or $x_2 = 0.716$ and $h_2 = 191.8 + 0.716 \times 2392.8 = 1905$ kJ. Steam is produced because heat $\{= 20 \times 1 \times (2000 - 600) = 28000 \text{ kJ}\}$ is transferred to steam during combustion from hot gases, which in turn cools from 2000 K to 600 K. If $m_s$ is mass of steam produced per kg of fuel burnt, then making the energy balance for heat transfer to steam, we get

$$m_s \times (h_1 - h_4) = m \times c_p \times (T_g - T_{st})$$

$$\Rightarrow m_s = 28000/(2797.7 - 191.8) = 10.75 \text{ kg}$$

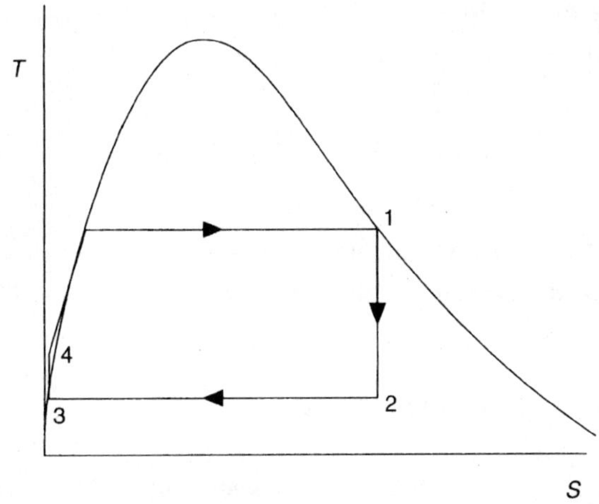


**Fig. 3.13** Rankine cycle

Here, $T_{st}$ is the temperature of the stack gas and state of saturated water at 3 is changed to dry and saturated steam at 1. In this process, entropy of steam is also increased from $s_3$ to $s_1$ per kg of steam.

$\therefore$ availability of heat supplied to steam = 28000 – 10.75 × 300 × $(s_1 - s_3)$ = 10681.43 kJ

Unavailable energy during steam formation = 28000 – 10681.43 = 17318.57 kJ.

**Availability of heat rejected in condenser.**

Heat rejected in condenser = $m_s (h_2 - h_3)$

$$= 10.75 \times (1905 - 191.8) = 18416.9 \text{ kJ}$$

Change in entropy during heat rejection in condenser = $m_s (s_2 - s_3)$ = 57.7286

$\therefore$ availability of heat rejected = 18416.9 – 300 × 57.7286 = 1098.33 kJ

In fact, such a small loss in availability in a heat condenser was expected because of low temperature of steam in the condenser. We may appreciate it from the following fractions.

Heat rejected in condenser/Heat transferred to steam = 18416.9/28000 = 0.66, and

availability of heat rejected in condenser/availability of heat transferred to steam

$$= 1098.33/17318.57 = 0.063$$

That is, while 66% of energy supplied to steam is rejected in the condenser, which accounts to only 6.3% of exergy loss. Both the figures are of importance, since the former tells how much the power manufacturer is unnecessarily paying for the fuel, whereas the latter tells the designer where to concentrate to get better utilisation of available energy.

**Work done in turbine.** Flow through the turbine is isentropic and steady. The steady flow energy equation therefore gives:

$$\begin{aligned}\text{Work from turbine} &= m_s (h_1 - h_2)\\ &= 10.75 \times (2797.7 - 1905)\\ &= 9596.53 \text{ kJ.}\end{aligned}$$

Since work is done by steam, it loses this much amount of availability. Table 3.2 summarises the availability in the given power plant.

**Table 3.2** Availability in a given power plant

| Process<br>1 | Energy exchange, kJ<br>2 | Availability, kJ<br>3 | Unavailable energy, kJ<br>4 = 2 – 3 | %<br>5 = 4/2 |
|---|---|---|---|---|
| *Availability credit* | | | | |
| Combustion in boiler | 34000.00 | 22617.28 | 11382.72 | 33.48 |
| *Availability debit* | | | | |
| Heat available from fuel gases | 6000.00 | 1841.10 | 4158.90 | 69.32 |
| Heat transfer to steam | 28000.00 | 10681.10 | 17318.90 | 61.85 |
| *Total of availability debit* | 34000.00 | 12522.40 | 21477.80 | |
| *Rankine cycle* | | | | |
| Work production by turbine | 9596.53 | 9596.53 | 0.00 | 0.00 |
| Heat rejection in condenser | 18416.90 | 1098 | 17318.57 | 94.00 |

From Table 3.2, we conclude the following.

1. Unavailable energy in the combustion process and stack gases amounts to 33.48% and 12.43%, respectively, of the total available energy of the fuel and air mixture, which sums up to 45.71%. It is a great loss and needs serious re-examination of the combustion process.
2. Unavailable energy of heat in the stack gases is 69.32% and that in the condenser 94%. This is due the to the fact that temperature of the gases and steam is low. This was expected since heat is low grade of energy.
3. Loss of availability in the furnace is = 22617.28 – 10681.1 = 11936.28 kJ. Of this 4158.9 kJ loss is by internal energy carried by fuel gases, and the remaining from irreversible heat transfer of heat from the furnace to the water in the boiler due to the large temperature difference, which comes out to

$$11936.28 - 4158.9 = 7777.38 \text{ kJ.}$$

4. Availability loss in the combustion process itself is

   34000 – 22617.28 = 11382.7 = 33.48%, which is a very significant loss.

5. Rankine cycle produces work = 9596.53 kJ from the heat supplied to it is 28000 kJ, having availability = 10681.1 kJ.

   ∴ exergetic efficiency = 9596.53/10681.1 × 100 = 89.85%

   and thermal efficiency = 9596.53/28000 × 100 = 34.27%.

6. Similarly, for the plant, we may find out the overall exergy and thermal efficiencies. These come out to

   exergetic efficiency = 9596.53/22617.28 × 100 = 42.63%

   and thermal efficiency = 9596.53/34000 × 100 = 28.23%.

7. Comparison of results in para 5 and 6 reflects that the combustion process and heat transfer to the working substance in the Rankine cycle from the combustion products are the main culprits for poor exergy and thermal efficiency of the power plant.

8. Availability of energy in the stack gases and steam condensed in the condenser is low. In fact, it is a complete waste, as the available energy in these two fluids is left to the atmosphere. Therefore, our efforts should be to minimise these losses to enhance the overall thermal and exergy efficiencies of the system, either by improving the methods or utilising it in some alternative ways. The concept of cogeneration is based on it.

## SUMMARY

In Chapter 2, we studied the first law, and in this chapter, we discussed different aspects of the second law of thermodynamics, both of which are the law of nature, and of which the only proof is that these have not been proved to fail in a single application. Both the laws have their importance, and both are needed for the analysis of any system. But still, the second law has an edge over the first law, since it can analyse an individual process of a cycle. Therefore, a detailed analysis of a system undergoing different processes is possible and the causes of poor performance of the system can be pin-pointed.

One of the applications, which is of importance to mechanical engineers, are the power and refrigeration cycles. A device that produces work is called an engine, and has to undergo cyclic change in its state in order to produce work continously. These cycles are called power cycles and there are a number of such cycles, which differ from each other in the manner of heat addition and heat rejection. The type of cycle to be used also depends on the nature of the working substance. Power cycles have been discussed at a later stage. Basically, an engine extracts heat from a thermal reservoir at higher temperature, and rejects to another at lower temperature to produce work. Thus, for a power cycle, heat is the input and work is its output. Its performance is measured by the ratio of output to input and is called efficiency. If the reversible cycle operates in other direction, then it results in the refrigeration cycle used for pumping heat from lower temperature to higher temperature, which is not possible as per the Clausius statement. In this cycle, work is required to be done. Coefficient of performance is a measure of performance of this cycle, and is defined as the net refrigeration produced per kJ of work input. The term source and sink of heat is used for thermal reservoirs, which give heat to a cyclic device or receive heat from it, respectively, irrespective of the temperature of these reservoirs. For an engine, a source is at higher temperature, whereas for a refrigerator, it is at lower temperature.

An interesting feature of the first and second law analysis is that $\delta q$ and $\delta w$, which are path functions, are made point functions $(\delta q - \delta w)$ and $(\delta q/T)$. The quotient $(\delta q/T)$ is to be determined for a reversible process. If the process is not reversible, then a reversible process has to be imagined between the given states to determine it. We named these properties as energy and entropy.

Both the first and the second law of thermodynamics are a product of our day-to-day observations, and do not have any mathematical proof. The second law has importance, as it leads to explain the direction of a process and examining the kind of system stability (not discussed in this book). Not only this the second law analysis has been a milestone in prediction and making improvement in a process, so that efficiency of the system could be improved. The importance of this law is because of its following conclusions.

1. Performance of a reversible cycle is maximum, and is same for all cycles working between the same source and sink, and is independent of the working substance. However, there exists no reversible cycle, because of the presence of internal and external irreversibility.
2. The second law gives the basis for a thermodynamic temperature scale. Lord Kelvin gave a linear temperature scale based on Eq. 3.1, with its zero as the absolute zero of the scale and dividing the interval between the triple point temperature of water and absolute zero linearly into 273.16 parts. The unit of temperature on this scale is called Kelvin, written as K (and not °K). However, for practical purposes, another scale, called the international practical temperature scale has been used and correction factors were determined to obtain reasonable accurate thermodynamic temperature, since the Kelvin scale is theoretical only.
3. Clausius gave the famous inequality principle that $\int \delta Q/T$ over cycle is $\leq 0$; equal for reversible and less than for an irreversible cycle. This then led to the law of increase in entropy of the universe in a given irreversible process and entropy production. Thus, no process is possible in which entropy of the universe will decrease. Latter it was linked to irreversibility too Eq. 3.16.

4. Available sources were classified as high-grade and low-grade sources, depending on whether a given amount of interaction between the system and the surrounding is convertible to mechanical work. In this respect, electrical and all other mechanical or non-mechanical form of work, kinetic and potential energy, etc. fall under the former category. However, the presence of internal and external irreversibility decreases the amount of mechanical work obtainable from these sources of energy, and this fact is expressed by the law of degradation of energy.
5. Another property of the system, the availability, was defined, which is the amount of maximum work obtained from a system in bringing its state to a dead state through a reversible process. The dead state is fixed by fixing the reference frame to the ground and taking the temperature and pressure equal to 300 K, 1 bar, respectively. From this, the concept of exergy and the second law analysis, which is now mandatory in many countries to be incorporated as part of the project report, emerged. Different reference states used for measurement of properties of different working substances (discussed in Chapter 4) is circumvented by adding a constant, as in Eq. 3.15, and determining it by making exergy of all substances zero at the dead state. Like efficiency, exergetic efficiency is defined, as in Eq. 3.22.

In fact, the second law is much more than this.

## EXERCISES

### Review Questions

3.1 Show that a decrease in temperature of a sink by a differential amount $\delta T$ will be more effective to increase the cycle efficiency as compared to an increase in source temperature by the same amount.

3.2 Carnot engines, when operating between adjacent temperatures anywhere on the thermodynamic scale, produce the same amount of work. Prove the statement.

3.3 What are the different mechanisms for transferring energy to a control volume?

3.4 An experimentalist claims to raise the temperature of water to 150°C by transferring heat to it from steam at 120°C. Examine his claim.

3.5 What is a thermal reservoir? Give examples of finite and infinite reservoirs.

3.6 What are the characteristics of a heat engine?

3.7 A homeowner claims that her 5-year-old room heater has 100% efficiency of energy conversion. Is her claim right or does it violate some law?

3.8 Consider a thermal 1000 MW power plant having thermal efficiency 40%. Determine the coal consumption per hour if its heating value is 28000 kJ/kg.

3.9 Show how the ideal gas temperature scale is developed. State its connection with the thermodynamic scale.

3.10 What are primary fixed points? What are their values in Fahrenheit and Celsius scales?

3.11 Show that a device violating the Clausius statement of heat transfer will also violate the Kelvin–Planck statement.

3.12 Show that all the reversible engines operating between the same source and sink will have the same efficiency.

3.13 Does a cycle for which $\int \delta q$ over a cycle > 0 violate the Clausius inequality? Why?

3.14 Is the $\int \delta q/T$ same from state 1 to 2 for all processes between these two states? Explain.

3.15 To determine entropy change between points 1 and 2, should $\int \delta q/T$ be evaluated along the actual path or an imaginary reversible path?

3.16 Entropy of a hot potato decreases as it cools. Does it violate the second law?

3.17 The entropy of steam or any other gas will not change as it flows through an adiabatic turbine. Comment.

3.18 Is a process that is internally reversible and adiabatic necessarily isentropic? Explain.

3.19 Enthalpy and internal energy for a perfect gas vary with temperature alone. Is it true for entropy too?

3.20 How does reversible work differ from useful work?

3.21 Is the exergy different in different environments?

3.22 Consider two geothermal wells whose energy contents are estimated to be the same. Will the exergy of these wells be essentially the same? Explain.

3.23 A researcher suggests to cool steam while it is expanding in a turbine. His argument is that it will cool the steam and hence entropy of steam will decrease, which will reduce irreversibility. Check his argument.

3.24 Compare and contrast the characteristic of control mass and control volume.

3.25 Name three types each of reversible and irreversible processes and state the criterion in terms of the second law of thermodynamics to recognise an irreversible process. A system is taken from state 1 to 2 by (a) reversible, and (b) irreversible process. Is change in entropy of the system same or different in the two processes?

3.26 Suppose, a device working on a reversible cycle is operated on the reversed cycle, what is the relation of heat and work transfer between the two directions of operation of the cycle?

3.27 Does the availability of the system depend on the condition of the system and the surrounding? Explain.

## Problems for Practice

3.1 An air conditioner removes heat steadily from a house at a rate of 750 kJ/min while drawing electric power at a rate of 6 kW. Determine (a) COP of the air conditioner, and (b) the rate of heat transfer to the outside air.

[*Ans.* 2.08, 1110 kJ/min].

3.2 A compressed air bottle of volume 0.15 $m^3$ contains air at 40 bar and 27°C. It is used to drive a turbine, which exhausts to atmosphere at 1 bar. If the pressure in the bottle is allowed to fall to 2 bar, determine the amount of work that could be delivered by the turbine. [*Ans.* 585 kJ.] (IES 98).

3.3 A system absorbs 100 kJ of heat from a body at 500°C and does 50 kJ of work. If it is required to restore the system to its initial state, is it possible to do so by an adiabatic process? [*Ans.* no].

3.4 It is recommended that food should be cooled to room temperature before putting it in the refrigerator as it saves energy. This can be seen from the example of a pan containing food having total mass of 5 kg. The average temperature of the kitchen is 20°C and the average temperature of food is 95°C when it is taken off the stove. The refrigerator space is maintained at 3°C, and the average specific heat of the food and the pan can be taken to be 3.9 kJ/kg K. If the refrigerator has a coefficient of performance of 1.2 and the cost of electricity is Rs 3 per unit, determine how much saving will be made per year by waiting for the food to cool to room temperature before putting it into the refrigerator.

3.5 A steam power plant works between a constant temperature boiler at 630 K and a constant temperature condenser at 310 K. For each kg of steam passing through the plant, 2100 kJ of heat is required to be transferred to the boiler, and during the cyclic process, 150 kJ of heat is lost from the condenser. Verify if the process is reversible, irreversible or impossible.

3.6 Making a not-much-deep lake with salt spread in its bottom may create a source of heat. Water in the lake absorbs sun radiation and gets heated. The top water layer, however, gets cooled to the atmospheric temperature by loosing heat to it. Adding salt at the bottom increases the density of water, and therefore, convection currents are not set up, as the density of water particles of the bottom layer is more. Also, the cold top surface prevents any loss of heat from the stored water. Now, a power plant can be run using an organic working substance while taking heat from the bottom water and rejecting it to the water at the top. If the water temperature is 35 °C near the surface and 80°C near the bottom of the lake, determine the maximum thermal efficiency that this power plant can have. Is it realistic to use 35 and 85°C for temperature in the calculations? Explain.

3.7 A reversible engine A absorbs energy from a reservoir at $t_1$ and rejects heat to a reservoir at temperature $t_2$. A second reversible engine *B* absorbs the same amount of heat as rejected by *A* to a reservoir at $t_2$ and rejects heat to a reservoir at $t_3$. What is the relation among $t_1$, $t_2$ and $t_3$ if: (a) the efficiencies of the two engines are the same, (b) work done by the two engines is the same.

[*Ans.* $t_2 = \sqrt{t_1 t_3}$, $t_2 = 1/2\,(t_1 + t_3)$].

3.8 In tropical climates, the water near the surface of the ocean remains warm throughout the year as a result of solar energy absorption. In the deeper part of the ocean, however, the water remains at a relatively low temperature, since the sun's rays cannot penetrate very far. It is proposed to take advantage of this temperature difference and construct a power plant that will absorb heat from the warm water near the surface, and reject the waste heat to the cold water. Determine the maximum thermal efficiency of such a plant if the water temperatures at the two respective locations are 24 and 3°C.

3.9 Consider two Carnot heat engines operating in series. The first engine receives heat from the reservoir at 2400 K and rejects the waste heat to another reservoir at temperature *T*. The second engine receives this energy rejected by the first one, converts some of it to work, and rejects the rest to a sink at 300 K. If the thermal efficiency of both engines are the same, determine the temperature *t*. [*Ans.* 849 K].

3.10 An electric stove has one heating element at 300°C getting 500 W of electric power. It transfers 90% of power to 1 kg of water in a kettle initially at ambient state 20°C, 1 bar; the remaining 10% leaks to the room air. The water at uniform temperature is brought to the boiling point. At the start of the process, what is the rate of availability transfer by (a) electric input, (b) from the heating element, and (c) into the water at $t_{water}$? [*Ans.* 00 W, 244 W, 0].

3.11 0.5 kg of air (argon gas) executes a Carnot power cycle having a thermal efficiency of 0.5. The heat transfer to the air during the isothermal expansion is 40 kJ. At the beginning of the isothermal expansion, the pressure is 7 bar and volume is 0.12 m$^3$. Determine the maximum and minimum temperature for the cycle in K, the volume at the end of isothermal expansion, and the work and heat transfer for each of the four processes in kJ. For air, $c_v = 0.721$ and $c_p = 1.008$ kJ/kg – K. [*Ans.* 585 K, 292.5 K] (IES 1993).

3.12 A reversible engine is connected to three thermal reservoirs at 200, 300 and 400 K, respectively. It draws 5 MJ of heat from a 400 K reservoir during a cycle of operation, and does 840 kJ of work. Find the amount and direction of heat interaction with other reservoirs. [*Ans.* –4980 kJ and 820 kJ with 300 and 200 K reservoirs] (IES 99).

3.13 Exhaust gases leave an internal combustion engine at 800°C and 1 bar, after having done 1050 kJ of work per kg of gas in the engine ($c_p$ = 1.1 kJ/kg K). The temperature of the surroundings is 30°C. (i) How much available energy per kg of gas is lost by throwing away the exhaust gas? (ii) What is the ratio of the lost available exhaust gas energy to the engine work? [*Ans.* 425.58 kJ, 0.405] (IES 99).

3.14 A steam turbine receives steam at 20 bar, 300°C with a velocity of 60 m/s, and leaves the turbine at 2 bar and 150 m/s. There is no heat transfer from the turbine. Assuming the temperature of the atmosphere at 25°C, determine the maximum and actual work delivered by the turbine per kg of steam.

3.15 A closed system contains 0.5 kg of air. It expands from 2 bar, 60°C to 1 bar, 40 °C. During expansion, it receives 2 kJ of heat from a reservoir at 100°C. Assuming atmospheric conditions to be 0.95 bar and 30°C, calculate (i) the maximum work, (work done on the atmosphere), and (iii) change in availability. [*Ans.* 28.26 kJ, –19.97 kJ, 8.288 kJ.] (IES 96).

3.16 Steam enters an isothermal steam turbine at a pressure of 12.5 bar, 250°C and velocity 50 m/s, and leaves the turbine at a pressure of 1.2 bar and velocity 130 m/s. Determine the heat transferred to the turbine per second, and the power developed by it when steam flows through the turbine at the rate of 0.1 kg/s. Assume that the process is reversible. [Hint: Heat transfer can be obtained for isothermal an process using the Clausius relation.].

3.17 One technique for operating a steam turbine in part-load power output is to throttle the steam to a lower pressure before it enters the turbine. The steam line conditions are 20 bar, 400°C, and the turbine exhaust pressure is 0.1 bar. Assume the expansion inside the turbine to be reversible adiabatic. (a) Determine the full-load work output of the turbine. (b) Find the pressure of the steam to which it should be throttled to force specific work output of 80% of full load.

3.18 A household refrigerator has a freezer at $T_f$ and a cold space at $T_c$. Heat is withdrawn from both the spaces and rejected to the atmosphere at $T_a$. Assuming that the rate of heat transfer from the cold space, $Q_c$, is the same as from the freezer, $Q_f$, find an expression for the minimum power into the heat pump. Evaluate this power when $T_a$ = 20, $T_c$ = 5 and $T_f$ = – 10°C and $Q_f$ = 3 kW.

3.19 An air compressor receives atmospheric air at $T_0$ = 17°C, 1 bar and compresses it up to 14 bar. The compressor has isentropic efficiency of 88%, and it loses energy by heat transfer to the atmosphere as 10% of the isentropic work. Find the actual exit temperature and the reversible work.

3.20 A rock bed consists of 6000 kg granite and is at 70°C. A small house with lumped mass of 12000 kg wood and 1000 kg iron is at 15°C. They are now brought to a uniform final temperature with no external heat transfer, by connecting the house with the rock bed through some heat engines. If the process is reversible, find the final temperature and the work done during the process.

3.21 Air enters the turbocharger compressor of an automotive engine at 1 bar, 30°C and exits at 1.7 bar. The air is cooled by 50°C in an intercooler before entering the engine. The isentropic efficiency of the compressor is 75%. Determine the temperature of the air entering the engine and the irreversibility of the compression–cooling process. [*Ans.* 46.3°C, 19.8 kJ/kg]

3.22 A rigid container with volume 200 *l* is divided into two equal volumes by a partition. Both sides contain $N_2$; one side is at 20 bar, 300°C, and the other side at 10 bar, 50°C. The partition ruptures, and the $N_2$ comes to uniform state at 100°C. Assuming the surroundings are at 25°C, find the actual heat transfer and the irreversibility in the process.

3.23 A steady state of R-22 at ambient temperature, 10°C and at 7.5 bar enters a solar collector. The stream exits at 80°C and 7 bar. Calculate the change in availability of the R-22 between these states. [*Ans.* 6.237 kJ/kg].

## Multiple-choice Questions

3.1 In a power plant, Turbine work = 10,000 kJ, pump work = 10 kJ, heat supplied by boiler = 30000 kJ. Thermal efficiency of the plant will be

(a) 27%. (b) 33.333%. (c) 33.33%. (d) 33.30%.

3.2 A refrigerator and heat pump operate between the same temperature limits. If COP of the refrigerator is 4, the COP of the pump would be

(a) 3. (b) 5. (c) 4. (d) cannot predict.

3.3 Energy requirement to heat a fully insulated enclosure will be fulfilled if

(a) a heat pump is used for heating. (b) an electric strip heater is used for heating.
(c) steam is used for heating. (d) hot water is used for heating.

3.4 It is impossible to construct an engine, which while operating in a cycle, produces no other effect to extract heat from a single reservoir and do equivalent amount of work.

(a) This refers to the Clausius statement.
(b) This refers to the Kelvin–Plank statement.
(c) This refers to the Carnot theorem.
(d) This does not refer to any of the above.

3.5 The Carnot cycle operates between temperatures of 1000 K and 500 K. The Carnot efficiency will be 50% if the working substance is
(a) air (b) nitrogen. (c) ammonia. (d) any substance.

3.6 The Carnot cycle comprises
(a) two isothermal processes and two isentropic processes.
(b) two constant volume processes and two isentropic processes.
(c) one constant volume, one constant pressure and two isentropic processes.
(d) two constant pressure and two isentropic processes.

3.7 A heat engine is supplied with 800 kJ/sec of heat at 600 K, and heat rejection takes place at 300 K. Which of the following results report a reversible cycle?
(a) 200 kJ/s are rejected (b) 400 kJ/s are rejected
(c) 100 kJ/s are rejected (d) 500 kJ/s are rejected.

3.8 A heat engine is supplied with 300 kJ/s of heat at 600 K, and rejects 100 kJ/s at 300 K. The data refers to
(a) reversible cycle. (b) irreversible cycle.
(c) impossible cycle. (d) none of the above.

3.9 The entropy of a fixed amount of incompressible substance,
(a) decreases in every process in which the temperature increases.
(b) remains the same in every process in which the temperature increases.
(c) increases in every process in which temperature increases.
(d) is not affected in the process by increase or decrease of temperature.

3.10 A closed system undergoes a process in which the work done on the system is 5 kJ and heat transfer $Q$ occurs only at temperature $T_0$.
(i) The change in entropy for $Q = 5$ kJ and internally reversible process is
(a) positive. (b) negative. (c) zero. (d) indeterminate.
(ii) For $Q = 0$ and internally reversible process. The change in entropy is
(a) positive. (b) negative. (c) zero. (d) indeterminate.
(iii) for $Q = -5$ kJ and internally reversible, the change in entropy is
(a) positive. (b) negative. (c) zero. (d) indeterminate.
(iv) for $Q = +5$ kJ with internal irreversibility present, the entropy change is
(a) positive. (b) negative. (c) zero. (d) indeterminate.
(v) for $Q = 0$ with internal irreversibility present, the entropy change is
(a) positive. (b) negative. (c) zero. (d) indeterminate.
(vi) for $Q = -5$ kJ with internal irreversibility present, the entropy change is
(a) positive. (b) negative. (c) zero. (d) indeterminate.

3.11 Which of the following statements is true?

(a) No process is allowed in which the entropy of both the system and the surroundings increase.

(b) During a process, the entropy of the system might decrease, while the entropy of the surroundings increases, and conversely.

(c) No process is allowed in which the entropies of both the system and the surroundings remain unchanged.

(d) A process can occur in which the entropies of both the system and the surroundings decrease.

3.12 A heat source at temperature $T_1$ transfers heat to a system at temperature $T_2$ ($T_1 > T_2$). Which of the following statements is not true?

(a) $\Delta S_{source}$ decreases. (b) $\Delta S_{system}$ increases.

(c) $(\Delta S_{system} + \Delta S_{source})$ increases. (d) $(\Delta S_{system} + \Delta S_{source})$ decreases.

3.13 In order to increase work capacity of energy transferred by heat transfer from high temperature to low temperature

(a) temperature difference should be increased.

(b) temperature difference should be decreased.

(c) higher temperature should be increased keeping temperature difference the same.

(d) lower temperature should be lowered keeping temperature difference the same.

3.14 Helmholtz function is expressed as

(a) $(u - Ts)$. (b) $h - Ts$. (c) $(u + pv)$. (d) $(-sdT + vdp)$.

3.15 Gibbs function is expressed as

(a) $(U + pV - TS)$. (b) $(U + pV - TdS)$.

(c) $(U + pV - TdS)$. (d) $(U + pv - SdT)$.

3.16 Availability function is expressed as

(a) $b = U + PoV - T_0S$. (c) $b = U + P_i\,dV + To\,dS$.

(b) $b = dU + p_0\,dV - To\,dS$. (d) $b = U + poV + T_0\,dS$.

# CHAPTER FOUR

# Working Substance



## 4.1 INTRODUCTION

In Chapter 1, we learnt the tools used for analysis of thermal systems. In this chapter, we shall study behaviour of substances which are used in thermal systems. Based on its chemical properties, a substance may be classified as an element, compound or a mixture. A substance is called a mixture if (i) its constituents can be separated by suitable physical or mechanical means, (ii) if its constituents can occur in all proportions, (iii) its heat of formation[1] is zero, and (iv) the overall properties are weighted average properties. A mixture is homogeneous if its composition is the same throughout the system, else it is heterogeneous. A substance may exist in any one or more of the phases, namely solid, liquid or gas. The term pure substance is used for a substance that has the same chemical

---

1. For the analysis of the first law, we need enthalpy or internal energy that are extrinsic properties. For a single component system, there is no problem as to what reference is being used for their measurement. However, consider a system made of a mixture of different substances (having different references for measurement of these quantities) undergoing a chemical reaction, changing the chemical composition. In such a case, it is not possible to find out change in these properties. Therefore, a common reference needs to be prescribed. Heat of formation is the result of such attempts, and is defined as the heat produced in an exothermic (or absorbed in endothermic) reaction when a reaction occurs between its constituent elements at 1 bar and 288 K standard reference conditions.

composition throughout the system, and which does not change during the process. Examples of a pure substance are (i) nitrogen, oxygen, water, etc.; (ii) a mixture of two or more phases of a pure substance, like wet steam; (iii) mixture of single phases of compounds, like air; (iv) mixture of lubricant oil miscible with the refrigerant, etc. However, a mixture of water and oil cannot be considered as a pure substance, as the chemical composition is different in different parts of the system.

We know steam is used in power plants, air is used in air compressors and gas turbines, and different kinds of refrigerants are used in refrigeration plants, and so on. Therefore, it is worthwhile to devote some time to study the behaviour of the working substance under different conditions of pressure and temperature. The extent of pressure and temperature range generates innumerable states, and hence makes presentation of such vast data on paper a hurricane job. Methods like charts, tables and mathematical relations have been used for the purpose; charts include two and three-dimensional diagrams, which are discussed in Section 4.2. Section 4.3 comprehends the tabular way of representing functions of one and two independent variables, and the concept of linear interpolation. For mathematical form of representation of properties, relation among various properties may be in differential, algebraic or parametetric forms. Section 4.4.2 analyses the various useful differential relations, and how one relation may be changed to another form using the available differential relations among the various partial derivatives we learn in a course on partial. For convenience, some of them used in the analysis have been reviewed, too. Knowledge of these relations help us in drawing the various processes on 2D charts as well. Behaviour of gases is covered in Section 4.4.1, wherein the concepts of compressibility factor and pseudo-pressure, known as fugacity, are introduced to take care of deviation of real gases from perfect gas.

## 4.2 DIAGRAMMATIC REPRESENTATION

By Gibbs phase rule, we know that a system consisting of a single component pure substance may need 0, 1 or 2 independent intensive properties to describe it, depending upon the number of phases present. To encompass the whole canvas, we have to fix any two of the many available properties, and then the third can be taken in turn. This leads to a three-dimensional plot. Three-dimensional plots such as $p$-$v$-$T$, $p$-$v$-$u$, etc. have been developed to show various states of a pure substance, including its many phases. We shall discuss only a p-v-T surface, as these are most relevant in applications. Each point on such a surface represents a unique state of the substance, and each substance has its unique 3D model, which resemble and differ in many ways for different substances; hence a single model is not enough for all substances.

We are familiar with water ($H_2O$). $H_2O$ exists in three single phases, namely ice (solid), water (liquid) and steam (vapour or gas). Did you ever notice that when we heat a given amount of ice, it first converts into water without change in temperature, till phase conversion is complete? Further heating increases the temperature of water, and at some point of time, water starts changing into steam slowly and slowly again at constant temperature, till the whole of water is transformed into steam, and thereafter, temperature of the steam, like water, starts increasing. It never happens that the whole of ice changes into water or water into steam at once or temperature changes during phase change. During the change of phase, the two phases remain in equilibrium with each other. As water converts into steam, its entropy, volume, and enthalpy increase. That is, during this phase change, the extensive properties change and the intensive properties remain constant. Such a change is known as *first order phase change.* On the contrary, liquid helium jumps to gas without any change in extensive property, which is a different behaviour from water. However, in this case, there is jump in value of specific heat. This characterises a *second order phase change.*

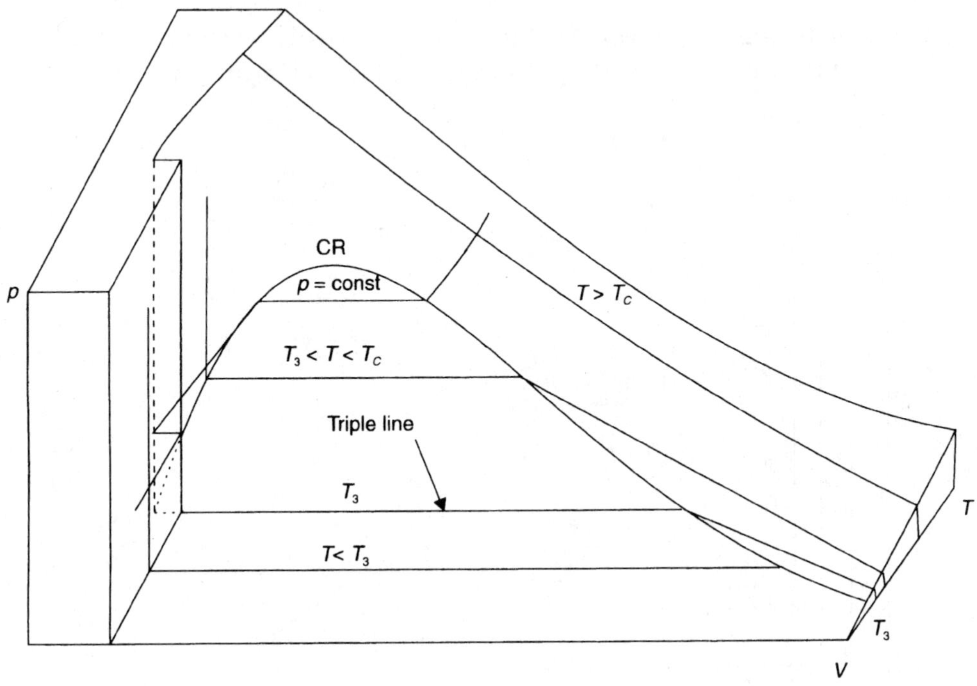


**Fig. 4.1** *p-v-T* surface for $H_2O$

Let us consider another aspect of water. When ice melts, its volume decreases. This behaviour is different from the behaviour of normal substances like carbon-dioxide, oxygen, ammonia, etc., in which case the volume increases like other extensive properties. This different behaviour of water during melting is called the *anomalous behaviour of water.* Bismuth is another substance which has similar behaviour. Figure 4.1 shows the p-v-T surface for $H_2O$. We see a dotted line at the interface of solid and liquid, representing decrease in volume of ice. But in Fig. 4.2, a p-v-T surface for $CO_2$, volume is increasing under similar conditions. Figures. 4.1 and 4.2 also show a few isobaric and isothermal lines.

Now, let us talk about a few similarities found in the two p-v-T surfaces shown in Figs. 4.1 and 4.2. We observe the following.

(i) There are three single-phase regions; each has double curvature at all points on it. Slope of the surface changes both along the temperature and volume axes, which is positive. That is $(\partial p/\partial T)_v$ and $(\partial p/\partial v)_T$ are > 0, though their values will be different at different states.

(ii) Single phase surfaces are separated from each other by two phase regions, which are contiguous with and join them along the crease called the phase boundary or boundary line. Different names are given to these boundaries, as shown in Fig. 4.5 on a phase diagram. The two-phase surfaces have single curvature, and their slope changes along the temperature axis. That is, curvature of the boundary lines will be seen on the p-T diagram, which is known as the *phase diagram.*

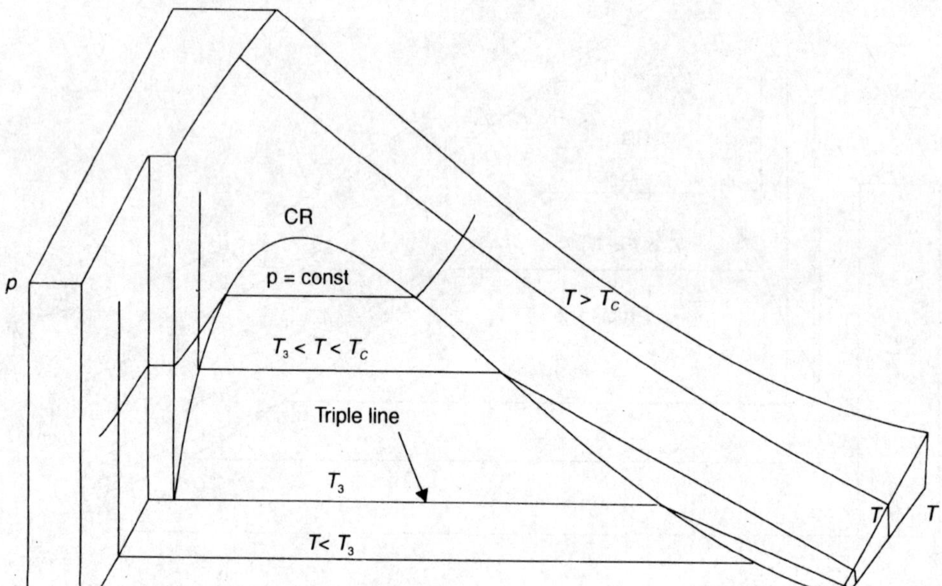


**Fig. 4.2** *p-v-T* surface for $CO_2$

(iii) The three two-phase surfaces intersect each other at boundaries on which the substance will exist in all its three phases. Such lines of intersection are known as triple line, and this occurs only at a particular value of temperature and pressure referred to as the triple point temperature and pressure, respectively. Therefore, on the p-T diagram, the triple line will appear as a point and is called the triple point. Note that at the triple point, only the pressure and temperature are fixed and not the extensive properties.

(iv) Lastly, there is a point of inflexion only on the liquid–vapour phase surface at which point $(\partial p/\partial v)_T = (\partial^2 p/\partial v^2)_T = 0$. This point is known as the critical point, and the properties are referred to as critical property, like critical pressure $p_c$, critical volume $v_c$, critical temperature $T_c$, etc. It is difficult to determine properties at this state experimentally because (i) isothermal compressibility[2] becomes infinite and this results in setting up of density gradients in the presence of gravitational field, (ii) the heat capacity becomes too large and therefore, obstructs in setting of thermal equilibrium quickly, (iii) scattering of light occurs, which hampers the visual observation, and (iv) the liquid and vapour become indistinguishable.

### 4.2.1 Two-dimensional Charts

It is not that easy to draw three-dimensional diagrams on a paper and explain the various processes on it. Therefore, a two-dimensional chart is often used in practice. We obtain a p-v diagram from a p-v-T plot by intersecting an isothermal plane with a p-v-T surface. Isothermal planes for $T \leq T_3$ (triple point temperature), $T_3 \leq T \leq T_c$ and $T > T_c$ are shown on surfaces for $H_2O$ and $CO_2$. The intersection shows changes in the state during an isothermal process. An orthographic projection, when seen in the direction of $T$, gives a p-v diagram. We may also similarly show intersections made by constant pressure or constant volume planes and then plot on a two-dimensional plot that yields a v-T or p-T diagram, respectively. However, the most commonly used charts are p-v and T-s charts. Other charts are also in use. For example, h-s chart, developed by Mollier and named after him, is used in steam power plants, and the p-h chart for refrigeration systems, as these are handy for analysis of thermal systems using steam or a refrigerant.

p-v and T-s charts for $H_2O$ are shown in Fig. 4.3 for the liquid–vapour region only. We observe a dome formed by *saturated liquid* and *saturated vapour lines* meeting at the critical point at its top. The dome encompasses wet steam, which is a homogeneous mixture of water and steam. Line 'fg' in the wet dome region represents an isobaric process as well as an isothermal process. For every pressure, there is a unique temperature

2. This is a property of the substance, which measures change in its volume with increase in pressure when it is subjected to an isothermal process. At the critical point, even a small change in pressure brings a very large change in volume during the isothermal compression or expansion process, (see Eq. 4.15).

at which water starts boiling, known as boiling temperature in day-to-day language. However, pressure and temperature (during change of phase) are referred as saturation pressure and saturation temperature, respectively. Water changes during boiling from saturated liquid at *f* to saturated steam at *g*; subscript *f* and *g* with any property are used to show its value corresponding to these two states. Regions right of the saturated vapour line or left of the saturated liquid line are single-phase regions, called *superheated steam* and *compressed or undercooled liquid* regions. In these regions, any two intensive properties are required to define a state. Nonetheless, *internal energy, enthalpy and volume for liquids are very weak functions of pressure* to the extent that normally this dependence is often ignored. Constant temperature expansion process 1fg2 and isobaric process 3f'g'4 are shown in Fig. 4.3 on p-v and T-s charts.[3] We may show other processes also in the same manner and leave it to the reader to do so.

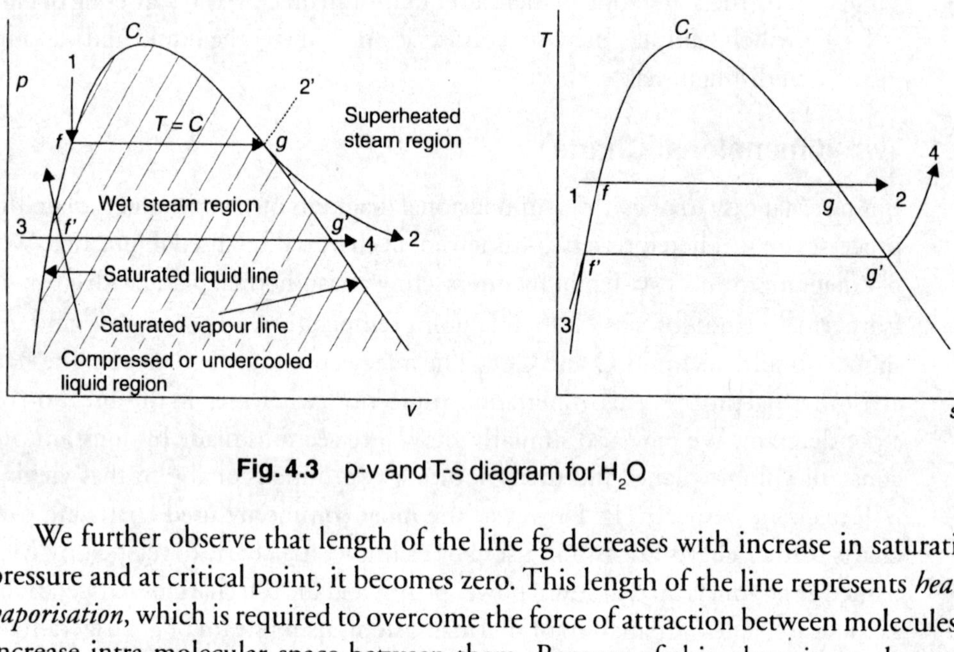


**Fig. 4.3** p-v and T-s diagram for $H_2O$

We further observe that length of the line fg decreases with increase in saturation pressure and at critical point, it becomes zero. This length of the line represents *heat of vaporisation*, which is required to overcome the force of attraction between molecules to increase intra-molecular space between them. Because of this, there is no change in temperature during the phase change, and the heat supplied is known as *latent heat*. Thus energy, enthalpy or entropy of $H_2O$ increases during the phase change. But change in their magnitude decreases with pressure. These observations help us to understand the performance of thermal systems using vapours as the working substance.

---

3. It is the author's experience that many students normally show constant temperature process in superheated region on the p-v diagram by the dotted line *g*2', as shown in figure. This confusion is the result of constant pressure process on the T-s diagram has this shape and student fails to realise it.

Mollier chart, shown in Fig. 4.4, is frequently used in the analysis of the steam power cycle. This chart has constant pressure, constant dryness fraction, constant temperature and constant volume lines; constant pressure lines are more steep compared to constant volume lines. Knowing any two properties of steam,[4] its state point can easily be located on the chart using the graphic interpolation technique. Importance of the h-s diagram is felt while making analysis of a power plant. A power plant may have 1 to 4 turbine modules depending on its power production rating. Each turbine module has many stages over which steam expands. If detailed analysis of each stage is required, then the use of steam table makes it very time consuming and irritating. But it is handy to draw various processes on the h-s chart. Once the process is shown on a diagram, we read the properties immediately from it. On the other hand, if the steam table, discussed in the next section, is used, then the process equation is to be solved to find out the end state. (Students may try some problems to have a first hand feeling of how time consuming is the process, in order to see importance of this chart.)

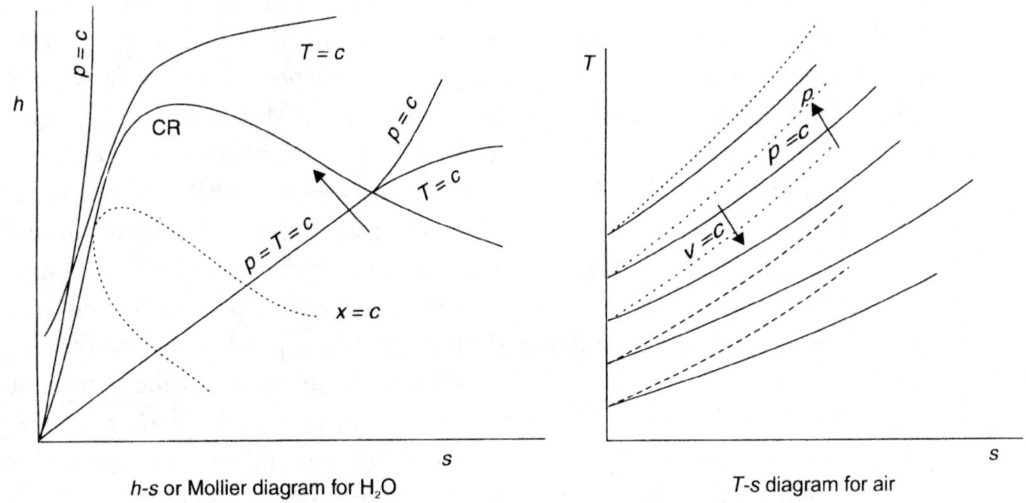


**Fig. 4.4** Useful two dimensional charts

Figure 4.4 also displays the T-s diagram for air. Recall that for air, assumed to behave like a perfect gas, $h \propto T$. Therefore, the h-s chart is also the T-s chart to a different scale. For this reason, the T-s and not the h-s chart is used for air and gases. The arrow on $p = C$ or $v = C$ curves shows that these quantities are increasing in that direction.

4. One of these properties has to be either pressure or temperature. See solved Example 4.1.

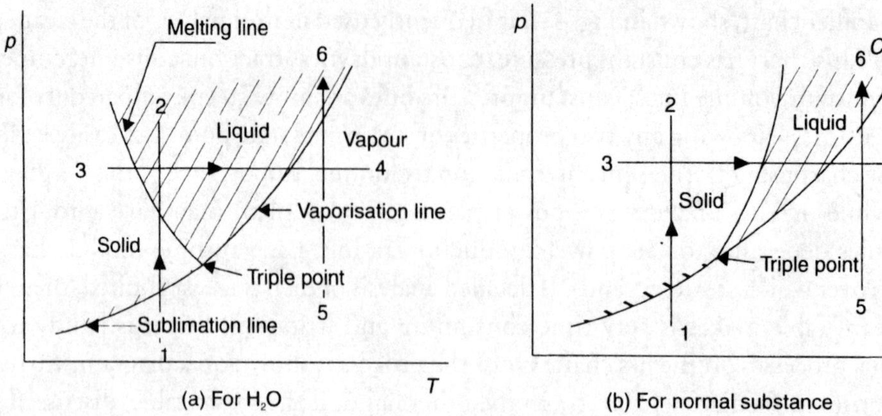


**Fig. 4.5** Phase diagram

One another chart that needs to be mentioned here is the phase diagram. It is p versus T, or simply the p-T chart. It is shown in Fig. 4.5 for a normal substance and $H_2O$. We see that the two-phase regions are shown as a curved line, since these surfaces have curvature along the temperature axis only, as mentioned before. For a normal substance, slope of all two-phase boundaries is positive, but for water, the solid–liquid line (melting or solidification line) has a negative slope. The phase diagram is important because it tells us how a phase would change during a process. An isothermal compression process 12 is shown in the figure for both the substances from its gaseous phase. $H_2O$ transforms first to solid and then to liquid, whereas a normal substance is changed to solid and stays in this phase for any pressure thereafter. Also shown is an isobaric process 34, and we may see from the figure how the phase change during a process depend much upon the initial point. The discussion applies for the initial state of the substance. If the initial state is different, then the story will be different as seen for the process 56 on these two phase diagrams; for both types of substances, phase changes to liquid with the application of pressure, and stays in that phase for any amount of pressure thereafter.

## 4.3 TABULAR PRESENTATION

Another convenient way of presenting the data is to organise them in a tabular manner, which is explained taking the example of steam; in fact, this organisation is universal for any substance for its liquid–vapour or any other two-phase region. Let us take a system consisting of m kg of any liquid such as water at room temperature, and start heating it at constant pressure. When heat is supplied, the system temperature increases, and at some temperature, the liquid starts changing into vapour, which occupies space above the liquid surface; more the heat supplied, more is the vapour formation. During this phase change process, liquid and vapour remain in equilibrium at constant temperature and pressure. Here, the two phases exist separately. If water and steam are mixed to give a uniform mixture, then the mixture is called wet vapour, whose composition is defined by the mass fraction formed by the ratio of mass of dry vapour to the total mass of the mixture. If $m_w$, $m_v$ are the mass of saturated liquid and vapour in the wet mixture, then the mass fraction is $m_v/(m_w + m_v)$. This ratio is called *dryness fraction* '$x$' of the wet vapour.

Continuous heating of water will convert it completely into vapour at one stage. Then, there shall be a single phase and temperature will once again start increasing; this phase is known as the superheated vapour phase. Therefore, for steam, we need two steam tables, one for wet and another for superheated steam. We know that for wet steam both pressure and temperature cannot be taken as independent variables. Therefore, we have the choice to arrange data as a function of temperature or pressure. That is why there are two tables for wet steam—in one temperature, and in another pressure is taken as the independent variable.

Since the data is vast, properties at all the temperatures or pressures cannot be tabulated. It was observed from the vast data bank that properties of saturated liquid, saturated vapour and superheated steam change linearly with independent variables up to some change in their value. For wet vapour, a property changes linearly with $x$ from $f$ to $g$. In fact, these observations hold true for all the substances belonging to the first order phase change. Thus, we may get any property for given vapour state if we know it for the extreme end of independent variable by using linear interpolation.

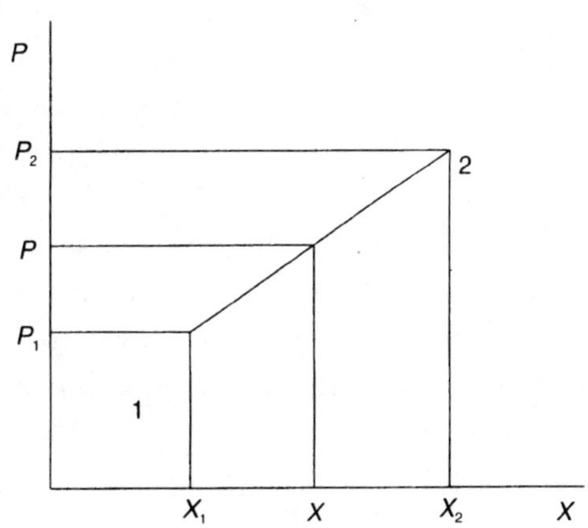


**Fig. 4.6** Linear interpolation

To understand linear interpolation, let us consider $P$ is dependent on an independent property $X$, that $P$ has values $P_1$ and $P_2$ for values $X_1$ and $X_2$ of $X$, as shown in Fig. 4.6, and that points 1 and 2 are extreme ends in which $P$ varies linearly with $X$. From this figure, we can write $P = P_1 + (P_2 - P_1)/(X_2 - X_1) \times (X - X_1)$. This is called linear interpolation. For wet steam, interpolation may be required on the independent variables pressure, temperature or dryness fraction; and for superheated steam on pressure, temperature or both. In case points 1 and 2 correspond to saturated states $f$ and $g$, respectively of wet steam then it is usual to write $P_{fg}$ for $(P_2 - P_1)h_{fg}$ and $s_{fg}$ are also tabulated in the wet steam table to save time consumed in subtraction and to eliminate any chances of error in subtraction.

This background makes it possible to put the vast data conveniently in tabular form for any wet or superheated vapour. See wet steam table (or any other wet vapour table) and note that properties such as specific volume, specific internal energy, specific enthalpy and specific entropy are tabulated only for the saturated liquid and saturated vapour states for discrete values of pressure, and that the gap between pressure values also varies with increase in pressure. However, all these properties vary linearly in the pressure interval between two successive rows for which properties are tabulated. The structure of a wet steam table is given in Table 4.1.

**Table 4.1** Wet steam table

| $p$ bar | $t_s$ °C | $v$, m³/kg | | $u$, kJ/kg | | $h$, kJ/kg | | | $s$, kJ/kg K | | |
|---|---|---|---|---|---|---|---|---|---|---|---|
| | | $v_f$ | $v_g$ | $u_f$ | $u_g$ | $h_f$ | $h_{fg}$ | $h_g$ | $s_f$ | $s_{fg}$ | $s_g$ |
| | | | | | | | | | | | |

We want to find out, say, enthalpy of saturated vapour, for example, at some given pressure by using the wet steam table. We first move down the pressure column till we come across that value and then along the horizontal line to the column for $h_g$ and read the value. This is the standard procedure to look up a table. There is also a possibility that properties are not listed for this pressure. In that case, we locate two adjacent rows such that in one pressure is less and in another more than the pressure we are looking for. We read $h_g$ at these two located conditions and then use interpolation for $h_g$ on pressure. We explained the procedure taking pressure as an example, but, in fact, it can be any property whose values are tabulated in the table.

A superheated vapour table has the structure as shown in Table 4.2, which is a standard way of tabulating the equation $z = z(x, y)$, taking $x$ along the vertical line and $y$ along the horizontal.

The procedure to read any property from the superheated steam table is the same as that used for the wet steam table. For example, if steam pressure is 5 bar and the temperature of steam is 300°C, then the value of $h$ will be obtained at the intersection point of the row for 5 bar pressure and 300°C column. But if $p$ is 3 bar, then there are rows for 2 and 5 bar in the table and linear interpolation on pressure is made to find out $h$ at 300°C. The same procedure is followed if temperature is different from that given in the table.

**Table 4.2** Superheated table

| $T$,°C → <br> $p$, bar ↓ | | 50 | 100 | 150 | ... |
|---|---|---|---|---|---|
| $v$ <br> $u$ <br> $h$ <br> $s$ | $v_g$ <br> $u_g$ <br> $h_g$ <br> $s_g$ | | | | |

Lastly, it is observed that specific enthalpy, specific volume or specific entropy for superheated steam is more than the corresponding value for saturated steam. This observation is very useful, as after the vapour is subjected to a process, it is not known whether vapour is wet or superheated, and it is determined by comparing the property value with that of saturated vapour. Similarly, values of these properties for compressed liquid is less than that for saturated liquid.

Table for compressed liquids is also available. However, we may do away with it, as it is explained in the Appendix 2 at the end of book.

**Example 4.1** Find the entropy of steam when (a) $p = 5$ bar and $t = 350$°C (b) $h = 3167.7$ kJ/kg and $v = 0.570$ m$^3$/kg.

***Solution*** **Pure substance:** Steam.

**To find:** Specific entropy.

(a) Steam is superheated because its temperature is more than saturation temperature at 5 bar. We move down the pressure column and stop where the pressure is 5 bar. Then, we move along the row to the right up to the column for 350°C; we read from the superheated steam table

$$s = 7.633 \text{ kJ/kg K}$$

Any other property in this row may be read like this.

(b) Since the values have been chosen from the superheated steam table, we know that pressure and temperature for this state are 5 bar and 350°C, at which $s = 7.633$ kJ/kg K, as determined in (a). But how to arrive at this result in the absence of this knowledge is the real problem, though two properties are given, which is what is required for any single phase by the Gibbs phase rule. Do we see the whole table for all the pressure and temperature to locate the combination of given values of $h$ and $v$? Are we sure that this state is tabulated in the table, or is to be obtained by linear interpolation? Is it possible for steam to have some other pressure and temperature to give these vales of $h$ and $v$? No, that is not possible, since each state is unique. For example, if we look at 6 bar, then temperature of steam between 350 and 400°C may be obtained by linear interpolation to give $h$ = 3167.7. But in this case, the value of v will not match.

Should we conclude that entropy in this case cannot be determined? Definitely not, because if we use Mollier chart then the point of intersection of $h$ = 3167.7 kJ/kg and $v$ = 0.570 m³/kg lines define the condition of steam enthalpy now can be read. So we conclude that the difficulty is with steam table as it has been organised w.r.t. $t$ or $p$. Moreover, a little insight into how we arrive at the Gibbs phase rule reveals that potentials such as pressure, temperature, and chemical potential, are responsible for the thermodynamic equilibrium of a system. Therefore, at least one of these will be desirable to define the system state. For example, if we were given $t$ or $p$ instead of $h$ or $v$ in this case, then we would know how to proceed.

**Example 4.2** 0.5 kg of steam at 10 bar and dryness fraction 0.8 is heated at constant pressure in a container until its volume is doubled. Determine the final condition ot steam, work done, heat transferred and change of entropy.

***Solution*** Given: $m$ = 0.5 kg, $p_1$ = 10 bar, $x_1$ = 0.8, $p_2 = p_1$ = 10 bar and $v_2 = 2v_1$.

**System:** Steam in a container is the system and is wet, as the dryness fraction is given. Since there is no mass flow, the system is a closed system. We read from the steam table at 10 bar $v_g$ = 0.194, $u_f$ = 761.7, $u_g$ = 2583.6, $s_f$ = 2.139 and $s_{fg}$ = 4.448

$\therefore\ v_1 = x_1 \times v_g = 0.8 \times 0.194 = 0.1552$ m³/kg

$u_1 = u_f + x_1 (u_g - u_f) = 2219.22$ kJ/kg and

$v_2 = 2\,v_1 = 0.3104$ m³/kg

which is more than $v_g$ (= 0.194 m³/kg) at 10 bar. Therefore, steam is superheated after heating. From the superheated steam table, the temperature of steam lies between 400 and 450°C, at which temperatures the specific volume is 0.3066 and 0.3304 m³/kg, respectively. By linear interpolation with respect to (hence after abbreviated as w.r.t.) temperature gives $t_2$ = 400 + 50 (0.3104 – 0.3066)/(0.3304 – 0.3066) = 408°C.

The condition of steam after heating is superheated at temperature 408°C.

From the superheated steam table, u at 400°C and 450°C is 2957.3 and 3040.3 kJ/kg. Again using linear interpolation w.r.t. $T$, we get:

$$u_2 = 2957.2 + ((3040.3 - 2957.3)/50) \times (408 - 400)$$
$$= 2970.58 \text{ kJ/kg.}$$

For a constant pressure process, taking care of using proper units[5] for different quantities, we get

$$W = m\,p\,(v_2 - v_1) = 0.5 \times (10 \times 10^5)\,(.3104 - .1552)/10^3 \text{ kJ}$$
$$= 77.6 \text{ kJ.}$$

The positive sign indicates that work is done by the system. Using nfee, we get

$$Q = W + dU = 77.6 + 0.5 \times (2970.58 - 2219.22)$$
$$= 453.28 \text{ kJ.}$$

**Example 4.3** A 40-litre vessel contains 2 kg of Freon-12 at a pressure of 1.6 bar. Determine: (a) temperature, quality and enthalpy of the refrigerant, (b) the volume occupied by the vapour phase.

***Solution*** Given: $m = 2$ kg, $v = 40\ l = 40 \times 10^{-3} = 0.04$ m$^3$, $p = 1.6$ bar *working substance*: Freon-12.

(a) We get from the given data, $v$ = specific volume = .04/2 = 0.02 m$^3$/kg and from the Freon-12 table, $v_g$ at 1.6 bar = 0.1031 m$^3$/kg. This means that the vapour in the vessel is *wet* as $v < v_g$. Again, from the table, we read $t = -18.49$°C, $h_f = 19.18$ kJ/kg, $h_{fg} = 160.23$ kJ/kg and

$v_f = 0.0006876$ m$^3$/kg. Since $v = v_f + x\,v_{fg}$, we get:

$x = (0.02 - 0.0006876)/(0.1031 - 0.0006876) = 0.188.$

Here, it is assumed that the system consists of a homogeneous mixture; that is, liquid and vapours are thoroughly mixed. Since $x$ is known, enthalpy can now be obtained by linear interpolation to give:

$$h = h_f + x\,h_{fg} = 19.18 + 0.188 \times 160.23 = 49.3 \text{ kJ/kg.}$$

(b) However, this question implies as if the liquid and vapour are not mixed with each other. Rather, the liquid is settled at the bottom, and the space above it is occupied by vapours. But the mass of dry vapour in a wet mixture = $x \times m = 0.188 \times 2 = 0.376$ kg.

---

5. While making substitution of numerical values of the variables, one must take care of proper units. In this case pressure is given in bar, whereas the proper unit is N/m$^2$; thus, the factor of $10^5$ is used, since 1 bar = $10^5$ N/m$^2$. The result is obtained in J and is to be divided by the factor $10^3$ to get it in kJ. That is, for units normally used in practice, one multiplies by a factor $10^2$.

$\therefore V_g$ = volume of vapour phase

$= m_g v_g$

$= 0.376 \times 0.1031$

$= 0.03875 \text{ m}^3.$

**Example 4.4** Find out the change in entropy and heat supplied when 2 kg of saturated water in a closed vessel at 2 bar is converted to saturated vapour at (a) constant pressure, or (b) constant volume by heating. Assume missing data suitably.

***Solution*** Given: Water in a vessel. It is closed system as there is no mass flow across the vessel.

$m = 2$ kg, $p_1 = 2$ bar, and $x_1 = 0$

Process: Heating water at (a) constant pressure or (b) constant volume.

From the steam table: At 2 bar $v_f = 0.001061$, $v_g = 0.885$ m³/kg, $s_f = 1.5301$, $s_g = 7.1268$ kJ/kg K, $u_f = 504.4$, $u_g = 2529.5$ and $h_f = 504.7$ kJ/kg.

(a) Water can be heated at constant pressure in a container fitted with a piston having weight per square metre of its surface area = $2 \times 10^5$ N. When heat is supplied, water starts changing into vapour without any change in temperature, which means that heat supplied is latent heat. In this process, the specific volume changes from $v_f$ to $v_g$ at 2 bar, and therefore, the piston moves up to keep the pressure constant. Therefore, work done in moving piston,

$W = m \times p \times (v_g - v_f) = 2 \times 2 \times (v_g - v_f) \times 10^2$

$= 353.58$ kJ. Using nfee, we get heat transfer,

$Q = W + \Delta U = 353.58 + 2\,(2529.5 - 504.4)$

$= 4403.78$ kJ.

We may note in this case, that $q = (u_g - u_f) + p\,(v_g - v_f) = h_g - h_f$. That is, though the system is closed, but *heat transfer during change of phase is given by change in enthalpy, as if the system is open having no work interaction with the surrounding*. This is so because work is done in overcoming forces among the water molecules.

Also, change in entropy = $2 \times (7.1268 - 1.5301)$

$= 11.1934$ kJ/K

*Note*: (1) A boiler is a device in which steam is produced at constant pressure, wherein mass of water entering into it is equal to that of steam leaving, so that pressure remains constant. This example of constant pressure heating in a closed system differs from it.

(2) While using $\Delta U$ for $\Delta E$, we assumed that change in potential energy is zero. But this is not true as there is movement of piston, and hence the centre of gravity of the mass inside the cylinder is raised. However, the magnitude of change in potential energy is infinitesimal compared to the changes in internal energy, and hence it justifies our assumption.

(b) Data given in the question is incomplete. We shall now explore what solutions are possible depending on size of the vessel if the vessel is initially filled with saturated water at 2 bar. The following situations are likely to arise

*Case 1.* If the volume of the container is $2 \times v_g = 1.77$ m$^3$ (where $v_g$ is specific volume at 2 bar), then water occupies only $2 \times v_f = 0.002122$ m$^3$ of the container space in the beginning, and the rest is occupied by its vapour, which is in equilibrium with water. As heating proceeds, more and more vapours fill the space above the water, and ultimately, the whole of the space is filled with saturated vapours. In this process, temperature and pressure remain constant, and heat transfer is the same as in case (a), since it is constant pressure heating.

*Case 2.* If the volume of the container is just equal to the volume of water, that is 0.002122 m$^3$, then there is no space for vapours to occupy. Hence, heating will increase the pressure and temperature of water, and there will be no change in phase (see p-v diagram for $H_2O$), since the saturated liquid line is inclined to the right of the vertical line passing through the initial point.

*Case 3.* Now, assume that the volume of container is somewhere between the above two extreme limits, say 1.62 m$^3$. In this case, heating will increase both the pressure and the temperature, so that the saturated vapours will occupy the whole space and shall have $v_g = 1.62/2 = 0.81$ m$^3$/kg. We look down the $v_g$ column in the wet steam table, and find that under this condition, pressure $p_2 = 3.2$ bar, at which $s_{g2} = 7.0949$, $u_{g2} = 2532.8$, $v_{f2} = 0.001063$ and $v_{g2} = 0.810$ from the steam table. Since pressure is changing during building up of pressure, it is not straightforward to calculate the work done to overcome the molecular forces of water. We may however simplify the analysis by considering contents of vessel as system rather than water, as mentioned earlier. Vessel being rigid, the evaporation occurs at constant volume. Heat transfer is then obtained using nfee.

$$\therefore\ Q = m\,[(u_{g2} - u_{f1})]$$
$$= 2\,[(2532.8 - 504.4)]$$
$$= 4057 \text{ kJ and}$$
$$\Delta S = m\,(s_{g2} - s_{f1}) = 11.1296 \text{ kJ/K}$$

*Case 4.* Lastly, assume that the volume of the container is more than 1.77 $m^3$, say it is 2 $m^3$. In this case, water will evaporate till it is completely changed into vapour. At the end of evaporation, the vessel volume is more than the volume that saturated vapours at 2 bar would occupy. Hence, pressure will be less than 2 bar, which can be calculated by looking at the table for $v_g = 2/2 = 1$ $m^3$. It is a little less than 1.8 bar, say, it is approximately 1.8 bar. Heat transfer is obtained as in Case 3 by treating contents of vessel as system and is left is students to find it out.

**Discussion.** It is worth examining how water is brought to the given initial state. We read from steam table that saturation temperature at 2 bar pressure is 120.2°C. So, water pressure may be raised in a pump to a pressure much more than 2 bar, and then supplied to the vessel. Since volume of the vessel is more than that of water, water will expand after entering into it, resulting in lowering of pressure and temperature. So, pressure and temperature of water at the inlet to the vessel is to be decided, so that we are able to have saturated water at 2 bar pressure. This is however not implemented in practice.

Actually, this problem is fictitious to demostrate how one can explore various possible solutions, as given above, which assume that *somehow* the water in tank is saturated at 2 bar. And that is the snag in all these solutions. From this point of view, ignoring a possibility given in (1), Case 2 in the only solution. Water can be compressed to 2 bar in a vessel provided it is filled with water.

**Example 4.5** A steam boiler has a total volume of 3.2 $m^3$. The boiler initially contains 1.7 $m^3$ of water and 0.5 $m^3$ of vapour in equilibrium at 1 bar pressure. The boiler is fired resulting in increase in pressure to 50 bar. Find out the amount of heat supplied and steam formed.

***Solution*** Given: $p_1 = 1$ bar, $V_{f1} = 1.7$, $V_{g1} = 0.5$ $m^3$ and $p_2 = 50$ bar.

System: Contents of boiler, which is a closed system.

From the steam table: At 1 bar $v_{f1} = 0.001043$ and $v_{g1} = 1.694$ $m^3$/kg. This gives $m_{w1} = 1.7/0.001043 = 1629.91$ kg, $m_{s1} = 0.5/1.694 = 0.2952$ and $m = m_{w1} + m_{s1} = 1630.28$ kg.

We assume that the discharge valve is closed and water is heated. Heating will evaporate more and more water, which would increase the pressure inside the boiler. Let $m_{w2}$ and $m_{s2}$ be the mass of water and steam at the time the boiler pressure has reached 50 bar, at which pressure

$v_{f2} = 0.001286$ and $v_{g2} = 0.0394\ m^3/kg$ from the steam table.

In this problem, the total volume and mass shall remain constant, as the system is closed and its boundary is rigid.

$$\therefore\ m_{w2} + m_{s2} = 1630.28 \text{ and } m_{w2}\, v_{f2} + m_{s2}\, v_{g2} = 3.2$$

Solving these two simultaneous equations, we get $m_{s2} = 3.717$ and $m_{w2} = 1627.56$ kg

$\therefore$ Amount of steam formed $= m_s = 3.717 - 0.2952 = 1.8314$ kg

When valve is closed, the evaporation process occurs such that total volume remains constant. Applying nfee, we get, since $W = 0$, from steam table $u_{f2} = 1147.8$, $u_{g2} = 2597.1$, $u_{f1} = 417.4$ and $u_{g1} = 2506.1$ kJ/kg.

$$Q = (m_{w2}\, u_{f2} + m_{s2}\, u_{g2}) - (m_{w1}\, u_{f1} + m_{s1}\, u_{g1}) = 1194106.46 \text{ kJ}$$
$$= 1194.106 \text{ MJ}$$

We note that a very large amount of heat is required to raise steam in the boiler during its starting up period and it takes a long time. If for any reason, the boiler has to be shut down, there is a colossal waste of energy and a long time is consumed in cooling the boiler. Therefore, it demands that maintenance techniques be developed for repairs of the boiler during its operation, and the boiler should rarely be shut down.

**Example 4.6** A cylinder contains steam at 1 bar and 150.3°C. The steam is compressed reversibly and isothermally to a state where the specific volume is 0.28 m³/kg. Find the change in internal energy, entropy and enthalpy per kg of steam. Also, find out work and heat transfer during the process.

***Solution*** Given: $p_1 = 1$ bar, $t_1 = t_2 = 150.3$°C, $v_2 = 0.28$

System: Steam, which is a closed system.

At 1 bar pressure and 150.3°C, steam is superheated. From the superheated steam table, we observe that the property is listed at 150 and 200°C and not at 150.3°C. Therefore, linear interpolation is to be carried out to find out internal energy and entropy of steam. However, a little consideration may save our time and energy. We observe that the difference in 150.3 and 150°C is so small, that we may take the values at 150°C without making any significant error, so that $u_1 = 2576.9$ and $s_1 = 7.279$; also $v_g = 1.694$. At state 2, temperature and specific volume are known, and therefore, we have to first find out the condition of the steam. Since $v_2 = 0.28$ is less than $v_g = 1.694$, steam is wet with $x_2 = V_2/V_g = 0.28/.694 = 0.1653$. Move down the wet steam table along the temperature column to locate 150.3°C; values at 150 and 152°C are listed, and therefore, entropy and internal energy of the saturated liquid and saturated vapour is to be obtained at both temperatures from where the desired result is then obtained by linear interpolation. The procedure yields $p_2 = 4.8$, $s_{f2} = 1.845$, $s_{fg2} = 4.990$, $u_{f2} = 633.1$ and $u_{g2} = 2559.8$. Using linear interpolation during change of phase, we get $s_2 = s_{f2} + x_2 \times s_{fg2} = 3.67$ and $u_2 = 951.58$. (*We may have a feel how cumbersome is the process of getting the desired values from the steam table. However, the values obtained from the steam table are accurate compared to those from the Mollier chart.*)

By the second law of thermodynamics, heat transfer during a reversible isothermal process is $T(s_2 - s_1)$.

$$\therefore \quad q = (273 + 150.3)\,(3.67 - 7.279) = -1951 \text{ kJ/kg.}$$

and $$w = q - du = -1951 - (951.58 - 2576.9) = -326 \text{ kJ/kg.}$$

Results show that heat is rejected in the compression process to keep the temperature constant, and work is done by the surroundings on the system.

**Example 4.7** 1 m$^3$ of steam at 2 bar and 40% wet is compressed to 10 bar according to the law $pv^{1.2}$ = constant. Find the final volume and d.f., work done during compression, change in specific entropy and heat transfer per kg of steam. Neglect the volume of water.

***Solution*** Given: $p_1 = 2$ bar, $V_1 = 1$, $x_1 = 0.4$, $p_2 = 10$ and $n = 1.2$

System: Steam in a piston cylinder arrangement, and hence a closed system.

From the steam table: At 1 bar, $v_g = 1.694$, $s_f = 1.303$, $s_{fg} = 6.057$

$$u_f = 417.4,\ u_g = 2506.1$$

$$\therefore \quad v_1 = 0.6776,\ u_1 = 1252.88 \text{ and } s_1 = 3.758$$

$$\Rightarrow m = V_1/v_1 = 1.4758 \text{ kg}$$

Process equation. $(V_2/V_1)^{1.2} = (p_1/p_2)$ gives $V_2 = 0.2615$, and $v_2 = V_2/m = 0.1772$, which is less than the specific volume of saturated vapour at 2 bar. Hence, the quality of steam is wet. From the steam table at 2 bar $v_g = 0.886$, $s_f = 1.530$, $s_{fg} = 5.597$, $u_f = 504.4$, $u_g = 2529.5$.

$$\therefore\ x_2 = V_2/V_g\ 0.2027, \qquad u_2 = u_f + x_2\, u_{fg} = 914.98 \text{ kJ/kg}$$

$$S_2 = s_f + x_2\, s_{fg} = 3.664 \text{ kJ/kg (using linear interpolation)}$$

Also, $W = m \times (p_1v_1 - p_2v_2) \times 100/(n-1)\ (2 \times 0.6776 - 10 \times 0.1772)$

$$\times\ 1.4758 \times 100/(1.2 - 1) \text{ kJ} = -307.56 \text{ kJ}$$

$$\Delta U = m\,(u_2 - u_1)1.4758\,(914.98 - 1252.88)$$
$$= -498.67 \text{ kJ}$$

Using nfee, we get

$$QW + \Delta U = -806.23 \text{ kJ}$$

The negative sign shows that heat is rejected during the process.

**Example 4.8** Two-and-a-half kilogram of saturated liquid water at 10 bar is mixed with 1 kg superheated steam at 10 bar, 250°C. The mixing is adiabatic and at constant pressure. Determine the dryness fraction of the final state, the change in volume, the work done and the change in internal energy.

***Solution*** Let us represent saturated and superheated steam as A and B, respectively.

System: We may assume that the two fluids are contained in two partitions of an insulated vessel separated by a membrane. The system consists of the contents of the insulated vessel. The membrane is punctured to mix fluids at constant pressure. Thus, there is change in volume. The system is a closed system, as there is no mass flow across its boundary.

From the steam table: $h_A = 763.6$, $h_B = 2943.6$ kJ/kg. Using nfee, we get for a constant pressure process, $0 - pdv = dU$, or $d\,(U + pV) = 0$, or $dH = 0$. That is, the mixing process is isenthalpic. Using this process equation, we may write $m_A h_A + m_B h_B = (m_A + m_B)\, h_2$. In this equation all are known except $h_2$. Solving it, we get $h_2 = 1385.6$ kJ/kg. But at 10 bar, $h_f =$ 763.6 and $h_g = 2776.2$ kJ/kg. Since $h_2$ lies in between $h_f$ and $h_g$, steam after mixing is wet steam. Using equation $h = h_f + x\, h_{fg}$, we get $x_2 = 0.30939$.

Similarly, $v_2 = x_2\, v_g = 0.06089 \rightarrow V_2 = (m_A + m_B)\, v_2 = 0.2133115 \text{ m}^3$.

From steam table, we read specific volume for saturated liquid and superheated steam at initial conditions. This gives

$V_1 = 3.5 \times 0.001127 + 1 \times 0.23275 = 0.235568$. We observe that volume has decreased during mixing. Therefore, work is done on the system and is equal[6] to

$$pdV = 10 \times (0.2133 - 0.2356) \times 10^2 = -23.3 \text{ kJ}.$$

Since $\Delta H = 0$, $\Delta U = -$ work done $= 23.3$ kJ. Internal energy has increased.

---

6. Steam table gives accuracy up to four significant decimals. Therefore, a number should be rounded off accordingly.

## 4.4 EQUATIONS

The chart has a drawback, that it does not give accurate values whereas tables do give accurate values but consume time. A more desirable and convenient way rather, is to express the properties of a substance in the form of a mathematical function of independent properties. The other advantage is that it is easy to simulate it on a computer. A functional relation among pressure, volume and temperature of the type $f(p, v, T) = 0$ is called *equation of state*. This equation may also be written in explicit form such as $p = p\,(V, T)$, $v = v\,(p, T)$ or $T = T\,(p, v)$. This function may be a simple algebraic equation, parametric equation, or differential equation. Further, there may be one such equation for the entire range, or one for some range and another for another range of independent properties. Though such functions exist for all the phases of a pure substance and for different properties, but the relations are simple for gases, specifically for ideal gas, among pressure, temperature and volume. For this reason, it is the practice to use equation of state for gases, and to tabulate the property data for solids and liquids in tabular or chart form.

### 4.4.1 Algebraic Equations

Pressure, temperature and specific volume of any gas are related to each other through algebraic equations. A few equations for ideal and real gases are given below.

#### *4.4.1.1 Perfect Gas Equation*

We normally treat gases as perfect gas. In fact, a gas behaves as a perfect gas only at low pressure and high temperature, and therefore, error is likely to be introduced if this equation is used in a temperature and pressure range for which it is not applicable. For a perfect gas, we write

$$pV = mRT \text{ (mass form) or } pV = n\bar{R}\,T, \text{ (mole form)} \tag{4.1}$$

$$c_p - c_v = R \text{ (Mayer's relation), and} \tag{4.2}$$

$$dU = m\,c_v\,dT \tag{4.3}$$

where $p$, $V$, $T$ and $m$ have the usual meaning, $R$ is a gas constant and depends on the gas, $\bar{R}$ is the universal gas constant (same for all gases), $c_p$ and $c_v$ are the specific heats at constant pressure and volume, respectively, and $dT$ is the change in temperature of the gas during the process. For a perfect gas, the specific heats do not change with temperature and pressure, whereas internal energy and enthalpy depend only on temperature. However, for real gases, specific heat changes with temperature. For air, the gas constant, specific heats at constant pressure and volume, and specific heat ratio are 0.287 kJ/kgK, 1.005 kJ/kgK, 0.718 kJ/kgK, and 1.4, respectively at a temperature of 300 K. We may use these values at other temperatures also, if accurate analysis is not required.

It has been the practice to use kmole instead of kg in equation of state, where kmole is mass of 1 mole of gas expressed in kg, and is numerically equal to the molecular weight of the gas. Equation of state then becomes $pV = n\bar{R}T$, where $\bar{R}$ is the universal gas constant whose value is 8.3143 kJ/kgK for all the gases, $n$ = m/M is the number of mole, and M the molecular weight of the gas.

**Example 4.9** Show that the relation $pv^{\gamma}$ = constant exists between $p$ and $v$ when a gas in a cylinder is subjected to an isentropic process.

***Solution*** System: Gas inside cylinder. It is a closed system.

Process: Isentropic, for which $\delta q = 0$. Also, $\delta w = p\,dv$ and $du = c_v\,dT$. So, we get from nfee, $-pdv = c_v\,dT$. This equation can be integrated by writing for $p$ from Eq. 4.1, to give $R\ln v + c_v \ln T = \ln k$, where $\ln k$ is constant of integration. As the relation between $p$ and $v$ is required, $T$ can be eliminated using Eq. 4.1, and also rewriting Eq. 4.2 as $c_p\,(\gamma - 1)/\gamma = R$, where $\gamma = c_p/c_v$; we get $pv^{\gamma}$ = constant, after simplification.

#### 4.4.1.2 van der Waals Equation

van der Waals argued that at high pressure, molecules of the gases come close and attract each other. This reduces the pressure measured by any instrument by an amount inversely proportional to the square of the specific volume. Therefore, he increased pressure $p$ in the perfect gas equation by $a/v^2$, where '$a$' is a constant whose value changes from gas to gas. Further, volume '$b$' occupied by the molecules of the gas is not negligible in comparison to the volume of the space at high pressure, and he corrected the volume $v$ in Eq. 4.1 to $(v - b)$; here $b$ is a constant whose value also depends on the gas, and $v$ is the specific volume per k mol. Units of $a$ and $b$ are N m$^2$/k mol$^2$ and m$^3$/k mol, respectively. The perfect gas equation modifies to

$$(p + a/v^2)\,(v - b) = \bar{R}T, \tag{4.4}$$

which is called the van der Waals' equation. The constants are given to in Table A1.2. He obtained constants $a$ and $b$ in terms of the critical constant of the gas by noting that the critical point is the point of inflexion, where $(\partial p/\partial v)_T = (\partial^2 p/\partial v^2)_T = 0$. This yields two equations in two unknowns $a$ and $b$, and gives after simplification;

$$v_c = 3b,\ p_c = a/(27\,b^2),\ T_c = 8\,a/(27b\bar{R})\ \text{and}\ p_c v_c/(\bar{R}T_c) = 3/8. \tag{4.5}$$

As mentioned before, accurate measurement of critical constants is quite tedious, and hence the calculated values of $a$ and $b$ are likely to have error. Experimental results show $v_c$ is equal to $2b$ rather than $3b$ and $p_c v_c/\bar{R}T_c$ varies from 0.2 to 0.3, which is lower than 0.375 (= 3/8).

$a$ and $b$ were also obtained by differentiating van der Waals equation with $T$ at constant $v$ to give one equation, and simplifying it with van der Waals equation as second equation. This gives

$$b = v - \bar{R}(\partial p/\partial v)_T \text{ and } a = v^2\,[T\partial p/\partial T)_v - p]$$

When determined this way, the constants were found to depend on $T$, which agrees with experimental results.

Isotherms plotted from Eq. 4.4. for $CO_2$, as shown in Fig. 4.7 on the p-v diagram, were compared with Andrew's experimental results for $CO_2$. These curves resemble each other in general, except in the two-phase zone. The maximum and minimum points, shown in Fig. 4.7, are missing in the experimental results. The differences in theoretical and experimental results can be explained as follows.

- All states from $p$ to $n$ are unstable, because during compression of the gas, pressure of the gas starts decreasing after point $p$, and therefore, the piston will jump to point $r$ after it reaches point $p$.
- The portion $pq$ and $mn$ of the curve represent the supercooled vapour and superheated liquid states, and equilibrium in these regions can easily be disturbed in the presence of vibration, dust, etc. Therefore, during experimentation, path 'qom' is very likely to be followed by the gas, rather than exhibiting maxima and minima.

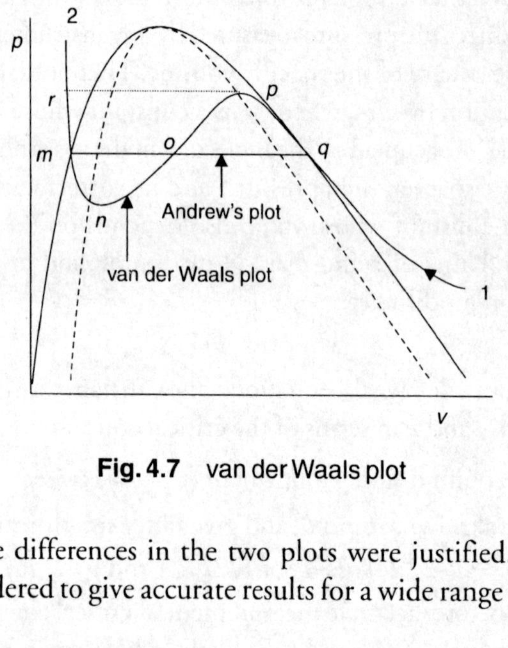


**Fig. 4.7** van der Waals plot

This way, the differences in the two plots were justified. As such, van der Waals equation is considered to give accurate results for a wide range of operating pressure and temperature.

#### 4.4.1.3 Other Equations of State

Though the van der Waals equation gave better results compared to the perfect gas equation, yet, we have seen its many limitations, especially in the high-pressure region. Therefore, many other equations were proposed by different scientists. A few of them, which have found prominent acceptance in use, are given below.

**Redlich and Kwong equation.** In 1949, Redlich and Kwong gave another equation involving two constants, which were determined in the same way as those for the van der Waals equation. The equation is

$$p = \bar{R}\,T/(\bar{v} - b) - a/\{\bar{v}(\bar{v} + b)^{3/2}\}, \text{ where}$$
$$a = 0.42748\ \bar{R}\,T_c^{5/2}/p_c \text{ and}$$
$$b = 0.08664\ \bar{R}\,T_c/p_c.$$

This equation also does not give accurate results, but has been used in gas mixture calculation and phase equilibrium with more success.

**Beattie–Bridgeman equation.** This is a pressure-explicit empirical relation involving five constants that are determined by graphical procedures from the experimental data. The equation is

$$p = \bar{R}T/v + \beta/v^2 + \gamma/v^3 + \delta/v^4$$

where $\beta = B_0\bar{R}\,T - A_0 - c\,R/T$, $\gamma = -B_0 b\bar{R}\,T + A_0 a - B_0 c\bar{R}/T^2$, and $\delta = B_0\,b\,c\,T/T^2$ where $A_0$, $B_0$, $a$, $b$ and $c$ are the five unknowns of the equation. This equation gives reasonably correct result for densities less than about 0.8 times that at critical point. Other equations involving 8 or 16 constants were also proposed and used. The additional terms take care of high densities.

*Note*: In all the above equations $R$ can be used in place of $\bar{R}$ modifying constants suitably, since tables given these values on kmole basis and not kg basis.

### 4.4.1.4 Generalised Compressibility Chart

Many equations of state were given for real gases by different research workers, which are accurate in different ranges of pressure and temperature. These equations do not give accurate results for all pressures and temperatures because of the difference in their compressibility behaviour. Therefore, a simple method was thought of to introduce a compressibility factor $Z$ in the equation of state for perfect gas, modifying it in the following manner

$$pv = Z\bar{R}\,T, \tag{4.6}$$

and to plot isothermals on chart $Z$ v/s p to give the value of $Z$. This chart is known as the *compressibility chart*. Naturally, there would be different charts for different gases. Nevertheless, reduction of pressure $p$, volume $v$ and temperature $T$ to what we call as *reduced values*, by dividing these values with their corresponding value at the critical point, helped in developing a single chart for all the gases. These reduced variables are

$$p_r = p/p_c,\ v_r = v/v_c,\ \text{and } T_r = T/T_c \tag{4.7}$$

$p_r$ was plotted against $T_r$ for constant $v_r$ to study the effectiveness of this approach for pressures and temperatures much higher than the critical values for a group of hydrocarbons. (Critical point data is given in A1.3). This study revealed that the entire *p-v-T* behaviour for all gases is satisfactory to a precision of 10 – 15%. However, at low pressures, the results were not satisfactory.

The family of isothermals on $Z$ versus $p_r$ chart for different gases were also drawn, and were found almost the same within an accuracy of 5% for $Z > 0.6$. Therefore, any such chart was found to be more suitable, and is widely used. This chart is known as the *generalised compressibility chart.*[7] Substituting from Eq. 4.7 in Eq. 4.6, we get

$$v_r = \frac{ZT_r}{p_r}\left(\frac{\bar{R}\,T_c}{p_c v_c}\right)$$

$$= \frac{ZT_r}{p_r} R_{cr} \tag{4.8}$$

where, $R_{cr}$ is the critical coefficient and has different values for different gases. This problem was solved by the introduction of the pseudo reduced volume $v'_r$ in 1901 by Kammerlingh Onnes as follows

$$v'_r = Z\,T_r/p_r \tag{4.9}$$

The value of $Z_c$, the compressibility factor at the critical point, is found to vary from 0.3 to 0.33 for different substances, and therefore, the generalised compressibility chart is not accurate in the vicinity of the critical point. More accurate generalised compressibility charts with three parameters, including $Z_c$ as one parameter, have been developed. However, the generalised compressibility chart with $Z = 0.27$, which is a good average value, are much in use. The generalised compressibility chart was drawn by Weber in 1933 from the data of Su, and was redrawn later by Nelson and Obert (A1.15). One such chart is shown in Fig. 4.8, where isothermals and constant pseudo volume lines are shown. Therefore, if two gases have the same reduced pressure and temperature, then they will also have the same reduced volume. This is the law of corresponding state, and we may reduce van der Waals equation in reduced co-ordinates to

$$(p_r + 3/v_r^2)\,(v_r - 1/3) = (8/3)\,T_r \tag{4.10}$$

Another observation about the behaviour of this chart worth noting is in the region of low pressure. Figure 4.9 shows isotherms for different values of $T$ on the compressibility chart in the region of very low pressure. We observe that in this region, the isothermals approach the same point at $Z = 1$ along a straight line with different slopes. However, the

---

7. In case of gases like hydrogen, helium and neon, reduced pressure and reduced temperature obtained by adding 8 to the corresponding critical value gives accurate results.

slope of these isotherms increases with $T$ from negative values at low values of $T_r$ to 0 at $T_r \approx 2.5$, maximum at $T_r \approx 5$, and then starts decreasing. This single temperature, about 2.5 times the critical temperature, is called the *Boyle temperature.* Only at this temperature

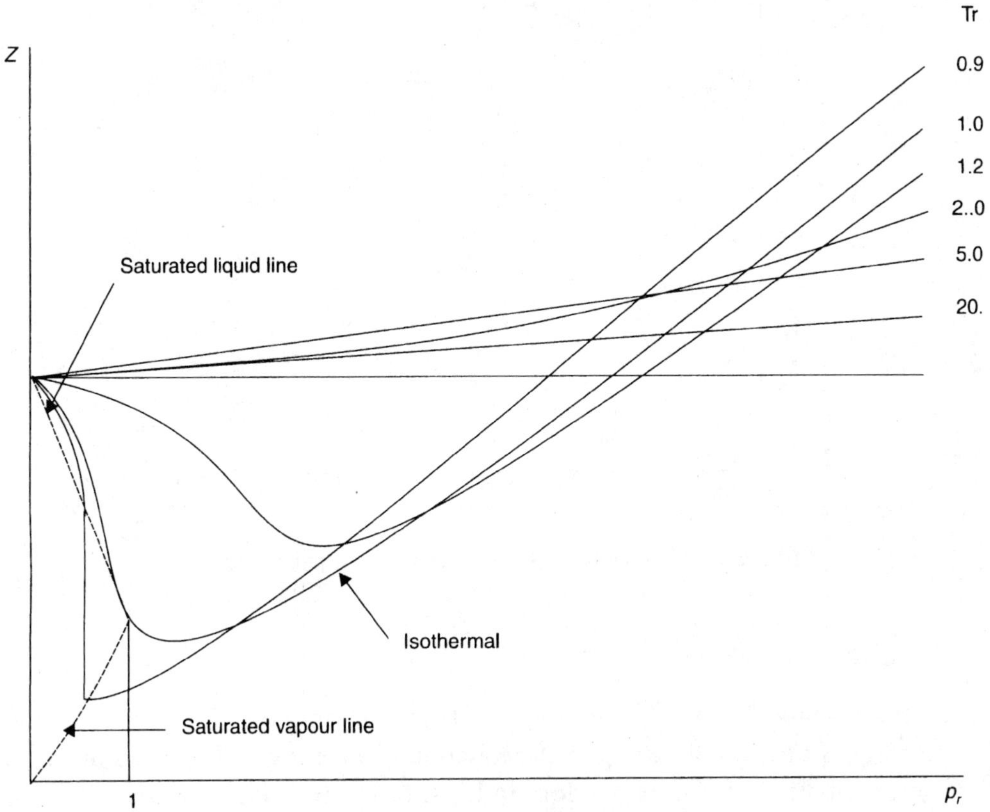


**Fig. 4.8** Generalised compressibility chart

$$\lim_{p \to 0}\left(\frac{\partial Z}{\partial p}\right) = 0 \tag{4.11}$$

So, the Boyle temperature is the only temperature at which a gas behaves exactly as a perfect gas at low finite pressure. To get more insight, let us consider 1 kg of gas having pressure, volume and temperature $p$, $v$ and $T$, respectively. The perfect gas equation gives volume of this gas to be $RT/p$. We define residual volume $v = RT/p - v$ or $pv' = RT - pv$. Now, as $p \to 0$, $pv' \to 0$ and hence $pv \to RT$. Since $p$ is tending to zero, it implies that $v'$ necessarily does not tend to zero under such circumstances. That is, the volume obtained experimentally and from the perfect gas equation need not be the same. Or, in other words, a low pressure does not ensure applicability of the perfect gas equation, as is normally accepted. It is only at the Boyle temperature and low finite pressures, that the ideal gas equation is exactly applicable.

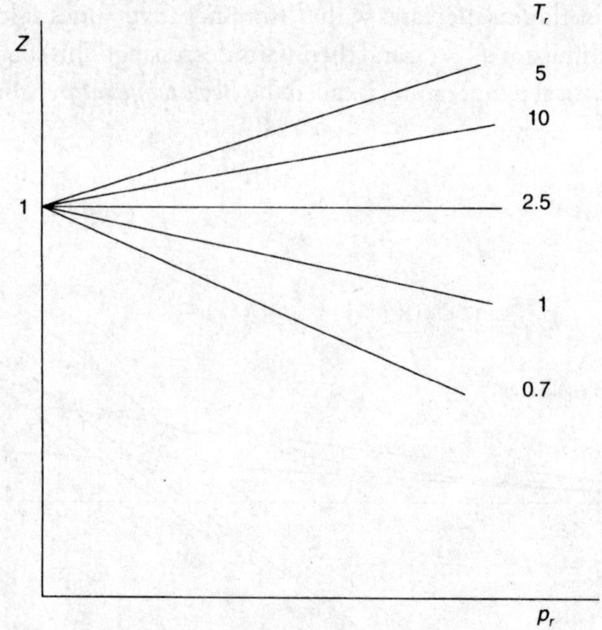


**Fig. 4.9** Behaviour of gases in the low pressure region

### 4.4.1.5 Fugacity

Another way to handle the deviation of a real gas from an ideal gas is to introduce pseudopressure $f$ in place of $p$ in the equation of state, Eq. 4.1, instead of adding the compressibility factor $Z$, as was done in Eq. 4.6. The pseudopressure is called fugacity, which is determined as follows, Integration of Eq. 4.17(c) for an isothermal process, replacing $v$ from Eqs. 4.1 and 4.6, gives

$$dg_T = RT/p\ dp_T = RT\ d(\ln p)_T \quad \text{(using Eq. 4.1)} \tag{4.12a}$$

and
$$dg_T = ZRT/p\ dp_T = ZRT\ d(\ln p)_T \quad \text{(using Eq. 4.6)} \tag{4.12b}$$

for an ideal and real gas, respectively. We define fugacity as:

$$dg_T = RT\ d(\ln f)_T \text{ such that } \lim_{p \to 0}(f/p) = 1 \tag{4.13}$$

Therefore, as $p \to 0, f \to 0$ to satisfy Eq. 4.13. Comparison of Eqs. 4.12(b) and 4.13 gives

$$d(\ln f)_T = Z\,d(\ln p)_T \text{ or } (Z-1)\,d(\ln p)_T$$
$$= d(\ln f)_T - d(\ln p)_T = d(\ln f/p)_T \tag{4.14}$$

Also, $dp/p = dp_r/p_r$ or $d\ln p = d\ln p_r$

$$\therefore\ (Z-1)\,d(\ln p_r)_T = d(\ln f/p)_T$$

Integration of Eq. 4.13 from $f = p = o$ to the given pressure $p$ using Eq. 4.14 is evaluated graphically taking the necessary data from the generalised compressibility chart to give fugacity of the gas at the given state. These values so obtained are plotted in the form of a generalised fugacity chart, which shows variation of $f/p$ against $p_r$ with $T_r$ as a parameter. So, from this chart, for given pressure and temperature of the gas, fugacity can be read and then using the ideal gas equation, the volume of the gas is obtained.

## 4.4.2 Differential Relations

In addition to pressure, temperature and volume, there are many other properties which are needed for analysis. These are internal energy, enthalpy, and entropy, which are obtained by integrating differential relations[8] formed by these variables with easily measurable variables like pressure, volume or temperature. We shall derive differential relations later. Other properties which are often needed are coefficient of volume expansion $\beta$, isothermal compressibily, $\kappa$, specific heat at constant pressure and volume (A1.4). These properties have importance because these can be measured experimentally in the laboratory at different pressures and temperatures, and their variation with temperature or pressure is slow. Variation of specific heats for solids and liquids is found to depend only on temperature and not pressure. We define these as follows

$$c_p = \left(\frac{\partial h}{\partial T}\right)_p, \; c_v = \left(\frac{\partial u}{\partial T}\right)_v, \; \beta = \frac{1}{v}\left(\frac{\partial v}{\partial T}\right)_p \quad \text{and} \quad \kappa = -\frac{1}{v}\left(\frac{\partial v}{\partial p}\right)_T \tag{4.15}$$

A negative sign in the definition of $\kappa$ is used to make $\kappa$ positive, since the slope of the isothermal on the p-v diagram is negative for the entire range of pressure and temperature. Moreover, we observe from the equation that coefficient of volume expansion is analogous to the linear expansion coefficient, and inverse of isothermal compressibility to Young's modulus of elasticity. Young modulus is the ratio of stress to strain, defined as the ratio of change in length to the original length.

---

8. Since we are going to use differential relations, it is worthwhile to devote some time to recall some concepts from partial differential calculus. These are:
   1. If $x$, $y$, and $z$ are point functions related to each other, then the functional relation between them can be written as $z = z(x, y)$, $y = y(x, z)$ or $x = x(y, z)$. Also, the relation $(\partial x/\partial y)_z (\partial y/\partial z)_x (\partial z/\partial x)_y = -1$ exists among their partial derivatives, which is known as a cyclic relation, as $x$, $y$ and $z$ rotate in cyclic way in the successive derivatives.
   2. If $dz = m\,dx + n\,dy$ then it can be integrated only if $(\partial m/\partial y)_x = (\partial n/\partial x)_y$.
   3. If four variables are related such that $w = w(x, y)$, $w = w(y, z)$ and $x = x(y, z)$, then $(\partial x/\partial z)_y (\partial z/\partial w)_y (\partial w/\partial x)_y = 1$. This relation is known as a chain relation. Literally, it means that derivatives with the same fixed property may be multiplied as ordinary quotients. Also, $(\partial w/\partial y)_z = (\partial w/\partial y)_x + (\partial w/\partial x)_y (\partial x/\partial y)_z$.

Therefore, our effort will be to express differential change in any of the properties in terms of these derivatives, and then use any graphical or computational integration technique to evaluate the change. This requires knowledge of some relations like Gibbsian equations and Maxwell relations, which we shall discuss first in the next sections.

#### 4.4.2.1 Gibbsian Equations

While studying thermodynamic equilibrium of a system, we come across four functions, namely, internal energy, enthalpy, Helmholtz and Gibbs function. These functions are known as *Legendre Transform*. We know about the first two functions, the later two functions are defined as follows

$$g = h - Ts \text{ and } f = u - Ts \tag{4.16}$$

which represent a group of extensive properties and hence are a property of the system. For a closed system, $\delta w = p\,dv$, and heat transfer in a reversible process (by the second law of thermodynamics) is $\delta q = T\,ds$. Writing these values for work and heat transfer, non-flow energy equation gives

$$du = Tds - pdv \tag{4.17a}$$

Also, on differentiation, $h = u + pv$ gives $dh = du + pdv + vdp$. Writing for $du$ from Eq. 4.17 (a), we get

$$dh = T\,ds + v\,dp \tag{4.17b}$$

Similarly, we may express Helmholtz and Gibbs function in their differential form to give

$$df = -s\,dT - p\,dv \text{ and } dg = -s\,dT + v\,dp \tag{4.17c}$$

Eqs. (4.17) are known as *Gibbsian equations.*

Pressure, temperature, entropy and volume can be expressed as a derivative of the Legendre transforms. Consider the Gibbsian equation for $du$; here, the properties $s$ and $v$ appear on the right side of the equation. The variables $s$ and $v$ are called the *natural variable* for function $u$. Similarly, we can observe natural variables for other Legendre transforms from Eq. 4.17 and express these as a function of their natural variables. That is, we may write functions: $u = u(s, v)$, $h = h(s, p)$, $f = f(v, T)$ and $g = g(p, T)$. The functional form for $u$ gives $du = (\partial u/\partial s)_v\, \partial s + (\partial u/\partial v)_s\, dv$ which, when compared with the Gibbsian equation for $du$, results in $T = (\partial u/\partial s)_v$ and $p = -(\partial u/\partial v)_s$. This procedure is repeated for other functions. The results are summarised below

$$-p = (\partial u/\partial v)_S = (\partial f/\partial v)_T \qquad -s = (\partial g/\partial T)_p = (\partial f/\partial T)_v$$

$$v = (\partial h/\partial p)_S = (\partial g/\partial p)_T \quad \text{and} \quad T = (\partial u/\partial s)_v = (\partial h/\partial s)_p \tag{4.18}$$

We find it very difficult to remember these relations. We shall discuss later, the mnemonic method for remembering these and other relations, like Gibbsian and Maxwell.

#### 4.4.2.2 Maxwell Relations

Since $u$, $h$, $f$ and $g$ are property, their derivatives must therefore form the exact differential for which their right hand side in the Gibbsian equations must satisfy the necessary condition given as a footnote in Section 4.2. The application of the condition results in the following Maxwell relations, which have been found very useful in changing the derivatives in desirable form:

$$(\partial T/\partial v)_s = -(\partial p/\partial v)_v \quad (\partial T/\partial p)_s = (\partial v/\partial s)_p$$
$$(\partial s/\partial v)_T = (\partial p/\partial T)_v \quad (\partial s/\partial p)_T = -(\partial v/\partial T)_p \tag{4.19}$$

Maxwell relations have importance in changing various derivatives involving the derivative of $s$, or w.r.t. $s$ in terms of derivatives made from $p$, $v$ and $T$. For example, whenever the derivative $(\partial p/\partial s)_v$ comes, it can be changed to $(\partial T/\partial v)_s$, which can be determined for an isentropic process.

#### 4.4.2.3 Specific Heat Relations

When any substance is heated, its temperature increases. However, different substances require different amount of heat for increasing the temperature by one Kelvin or one °C from 14.5°C to 15.5°C. This property of the substance is called specific heat; the temperature range is defined here because the amount of heat required is different for different temperature ranges. For liquids and solids, heating at constant pressure or volume does not matter significantly. But consider air contained in a vertical cylinder, fitted with a piston, as shown in Fig. 4.10. If air is heated at constant pressure, then the piston will rise, as heat is added and work is done by the system (air) on the surrounding. On the other hand, if heating is done at constant volume, then no work will be done. This means that heat required for a constant pressure process will be more than that required for a constant volume process for the same temperature rise by an amount equal to work done in the former process. It also implies that for any other process, heat required should be different. Or, in other words, for compressible fluids, specific heat will be different for different processes. Nonetheless, specific heats at constant pressure $c_p$ and constant volume $c_v$ is mostly used. Specific heat at constant volume is the rate at which the internal energy of the system changes w.r.t temperature in a constant volume process.[9]

---

9. Relations for properties of a substance like entropy, and energy, are derived in terms of those which can easily be measured by instruments like pressure, temperature, or can be determined experimentally such as $\beta$, $\kappa$ or $(\partial p/\partial T)_v = \beta/\kappa$ (see Eq. 4.21). The procedure is always the same: to use the functional form, use cyclic or chain relation, and if required, use the Maxwell relation, and finally, to transform in terms of derivatives made from $p$, $v$ and $T$, as their value is known in terms of $\beta$ and $\kappa$. A useful functional form for Legendre transforms is in terms of their natural variables, and this choice is quite often made to get fruitful results.

Similarly, specific heat at constant pressure is the rate at which the enthalpy of a system changes w.r.t temperature during a constant pressure process.

Using the definition of specific heat at constant pressure and constant volume, chain relation and Eq. 4.18, we can write

$$C_v = (\partial u/\partial T)_v = (\partial u/\partial s)_v (\partial s/\partial T)_v = T\,(\partial s/\partial T)_v \quad (4.20a)$$

$$C_p = (\partial h/\partial T)_p = (\partial h/\partial s)_p (\partial s/\partial T)_p = T\,(\partial s/\partial T)_p \quad (4.20b)$$

and $$\therefore \quad C_p - C_v = T[(\partial s/\partial T)_p - (\partial s/\partial T)_v] \quad (4.20c)$$

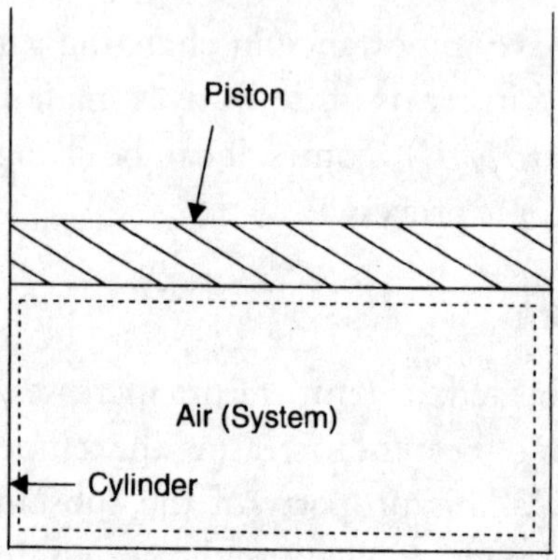


**Fig. 4.10** Piston cylinder arrangement

This way of defining specific heat has now been changed as discussed later. Specific heat at constant volume is the rate at which the internal energy of the system changes w.r.t temperature in a constant volume process. Similarly, specific heat at constant pressure is the rate at which the enthalpy of a system changes w.r.t. temperature during a constant pressure process.

Equation 4.20(c) suggests to take $s = s\,(T, v)$, since both the derivatives are w.r.t. $T$; the other variable could be taken as $v$ or $p$. Here, we took $v$.

$$\therefore \; ds = (\partial s/\partial T)_v\, dT + (\partial s/\partial v)_T\, dv$$

This gives change in $s$ for a given change $dT$ in temperature and $dv$ in volume in the process. If some property $z$ remains constant in the process, then we can divide the equation by $dT$ (or $dv$) and write $(\partial s/\partial T)_z = (\partial s/\partial T)_v + (\partial s/\partial v)_T (\partial v/\partial T)_z$.[10] Taking $z$ as $p$, we get the following equation after rearranging the terms and using the Maxwell relation

$$(\partial s/\partial T)_p - (\partial s/\partial T)_v = (\partial s/\partial v)_T (\partial v/\partial T)_p$$

$$= (\partial p/\partial T)_v (\partial v/\partial T)_p$$

10. Students are advised to understand the procedure because, from now on, the final equation will be written omitting the details.

$(\partial p/\partial T)_v$ can, however, be expressed as follows in terms of $\beta$ and $\kappa$ using the cyclic relation among $p$, $v$ and $T$ and noting that $(\partial v/\partial T)_p = \beta v$, $(\partial v/\partial p)_T = -\kappa v$.

$$(\partial p/\partial T)_v (\partial T/\partial v)_p (\partial v/\partial p)_T = -1$$

or,
$$(\partial p/\partial T)_v = -(\partial v/\partial T)_p/(\partial v/\partial p)_T = \beta v/\kappa v = \beta/\kappa \qquad (4.21)$$

$$\therefore \quad C_p - C_v = (T\,\beta/\kappa)\,\beta v = T\beta^2 v/\kappa$$

Since the right hand side is invariably positive, $C_p$ is always greater than $C_v$, except if $T = \beta = 0$ or $\kappa$ is $\infty$; in that case, both are equal. These conditions are, however, not satisfied in practice. Thus, for all practical purposes, $C_p > C_v$. We may also take the ratio of the two specific heats to get:

$$\gamma = C_p/C_v = (\partial s/\partial T)_p/(\partial s/\partial T)_v$$

and change it in useful form by writing the cyclic relation among the variables $s$, $T$, and $p$ or $s$, $T$, and $v$.

$$(\partial s/\partial T)_p (\partial T/\partial p)_s (\partial p/\partial s)_T = -1 \text{ and } (\partial s/\partial T)_v (\partial T/\partial v)_s (\partial v/\partial s)_T = -1$$

Dividing the first by the second and rearranging the terms, we get

$$\gamma = (\partial s/\partial T)_p/(\partial s/\partial T)_v = [(\partial p/\partial T)_s/[(\partial v/\partial T)_s] \times [(\partial s/\partial p)_T/[(\partial s/\partial v)_T]$$
$$= (\partial p/\partial v)_s/(\partial p/\partial v)_T$$

We showed above that $C_p > C_v$, it implies that $\gamma > 1$. Therefore, the *slope of an isentropic line is more than that of the isothermal on the p-v diagram.*

We can also find dependence of $C_p$ and $C_v$ on pressure, volume and temperature for a substance by differentiating Eq. 4.20 (a) and (b). Differentiating Eq. 4.20(a) for $c_v$ w.r.t. volume and for $c_p$ w.r.t. pressure at constant temperature, we get

$$\left(\frac{\partial C_v}{\partial v}\right)_T = \left[\frac{\frac{\partial}{\partial v}\left[T\left(\frac{\partial s}{\partial T}\right)_v\right]_T}{\partial v}\right]_T = T\left[\frac{\frac{\partial}{\partial v}\left[T\left(\frac{\partial s}{\partial T}\right)_v\right]_T}{\partial T}\right]_v = T\left(\frac{\partial^2 p}{\partial T^2}\right)_v$$

and,
$$\left(\frac{\partial C_p}{\partial p}\right)_T = \left[\frac{\frac{\partial}{\partial v}\left[T\left(\frac{\partial s}{\partial T}\right)_p\right]_T}{\partial p}\right]_T = T\left[\frac{\frac{\partial}{\partial v}\left[T\left(\frac{\partial s}{\partial T}\right)_v\right]_p}{\partial T}\right]_p = -T\left(\frac{\partial^2 v}{\partial T^2}\right)_p$$

using the Maxwell relation. For a perfect gas, both the second order derivatives are 0. This means that both $c_p$ and $c_v$ do not depend on pressure or volume. These, therefore, may either be constant or a function of temperature; for a perfect gas, specific heats are constant.

Further, from Eq. 4.20, we note that $(\partial T/\partial s)_p = T/c_p$ and $(\partial T/\partial s)_v = T/c_v$. This means that $(\partial T/\partial s)_p < (\partial T/\partial s)_v$, since $c_p > c_v$. That is, *constant volume lines are steeper than constant pressure lines on the T-s chart.* This knowledge is of great help in drawing constant pressure and constant volume processes on the T-s diagram at a given point. Figure 4.4(b) shows constant pressure and constant volume lines on the T-s diagram for air. We can see that the isometric lines are steeper than isobaric lines. Further for a point to the right of the current point on the isometric or isobaric line, $T$ would be more, as both have a positive slope. This means that the slope on the successive points to the right would increase. Thus, the curves will bend upward as is evident from Fig. 4.4 (b).

We also get an interesting result from the relation $T = (\partial h/\partial s)_p$ (Eq. 4.18). This shows that the slope of the constant pressure line on the h-s diagram depends on $T$. This means slope of the constant pressure line will not change if T is constant or the isobaric lines in the wet region are lines, whose slope increases with pressure. Pressure increases in the direction shown on the Mollier diagram in Fig. 4.4(a).

#### 4.4.2.4 Entropy Relations

Entropy relations are also called $T\,ds$ relations. These relations can be obtained by taking $s$ as a function of $(v, T)$, $(p, T)$ or $(p, v)$. This gives the first, second or third $Tds$ equation, respectively, which on integration gives the change in entropy of the system during the process. We shall demonstrate the procedure for the third $Tds$ equation; the other two are obtained in the same way. Since $s = s\,(p, v)$,

$$ds = (\partial s/\partial p)_v\, dp + (\partial s/\partial v)_p\, dv$$

The chain relation among the variables $(s, p, v, T)$ is

$$(\partial s/\partial p)_v (\partial p/\partial T)_v (\partial T/\partial s)_v = 1$$

Using Eq. 4.20 (a) and 4.21,

$$(\partial s/\partial p)_v = (c_v/T)\,(\partial T/\partial p)_v = c_v\,\kappa/(T\,\beta)$$

Similarly, $(\partial s/\partial v)_p (\partial v/\partial T)_p (\partial T/\partial s)_p = 1$

$$\therefore\ (\partial s/\partial v)_p = (c_p/T)\,(\partial T/\partial v)_p = c_p/(\beta v T)$$

using Eqs. (4.20) and (4.21). The three $T\,ds$ equations are

$$\begin{aligned} T\,ds &= c_v(\partial T/\partial p)_v\, dp + c_p(\partial T/\partial v)_p\, dv \\ &= c_v\,\kappa/\beta\, dp + c_p/(\beta v)\, dv \quad \text{(3rd } Tds \text{ Eq.)} \end{aligned} \tag{4.22a}$$

$$\begin{aligned} T\,ds &= c_v\, dT + T\,(\partial p/\partial T)_v\, dv \\ &= c_v\, dT + \beta T/\kappa\, dv \quad \text{(2nd } Tds \text{ Eq.)} \end{aligned} \tag{4.22b}$$

$$\begin{aligned} T\,ds &= c_p\, dT - T\,(\partial v/\partial T)_p\, dp \\ &= c_p\, dT + \beta T/\kappa\, dp \quad \text{(1st } Tds \text{ Eq.)} \end{aligned} \tag{4.22c}$$

The last two equations in Eq. 4.22 are obtained by taking $s = s\,(v, T)$ and $s = s\,(p, T)$, respectively.

#### 4.4.2.5 Energy Equations

Consider the function $h = h(T, p)$, so that $dh = (\partial h/\partial T)_p dT + (\partial h/\partial p)_T dp$. Taking $h$ as a function of its natural variables $s$ and $p$, i.e. $h = h(s, p)$, we may write using Eqs. 4.18 and 4.19

$$(\partial h/\partial p)_T = (\partial h/\partial p)_s + (\partial h/\partial s)_p (\partial s/\partial p)_T = v - T(\partial v/\partial T)_p = v - \beta v T \quad (4.23a)$$

and $$dh = (\partial h/\partial T)_p dT + (\partial h/\partial p)_T dp = c_p dT + v(1 - \beta T)\, dp \quad (4.23b)$$

Similarly, by taking $u = u(s, v)$, the following relation for $du$ can be obtained

$$(\partial u/\partial v)_T = -p + T(\partial p/\partial T)_v \quad \text{and} \quad du = c_v dT + (\beta T/\kappa - p)\, dv \quad (4.23c)$$

Like specific heat, we can show that $u$ and $h$ are either constants or functions of temperature alone for an ideal gas. For a perfect gas, $(\partial p/\partial T)_v = R/v$ and $(\partial v/\partial T)_p = R/p$. From the above equations, $(\partial u/\partial v)_T = (\partial h/\partial p)_T = 0$, which proves the statement. Experimental results show that $u$ and $h$ are a function of temperature alone for an ideal gas, and pressure also for real gas.

#### 4.4.2.6 Joule –Thomson Coefficient

Throttling is a process in which the flowing fluid is expanded adiabatically through a restricted passage. Since there is no energy interaction between the system and its surrounding, enthalpy of the fluid will not change. Such a process is called an isenthalpic process, and occurs when a fluid flows through a restricted passage in a valve or porous media, or suddenly expands in an insulated container from high to low pressure. Since enthalpy of a perfect gas is a function of temperature alone, the gas temperature will not change after expansion. However, the result is different for a real gas, which may be heated or cooled after expansion, depending on its Joule–Thomson coefficient.

The *Joule–Thomson coefficient* for a substance is defined as rate of change of its temperature with pressure during the isenthalpic process; a positive value means decrease in temperature (cooling) after expansion. It is represented by the symbol $\mu$. That is, $\mu = (\partial T/\partial p)_h$. It can be expressed in terms of $\beta$ by writing the cyclic relation among the variables $T$, $p$ and $h$, and using Eq. 4.23a. This gives

$$(\partial T/\partial p)_h (\partial p/\partial h)_T (\partial h/\partial T)_p = -1. \text{ But, } (\partial h/\partial T)_p = c_p \text{ (by definition)}$$

or, $$\mu = (\beta T - 1)\, v/c_p \quad (4.24)$$

We may determine the Joule–Thomson coefficient by using Eq. 4.6 and evaluating $(\partial T/\partial p)_p$ from it. This gives

$$\beta = \frac{1}{v}\left(\frac{\partial v}{\partial T}\right)_p = \frac{1}{pv}\left[ZR + RT\,(\partial Z/\partial T)_p\right] \text{ and}$$

$$\mu = RT^2 (\partial Z/\partial T)_p/(pC_p)$$

Recall our discussions of Section 4.4.1.4 that $Z$ tends to 1 as $p$ tends to 0 for different isotherms with different slopes. The slope is zero at the Boyle temperature, negative for $T_r < 2.5$, maximum at $T_r \approx 5$ and thereafter it starts decreasing, as seen in Fig. 4.9. This figure gives an idea of how $(\partial Z/\partial T)_p$ varies with $T$ at a constant pressure. To have some idea, draw a vertical line at any $p$. We observe that as $T$ approaches the isotherm for $T_r \approx 5, Z$ increases and then it decreases. That is, $(\partial Z/\partial T)_p$, and therefore, $\mu$ is positive up to some value of $T$, and negative after that temperature. A positive value of $\mu$ implies that as pressure decreases in an isenthalpic expansion process, temperature also decreases, and the gas is cooled after expansion provided the gas temperature before expansion is less than this value. This maximum temperature is called the inversion temperature. This fact has been used in refrigeration cycles. We shall discuss further about it in the chapter on the refrigeration cycle.

#### 4.4.2.7 Integration of Energy and Entropy Equations

The basic data for preparing the p-v-T surface is obtained from macroscopic measurements, or even at the microscopic level to cover the entire range of these variables. From this data, the equation of state is obtained using curve-fitting techniques. Recall our earlier discussion on how to choose the independent variables. We mentioned that temperature is normally taken as one variable. The other variable from $p$ and $v$ is decided, depending upon which one of these may be expressed explicitly. Therefore, for the van der Waals equation, the right choice for independent variables is $p$ and $T$. The choice of independent variables decides the differential equation to be integrated for determining a property at any state of the substance. This also requires assigning a reference state. *The reference state is so chosen that values of all the properties remain positive for the entire range of pressure and temperature.* For this reason, different substances have different reference states. For example, enthalpy for water is taken zero at 0.1°C, whereas for air at 0 K.

Integrating Eqs. (4.22), we get change in entropy for a perfect gas

$$s_2 - s_1 = c_v \ln (T_2/T_1) + R\ln(v_2/v_1), \quad s_2 - s_1 = c_p \ln (T_2/T_1) - R\ln(p_2/p_1)$$

and $$s_2 - s_1 = c_p \ln (v_2/v_1) + c_v \ln(p_2/p_1) \tag{4.25}$$

noting $(\partial p/\partial T)_v = R/v$, $(\partial v/\partial T)_p = R/p$ and $c_p$, $c_v$ are constant for a perfect gas. However, integration of the energy and entropy equations is not that simple, as we found in case of a perfect gas, because specific heats, $\beta$ and $\kappa$ are functions of temperature and pressure. It then requires graphical or numerical integration on a computer.

We may illustrate the procedure by taking the example of finding change in entropy of water during constant pressure heating from liquid state 1 to steam state 2, as shown in Fig. 4.11 on the v-T diagram. This is determined by considering the change in following three steps.

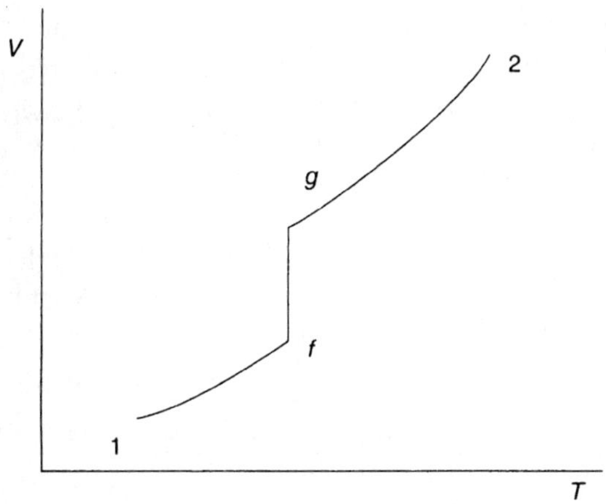


**Fig. 4.11** Integration

1. Sensible heating of water from state 1 to $f$ when it starts changing phase. Equation 4.22(c) gives change in entropy during this process, $s_f - s_1 = \int (c_p/T)\, dT$, since pressure is constant. The integral is evaluated by plotting $c_p/T$ against $T$ for temperatures from point 1 to $f$; the area under this curve is the required value.
2. Change in entropy during phase change from $f$ to $g$ during which latent heat, $h_{fg}$, is transferred reversibly at constant saturation temperature $T_{sat}$. Change in entropy by the second law of thermodynamics is $h_{fg}/T_{sat}$.
3. Lastly, sensible heating of vapour to state 2. Change in entropy can be evaluated as in step 1.

Total change is the sum of the three changes. This procedure can be used when $p$ and $T$ are independent variables. Equation 4.22(a) or 4.22(b) will be used for this purpose in case $p$ and $v$ or $v$ and $T$ are independent variables.

### 4.4.2.8 Clausius–Clapeyron Equation

We learnt before that change from one phase to another for most of the substances is of the first order, during which extensive properties like volume and energy change continuously, but the Gibbs function does not change. That is, in step 2 of Section 4.4.2.7, $g_f = g_g$. Now, consider another similar process, but at pressure $p + dp$ for which $(g + dg)_f = (g + dg)_g$, where $dg$ is change in $g$ for change in pressure from $p$ to $p + dp$. These two equations combined give $dg_f = dg_g$. Writing for $dg$ from Eq. 4.17(c), we get

$$(-sd_T + vd_p)\, f = (-sd_T + vd_p)_g$$

or

$$(s_g - s_f)\, dT = dp\, (v_g - v_f)$$

since $dp$ and $dT$ is same at $f$ or $g$

$$\therefore\ (dp/dT)_{sat} = s_{fg}/v_{fg} = (h_{fg}/T_{sat})/v_{fg} \qquad (4.26)$$

This equation was obtained by Clapeyron, and is named after his name. $v_{fg}$ may be replaced by $v_g$, since volume of liquid is negligible compared to that of a vapour. Also, treating vapours as perfect gas, $v_g = RT_{sat}/p$, Eq. 4.26 can be written as

$$(dp/dT)_{\text{sat}} = h_{fg}\, p/\left(RT_{\text{sat}}^2\right) \tag{4.27}$$

Equation 4.27 is the Clausius–Clapeyron equation. This equation may also be obtained from the Maxwell relation $(\partial s/\partial v)_T = (\partial p/\partial T)_v$. Recall that for a two-phase substance, $p = p\ (T)$, and therefore, derivatives like $(\partial T/\partial p)_s$ and $(\partial p/\partial T)_v$ may be written as $(dT/dp)_{sat}$ and $(dp/dT)_{sat}$, respectively. Therefore, Maxwell's third relation may be written as $(\partial s/\partial v)_T = (dp/dT)_{\text{sat}}$, in which the right hand side remains constant during phase change at a given pressure or temperature. This equation now can easily be integrated between saturated liquid and vapour states to give:

$$s_g - s_f = (dp/dT)_{sat}\,(v_g - v_f)$$

Since pressure is constant during the process, Eq. 4.17(b) reduces to $dh = Tds$, which gives on integration:

$$\int_f^g dh = \int_f^g Tds \quad \rightarrow \quad h_{fg} = Ts_{fg}$$

Combining this equation with the previous one, we get an important result

$$(dp/dT)_{sat} = h_{fg}/(Tv_{fg})$$

which is Eq. 4.26. For normal substances, $(dp/dT)_{sat}$ is positive for the entire range of pressure in all the three 2-phase regions, and therefore, for such substances both entropy and volume increase or decrease together during phase change. However, phase boundary from solid to liquid for $H_2O$ has negative $(dp/dT)_{sat}$ for its entire range of pressure. Since change in enthalpy during change from solid to liquid is positive, a negative slope means volume of $H_2O$ will decrease on melting, a result we know.

These equations with the phase diagram give useful information about the change in volume, entropy and latent heat during change of liquid to vapour phase with increasing pressure. From the phase diagram, we observe that the slope of the vaporisation line increases with pressure. This means that saturation pressure increases at a faster rate compared with the saturation temperature, and therefore, specific volume of saturated vapour decreases with increasing rate. The result is a decrease in latent heat with increasing pressure, to zero at the critical point (see Eq. 4.26).

Change in other properties like internal energy and enthalpy may be obtained similarly by graphical integration. Recall that whereas change in enthalpy and internal energy of a gas depends on temperature (dependence on pressure is negligibly small and ignored), the change in entropy cannot ignore pressure effects. Therefore, a function $\phi = \int (c_p/T)\, dT$ is defined, so that it is function of temperature alone, and care is taken to add the change in entropy due to pressure given by the second integral of Eq. 4.22(c). Now, $u$, $h$ and $\phi$ are a function of temperature only, and have been tabulated against temperature for different gases including water vapour in tables known as gas tables. For all these substances, reference is taken at absolute zero, that is, $u$, $h$ and $\phi$ are zero at 0 K.

### 4.4.3 Mnemonic Diagram

The biggest problem in thermodynamic relations is to remember the Maxwell, Gibbsian and other relations, and even if we remember them, we are always in doubt whether or not the relation has been written correctly. It, therefore, calls for some mnemonic as an aid to our memory. One such method is to draw a rectangle, as shown in Fig. 4.12. Place the properties TPSV (Tapasvi) in that order at the four corners of the rectangle (starting from the top left corner and moving clockwise), and Legendre transform on the side joining its natural variables at the corner of the rectangle. For example, $p$ and $s$ are natural variables for $h$, and hence $h$ has been placed on the side joining $p$ and $s$. We will now discuss the pattern observed among different relations discussed. A relation can be written following the observations.

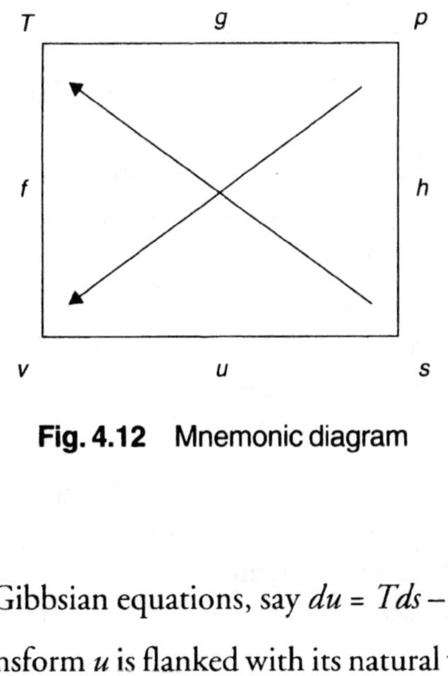


**Fig. 4.12** Mnemonic diagram

#### *Gibbsian Equations*

We observe from the Gibbsian equations, say $du = Tds - pdv$, the following.

(i) Legendre transform $u$ is flanked with its natural variables $s$ and $v$, which appear as derivative.

(ii) A negative sign comes with the $p\,dv$ term, where $v$ lies on arrowhead and + sign with $T\,ds$, where $s$ does not lie on arrowhead.

(iii) Properties $T$ and $s$ in $Tds$, or $p$ and $v$ in $pdv$ lie on opposite corners of the rectangle.

Say, we want to write such a relation for $dh$. We see that $h$ is flanked with natural variables $s$ and $p$, both of which do not lie on arrowhead. So, the differential will *be* $+ ds$ and $+ dp$. Also, diagonal properties in the rectangle corresponding to $s$ and $p$ respectively, are $T$ and $v$. So one writes $dh = Tds + vdp$, which is Eq. 4.17(b).

### Maxwell Relation

Consider Maxwell relations $(\partial T/\partial v)_s = -(\partial p/\partial s)_v$ and $(\partial T/\partial p)_s = (\partial v/\partial s)_p$ for example. We make the following observations, which applies for others, too.

(i) Derivatives involve properties $p$, $v$, $s$ and $T$, and not any Legendre transform.
(ii) Properties that remains constant and the property differentiated in the derivative lie at diagonally opposite corners of the diagram; for example $T$ and $s$ in derivative $(\partial T/\partial v)_s$ or $p$ and $v$ in $(\partial p/\partial s)_v$.
(iii) In the first relation, both $T$ and $v$ or $p$ and $s$ lie together on the vertical line, whereas in the second relation, these properties lie together on the horizontal line of the rectangle. That is, these properties lie together either on the vertical or on the horizontal line.
(iv) The property differentiated on one side of the relation does not appear on its other side, and is replaced with this missing property to form the right hand side derivative.
(v) The position of the property that is constant in the derivative is interchanged with that with respect to which the derivative is formed; for example, $v$ and $s$ in the first Maxwell relation have swapped their position on derivatives $(\partial T/\partial v)_s$ and $(\partial p/\partial s)_v$ ← look at position of $s$ and $v$.
(vi) A negative sign appears if the property forming the derivative lies on the vertical line, and a positive sign when on the horizontal line. For example, a negative sign is present in the first relation in which $p$ and $s$ lie on the vertical line.

The first three conditions must be checked to ensure that the Maxwell relation exists for the given derivative, and only then an attempt to change it by using the Maxwell relation should be made. Say, we want to write the Maxwell relation for the derivative $(\partial p/\partial s)_v$. This satisfies all the first three conditions, and hence the Maxwell relation can replace it. In this derivative, $T$ is missing and $p$ and $s$ lie on the vertical line, so a negative sign follows. Thus, the derivative is $-(\partial T/\partial v)_s$.

### Properties as a Derivative of Legendre Transform

In the relation $T = (\partial u/\partial s)_v = (\partial h/\partial s)_p$ we observe following pattern.

(i) The derivative is formed with $s$, where s lies on the corner (opposite to $T$) of the rectangle, and is flanked with Legendre transforms $u$ and $h$ along the horizontal and vertical line, respectively.

(ii) $v$ in $(\partial u/\partial s)_v$ is another natural variable of $u$.

(iii) A derivative of Legendre transform lying on the arrowhead of the rectangle has a positive sign.

Say, we want to express $p$ in terms of derivatives of the Legendre transforms. It occupies the right top corner, which does not lie on the arrowhead, and therefore, there will be a negative sign. The property lying on the opposite corner is $v$, which is flanked with Legendre transforms $f$ and $u$. For $f$ the other natural variable is $T$ and that for $u$ is $s$. Therefore, we may write

$$p = -(\delta u/\delta v)_s = -(\delta f/\delta v)_T$$

which is the same as Eq. 4.18. Similarly, we may change the derivative $(\delta f/\delta v)_T$ or $(\delta u/\delta v)_s$ to $-p$.

## 4.5 ANALYSIS OF A GAS SYSTEM

Consider a system is made up of a perfect gas and is subjected to some process. Figure 4.13 shows its initial state $i$, and the state after executing different processes. These processes can be represented by a mathematical equation (*process equation*),

$$pv^n = \text{constant} \tag{4.28}$$

where, $n$ is the index called the index of expansion or compression, depending on the type of process. In compression process, volume decreases while pressure increases, and in the expansion process, the reverse happens. The equation represents isobaric, isothermal, isentropic and isometric processes for values of $n$ = 0, 1, $\gamma$ and $\infty$, respectively. These processes lie either in the second or fourth quarter. Logarithmic differentiation of Eq. 4.28 gives $dp/dv = -n\,(p/v)$. Therefore, at point $i$ on the process line, the slope is more for that process for which $n$ is large, and is negative for all the processes. This explains the trend of the processes shown in Fig. 4.13.

There exists a process in which both volume and pressure increase. Consider a gas, contained in a cylinder fitted with a spring-loaded piston, is heated, increasing its pressure which pushes the piston upward. Compression of the spring increases the downward force on the piston, and hence pressure of the gas. Thus, in this process, heating will not only increase the volume but also the pressure.

By eliminating pressure or volume from the process equation using the perfect gas equation, we may also write the process equation in the forms $Tv^{(n-1)}$ = constant and $T/p^{(n-1)/n}$ = constant (that is, $T \propto p^{(n-1)/n}$). Note that while $pv^n$ = constant is applicable to any gas or vapours, *the other relations hold true for a perfect gas only,* as these have been obtained by eliminating $p$ or $v$ with the help of a perfect gas relation.

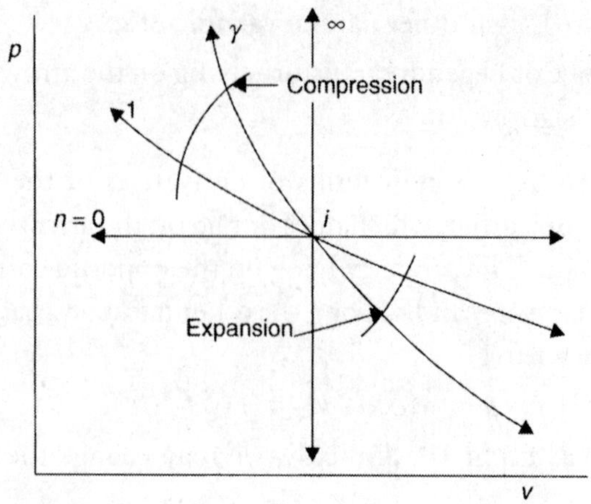


**Fig. 4.13** Process equation

We learnt in Chapter 1 that the amount of work transfer between the system and the surrounding is given by the integral $\int p\,dv$ from initial to final state using the functional relation between $p$ and $v$, which is a process equation. This is simple integration, which gives:

$$w = \frac{p_1 v_1 - p_2 v_2}{n-1} = \frac{R(T_1 - T_2)}{n-1}, \text{ if } n \neq 1 \text{ and}$$

$$= p_1 v_1 \ln(v_2/v_1) = RT_1 \ln(v_2/v_1), \text{ if } n = 1 \tag{4.29}$$

Equations 4.1 to 4.3, 4.29 and the energy equation, discussed in Chapter 1, are sufficient to analyse any system consisting of a perfect gas.

**Example 4.10** Find out the heat transfer for a gas contained in a cylinder subjected to different processes. Also, determine specific heat.

***Solution*** Gas in the cylinder is subjected to a process, and hence it is a closed system for which non-flow energy equation is applicable. Work transfer is only through boundary displacement given by $pdv$. We assume that the gas may be treated as a perfect gas, and hence $du = c_v\,dT$. We shall now consider isometric, isobaric, isothermal and general processes.

Constant volume process: $\delta w = 0$ and therefore $\delta q = du$. This gives

$$c_v = (\partial u/\partial T)_v$$

Constant pressure process: $\delta q = pdv + c_v\,dT$. Differentiation of $pv = RT$ at constant pressure yields $p\,dv = R\,dT$, and nfee then results in

$$\delta q = (c_v + R)\,dT = c_p\,dT \text{ (using Mayer's relation).}$$

Alternately, we may write $\delta q = pdv + du = d(u + pv) = dh$, since $p$ is constant. This gives $c_p = (\partial h/\partial T)_p$

Constant temperature process: For this process, $dT = 0 = du$, and hence $\delta q = pdv$. Therefore specific heat at constant temperature is infinite, as there is no change in temperature; heat supplied to the system is used in doing work.

General process: ($pdv$) term may be changed in terms of $dT$, noting from Eq. 4.29, that for an infinitesimal process $\delta w = R\,dT/(n-1)$. Writing this in nfee, we get: $\delta q = c_v\,dT + R\,dT/(n-1) = [(c_v + R/(n-1)]\,dT$. Therefore, specific heat for any general process is $c_v + R/(n-1)$.

**Example 4.11** 2 m$^3$ of air at 1.1 bar and 20°C is heated in a cylinder at constant pressure until its volume is 4.5 m$^3$. It is then compressed according to the law $pv^n$ = constant till pressure and volume change to 5.5 bar and 0.6 m$^3$. Calculate change in entropy, work and heat transfer during the processes and the value of $n$.

***Solution*** Given: Air is closed system and

$p_1 = 1.1$ $V_1 = 2$ $T_1 = 20 + 273 = 293$

$p_2 = 1.1$ $V_2 = 3.5$

$p_3 = 5.5$ $V_3 = 0.6$ $n = ?$

Arranging data in the above manner gives an idea of what is given and what not, by just having a glance over it.

To find: Heat and work transfer, change in entropy during two processes.

Assumption: Air is a perfect gas for which $R$ = 0.287 kJ/kgK and $c_v$ = 0.718 kJ/kg K.

$$m = \text{mass of air} = pv/RT = 1.1 \times 2 \times 100/(0.287 \times 293) = 3.616 \text{ kg}$$

Analysis: The system is subjected first to a constant pressure process 12 and then to a general compression process 23. Therefore, we shall analyse it process by process, as given below.

Process 23: We know pressure and volume at the initial and final state. The process equation gives

$$\frac{p_2}{p_1} = \left(\frac{v_2}{v_1}\right)^n \text{ or } 5.5/1.1 = (3.5/0.6)^n \rightarrow n = 1.1277$$

$W_{23}$ = work transfer in process 23, using Eq.(4.29)

$$= (1.1 \times 3.5 - 5.5 \times 0.6) \times 100/(1.1277 - 1) = -430.51 \text{ kJ}$$

The perfect gas equation, when applied to states 2 and 3 gives

$$T_2 = 1.1 \times 3.5 \times 100/(3.616 \times 0.287) = 366.28 \text{ K and}$$
$$T_3 = 439.54 \text{ K}$$
$$\therefore\ U_3 - U_2 = 3.616 \times 0.718\ (439.54 - 366.28) = 137.60 \text{ kJ, and}$$
$$Q_{23} = {}_2W_3 + U_3 - U_2 = -292.91 \text{ kJ (from nfee)}$$

Using Eq. 4.25, change in entropy is

$$S_3 - S_2 = 3.616 \times [1.005 \ln (439.54/366.28) - .287 \ln (5.5/1.1)]$$
$$= -0.729 \text{ kJ/K}$$

The negative sign comes because heat is rejected during the process.

Process 12: It is a constant pressure process for which

$W_{12} = p\,(V_2 - V_1) = 1.1\ (3.5 - 2) \times 100 = 55$ kJ. Work is done by the system. As above, we may find out

$$U_2 - U_1 = 173.64,\ Q_{12} = 193.64 \text{ kJ}$$

and, $S_2 - S_1 = 3.616 \times [1.005 \ln (366.28/293) - .287 \ln (1.1/1.1)]$

$$= 0.5959 \text{ kJ/kg K}$$

Heat is added during the process, and hence a positive sign in change in entropy.

**Example 4.12** Initial volume, pressure and temperature of a gas contained in a cylinder respectively, are 1.25 m³, 1 bar and 47°C. Air is compressed to 7 bar and 197°C, during which 6 kJ heat is rejected. Find specific heat at constant pressure and constant volume.

***Solution*** Given: gas in the cylinder is confined to it, and hence it is a closed system. Also,

$$p_1 = 1,\ V_1 = 1.25,\ T_1 = 47 + 273 = 320,\ p_2 = 7,\ T_2 = 470 \text{ and } Q = -6.0.$$

To find: Specific heat at constant pressure and constant volume.

Assumption: The gas is a perfect gas.

Analysis: From the given data, we can first find out the index of compression $n$, work in compression, and finally, change in internal energy (using nfee). Equation 4.3 gives $m\,c_v$ and Eq. 4.1 gives $m\,R$.

Since for a perfect gas, $T \propto p^{(n-1)/n}$, we get

$$470/320 = (7/1)^{(n-1)/n} \rightarrow n = 1.246$$

and $V_2 = (p_1/p_2)\,(T_2/T_1)\,V_1 = 0.2623$ (from the perfect gas equation)

$$\therefore\ W = (1 \times 1.25 - 7 \times 0.2623) \times 100/(1.246 - 1) = -238.19 \text{ kJ}$$

and $dU = Q - W = -6 - (-238.19) = 233.19 = mc_v\,(470 - 320)$

$$\Rightarrow m\,c_v = 1.548 \qquad \text{(A)}$$

Equation of state for a perfect gas results in

$$m\,R = 1 \times 1.25 \times 100/320 = 0.391 \qquad \text{(B)}$$

These two equations (A) and (B) give rise to

$R/c_v = 0.2523$

$\Rightarrow \gamma$ = specific heat ratio for gas

= 1.2523

Specific heat cannot be determined, as there are three unknowns, namely, $c_p$, $c_v$, and $R$ whereas there are only two equation A and B. However, we may try to get these values, assuming the gas to be air for which $R = 0.287$ kJ/kg K. Equation B gives, $m = 0.391/0.287 = 1.3624$ kg and then Eq. A gives, $c_v = 1.548/1.3624 = 1.136$ (which does not match with the value for air for which $c_v = 0.71$). So, we conclude that specific heat for this problem cannot be determined as complete information is not given. If $m$ is known, then specific heat can be determined.

**Example 4.13** Show, for a gas obeying the equation of state $PV = RT\,(1 + \beta' p)$ in which $\beta'$ is a function of temperature, that

$$c_p = -RTp\,\frac{d^2}{dT^2}(\beta' T) + (c_p)_0$$

where $(c_p)_0$ = value $c_p$ at very low pressure.

***Solution*** Differentiating equation of state twice with respect to $T$, while keeping $p$ constant, we get

$$p\left(\frac{\partial v}{\partial T}\right)_p = R + pR\,\frac{d}{dT}\,(\beta' T)$$

and $$p\left(\frac{\partial^2 v}{\partial T^2}\right)_p = pR\,\frac{d^2}{dT^2}\,(\beta' T)$$

or $$p\left(\frac{\partial^2 v}{\partial T^2}\right)_p = R\,\frac{d^2}{dT^2}\,(\beta' T)$$

$$\therefore \left(\frac{\partial c_p}{\partial p}\right)_T = -T\left(\frac{\partial^2 v}{\partial T^2}\right)_p = -RT\,\frac{d^2}{dT^2}\,(\beta' T)$$

(See Section 4.23)

Integrating this equation with $p$ keeping $T$ constant gives

$$c_p = -RpT \frac{d^2}{dT^2}(\beta' T) + \phi(T)$$

But at low pressure

$$p = 0, c_p = (c_p)_0$$

$$\therefore \quad \phi(T) = (c_p)_0$$

and $$c_p = (c_p)_0 - RpT \frac{d^2}{dT^2}(\beta' T)$$ Q.E.D.

**Example 4.14** Show that $\mu$ for van der Waals gas is

$$\mu = \frac{v}{c_p} \times \frac{2a(v-b)^2 - RTv^2 b}{RTv^3 - 2a(v-b)^2}$$

and determine the equation of inversion curve.

***Solution*** It has been proved before that

$$\mu = \frac{v}{c_p}[\beta T - 1] \text{ (Eq. 4.24)}$$

Differentiating van der Waals equation

$$p = \frac{RT}{v-b} - \frac{a}{v^2}$$

with respect to $T$, keeping $p$ constant, we get

$$0 = -\frac{RT}{(v-b)^2}\left(\frac{\partial v}{\partial T}\right)_p + \frac{R}{v-b} + \frac{2a}{v^3}\left(\frac{\partial v}{\partial T}\right)_p$$

or $$\left(\frac{\partial v}{\partial T}\right)_p = \frac{v^3 R(v-b)}{v^3 RT - 2a(v-b)^2}$$

$$\therefore \beta T - 1 = \frac{v^2 R(v-b)T}{v^3 RT - 2a(v-b)^2} - 1 \text{ (using Eq. 4.15)}$$

$$= \frac{2a(v-b)^2 - RTbv^2}{RTv^3 - 2a\,(v-b)^2}$$

and $$\mu = \frac{v}{c_p} \times \frac{2a(v-b)^2 - RTbv^2}{RTv^3 - 2a\,(v-b)^2}$$ Q.E.D.

At inversion point $\mu = 0$

This gives $T_i = \dfrac{2a(v-b)^2}{Rbv^2}$

or $$\frac{v-b}{v} = \left(\frac{RbT_i}{2a}\right)^{1/2}$$

$$\therefore\ v = \frac{b}{1-\left(\dfrac{RbT_i}{2a}\right)^{1/2}}$$

Substituting $v$ in van der Waals equation

$$p_i = \frac{RT_i}{b\left[\dfrac{2}{1-\left(\dfrac{RbT_i}{2a}\right)^{1/2}} - 1\right]} - \frac{a}{\left[\dfrac{b}{1-\left(\dfrac{RbT_i}{2a}\right)^{1/2}}\right]^2}$$

$$= \frac{RT_i\left[1-\left(\dfrac{RbT_i}{2a}\right)^{1/2}\right]}{b\left(\dfrac{RbT_i}{2a}\right)^{1/2}} - \frac{a\left[1-\left(\dfrac{RbT_i}{2a}\right)^{1/2}\right]^2}{b^2}$$

or $$b^2\,p_i\left(\frac{RbT_i}{2a}\right)^{1/2} = \left[1-\left(\frac{RbT_i}{2a}\right)^{1/2}\right] \times$$

$$\left[RbT_i - a\left\{1-\left(\frac{RbT_i}{2a}\right)^{1/2}\right\}\left(\frac{RbT_i}{2a}\right)^{1/2}\right]$$

$$= \left[1 - \left(\frac{RbT_i}{2a}\right)^{1/2}\right] \times \left[\frac{3}{2} RbT_i - a\left(\frac{RbT_i}{2a}\right)^{1/2}\right]$$

or $$b^2 p_i = a\left[1 - \left(\frac{RbT_i}{2a}\right)^{1/2}\right] \times \left[3\left(\frac{1}{2RbT_i a}\right)^{1/2} - 1\right]$$

which is the equation of inversion curve.

**Example 4.15** Prove that $u$ and $h$ are functions of temperature only for the perfect gas.

***Solution*** We have already proved that

$$\left(\frac{\partial u}{\partial v}\right)_T = T\left(\frac{\partial p}{\partial T}\right)_v - p \qquad \text{(Eq. 4.23c)}$$

and $$\left(\frac{\partial h}{\partial p}\right)_T = v - T\left(\frac{\partial v}{\partial T}\right)_p \qquad \text{(Eq. 4.23a)}$$

For perfect gas, $pv = RT$

$$\therefore \left(\frac{\partial p}{\partial T}\right)_v = \frac{R}{v}$$

and $$\left(\frac{\partial v}{\partial T}\right)_p = \frac{R}{p}$$

or $$\left(\frac{\partial u}{\partial v}\right)_T = 0$$

and $$\left(\frac{\partial h}{\partial p}\right)_T = 0$$

We conclude variation of $u$ and $h$ does not depend on $v$ or $p$. This implies that their variation may or may not depend on temperature. It is found that for a perfect gas, $u$ and $h$ are functions of temperature alone.

**Example 4.16** Determine the change in internal energy of a gas obeying van der Waals equation. Also prove that internal energy is not a function of temperature only for such gases evalute $(\Delta h)_T$ for van der Waals gas.

***Solution*** The van der Waals equation is

$$\left(p + \frac{a}{v^2}\right)(v - b) = RT$$

$$\therefore \quad \left(\frac{\partial p}{\partial T}\right)_v = \frac{R}{(v - b)}$$

and $$\left(\frac{\partial u}{\partial v}\right)_T = \frac{RT}{v - b} - p = \frac{a}{v^2} \quad \text{(using Eq. 4.23c)}$$

which shows that $u$ is not a function of temperature alone.

Further, $$du = c_v\, dT + \left(\frac{\partial u}{\partial v}\right)_T dv$$

On substitution

$$du = c_v\, dT + a\,\frac{dv}{v^2}$$

Now, integrating this equation between states 1 to 2, we get

$$u_2 - u_1 = \int_{T_1}^{T_2} c_v\, dT + a\left(\frac{1}{v_1} - \frac{1}{v_2}\right)$$

$$= c_v(T_2 - T_1) + a\left(\frac{1}{v_1} - \frac{1}{v_2}\right)$$

if $c_v$ is assumed constant.

Since $$\left(\frac{\partial h}{\partial p}\right)_T = v - T\left(\frac{\partial v}{\partial T}\right)_p \quad (4.23a)$$

$$(\partial h)_T = \left[v - T\left(\frac{\partial v}{\partial T}\right)_p\right](dp)_T$$

Let $$p = p(v, T)$$

$$\therefore \quad dp = \left(\frac{\partial p}{\partial v}\right)_T dv + \left(\frac{dp}{\partial T}\right)_v dT$$

and $$(dp)_T = \left(\frac{\partial p}{\partial v}\right)_T dv,$$

since $$dT = 0$$

$$\therefore\ (dh)_T = \left[v - T\left(\frac{\partial v}{\partial T}\right)_p\right]\left(\frac{\partial p}{\partial v}\right)_T dv$$

$$= \left[v\left(\frac{\partial p}{\partial v}\right)_T - T\left(\frac{\partial v}{\partial T}\right)_p\left(\frac{\partial p}{\partial v}\right)_T\right]dv$$

But

$$\left(\frac{\partial v}{\partial T}\right)_p\left(\frac{\partial T}{\partial p}\right)_v\left(\frac{\partial p}{\partial v}\right)_T = -1$$

$$\therefore \left(\frac{\partial v}{\partial T}\right)_p\left(\frac{\partial p}{\partial v}\right)_T = -\left(\frac{\partial p}{\partial T}\right)_v$$

and $$(dh)_T = \left[v\left(\frac{\partial p}{\partial v}\right)_T + T\left(\frac{\partial p}{\partial T}\right)_v\right]dv$$

For van der Waals gas,

$$(p + a/v^2)\,(v - b) = RT$$

$$\therefore\ \left(\frac{\partial p}{\partial T}\right)_v = \frac{R}{v - b}$$

and $$\left[\left(\frac{\partial p}{\partial v}\right)_T - \frac{2a}{v^3}\right](v - b) + \left(p + \frac{a}{v^2}\right) = 0$$

or $$\left(\frac{\partial p}{\partial v}\right)_T = -\frac{p + a/v^2}{v - b} + \frac{2a}{v^3}$$

$$= -\frac{RT}{(v - b)^2} + \frac{2a}{v^3}$$

$$\therefore (dh)_T = \left[\frac{-RvT}{(v-b)^2} + \frac{2a}{v^2} + \frac{RT}{v-b}\right] dv$$

or $$(\Delta h)_T = -RT\int_{v_1}^{v_2} \frac{v\,dv}{(v-b)^2} + 2a\int_{v_1}^{v_2} \frac{dv}{v^2} + RT\int_{v_1}^{v_2} \frac{dv}{v-b}$$

$$= -RT\left[\ln\frac{v_2-b}{v_1-b} - b\left\{\frac{1}{v_2-b} - \frac{1}{v_1-b}\right\}\right]$$

$$-2a\left(\frac{1}{v_2} - \frac{1}{v_1}\right) RT \ln\frac{v_2-b}{v_1-b}$$

$$= bRT\left(\frac{1}{v_2-b} - \frac{1}{v_1-b}\right) - 2a\left(\frac{1}{v^2} - \frac{1}{v_1}\right)$$

**Example 4.17** On a Mollier diagram prove that

(a) the slope of the curve representing a reversible isothermal process is equal to $T - 1/\beta$

(b) for a isochoric process (constant volume) the slope is $T + (\gamma - 1)/\beta$ and

(c) the constant pressure lines have equal slopes at the same temperature.

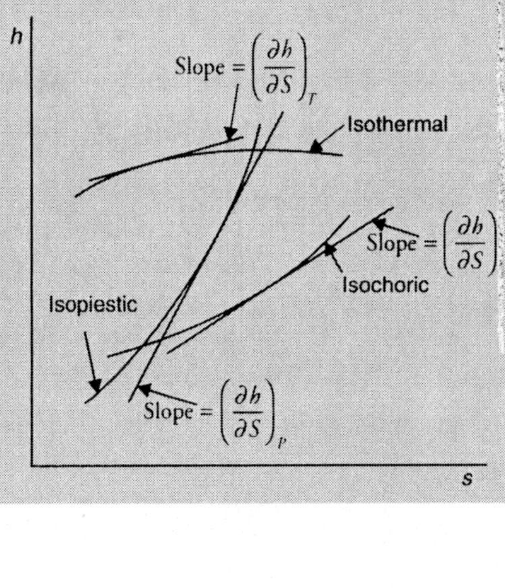

***Solution*** (i) Let $h = h(s, p)$

$$\therefore \left(\frac{\partial h}{\partial s}\right)_T = \left(\frac{\partial h}{\partial s}\right)_p + \left(\frac{\partial h}{\partial p}\right)_s \left(\frac{\partial p}{\partial s}\right)_T$$

where $$\left(\frac{\partial h}{\partial s}\right)_p = T \text{ (Gibbsian relations)}$$

$$\left(\frac{\partial h}{\partial p}\right)_s = v$$

and $$\left(\frac{\partial s}{\partial p}\right)_T = -\left(\frac{\partial v}{\partial T}\right)_p \text{ (Maxwell relation)}$$

$$\therefore \left(\frac{\partial h}{\partial s}\right)_p = T - \frac{v}{\left(\dfrac{\partial v}{\partial T}\right)_p}$$

$$= T - \frac{1}{\beta}, \text{ since} \left(\frac{\partial v}{\partial T}\right)_p = \beta v$$

or slope of isothermal line on $h$-$s$ plane $= T - \dfrac{1}{\beta}$

(ii) Let $h = h(s, p)$

$$\therefore \left(\frac{\partial h}{\partial s}\right)_v = \left(\frac{\partial h}{\partial s}\right)_p + \left(\frac{\partial h}{\partial p}\right)_s \left(\frac{\partial p}{\partial s}\right)_v$$

But $$\left(\frac{\partial h}{\partial s}\right)_p = T \text{ (Gibbsian relations)}$$

$$\left(\frac{\partial h}{\partial p}\right)_s = v$$

$$\therefore \left(\frac{\partial h}{\partial s}\right)_v = T + v\left(\frac{\partial p}{\partial s}\right)_v$$

$$= T + v\left(\frac{\partial p}{\partial T}\right)_v \left(\frac{\partial T}{\partial s}\right)_v$$

Substitute $\left(\frac{\partial s}{\partial T}\right)_v = \frac{c_v}{T}$

and $\left(\frac{\partial p}{\partial T}\right)_v = \frac{\beta}{K}$

$$\therefore \left(\frac{\partial h}{\partial s}\right)_v = T + v\left(\frac{\beta}{K}\right)\left(\frac{T}{c_v}\right)$$

Also $c_p - c_v = \beta^2 Tv/K$

or $c_v(\gamma - 1) = \beta^2 Tv/K$

Substituting $c_v$, the slope of isochoric line on Mollier diagram is

$$= \left(\frac{\partial h}{\partial s}\right)_v$$

$$= T + \frac{\beta v T}{K} \times \frac{\gamma - 1}{\beta^2 Tv/K}$$

$$= T + \frac{\gamma - 1}{\beta} \qquad \text{Q.E.D.}$$

The slope of constant pressure process of Mollier diagram is $\left(\frac{\partial h}{\partial s}\right)_p$

But $\left(\frac{\partial h}{\partial s}\right)_p = T$

That is, constant pressure lines have same slope on h-s chart at the same temperature.

## SUMMARY

The term pure substance is used for a substance that has the same chemical composition throughout the system, and which does not change during the process. A pure substance may be a single chemical component (molecule or a compound) or a mixture of different compounds, so that the chemical composition remains the same throughout the system. It may exist in three single phases (namely solid, liquid and gas phase), two phases or three phases; different phases in a two- or three-phase system are in thermodynamic equilibrium with each other. (This means all the phases have the same potentials.) The state of a thermodynamic system is defined by a set of properties like pressure, temperature, volume, internal energy, and entropy. The Gibbs phase rule gives the degree of freedom of a system, that is, the minimum number of properties required defining the state of the system. The choice of independent properties requires an in-depth study of the physics of the given problem. A thumb rule is to choose temperature as one such property.

To determine all the properties of a single component for the entire range of pressure and temperature, algebraic and differential equations have been developed. Differential equations can be integrated using graphical or computational integral procedure, described in Section 4.4.2.7. However, the choice of equation to be integrated depends on the independent variable used. The determination of all the properties also requires knowledge of specific heat at constant pressure and volume, coefficient of volume expansion $\beta$ , and isothermal compressibily $\kappa$ , which are determined experimentally in the laboratory. However, determination of entropy poses a problem, as it also depends on pressure. Therefore, a function $\phi$ has been defined, which is a function of temperature only. Section 4.4.2.7 discusses how to incorporate effects of change in pressure.

Since differential relations give change in the property, it requires to fix a reference condition or state, also called datum. In gas tables, this reference has been fixed at absolute zero of the thermodynamic temperature scale. However, for different substances, different datum has been fixed, keeping in mind to avoid negative numbers for all extensive properties. For water, the reference is 0.01°C at which internal energy, entropy and entropy are zero. The difficulty with differential relations is that these are made from a partial derivative involving three properties formed from eight properties ($p$, $v$, $T$, $u$, $h$, $s$, $f$, and $g$), and there may be two, three or more such derivatives present in a relation. The number of ways in which the three derivatives may be formed from eight properties is $8P_3 = 336$. Similarly, the number of ways in which equations having four or three derivatives may be formed from these 336 partial derivatives is $336\,C_4 \approx 5 \times 10^8$ and $336\,C_3 \approx 5.6 \times 10^6$ ways. It becomes a difficult job to remember these relations. A nemonic diagram becomes very handy, which is explained in Section 4.4.3.

Thermodynamic relations developed for specific heat, enthalpy, internal energy and entropy also help in understanding the behaviour of substances. For example, relations for specific heat tell us that constant pressure lines diverge on the T-s chart for air. It is this characteristic that a gas turbine cycle could be developed. Or, that constant volume lines are steeper than constant pressure lines, and this information helps a lot in drawing the two processes through the point, as has been used in gas power cycles later. Take another example. $(\partial h/\partial s)_p = T$ shows that constant pressure lines will be a straight line in a two-phase region having the slope $T$, which means constant pressure lines will diverge along the right on the Mollier diagram. This knowledge is very useful to draw the cycles and to draw conclusions from these. Like this the relations have immense information contained in them.

There are infinite states in which a pure substance may exist, and hence representation of its property becomes a hurricane job. Different methods such as three-dimensional chart, two-dimensional chart, tables, algebraic and differential equations have been developed. However, in day-to-day practice, three-dimensional chart is not used. Two-dimensional charts are useful as these give a picture of how the process will proceed, and what are the possible final states attainable by the system in accordance with nature's laws, namely, the first and second law of thermodynamics, and the presence of irreversibility in the system. Nonetheless, we would prefer an algebraic relation and that too in explicit form, as these can easily be solved and simulated on a computer. Unfortunately, the equations for solid and liquid are quite complex and chart or tables are used. However, for gases, the equations are comparatively simple, but a single equation is not applicable for the entire range of temperature and pressure, especially at high pressure and low temperature. The main problem is the compressibility of a gas. The problem is circumvented by developing a number of gas equations, which are applicable for different pressure and temperature ranges. A common technique is to introduce more and more unknowns to take care of the compressibility effect. These equations involve some unknowns, which are determined for different gases. On the other hand, Weber in 1933, and later Nelson and Obert, plotted the property data of Su on the generalised compressibility chart, which is a plot of the compressibility factor (defined by Eq. 4.6) against reduced pressure with reduced temperature as parameter; reduced parameters are defined in Eq. 4.7. This chart, though gives error close to critical point is still preferred for accurate work.

## EXERCISES

### Review Problems

4.1 What is the difference between a compound and a mixture?

4.2 Does the boiling temperature also increase if pressure is increased during boiling?

4.3 In what way does a pressure cooker help? Will cooking time reduce if another heavier one replaces the dead weight? Discuss its side effects.

4.4 Is it possible to boil water contained in a container at 20°C without heating it? Explain how.

4.5 Is it possible to have water vapour at –10°C?

4.6 Why is temperature or pressure required in the wet mixture region and not both?

4.7 What is the difference between the critical point and triple point? Do all phases have a critical point?

4.8 Vegetable is cooked in a pan that is (a) uncovered, (b) covered with a light lid, and (c) covered with a heavy lid. For which case will the cooking time be the shortest? Why?

4.9 What is the difference between saturated vapour and superheated vapour?

4.10 How does the boiling process in a super-critical pressure boiler differ from that in a sub-critical pressure boiler?

4.11 Does the reference point selected for the properties of a pure substance have any effect in thermodynamic analysis? Mention the criterion for selection of the reference point.

4.12 What is quality of vapour? Does it have any significance for superheated steam?

4.13 How do various properties change during change of phase? Justify your answer using charts.

4.14 For $H_2O$, pressure is given. How will you ascertain whether it is liquid, wet steam or superheated steam if its volume, internal energy or enthalpy is given as another property?

4.15 Is it true that as we go up the mountain, water starts boiling earlier? Why?

4.16 What happens if a saturated liquid is heated at constant (a) pressure, and (b) volume?

4.17 Why is skating on ice easy as compared to skating on cement floor?

4.18 How will you specify a mixture of a pure substance at its triple point pressure?

4.19 Why is the Mollier chart usually used in the analysis of a power cycle, and the p-h chart in the refrigeration cycle?

4.20 What does the Joule–Thomson coefficient represent? How is its knowledge helpful in cooling of gases? Does it change with temperature at a given pressure? Estimate the Joule–Thomson coefficient for *R*-12 at 6 bar, 1000°C.

4.21 A gas follows equation of state $p\,(v - a) = RT$, where a is a positive constant. Is it possible for this gas to cool by throttling?

4.22 Estimate the specific heat difference, $\beta$ and $\kappa$ for *R*-12 at 1.4 bar and 200°C.

4.23 Is it possible to find out variation of specific heat, enthalpy and internal energy with temperature from the knowledge of *p-v-T* data alone? Justify your answer.

4.24 Show that the slopes of the three two-phase lines on a phase diagram is different at triple point.

4.25 What is the importance of the Clapeyron equation? Are we justified in making assumptions used for deriving the Clausius–Clapeyron equation?

4.26 What do you understand by the order of phase difference and how does it help us in developing the steam table?

4.27 Prove the Meyer's relation for a perfect gas.

4.28 Show that the process equation for a system made up of an ideal gas subjected to isentropic process is given by $pv^{\gamma}$ = constant, when specific heat are constant.

4.29 Find out the relation between $p$ and $v$ for a gas obeying the van der Waals equation subjected to an isentropic process.

4.30 For a homogeneous single-phase pure substance, pressure and temperature are taken as independent properties. Show that change in volume is given by $dv/v = \beta\ dT - \alpha\ dp$, and find out the ratio $v_2/v_1$ for the process to which this system is subjected to.

4.31 Determine change in entropy, internal energy and enthalpy of a gas obeying van der Waals equation, and show that internal energy and enthalpy does not depend only on temperature.

4.32 Verify Eqs. 4.5 and 4.10.

4.33 What is an equation of state? Discuss the most commonly used equations.

4.34 Determine the specific heat for any process in terms of specific heat for a constant pressure and constant volume process.

4.35 Show that for a perfect gas, the slope of the constant volume line on the T-s diagram is steeper than that of the constant pressure line. Sketch the constant volume and constant pressure lines on this diagram.

4.36 Sketch the inversion curve and indicate the regions of cooling and heating. The maximum inversion temperature for He is 24 K. What should be its initial temperature if it is required to be cooled by throttling?

## Problems for Practice

4.1 A 0.5 $m^3$ tank contains a saturated mixture of *R*-12 at 2 bar. If the liquid occupies 10% of the space, determine the quality and total mass of the refrigerant in the tank.

4.2 A rigid tank of volume 3.5 $m^3$ contains 5 kg of saturated liquid-vapour mixture of water at 75°C. It is now heated slowly. Determine the temperature at which the liquid in the tank is completely vaporised and the mass of water in the initial state. How will this temperature change if (a) the initial mass of mixture is 4 kg or (b) the initial temperature is 60°C.

4.3 A piston-cylinder device contains 0.1 m$^3$ of liquid water and 0.9 m$^3$ of vapour in equilibrium at 8 bar. It is heated at constant pressure till the temperature increases to 300°C. Find out the amount of heat transfer and the final volume.

4.4 A rigid tank is divided into two equal partitions. Initially, one side contains 5 kg of water at 2 bar and 25°C, and the other side is evacuated. The partition is then removed, and the water expands into the entire tank. The water is allowed to exchange heat with its surrounding until the temperature in the tank returns to the initial value of 25°C. Determine (a) the volume of tank, (b) the final pressure, and (c) the heat transfer for this process. [*Ans.* 0.01 m$^3$, 0.0317 bar, 0.25 kJ]

4.5 A throttling calorimeter is used to measure the dryness fraction of steam in a steam main. The calorimeter reading after throttling are 1.00 bar and 120°C; pressure in the main is 6 bar. Find d.f.

4.6 Steam expands isentropically from 30 bar 300°C in a turbine to 0.07 bar. Find out the quality of steam after expansion, and the amount of work produced in the turbine.

4.7 For the following processes, determine changes in $h$, $s$, $v$, $T$ and $x$ (appropriate) of $H_2O$. Assume $p_1$ = 5 bar. (a) Constant volume process at $v$ = 0.3 m$^3$/kg and $p_2$ = 3.0 bar; (b) constant entropy, $s_1$ = 6.3 kJ/kg K, $p_2$ = 1.5 bar; (c) constant volume, $h_1$ = 2500 kJ/kg, $p_2$ = 2.0 bar; (d) constant enthalpy, $s_1$ = 6.4 kJ/kg K, $p_2$ = 2.0 bar; (e) constant temperature, $T_1$ = 200°C, $p_2$ = 2.0 bar; (f) constant pressure, $x_1$ = 0.5, $T_2$ = 400°C.

4.8 Refrigerant-134a at 10 bar and 80°C is to be cooled to 10 bar and 30°C in a condenser by air. The air enters at 1 bar and 27°C with a volume flow rate of 800 m$^3$/min and leaves at 0.95 bar and 60°C. Determine the mass flow rate of the refrigerant. [*Ans.* 139 kg/min]

4.9 A 1 m$^3$ tank contains 2.841 kg of steam at 6 bar. Determine temperature of steam using (a) ideal-gas equation, (b) van der Waals equation, and (c) steam table.

4.10 On Mollier diagram show that slope of the isothermal and isometric processes is given by $T - 1/\beta$ and $T + (\gamma - 1)/\beta$, respectively. Also, show that isobaric lines on this chart have the same slope at the point of their intersection with a constant temperature line.

4.11 The relation between saturation pressure and temperature for mercury in the MKS system of units is given by $\log p = 6.7173 - 3277/T - 0.6520 \log T$. Calculate the specific volume of saturated mercury vapour at 0.1 bar. Take latent heat of vaporisation at this pressure as 69 kcal/kg.

4.12 Using the Clapeyron equation, estimate the enthalpy of vaporisation of $R$-12 at 30°C and compare it with the tabulated value. Justify the difference.

4.13 Argon gas enters a turbine at 0.7 bar, 600 K with velocity 100 m/s, and leaves at 0.1 bar with a velocity of 150 m/s at a rate of 5 kg/s. Heat is lost to the surroundings at 25°C at a rate of 60 kW. Using the generalised compressibility chart, determine the power output.

4.14 An insulated tank having volume 0.1 $m^3$ contains $CO_2$ at 15 bar, 60°C. It is pressurised from a line carrying gas at 90 bar, 60°C, and the valve is closed when pressure reaches 90 bar. Using the generalised compressibility chart, determine the amount of the gas filled in.

4.15 A piston-cylinder device contains 50 kg of water at 1.5 bar and 25°C. The cross-sectional area of the piston is 0.1 $m^2$. Heat is now transferred to the water, causing part of it to evaporate and expand. When the volume reaches 0.2 $m^3$, the piston reaches a linear spring, whose spring constant is 100 KN/m. More heat is transferred to the water until the piston rises 20 cm more. Determine (a) the final pressure and temperature, and (b) the work done during this process. Also, show the process on a *pv* diagram.

4.16 A piston-cylinder device with a set of stops contains 10 kg of the refrigerant-134a. Initially, 8 kg of the refrigerant is in the liquid form, and the temperature is – 8°C. Now, heat is transferred slowly to the refrigerant until the piston hits the stops, at which point the volume is 400 litres. Determine (a) the temperature when the piston first hits the stops, and (b) the work done during this expansion process. Also, show the process on a pv diagram. [*Ans.* –8°C, 45.6 kJ]

4.17 A well-insulated rigid tank contains 5 kg of saturated liquid–vapour mixture of water at 1 bar. Initially, three-quarter of the mass is in the liquid phase. An electric resistor placed in the tank is connected to a 220 V source, and a current of 4 A flows through the resistor when the switch is on. Determine how long it will take to vaporise all the liquid in the tank. Also, show the process on a T-v diagram with respect to saturation lines.

4.18 An insulated vessel is divided into two compartments by a valve. Initially, one compartment contains steam at 10 bar, 500°C, and the other is evacuated. The valve is opened and the steam is allowed to fill the entire volume, achieving a final pressure of 1 bar. Determine the final temperature, the % vessel volume initially occupied by steam, and the amount of entropy produced.

[*Ans.* 495.2°C, 10%, 1.11 kJ/kg – K] (IES 1993).

4.19 A piston-cylinder contains 3 kg of wet steam at 1.4 bar. The initial volume is 3.25 $m^3$. The steam is heated till its temperature reaches 400°C. The piston is free to move up to the top, where it is stopped by a stopper; the cylinder volume at that instant is 4.65 $m^3$. Determine the amount of work and heat transfer. [*Ans.* 614.5 kJ, 4390.6 kJ] (IES 1994). {Hint: $v_2$ and $T_2$ are given. This gives $p_2$ and $h_2$.}

4.20 0.14 $m^3$ of steam at 20 bar and 250°C is expanded reversibly and polytropically to 2 bar. Find the final temperature, heat transfer and the change in entropy, if the index of expansion is 1.25. [*Ans.* $x = 0.79$, 413.6 kJ, 313.7 kJ, –0.8171 kJ/K} (IES 96).

4.21 Air at 0.9 bar and – 3°C enters an adiabatic diffuser steadily with a velocity of 183 m/s, and leaves with a low velocity at a pressure of 1 bar. The exit area of the diffuser is five times the inlet area. Determine (a) the exit temperature, and (b) the exit velocity of air.

4.22 Air is compressed from 1.01 bar and 107°C to a pressure of 10 bar while being cooled at a rate of 25 kJ/kg by circulating water through the compressor casing. The volume flow rate of the air at the inlet conditions is 142 $m^3$/min, and the power input to the compressor is 522 kW. Determine (a) the mass flow rate of the air, and (b) the temperature at the compressor exit. [*Ans.* 2.88. kg/s, 441.7 K]

4.23 Specific heat of a gas are of the form $cp = a + kT$ and $cv = b + kT$, where $a$, $b$ and $k$ are constants. Derive the formula (process equation) for adiabatic expansion of the gas. [*Ans.* $Tb\ v\ (a - b)\ ekT = C$] (IES 96).

4.24 A 3.27 $m^3$ tank contains 100 kg of nitrogen gas at 225 K. Determine pressure in the tank using (a) perfect gas equation, (b) van der Waal's equation, and (c) generalised compressibility chart.

4.25 For an ideal gas, specific heat varies linearly with temperature as $c_p = \beta + \lambda T$, and $c_v = \alpha + \lambda T$, where $\alpha, \beta$ and $\gamma$ are constants. Find out the process equation for an adiabatic process.

## Multiple-choice Questions

4.1 Universal gas constant of a perfect gas
- (a) increases with temperature.
- (b) decreases with temperature.
- (c) increases with increase in molecular weight.
- (d) is always constant.

4.2 Characteristic gas constant of any perfect gas
- (a) increases with increase in temperature.
- (b) increases with increase in pressure.
- (c) is a function of pressure and temperature.
- (d) is a constant.

4.3 The ratio of specific heat of gas at constant pressure and at constant volume
- (a) is always constant.
- (b) always varies with pressure.
- (c) always varies with temperature.
- (d) none of the above.

4.4 Characteristic equation of gas is given by ($v$ = specific volume, $m$ = mass of gas)
(a) $pv = RT$. (b) $pv = mRT$. (c) $pv^n = C$. (d) $pv^\gamma = C$.

4.5 Equation $TdS = dU + pdV$ can be applied to processes which are
(a) only reversible. (b) only irreversible.
(c) reversible or irreversible. (d) none of the above.

4.6 The polytropic index of expansion $n$ in the equation $pv^n = c$ for constant volume process is
(a) 1. (b) 1.4. (c) $\infty$. (d) 0.

4.7 The change of entropy of a closed system
(a) is the same for every process between two specified states.
(b) is not the same for every process between two specified states.
(c) is the same only for an isothermal process between two specified states.
(d) is the same only for a reversible adiabatic process between two specified states.

4.8 The entropy of fixed amount of an ideal gas
(a) decreases in every isothermal compression.
(b) increases in every isothermal compression.
(c) remains the same in every isothermal compression.
(d) may increase or decrease in every isothermal compression.

4.9 The specific internal energy, enthalpy and entropy of an ideal gas:
(a) each is a function of temperature alone.
(b) each is a function of pressure alone.
(c) each is a function of volume alone.
(d) All of the above are false.

4.10 One of the $T\,ds$ equations has the form
(a) $Tds = dh + u\,dp$. (b) $Tds = dh - u\,dp$.
(c) $Tds = du - pdu$. (d) $Tds = dh + pdu$.

4.11 At triple point
(a) ice on heating becomes super-heated vapour.
(b) solid, liquid and vapours co-exist.
(c) ice occupies maximum specific volume.
(d) liquid water and dry saturated steam co-exist.

4.12 The specific volume of water when heated from 0°C
(a) first increases and then decreases. (b) first decreases and then increases.
(c) increases steadily. (d) decreases steadily.

4.13 For water at triple point the properties are
   (a) pressure = 1.0332 kgf/cm$_2$ and temperature = 273.16 K.
   (b) pressure = 0.00622 kgf/cm$_2$ and temperature = 373.16 K.
   (c) pressure = 0.00622 kgf/cm$_2$ and temperature = 273.16 K.
   (d) pressure = 0.1 bar and temperature = 273. 16 K.

4.14 With the increase in pressure
   (a) boiling point of water increases and enthalpy of evaporation remains constant.
   (b) boiling point of water increases and enthalpy of evaporation decreases.
   (c) boiling point of water increases and enthalpy of evaporation increases.
   (d) None of the above.

4.15 At critical point for water
   (a) $p$ = 221.2 bar, $T$= 647.3 K, $v$ = 0.00317 m$^3$/kg.
   (b) $p$ = 221.2 bar, $T$ = 650 K, $v$ = 0.00312 m$^3$/kg.
   (c) $p$ = 225.65 kgf/cm$^2$, $T$ = 773 K, $v$ = 0.00317 m$^3$/kg.
   (d) $p$ = 225.65 kgf/cm$^2$, $T$ = 673 K, $v$ = 0.00317 m$^3$/kg.

4.16 With increase in pressure, for water
   (a) enthalpy of dry saturated steam increases.
   (b) enthalpy of dry saturated steam decreases.
   (c) enthalpy of dry saturated steam remains the same.
   (d) enthalpy of dry saturated steam first increases and then decreases.

4.17 The steam whose dryness fraction is to be determined is at a pressure of 40 bar. The expected dryness fraction is approximately 0.85. The results can be obtained by
   (a) employing only a throttling calorimeter.
   (b) employing a separating and throttling calorimeter.
   (c) employing two throttling calorimeters.
   (d) none of the above, as dryness fraction is too low to be determined by a calorimeter.

4.18 Only a throttling calorimeter is used for measuring
   (a) very low dryness fraction up to 0.7.
   (b) very high dryness fractions 0.98.
   (c) dryness fraction of only low pressure steam.
   (d) dryness fraction of only high pressure steam.

4.19 Normal boiling point of Freon 12 is:
   (a) greater than that of Freon 22. (b) less than that of Freon 22.
   (c) equal to that of Freon 22. (d) not predictable.

4.20 The normal boiling point among the substances helium, hydrogen, oxygen and nitrogen is the lowest for

(a) hydrogen. (b) helium. (c) oxygen. (d) nitrogen.

4.21 The normal boiling point among the substances Freon 134a, Freon 12, Freon 22 and ammonia is the highest for:

(a) Freon 12. (b) Freon 134a. (c) Freon 22. (d) Ammonia.

CHAPTER FIVE

# Gas and Vapour Mixtures



## 5.1 INTRODUCTION

In Chapter 3, we studied the behaviour of a single component substance. However, there exists many thermal systems, notably where mixture of pure substances is used as working fluid. We would like to develop methods so that these can be analysed, like single component systems, without complexity. Chemical reaction is another field where the chemical composition of mixture of a number of single components changes the process during which heat is liberated to or extracted from the surroundings. This shall be discussed in the next chapter. We shall devote this chapter to the mixtures of gases, or gases and vapours where no chemical reaction occurs and the chemical composition does not change.

A gas mixture is formed when different gases are mixed with each other. Air is an example of a gas mixture which contains gases like oxygen, nitrogen, and carbon-dioxide, as well as moisture in very small amounts. But it contains mostly oxygen and nitrogen, both of which are diatomic gases having almost equal molecular weight, and behave as perfect gases in most operating ranges of pressure and temperature. But all the gases in a system need not necessarily be perfect gases or non-reactive. It is always possible that some of the constituents in a mixture may react chemically, changing the composition of the system as in combustion. Also, there may be a mixture of two phases of different pure substances as in vapour absorption refrigeration systems.[1]

---

1. In ammonia–water absorption refrigeration, ammonia gas is absorbed in water in absorber and liberated as gas in generator from its solution formed in absorber. On the other hand, lithium bromide is the solid and water is the liquid in a lithium-bromide refrigeration system.

For analysis of gas mixtures, what is needed is the determination of properties like gas constant, molecular weight and specific heat of the mixture from the knowledge of their values for constituent gases. Once this is known, then the gas mixture can be analysed by treating it as a single component gas.

We know that if two gases are mixed, then their total mass is the sum of the mass of the individual gases (mass conservation). Mass and the number of moles are used to represent the quantity of gas, and thus we can form the mass fraction or the mole fraction to represent the composition of the mixture. *Mass fraction* is defined as the ratio of the mass of a gas to the total mass of the gas. Similarly, *mole fraction* is defined replacing mass with mole. If $m_i$ and $n_i$ are the mass and the number of moles of the i$^{th}$ constituent of the mixture, then

$$m_{fi} = \text{mass fraction} = m_i/\Sigma m_i \quad \text{and} \quad x_i = \text{mole fraction} = n_i/\Sigma n_i \tag{5.1}$$

where, summation is for all the gases in the mixture. Another fraction that has been used is volume fraction, defined in the same way. But all the gases occupy the same volume—the volume of the container holding the gases. So, what is the volume fraction of a constituent gas? We note that the gases at the same $V$ and $T$ will have different pressure. Therefore, volume in volume fraction is to be determined, so that all the constituents are at the temperature and pressure of the mixture. This volume $V_i$ is called *partial volume.* Thus, volume fraction at mixture pressure and temperature is obtained by determining the partial volume of all its constituent gases at mixture pressure and temperature. In mathematical form, volume fraction is defined as

$$V_{fi} = V_{fi}]_{p,T} / \Sigma V_{fi}]_{p,T} \tag{5.2}$$

The notation $V_{fi}]_{p,T}$ is used to represent partial volume to remind that this is determined at mixture temperature and pressure. However, we shall henceforth use $V_{fi}$ for partial volume, keeping in mind that partial volume is obtained at mixture pressure and temperature to write Eq. 5.2 in a simplified way.

The above fractions hold for mixture of gas and vapour as well. The difference lies in the fact that the vapours may or may not obey the gas laws, especially the perfect gas equation. But in situations where the amount of vapour is so small that its vapour pressure is negligible compared to the mixture pressure, then we may treat vapour as perfect gas without introducing much error that is permissible in most of such applications. Such mixtures which we often deal with are a mixture of air and water vapour in air conditioning, the space above the water level in a boiler, or steam condenser, etc. The other area is automobile engineering, where a mixture of fuel vapours and air is formed to give an uniform mixture. Section 5.1 deals with a mixture of ideal and real gases, while Section 5.2 with the mixture of air and water vapour, which is studied under psychrometry. A mixture of air and fuel vapours has not been discussed, but can be tackled in a similar way.

Mixtures of solid and liquids or liquid and gas are also used in some applications. Such mixtures pass through different kind of processes, the most significant is the rectification process. Thermodynamic analysis of such mixtures has been carried out in Section 5.3, and forms the perquisite for the chapter on refrigeration.

### 5.1.1 Dalton–Gibbs and Amagat–Leduc Laws

Let '*s*' number of perfect gases at temperature *T*, each occupying volume *V*, be mixed to form a mixture at volume *V*. In this case, the pressure of individual gases in the mixture will be different, which can be determined using equation of state. Pressure of individual gases in a mixture is called its *partial pressure*. For an individual constituent '*i*' in a mixture of perfect gases, we can write $p_i V = n_i \bar{R} T$, and similarly for the mixture itself $pV = n_i \bar{R} T$, where $n$ = total moles of mixture = $\Sigma n_i$. If the former equation is divided by the latter, we get:

$$p_i/p = n_i/n = x_i \tag{5.3}$$

That is, the ratio of partial pressure of a gas to the mixture pressure is equal to its mole fraction in the mixture. We may also write $p_i V = n_i \bar{R} T$ for *i*th gas of the mixture. Summing equations for all the gases, we get $\Sigma p_i V = \Sigma n_i \bar{R} T$, or $V \Sigma p_i = n_i \bar{R} T$, since $V$, $\bar{R}$ and $T$ are the same for all the constituent gases, and can be taken out of the sigma sign. Comparing this with similar equations for mixtures, we get an important result for a mixture of *perfect gases*

$$p = \Sigma p_i \tag{5.4}$$

in which $p_i$ is the partial pressure of an individual gas measured at mixture volume and temperature. This is known as Dalton's law of partial pressure. Gibbs extended it further to *extensive properties that are a function of temperature alone,* and is known as the Dalton–Gibbs law. It states that "*The total property of a mixture of ideal gases that is a function of temperature is equal to the sum of the property that the individual gas will have, if each occupied the total mixture volume alone at the mixture temperature*". That is,

$$U = n\bar{u} = \Sigma n_i \bar{u}_i, \qquad H = n\bar{h} = \Sigma n_i \bar{h}_i, \qquad c_p = n\bar{c}_p = \Sigma n_i \bar{c}_{pi}, \text{ and}$$
$$c_v = n\bar{c}_{vi} = \Sigma n_i \bar{c}_{vi},$$

or

$$\bar{u} = \Sigma x_i \bar{u}_i, \ [\bar{h} = \Sigma x_i], \qquad \bar{h} = \Sigma x_i \bar{h}_{pi}, \qquad \bar{c}_p = \Sigma x_i \bar{c}_{pi}, \text{ and}$$
$$\bar{c}_v = \Sigma x_i \bar{c}_{vi} \tag{5.5}$$

where (–) over a property represents its value for one mole of the gas. We may similarly proceed to develop such relations in terms of mass fraction. We get exactly the same relations as in Eq. 5.5, obtained by replacing mole fraction by mass fraction $m_{fi}$.

The Dalton–Gibbs law is applicable to those properties, which are a function of temperature only. This means that when gases are mixed, the total entropy of the mixture will not be equal to the sum of the entropy of individual gases, as entropy depends on pressure, too. To find out the change in entropy due to change in pressure during mixing of gases, consider two gases *a* and *b*, for example, at temperature *T* and pressure *p* mixed to form a mixture at temperature *T* and pressure *p*. This means that the two gases, initially occupying their partial volumes, now occupy the mixture volume *V*. In this process, pressure of individual gas changes to $p_a$ and $p_b$, respectively. This decrease in pressure will increase the entropy of the mixture by the amount obtained by integration of the second *Tds* equation. This gives

$$(\Delta S)_a = -n_a \bar{R} \ln(p_a/p) = -n_a \bar{R} \ln(x_a) \text{ and}$$

$$(\Delta S)_b = -n_b \bar{R} \ln(p_b/p) = -n_b \bar{R} \ln(x_b)$$

$$\therefore \; \Delta S = (\Delta S)_a + (\Delta S)_b = -\sum n_i \bar{R} \ln(x_i) = -\bar{R} \sum n_i \ln(x_i)$$

or

$$\Delta \bar{s} = -\bar{R} \sum n_i \ln(x_i)/n$$

$$= -\bar{R} \sum x_i \ln(x_i)$$

Thus change in entropy depends on the mole fraction of the constituent gases, and does not depend on the nature of gas. This implies that if we mix the same gas at the same temperature, then there shall be no change in entropy during mixing of the gas. Or, change in entropy may be interpreted as a measure of similarity or dissimilarity among the gases mixed.

Further, writing the perfect gas equation for the individual component *i* and the mixture, and dividing the former equation with the latter results in

$$v_{fi} = V_i/V = n_i/n = x_i \tag{5.6}$$

Further, summing all the equations for individual gases in the mixture at mixture *p* and *T*, we get the Amagat–Leduc law, $V = \sum V_i$ That is, the mixture volume is a summation of the partial volume of constituent gases at mixture pressure and temperature.

### 5.1.2 Molecular Weight and Gas Constant of the Mixture

Relations for these gas constants may be derived in terms of mass or mole fraction. Writing the gas equation $p_i V = m_i R_i T$ for individual constituents occupying mixture volume *V* at temperature *T*, and then summing over all the gases, we get

$$\therefore \; \sum p_i V = \sum m_i R_i T, \text{ or } V \sum p_i = T \sum m_i R_i, \text{ or}$$

$$pV = T \sum m_i R_i \;\; (\text{since } p = \sum p_i)$$

But for the mixture, $pV = m\,R\,T$, where *R* is the gas constant for mixture bases on mass. Equating the two expressions for *pV*, we get

$$\therefore \; R = \sum m_{fi} R_i / m = \sum m_{fi} R_i \text{ and } M = \text{Molecular weight of mixture} = \bar{R}/R \quad (5.7)$$

Also, the mass of the mixture will be equal to the sum of the mass of constituent gases. This means, if $M$ is the molecular weight of the mixture,

$$nM = \sum n_i M_i \;\rightarrow\; M = \sum x_i M_i \text{ and } R = \bar{R}/M \quad (5.8)$$

Equation 5.7 is useful when the mass fraction is given, and Eq. 5.8 if the mole fraction is given. So, all the constants that we need in the first and second law analysis are known.

**Example 5.1** A mixture of 3 moles of helium, 4 moles of neon and 5 moles of argon is at 1 bar and 300 K. Calculate (a) volume, (b) mole fraction and partial pressure of gases, (c) change of entropy due to mixing.

***Solution*** $n = \sum n_i = 3 + 4 + 5 = 12$. Using the perfect gas equation $pV = n\bar{R}T$, we get

$V$ = mixture volume

$= 12 \times (8.3143 \times 10^3) \times 300/(1 \times 10^5)$

$= 29.93 \text{ m}^3$

(b) $x_i$ = mole fraction of gas $i = n_i/n$ and $p_i = x_i p$ (Eq. 5.3). This gives

$x_{He} = 3/12$, $x_{O2} = 4/12$ and $x_{CO2} = 5/12$

$p_{He} = 3/12$, $p_{O2} = 4/12$ and $p_{CO2} = 5/2$ bar, since $p = 1$ bar

(c) Change in entropy after mixing is due to change in pressure of gases from 1 bar to their respective partial pressure, and is given by

$$\Delta\bar{s} = -\bar{R}\sum x_i \ln x_i = 8.3143\,[3/12 \ln (3/12) + 4/12 \ln (4/12) + 5/12\,(5/12)]$$

$= 0.51$ kJ/kmol K

**Example 5.2** A mixture of gases has the following composition by mass: $N_2$ = 60%, $O_2$ = 30% and $CO_2$ = 10%. 1 kg of mixture at 1 bar and 40°C is compressed to 5 bar; the index of compression may be taken as 1.2. Find the work done, heat transfer and change in entropy during the process. Also, find the molecular weight, gas constant and specific heat at constant pressure for the gas. Take $c_p$ for constituent gases as 1.04, 0.924 and 0.85 kJ/kg K for nitrogen, oxygen and carbon-dioxide, respectively.

***Solution*** Since the mass fraction is given and the molecular weight of nitrogen, oxygen and carbon dioxide, respectively, is 28, 32 and 44, gas constant of the mixture is

$R = \sum m_{fi} R_i = \sum m_{fi} \overline{R} / M_1$

$= 8.3143\,(0.6/28 + 0.3/32 + 0.1/44)$

$= 0.275$ KJ/kg K

$M$ = Molecular weight of the mixture = $\overline{R}/R = 30.23$

$c_p = \sum m_{fi} c_{pi} = 0.6 \times 1.04 + 0.3 \times 0.924 + 0.1 \times 0.85$

$= 1.0486$ kJ/kg K

$c_v = \sum m_{fi} c_{vi} = c_p - R = 0.7726$ kJ/kg K

$T_2 = T_1\,(p_2/p_1)^{(n-1)/n} = (273 + 40)\,(5/1)^{0.2/1.2} = 408$ K

$W$ = work transfer = $m\,R\,(T_1 - T_2)/(n - 1)$

$= 1 \times 0.275 \times (313 - 408)/0.2 = -131.1$ kJ

$Q = W + \Delta U$ (nfee)

$= -131.1 + 1 \times 0.7726 \times (408 - 313)$

$= -57.7$ kJ

$\Delta s = c_p \ln (T_2/T_1) - R \ln (p_2/p_1)$

$= 1.0486 \ln (408/313) - 0.275 \ln (5)$

$= -0.1646$ kJ/kg K

A negative sign with work and heat transfer indicates that work is done on the system while the system is rejecting heat during compression. Also, entropy is decreasing during this process because heat is rejected; by no means does it amount to violation of the second law of thermodynamics.

**Example 5.3** $n_1$ moles of an ideal monoatomic gas at temperature $T_1$ and pressure $p$ are in one compartment of an insulated container. In an adjoining compartment, separated by an insulating partition, are $n_2$ moles of another ideal monoatomic gas at temperature $T_2$ and pressure $p$. When the partition is removed, find out the final temperature and pressure and also change in entropy if the two gases are (a) the same, and (b) different.

***Solution*** Gases in insulated container form the system, which is a closed system, since there is no mass flow across its boundary. Further, the container is insulated, and hence there is no heat transfer. Neglecting work done in removing the partition, nfee gives: $dU = 0$. If $T$ is the final temperature, then $dU = (n_1 + n_2)\,C_v\,T - C_v\,(n_1\,T_1 + n_2\,T_2) = 0$; here, it has been assumed that internal energy is zero at a reference temperature of 0, and specific heat at constant volume is the same for both the gases, as both are monoatomic. This gives on simplification

$T = (n_1\,T_1 + n_2\,T_2)/(n_1 + n_2)$

If $p'$ is the mixture pressure, then $V$ = mixture volume = $(n_1 + n_2)\ \bar{R}\ T/p'$. But $V = V_1 + V_2$, $V_1$ and $V_2$ are volume occupied by the gases before mixing. These are: $V_1 = n_1\ \bar{R}\ T_1/p$ and $V_2 = n_2\ \bar{R}\ T_2/p$. Substituting for $V_1$, $V_2$ and $T$, we get $p' = p$.

**Entropy change.** When both gases are the same, then change in entropy is due to change in temperature alone. This change, using Eq. 4.25, is given by

$$\Delta S = \bar{C}_p\ n_1 \ln T/T_1 + \bar{C}_p\ n_2 \ln T/T_2$$

$$= \bar{C}_p (n_1 + n_2)\ [x_1 \ln T/T_1 + x_2 \ln T/T_2)]$$

However, when the gases are different, then the two gases will exert partial pressure in the mixture, and thus there is change in pressure of each gas, which will cause additional increase in entropy

$$= -\bar{R} \sum n_i \ln x_i$$

$$= -\bar{R}\ (n_1 + n_2) \sum x_i \ln x_i$$

$$= -\bar{R}(n_1 + n_2)(x_1 \ln x_1 + x_2 \ln x_2)$$

which is positive, since the mole fraction is less than one and its log is negative.

### 5.1.3 Gravimetric and Volumetric Analysis

In industries and laboratories, volumetric analysis is obtained using gas chromatographs or Orsat apparatus, which give the volume fractions of the constituent gases. However, we quite often need the mass fraction of the individual gases, instead of their mole or volume fraction for analysis. This, therefore, requires the conversion of volume fraction to mass fraction. This can be done systematically in the following manner (noting that $m = n \times m$)

$$m_{fi} = m_i/\Sigma m_i = n_i M_i/\Sigma n_i M_i = x_i M_i/\Sigma x_i M_i$$

Equation of state for ith component at the mixture $p$ and $T$ is

$$pV_i = m_i R_i T = m_i \left(\bar{R}/M_i\right) T, \text{ and } V_{fi} = V_i/v$$

$$\rightarrow m_i = \left(pV/RT\right) v_{fi}\ M_i$$

$$\therefore\ m_{fi} \text{ is also} = \frac{m_i}{\Sigma m_i} = \frac{v_{fi}\ m_i}{\Sigma v_{fi}\ M_i}$$

so

$$m_{fi} = \frac{x_i\ M_i}{\Sigma x_i M_i} = \frac{v_{fi}\ M_i}{\Sigma v_{fi}\ M_i} \tag{5.9}$$

Similarly, we may obtain volume or mole fraction if mass fraction is given.

$$v_{fi} = x_i = n_i/n = (m_i/M_i)/\Sigma(m_i/M_i) = (m_{fi}/M_i)/\Sigma(m_{fi}/M_i) \quad (5.10)$$

Equations 5.9 or 5.10 are implemented in tabular form as explained in the following example.

**Example 5.4** Dry products of combustion of a coal have the following volumetric composition 11% $CO_2$, 3.5% $O_2$ and 83.5% $N_2$ at atmospheric pressure 1 bar and temperature 27°C. Calculate (a) the gravimetric analysis, (b) the gas constant for the mixture, (c) the equivalent molecular weight, (d) partial pressure of each constituent of, (e) of the mixture specific heat (f) molar specific heat at constant pressure using the results in (e). Take values of specific heat from the tables.

***Solution*** Gravimetric analysis is obtained by using Eq. 5.9 in a tabular way (Table 5.1); the procedure is self-explanatory and involves determination of the numerator and denominator in a tabular manner.

**Table 5.1** Gravimetric analysis from volumetric analysis

| Gas | $V_{fi}$ | $M_i$ | $V_{fi}M_i$ | $m_{fi} = V_{fi}M_i/\Sigma V_{fi}M_i$ | | $P_i$, bar | $m_{fi}/M_i$ | $c_{pi}$ | $m_{fi}c_{pi}$ |
|---|---|---|---|---|---|---|---|---|---|
| | 1 | 2 | 3 = 1 × 2 | 4 = 3/Σ3 | 5% | 6% | 7 = 4/2 | 8 | 9 = 4 × 8 |
| $CO_2$ | 11.0 | 44 | 484 | .1620 | 16.20 | 0.110 | 0.003682 | 0.846 | 0.1371 |
| $O_2$ | 3.5 | 32 | 112 | .374 | 3.74 | 0.035 | 0.001168 | 0.918 | 0.0333 |
| $N_2$ | 83.5 | 28 | 2394 | .8006 | 80.06 | 0.885 | 0.028593 | 1.039 | 0.8318 |
| Total | 100.0 | | 2990 | 1.000 | 100 | 1.03 | 0.03344 | | 1.0032 |

Assume 1 kg of the mixture that is $m$ = 1 kg

(a) From column 5, the gravimetric analysis is 16.20% $CO_2$, 3.74% $O_2$ and 80.06% $N_2$.

(b) From Eq. 5.7

$$R_{\text{mix}} = \Sigma m_{fi} R_i = \bar{R}\,\Sigma m_{fi}/M_i$$

$$= 8.3143 \times 0.03344 = 0.278 \text{ kJ/kg K}$$

(c) $M_{\text{mix}} = \bar{R}/R_{\text{mix}} = 8.3143/0.278 = 29.9$

(d) Partial pressure of constituent gases in bar is given in column 6.

(e) $c_{p,\text{mix}} = \Sigma m_{fi} c_{pi} = 1.0032$ kJ/kg K

(f) $C_{p,\text{mix}}$ = Molar specific heat at constant pressure = $M_{\text{mix}}\, c_{p,\text{mix}}$

= 30.0 kJ/kmol K

**Example 5.5** A mixture of 0.5 kg of air and 0.5 kg of $H_2O$ is heated at a temperature of 150°C; quality of $H_2O$ is 0.5. (a) Find out the work done if this system is subjected to an isothermal process increasing its volume to five times. (b) Also, find out the temperature to which the mixture should be raised in order to have saturated vapours at the end of the process.

***Solution*** Given: mass of air = $m_{a1}$ = 0.5, mass of wet steam = $m_{v1}$ = 0.5, $x_1$ = 0.5 and $t_1$ = 150°C.

System: Vessel containing mixture of water and vapour in equilibrium. It is a closed system as there is no mass flow across it.

Assumptions: (i) Water and vapour are thoroughly mixed with each other and air, and (ii) vapour behaves as a perfect gas.

From the wet steam table: At 150°C, $p_1$ = 4.758 bar, $v_{f1}$ = 0.001091,

$v_{g1}$ = 0.3928, $u_{f1}$ = 631.7, $u_{g1}$ = 2559.5, $s_{f1}$ = 1.842 and $s_{fg1}$ = 4.996.

$\therefore s_1 = 1.842 + 0.5 \times 4.996 = 4.34$ kJ/kg K and

$v_1 = 0.5 \times 0.3928 = 0.1964$ m³/kg

$V$ = vessel volume = $m_{v1} \times v_1$ = 0.0982 m³ = volume of air

Using the perfect gas equation of air, we get air pressure

$$p_{a1} = [(0.5 \times 0.287 \times 423)/0.0982]/100$$
$$= 6.16 \text{ bar}$$

and $p_1$ = initial mixture pressure = $p_1 + pa_1$

$$= 10.918 \text{ bar}$$

(a) Given: $V_2 = 5 \times 0.0982 = 0.4921$ m³

$$\Rightarrow v_2 = 0.491/0.5$$
$$= 0.982 \text{ m}^3/\text{kg}$$

Since $v_2$ is greater than $v_g$ at 150°C, steam is superheated at the end of the process. From the superheated steam table, we note that pressure of steam lies between 1.5 and 2 bar. At these pressures, internal energy is 2579.8 and 2576.9, volume is 1.285 and 0.9596, whereas entropy is 7.419 and 7.279. Linear interpolation yields

$$p_2 = 1.5 + (2 - 1.5) \times (0.982 - 1.285)/(0.9596 - 1.285) = 1.962 \text{ bar}$$

$$u_2 = 2579.8 + (2576.9 - 2579.8) \times (1.962 - 1.5)/0.5$$
$$= 2577.12 \text{ kJ/kg, and}$$

$$s_2 = 7.419 + (7.279 - 7.419) \times (1.962 - 1.5)/0.5$$
$$= 7.2896 \text{ kJ/kg K}$$

$$\Delta S_v = \text{change in entropy of vapour} = 0.5\ (7.2896 - 4.34)$$
$$= 1.4748$$

$\Delta S_a$ = change in entropy of air (using Eq. 4.25) = $m_a R \ln(V_2/V_1)$

= 0.5 × 0.287 ln (5) = 0.231

and $\Delta S$ = change in entropy of mixture = $\Delta S_a + \Delta S_v$

= 1.7058 kJ/K

We assume that heat transfer is reversible at constant temperature, the amount of heat transfer

$Q = 423 \times \Delta S = 721.53$ kJ

Discussion: We observe that pressure decreases in the process. This can be verified from the T-s diagram, as shown in Fig. 5.1. It can be concluded from this diagram that isothermal heating is possible at constant pressure till steam is wet, and after that, only if pressure is decreased. During this process, volume is increasing, as given in the problem.

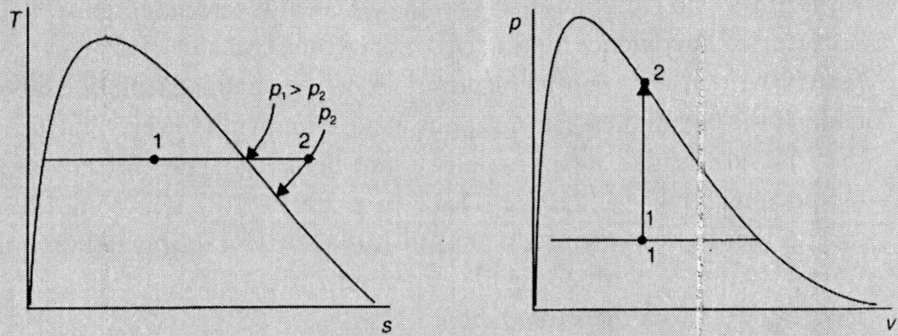


**Fig. 5.1** Wet steam subjected to (a) T = c (b) v = c process

(b) We are given that water is converted completely into vapour at the end of the process. This is possible by heating at constant pressure till the liquid is converted into vapour. The amount of heat required for the process can be obtained in the same way as we did in part (a) since in the final state, vapours are saturated.

Another way of doing it is to assume that the vessel is rigid. In that case, heating would occur at constant volume. Therefore, $v_2$ = 0.0982/0.5 = 0.19664 m$^3$/kg. From the steam table, pressure of saturated steam will lie between 9.8 and 10 bar. The pressure of steam by linear interpolation is 9.8 + 0.2 × (0.19664 – 0.198)/(0.194 – 0.198) = 9.858 bar. Similarly, $s_2$ = 6;5853, $u_2$ = 2583.25 and $t_2$ = 179.26°C. Therefore, temperature should be raised to 179.26°C.

## 5.2 MIXTURE OF AIR AND VAPOUR—PSYCHROMETRY

Water is always present in atmospheric air at a low pressure, such that it exists as superheated vapour at atmospheric temperature and can be treated as a perfect gas for thermodynamic analysis. In fact, partial pressure of vapour in air is so small, that even saturated vapour may be considered as a perfect gas with an error of less than 0.2 per cent. Therefore, any such mixture can be handled as a mixture of perfect gases.

The air–vapour mixture is studied under the umbrella of psychrometrics, wherein mathematical relations, psychometric tables and psychrometric charts have been developed for different thermodynamic properties of atmospheric air. For this purpose, air is treated as a mixture of dry air (made from different gases like oxygen and nitrogen) and water vapour. Whereas the composition of dry air does not change, the amount of moisture changes from season to season due to variation in evaporation of water from different water sources or its condensation from air. To differentiate atmospheric air from dry air, the term moist air is used for it, but properties of moist air are given for *one kg of dry air* with enthalpy and internal energy for dry air taken as zero at 0°C; this reference is chosen because it is the reference for water also. Therefore, enthalpy of dry air at any temperature $t$ can be written as $c_p t$. Enthalpy of vapour can be obtained from the superheated steam table, if the partial pressure of vapour in air is known. However, the pressure is so low, that the superheated steam table may not be useful. The enthalpy of vapour at any temperature is taken as $2501.3 + 1.82 \times t \cong 2501 + 1.82 \times t$, where 1.82 kJ/kg K is the average specific heat at constant pressure for vapours for the operating temperature range from −10 to 50°C with negligible error, and 2501.3 kJ/kg is the enthalpy of saturated steam at 0°C (from the steam table).

### 5.2.1 Terms Used in Air Conditioning

**Humidity ratio.** Moisture present in air may be expressed as its absolute amount per kg of dry air or per unit volume. It is called the *humidity ratio* when expressed on kg basis of dry air and is denoted by $\omega$. (*Note that it is not the mass fraction of vapour in the mixture, since the ratio is made with mass of dry air and not that of the mixture.*) Using the perfect gas equation for both vapour and dry air, we may express $\omega$ in terms of partial pressure $p_v$ of vapour and total pressure $p$ of moist air, as follows:

$$\omega = \frac{m_v}{m_a} = \frac{p_v V/(R_v T)}{p_a V/(R_a T)} = \frac{p_v / R_v}{p_a / R_a} = 0.622 \frac{p_v}{p_a} = 0.622 \frac{p_v}{p - p_v} \text{ kg}_\text{v}/\text{kg}_\text{a} \tag{5.11}$$

where, subscripts $v$ and $a$ refer to vapour and dry air, $V$ and $T (= 273 + t)$ are volume and temperature of moist air at mixture pressure $p$, $R_a$ is the gas constant for dry air based on mass basis, $R_v$ is the gas constant for vapour and $p = p_a + p_v$, by Dalton's law of partial pressure.

The amount of moisture present in air, expressed on per unit volume basis, at pressure and temperature of moist air is called *absolute humidity*. This is used for defining relative humidity of air, discussed later.

**Relative humidity and degree of saturation.** Air at atmospheric pressure has limited capacity to hold moisture, and if, for any reason, its moisture content increases beyond this capacity, then it separates out in the form of drops. Or, in other words, if water is added to moist air, then moisture in the air increases up to a certain limit, after which no more water can be added to air. Air at this condition is called saturated[2] air. During moisture addition, temperature of air would not change, and hence the moisture state on the T-s diagram for $H_2O$ would move along the horizontal line 1g towards the saturated vapour line from initial point 1, as shown in Fig. 5.2; here, point 1 is shown on the constant pressure line for partial pressure $p_v$ of moisture in the air. Addition of water at constant temperature increases the partial pressure of moisture in air to pressure $p_g$, as is seen from the figure, corresponding to point $g$ on the saturation curve, and can be determined from the steam table as the temperature is known. This implies that in the actual situation, air is partially saturated. The degree of saturation of air is expressed by the terms relative humidity and degree of saturation, as defined below

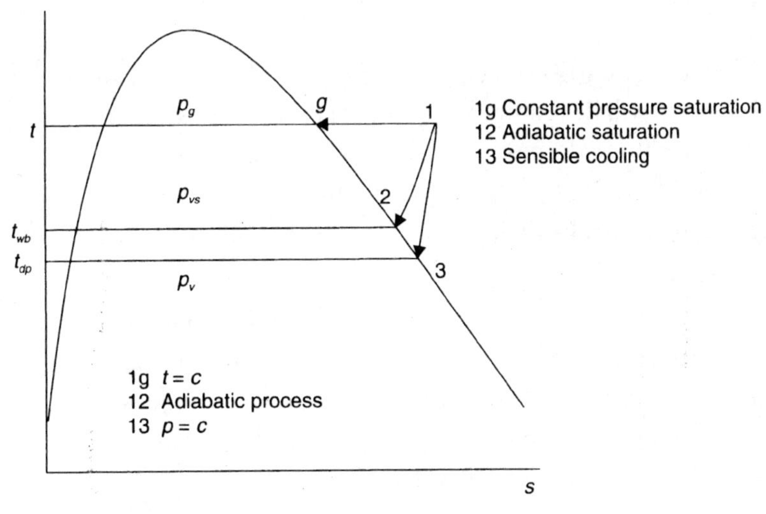


**Fig. 5.2** Air saturation

*Relative humidity* is defined as the ratio of absolute humidity of air to that of air saturated at mixture temperature $T$ and volume $V$. It is denoted as $\phi$. Using perfect gas equation, we may write

2. Here, saturated refers to a solution and not to the states of a pure substance lying on phase boundaries.

$$\varphi = \frac{m_v}{m_g} = \frac{p_v V/(R_v T)}{p_g V/(R_v T)} = \frac{p_v}{p_g} \tag{5.12}$$

Relative humidity is, therefore, also the ratio of partial pressure of water vapour to its pressure in saturated air, when saturated at moist air temperature. Since the ratio of partial pressure is also mole fraction, $\phi$ *may be thought of as the ratio of mole fraction of vapour in moist air to that in saturated air.*

On the other hand, the *degree of saturation* is defined on the basis of the ratio of the mass of vapour in air to that in saturated air when measured on the basis of unit kg of dry air, that is, on the basis of humidity ratio in contrast to relative humidity that is based on absolute humidity. Thus, the degree of saturation is the ratio of the humidity ratio of air at state 1 and $g$ in Fig. 5.2. It is represented as $\mu$ . Or,

$$\mu = \frac{\omega_1}{\omega_g} = \frac{0.622\left(\dfrac{p_v}{p - p_v}\right)}{0.622\left(\dfrac{p_g}{p - p_g}\right)} = \frac{p_v}{p_g}\left[\frac{p - p_g}{p - p_v}\right] = \varphi\frac{1 - p_g/p}{1 - \varphi\, p_g/p} \tag{5.13}$$

Another interesting observation made from Fig. 5.2 is that if air temperature increases, then more water will be required to be added to make air saturated, since point 1 moves away from the saturated vapour line. That is, during summer, more water will evaporate from a wet surface as compared to the amount in winter, since air temperature is high.

**Enthalpy of moist air.** Enthalpy is obtained using the Dalton–Gibbs law, and is

$$h = 1.005\, t + \omega\ (2501 + 1.82t) \tag{5.14}$$

where, $t$ and $\omega$ are dry bulb temperature and humidity ratio of the moist air.

**Sensible heat factor.** Heat may be added to air as sensible and/or latent heat. *Sensible heat factor* is the ratio of sensible heat to the total heat added to air, and is a very useful factor in designing of air conditioning systems. If $Q_S$ and $Q_L$ are the rate of sensible and latent heat addition to air in a process, then

$$\text{SHF} = Q_S/(Q_S + Q_L) \tag{5.15}$$

**Dry bulb, wet bulb, and dew point temperature.** Consider water being collected in an insulated container through which air flows over the water surface, as shown in Fig. 5.3. The container is assumed to be long enough, so that air leaving it is saturated. When air moves over the water surface, water molecules leave the water surface and join it as vapour; the heat of vaporisation partly comes from sensible heat transfer from air to water, and the rest from the water itself. Thus, the temperature of water decreases, which in turn increases heat transfer from air. At some stage, the whole heat of vaporisation will come from air and a steady state is then reached. Under this condition, water in the container will attain a constant temperature, which is the same as that of air at the exit from the container. Vapour added to the air carries with it the enthalpy of vapour at water temperature, in which the enthalpy of vaporisation comes from air under steady state. Therefore, the process of saturation increases the enthalpy of air by amount of enthalpy of saturated water being added in the process. Since humidity ratio of air at exit and enthalpy of saturated water are less, there is insignificant change in enthalpy during the process and it is even considered as isenthalpic process for all practical purposes without introducing any significant error.

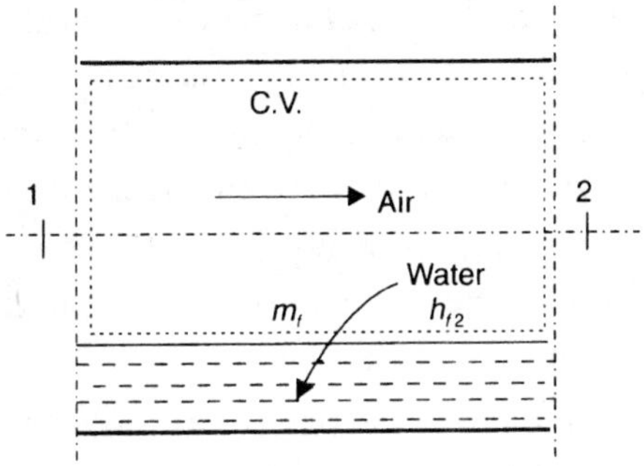


**Fig. 5.3** Adiabatic saturation

This process of saturation of air is adiabatic saturation and is shown by curve 12 in Fig. 5.2; temperature $t_{wb}$ at point 2 is called *wet bulb temperature*. This temperature is less than temperature $t_1$ at point 1, which is measured by any thermometer and is known as *dry bulb temperature*; the names dry or wet indicate that temperature is measured by a thermometer with its bulb dry or wet, respectively. In practice, wet bulb temperature is measured by a Sling thermometer, which is a thermometer whose bulb is masked with a wick dipped in water so that the wick is always wet. The thermometer is rotated till the temperature reading is constant. This temperature is the temperature of water attained after saturation of air surrounding the bulb, and hence it is a wet bulb temperature. However, in open space, there may be some heat transfer from other sources present around the thermometer bulb, and therefore, care should be taken to ensure their absence from the place of measurement. In literature, steady state temperature attained by water in Fig. 5.3 is referred to as thermodynamic wet bulb temperature, when air is completely saturated adiabatically, to differentiate it from the temperature measured by a Sling thermometer. It is represented as $t^*$.

Representing the inlet and outlet sections in Fig. 5.3 by subscripts 1 and 2, the mass and energy balance, when applied to the system shown in the figure by dotted lines gives:

$m_{a1} = m_{a2} = m_a$, because mass flow of dry air remains constant.

$m_a\omega_1 + m_f = m_a\omega_2$, mass of water added to air from water in the container is $m_f$.

We learnt above that with the addition of water during saturation of air, air enthalpy is increased by the enthalpy of saturated water only.

$$m_a h_1 + m_f h_{f2} = m_a h_2, \text{ since flow is steady, } Q = W = 0 \text{ and } d\,(\text{ke}) = d\,(\text{pe}) = 0.$$

or, $$m_a h_1 + m_a(\omega_2 - \omega_1)h_{f2} = m_a h_2 \;\rightarrow\; h_1 + (\omega_2 - \omega_1)h_{f2} = h_2$$

But for air, $h = h_a + \omega h_v = c_p t + \omega h_v$ per kg of dry air. This gives

$$\left(c_p t_1 + \omega_1 h_{v1}\right) + (\omega_2 - \omega_1)\, h_{f2} = \left(c_p t_2 + \omega_2 h_{g2}\right)$$

or $$\omega_1 = [c_p(t_2 - t_1) + \omega_2 h_{fg2}]/[h_{v1} - h_{f2}] \tag{5.16}$$

Since dry bulb and wet bulb temperatures can be measured at the inlet and outlet of the container, we read $h_{f2}$, $h_{fg2}$ and $p_{v2}$ at temperature $t_{w2}$ (= $t_2$) from steam table. Equation 5.11 gives $\omega_2$ and then Eq. 5.16 gives $\omega_1$ since all the other quantities in the equation are known. Once $\omega_1$ is known, vapour pressure in air at the inlet can be obtained using Eq. 5.11 and relative humidity from Eq. 5.12. However, this procedure is very time consuming and frustrating. The problem is circumvented by the use of psychrometric charts discussed in Section 5.2.2.

The main problem is to know the partial pressure of vapour $p_v$ in actual condition. Many empirical relations relating vapour pressure with wet bulb and dry bulb temperature were developed. These are

Apjohn equation: $p_v = p_{vs} - 1.8\, p(t - t_{wb})/2700$

Ferrel equation: $p_v = p_{vs} - 0.00066\, p(t - t_{wb})\, \{1 + 1.8\,(273 + t)/1571\}$

Carrier equation: $p_v = p_{vs} - (p - p_{vs})(t - t_{wb})/\{1547 - 1.44\,(273 + t_{wb})\}$

where, $p_{vs}$ is the pressure of vapours in air saturated at wet bulb temperature. The carrier equation is often used.

We may also obtain relative humidity by passing air over a cold surface, so that air cools at constant vapour pressure along path 13 in Fig. 5.2. At a certain temperature, moisture in air will start condensing. *This temperature is known as the dew point temperature, $t_{dp}$*. From the steam table, vapour pressure at $t_{dp}$ can be read and then relative humidity is obtained from Eq. 5.12. It can be seen that $t > t_{wb} > t_{dp}$. However, for saturated air having $\phi = 1$, all the three temperatures will be equal, that is, $t = t_{wb} = t_{dp}$.

### 5.2.2 Psychrometric Chart

We shall determine air properties using equations later in Example 5.6 and shall experience that this procedure is very tiring and time consuming. This problem aggravates when these calculations are to be done again and again while designing an air conditioning system consisting of a number of components. Therefore, the best way will be to calculate these values for different conditions once and for all, and then either tabulate or plot the results so obtained. The two-dimensional chart used in air conditioning systems is called the *psychrometric chart*, drawn at different atmospheric pressures because humidity ratio, wet bulb temperature, etc. depend upon atmospheric pressure. Psychrometric chart in common use is, however, drawn at atmospheric pressure of 1.01325 bar. For small change in barometric pressure, corrections can be used, which are usually given in the chart for those properties which are atmospheric pressure dependent.

A psychrometric chart is a plot of enthalpy $h$ versus humidity ratio $\omega$. It is different from other $X$-$Y$ plots in the sense that the two axes $X$ and $Y$ are taken perpendicular to each other. But, in this chart, these two axes are oblique in which the constant enthalpy line is inclined at an angle $\beta$ with negative $X$-axis and $\omega$ (is along ordinate $Y$ of $X$-$Y$ plot, as drawn by Goodman and shown in Fig. 5.4 (a). We will skip the details of the derivation of the equation used for drawing different lines at their constant values. Applying the law of triangles to the triangle 123 in Fig. 5.4 (a), Goodman derived the relation: $\cot\theta + \cot\beta = L_h/L_w$ where $L_h$ and $L_w$ are shown in the figure.

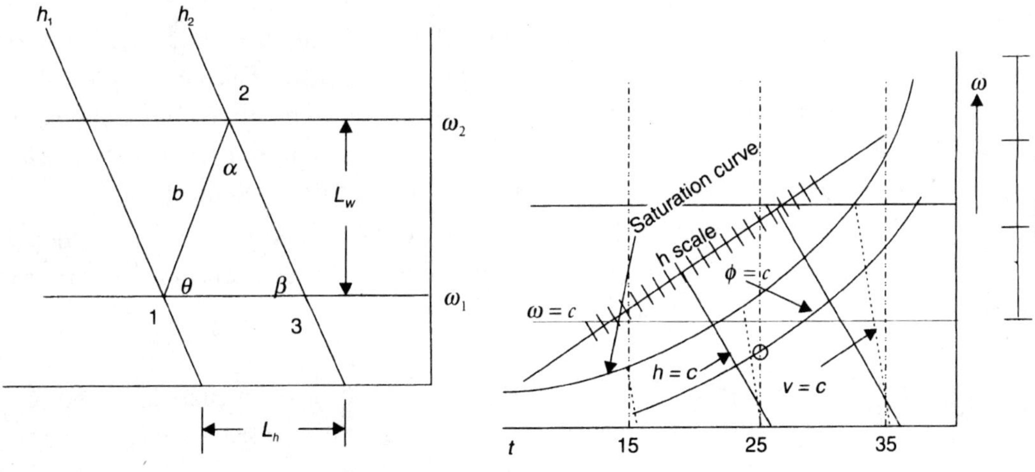


(a) Development of chart (b) Actual chart

**Fig. 5.4** Psychrometric chart

The psychrometric chart is shown in Fig. 5.4 (b). From its appearance, it looks as if this chart is a graph between the humidity ratio $\omega$ along the $Y$-axes and dry bulb temperature[3] $t$ along the negative $X$-axes. The chart also carries the constant enthalpy, volume and constant relative humidity lines. It also shows the saturation line for which $\phi = 1$. But constant wbt and dpt lines are not drawn. A little peep into the saturation

3. A look at the psychrometric chart shows that dbt lines are vertical. However, it is not so. For a constant dbt line on the chart, $\theta = 90$. This is derived from Goodman equation $\tan\beta = L_w/L_h = S/h_{g,t}$. Here, $S$ is the ratio of scale used to draw the chart for $\omega$ and $h$; $h_{g,t}$ is enthalpy of vapour at its dbt. Since $h_{g,t}$ depends on $t$, it is clear that all the constant dbt lines cannot be parallel; only one of them is taken vertical.

mechanism to define two temperatures will tell us the procedure. For example, consider point 1 on chart represents state of air, and we have to find out wbt and dpt at this state. Dew point temperature is obtained by sensible cooling of air; a horizontal line through this point represents the process. This line meets the saturation curve, at which point dbt of air is also the dew point temperature, since vapour pressure is not changing during the process.

Similarly, for adiabatic saturation, water is added to air at its wet bulb temperature. During this process, enthalpy of air increases, but this change is negligibly small. Therefore, a constant $h$ line passing through point 1 is also a constant wet bulb temperature line. (In more accurate charts, constant thermodynamic wet bulb temperature lines are also drawn.) This line intersects the saturation curve at some point; dbt at this point is then wbt of air at point 1, because for saturated air, the two temperatures will be the same. We may also see that if a point lies on the saturation curve, then for air at this state, $t = t_{wb} = t_{dp}$.

The chart also has a dot at the intersection of the 25°C dbt line with the 50% $\phi$ line and a sensible heat factor scale placed along the $\omega$-axis, such that the line joining any point on the sensible heat factor scale with this dot gives the direction of the process along which the condition of the cold air added to the air conditioned space, where heat comes from different sources in sensible and latent heat forms, would change.

Knowing any two properties of air, its state can be located on the chart. Once the state point 1 on the chart is known, the remaining properties of air may be read from chart.

**Example 5.6** Show the state of atmospheric air on psychrometric chart and read all those properties which can be read from this chart when air is at (a) $t$ = 32°C and $\phi$ = 0.60, (b) $t$ = 32°C and dew point temperature = 10°C, and (c) $t$ = 32 °C and $h$ = 50 kJ/kga. Also, compare these values with those obtained from mathematical relations. Assume barometric pressure is standard atmospheric pressure, which is 1.01325 bar.

***Solution*** Given: Moist air at different conditions.

(a) Dry bulb temperature and relative humidity of air is given. Therefore, the state point will lie on the intersection of vertical line through 32°C on the horizontal axis with 0.60 relative humidity line. We read from psychrometric chart: $\omega$ = 0.018 kg/kga, $h$ = 78.5 kJ/kga, $t_{wb}$ = 23.5°C, $t_{dp}$ = 23.5°C and $v$ = 0.8535 m$^3$/kg.

From equations. We note $\phi$ and $t$ are given. From the steam table, saturation pressure at 32°C, $p_g$ = 0.04759 bar. Therefore, Eqs. 5.12 and 5.11 give,

$p_v = \phi \times p_g = 0.028554$ bar

$\omega = 0.622 \times 0.028554/(1.01325 - 0.028554) = 0.018037$

(same as from chart)

Equation 5.14 gives

$h = 1.005 \times 32 + 0.018037\ (2501 + 1.82 \times 32) = 77.16$

(difference = (78.5 – 77.16)/77.16 = 1.7%)

From the steam table, saturation temperature at 0.028554 bar is somewhere between 23 and 24°C and is 23.74°C by linear interpolation. Therefore, the dew point temperature is 23.74°C.

Enthalpy may also be obtained from the equation, $h = 1.005\ t + \omega h_v$ where $h_v$ is obtained from the superheated steam table. Students may try it and see how difficult is the process. In fact, we prefer to use the chart only unless a high degree of accuracy is required. We observe that in this case, the error is not much.

(b) We are given $t$ = 32°C and $t_{dp}$ = 10°C. To locate point on the chart, find first point of intersection of 10°C dbt line with the saturation curve. Pass the horizontal line through this point, which meets the vertical line for 32 °C dbt at point 1. Properties now can be read from chart, which are $h$ = 52 kJ/kga, $\omega$ = 0.0076 kg/kga, $v$ = 0.875 m$^3$/kg and $\phi$ = 0.26.

From equation: Since $t_{dp}$ is given, we may obtain the vapour pressure from the steam table. This gives, $p_v$ = 0.0123 bar. Using Eq. 5.11, we get the humidity ratio,

$\omega = 0.622 \times 0.0123/(1.01325 - 0.0123) = 0.00764$ kgv/kga.

(From chart, $\omega$ = .0076)

Again, from the steam table, vapour pressure at 32°C is

$p_g$ = 0.0476 bar. Using Eq. 5.12, we get

$\phi = p_v/p_g = 0.0123/0.0476 = 25.84\%$ (From chart $\phi$ = 26%)

Specific enthalpy is obtained from Eq. 5.14. This gives

$h = 1.005 \times 32 + 0.00764(2501 + 1.82 \times 32)$

$= 51.71$

(from chart 52 kJ/kga)

Specific volume is obtained using the perfect gas equation, taking

$R = 8.3143/18 = 0.4619$ kJ/kg K. This gives

$v = \omega RT / p_v$

$= 0.00764 \times (0.4619 \times 10^3)(273 + 32)/(0.0123 \times 10^5)$

$= 0.875\ m^3/kga$

Comparison of values from the equations and chart shows that the two are very close to each other; any deviation may also be due to error made in reading the values from the chart as well.

(c) We are given $t = 32$ and $h = 50$. This point will lie on intersection of $h = 50$ line with $t = 32$ line. From the chart, we read $\omega = 0.0066$ kg/kga, $\phi = 0.20$, and $v = 0.875\ m^3/kg$. To find the dew point temperature, draw a horizontal line passing through the point meeting the saturation curve; read dbt at this point $\rightarrow t_{dp} = 7.3°C$. Similarly, draw a constant $h$ line through this point, and find dbt at the point of its intersection with saturation line $\rightarrow t_{wb} = 17.7°C$.

From equation: Since $h$ and $t$ are known, Eq. 5.14 is used to determine the humidity ratio. This gives:

$$\omega = (50 - 1.005 \times 32)/(2501 + 1.82 \times 32) = 0.00697$$

(from the chart 0.0066)

Vapour pressure is obtained from Eq. 5.11 to give

$$0.00697 \times (1.01325 - p_v) = 0.622\, p_v \quad \text{or} \quad p_v = 0.0112284 \text{ bar}$$

From the steam table, $p_g$ at 32°C = 0.0476 bar. Therefore,

$$\phi = 0.0112284/0.0476 = 23.6\%$$

We may also determine the saturation temperature at $p_v$ to give $t_{dp}$. In this question, values from the chart and equation are different. This is mainly because we have made an error in locating the point on the chart, since line for $h = 50$ does not exist on it and has to be drawn by linear interpolation.

### 5.2.3 Psychrometric Processes

The purpose of air conditioning is to control four parameters, namely dry bulb temperature, relative humidity, pollution level in the air and velocity of flow of air as per the needs of the application. For this, the condition of air is changed in the air conditioner and added to the space where air conditioning is required. The amount of heat and moisture added to the space from outside, or the occupants of the space, is therefore required to be removed to maintain steady conditions. For this, air is withdrawn from the space and passed to the air conditioner. A certain percentage of air from it is replaced by fresh air to maintain the pollutant level in the condition space. Any fresh air added will increase the heating load on the conditioner. Therefore, a minimum percentage of fresh air is taken to keep pollution at an acceptable level, while requiring least conditioning. This percentage depends on the level of pollution, and in practice, it varies from 20 to 25 per cent of air withdrawn from the space for conditioning in the air conditioner. (In case pollution is due to some activity, taking place in the space, simply replacement of air may not be sufficient. In that case, methods suitable to remove that pollutant may be used.) The temperature and humidity ratio of this mixed air is then modified to suitable values and sent to the conditioned space, so that the supply air maintains the desired temperature and humidity in the conditioned space.

Once temperature, relative humidity and pollution level in the conditioned space has been fixed as per the application of the air conditioning system and outside air conditions[4] are known, then a designer can combine any number and any of the following processes to achieve the goal—heating or cooling, humidification and heating, adiabatic heating and cooling, humidification or dehumidification, dehumidification and cooling, and adiabatic mixing of air streams.

The designer has to also calculate sensible and latent heat to determine the sensible heat factor, so that he knows in what direction the air condition would move once the conditioned air is supplied to space. He has to also ensure that air is supplied to space, so that it is uniformly distributed and the condition of air around the space is almost uniform, and that minimum heat and pressure loss occurs while air is transported from the air conditioning unit to the space.

We use different equipment for implementing the above processes. For example, a coil is used through which cold or hot fluid (steam) flows for cooling or heating purposes. Liquid is sprayed in a chamber from a nozzle through which air flows for adiabatic cooling or heating, water is used for humidification and a chemical for dehumidification. We shall analyse all the processes in the following paragraphs. Air conditioning equipment are steady flow open systems, and hence sfee is a suitable tool for energy analysis, wherein change in kinetic or potential energy is normally negligible.

### 5.2.3.1 Heating and Cooling Processes

Air can be heated or cooled by passing it over a coil carrying hot or cold fluid, shown in Fig. 5.5 (a), where point 1 represents the state of air at the entrance. The humidity ratio will remain constant, since no moisture is added in the process. The process is shown in Fig. 5.5 (b); path 12 represents sensible cooling, since temperature is decreasing and 12'

4. An air conditioning system is required round the year and the outside air conditions vary year to year. Therefore, the designer has to make the right choice; his experience in the area and local requirements guide him in this.

sensible heating. However, while cooling, it is possible that condensation of moisture may start if cooled beyond the dew point temperature. In that case, the humidity ratio will decrease. Therefore, for mathematical analysis, we shall take a general case in which there is moisture removal also. We take air inside rectangle of Fig. 5.5(a) as system and let $\pm Q$ kJ heat as added to (+ sign) or extracted (–ve sign) from air while air is moving over the coil then mass and energy balance gives:

$m_{a1} = m_{a2} = m_a$ mass flow of dry air remains constant.

$m_a\omega_1 - m_w = m_a\omega_2$ mass of water removed from air is $m_w$; it may be zero also.

and $m_a h_1 \pm Q = m_a h_2$ since work done = 0

or
$$\pm Q = m_a (h_2 - h_1) \text{ and } m_w = m_a (\omega_1 - \omega_2) \tag{5.17}$$

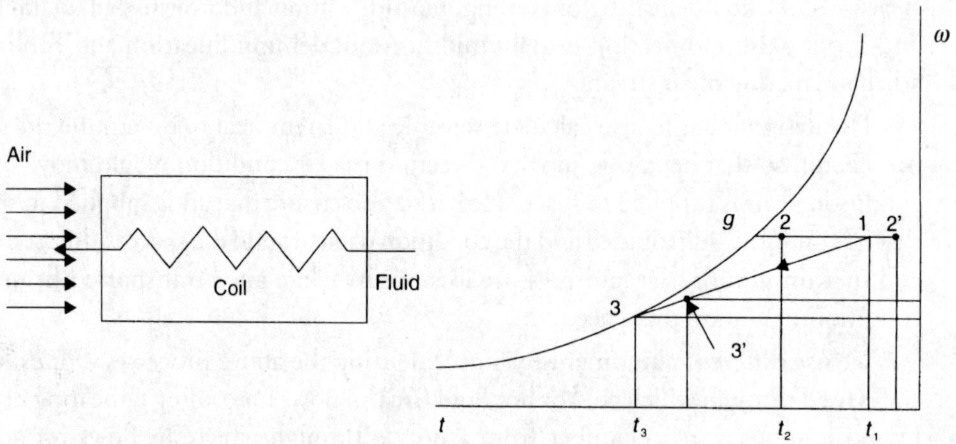


(a) Cooling coil (b) Cooling process on chart

**Fig. 5.5** Sensible heating and cooling

Sensible cooling continues till the constant $\omega$ line meets the saturation curve at $g$, after which moisture will start condensing and the condition of air then moves along the saturation curve, such that both $\omega$ and $t$ decrease. Air can cool theoretically only up to the temperature of the cooling coil, which is known as *apparatus dew point* (ADP) temperature. This point is shown as point 3 in the figure. The process of *cooling and dehumidification* of air is commonly used for removing moisture from air during the rainy season, and in industries where dry air is required for instrumentation, pneumatic control equipment, and many other applications.

The combined cooling and dehumidification process involves both heat and mass transfer. When warm air comes in contact with the cold coil, its temperature decreases. Moreover, vapour pressure in air will be more than that at ADP. This vapour pressure difference will cause diffusion of moisture from air to the coil, and the humidity ratio of air would decrease. Thus, in this process, both temperature and moisture decrease as more and more air comes in contact, even if air temperature is more than dpt. These two rates are such that the path of the state followed by air is actually not along 1g3, but it is along the straight line joining the point 1 with the point 3 on the saturation curve at ADP. This behaviour is observed when the Lewis number[5] is unity. Moreover, air coming out of the cooling coil will in fact be at 3' instead of 3, since the time available for cooling is not sufficient, and some of the air does not even come in contact with the cooling coil surface. This fact is expressed by the term *bye pass* factor, defined as follows

$$\text{B.F.} = \text{Length } 3'3/\text{Length } 13 = (h_{3'} - h_3)/(h_1 - h_3) = (\omega_{3'} - \omega_3)/(\omega_1 - \omega_3) \quad (5.18)$$

**Example 5.7** Moist air enters a refrigeration coil at 30°C dbt and 60% $\phi$ at the rate of 200 m$^3$/min at standard atmospheric pressure. It leaves the coil at 25°C. Calculate the tons of refrigeration required and the final $\phi$.

***Solution*** We are given: $t_1 = 30$ and $\phi_1 = 0.6$. A refrigeration coil is a unit in which a cooling coil is placed and air entering from one side flows over the coil and comes out from other side. The refrigeration coil is the system, which is a steady flow open system. Air state at inlet can be located on the psychrometric chart, since $t$ and $\phi$ are given. From the psychrometric chart, we note that the dew point temperature of air at point 1 (see Fig. 5.5b) is 21.2°C. Therefore, process 12 is a constant $\omega$ process, as $t > t_{dp}$. Since $t_2 = 25$°C is given, and the process is constant $\omega$ the final state 2 of air can now be located on the chart. Reading from the chart, we get $h_1 = 72.0$, $h_2 = 66.2$ kJ/kga and $v_1 = 0.88125$ m$^3$/kga. This gives

$$m_{a1} = 200/0.88125 = 226.95 \text{ kga/min}$$

and $Q$ = heat removed in the cooling coil = $m_{a1}\,(h_1 - h_2)$

$$= 1316.31 \text{ kJ/min} = 1316.31/210$$

$$= 6.268 \text{ tonne}$$

Since 1 tonne = 210 kJ/min (conversion factor)

### 5.2.3.2 Combined Humidification and Heating

In winter, not only heating, but humidification is also required. It is generally obtained by spraying steam vapour into air. Or, alternately, air may first be heated sensibly by passing air over a heating coil, and then humidified by spraying water into it. (It is also possible to humidify and then heat it.) Figure 5.6 (a) shows the schematic arrangement.

---

5. Lewis number is a non-dimensional group $h_c/(h_D c_{pa})$, where $h_c$ and $h_D$ are heat and mass transfer coefficients, and $c_{pa}$ is the specific heat of dry air at constant pressure.

Process 12 in the heating coil and 23 in the humidifier are shown on the psychrometric chart (Fig. 5.6 (b)). Joining points 1 and 3 give the direction of the process in the unit. If under steady state, $h_w$ is the specific enthalpy of water added to air, $m_w$ and $Q$ kJ the amount of water and heat added in the process per unit time, then mass and energy balance gives:

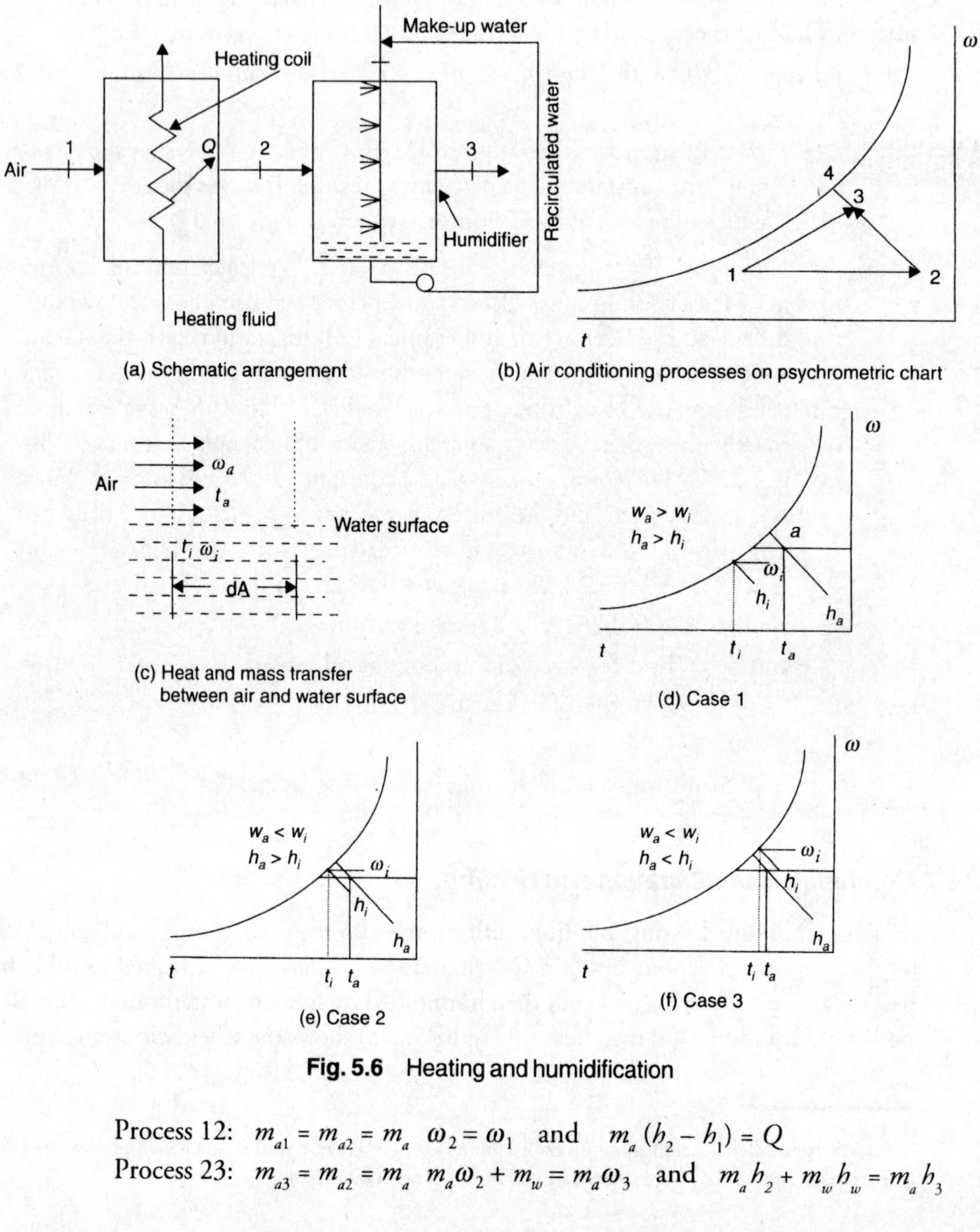


**Fig. 5.6** Heating and humidification

Process 12: $m_{a1} = m_{a2} = m_a \quad \omega_2 = \omega_1$ and $m_a(h_2 - h_1) = Q$

Process 23: $m_{a3} = m_{a2} = m_a \quad m_a\omega_2 + m_w = m_a\omega_3$ and $m_a h_2 + m_w h_w = m_a h_3$

These equations give $Q = m_a (h_3 - h_1) - m_w h_w$ and

$$(h_3 - h_1)/(\omega_3 - \omega_1) = Q/[m(\omega_3 - \omega_1)] + h_w \tag{5.19}$$

$(h_3 - h_1)/(\omega_3 - \omega_1)$ is the slope of the process line 13 with the vertical line through point 1, and should be positive for heating of air.

We may explain the process of sensible heating and mass transfer with the help of enthalpy difference. We know heat transfer is caused by temperature difference, whereas moisture transfer is due to vapour pressure difference. A combined heating and humidification process therefore involves both heat and mass transfer. Stoecker and Jones have derived the following relations by considering sensible and latent heat transfer (associated with vaporisation of vapours transferred) from the water surface to air flowing over the water surface, shown in Fig. 5.6 (c)

$$dq_s = \text{sensible heat transfer} = h_c\, dA\, (t_i - ta)$$

$$dq_L = \text{latent heat transfer} = h_D\, dA\, (\omega_i - \omega_a)$$

$$dq_t = \text{total heat transfer} = h_c\, dA\, (h_i - h_a)/c_{pm} \tag{5.20}$$

where, subscript *i* and *a* refer to water and air (at inlet) as shown in the figure, $h_c$ is the heat transfer coefficient between water and air, *dA* is the surface area from which heat transfer is taking place; other symbols have the usual meaning. So, the combined process of heat and mass transfer is due to some potential difference like any other thermodynamic process. This potential is enthalpy.

Whether air will be heated and humidified, or some other combination may occur depends on the difference of two enthalpies. This can be observed from figures *d*, *e* and *f* of Fig. 5.6, which show respectively, three situations in which the conditions are case 1, $\omega_a > \omega_i$ and $h_a > h_i$; case 2, $\omega_a < \omega_i$ but $h_a > h_i$; and case 3, $\omega_a < \omega_i$ and $h_a < h_i$. In all these situations, $t_a > t_i$. We may verify the following conclusions from the above heat transfer equations.

Case 1: Sensible, latent and total heat is from air to water.

Case 2: Sensible and total heat is from air to water, but latent heat transfer is from water to air.

Case 3: Sensible heat is from air to water, but latent and total heat transfer is from water to air.

At first sight in case 3, it looks as if heat flow is from lower temperature to higher temperature, contradicting the second law of thermodynamics. But it is not so, because this is the result of heat transfer associated with mass transfer also. If temperature of water is also more than that of air, then air will be heated as well as humidified. This suggests that steam may be injected into air in place of two different units for heating and humidification of air. If steam injection is used such that *it is saturated at dry bulb temperature of air* having enthalpy $h_g$, then making mass and energy balance, it gives after simplification.

$$(h_3 - h_1)/(\omega_3 - \omega_1) = h_g$$

Here, subscripts 1 and 3 refer to the beginning and end of the process. This process would change the state of air along the constant dbt line, for we may write $h = h_a + \omega h_g$ at the inlet and outlet. For a constant dbt process, there is no change in enthalpy of air, and change in enthalpy of vapour is due to change in $\omega$ of air only. Therefore, $(h_3 - h_1)/(\omega_3 - \omega_1) = h_g$, the same expression as above.[6] This implies that if steam is added at a little high pressure, then heating of air will take place. But if enthalpy of vapour added is less than this value, then cooling and humidification of air will occur; the limiting process is evaporative cooling, which is discussed in the next section.

#### 5.2.3.3 Evaporative Cooling

While defining wet bulb temperature, we discussed the mechanism of evaporation of water and how the equilibrium state is reached between water and flowing air. In this process, air is cooled because of evaporation of water, and is therefore known as evaporative cooling. It is more effective when the difference between dry bulb and wet bulb temperature of air is more. This difference is known as *wet bulb depression*, and is a measure of relative humidity of atmospheric air; a low depression means high relative humidity. In practice, fine drops of water are sprinkled through nozzles in cooler space, so that space is filled with water drops and air flows through it. Some water is evaporated to mix with air and the rest is collected in the bottom of the tank, which is recirculated through nozzles. After some time, water in the bottom of the tank will acquire wbt (wet bulb temperature) of air, and steady state is reached. Process 23 in Fig. 5.6 (b) shows the evaporative cooling process on the psychrometric chart.

Evaporative cooling has found commercial application for cooling of large and small buildings, because as compared to an air-conditioning system, it requires comparatively small amount of energy and its first cost is less. These are available as desert cooler or air washer in the market. In actual situations, air at the outlet of air washer is not saturated, since some of the air does not come in contact with water, and it mixes later with cooled air. The state of air coming out of evaporative cooler will therefore be at, say, 3. We express it by terms like 'bypass factor' and efficiency of air washer, defined as follows

$$\text{BF} = (t_3 - t_4)/(t_2 - t_4) = (\omega_4 - \omega_3)/(\omega_4 - \omega_2)$$

and air washer efficiency

$$= 1 - F \qquad (5.21)$$

---

6. Addition of steam into air increases vapour pressure. Therefore, enthalpy of vapour in air at the beginning and end of steam addition process will not be the same. However, this difference is very small and hence neglected.

**Example 5.8** Air initially at 9°C dbt and –3°C dpt is heated to 17°C. It then flows through an air washer in which water constantly recirculates without being heated or cooled. After leaving the air washer, the air is reheated to final condition of 19°C dbt and 12.0°C wbt. Find (a) wbt, $\phi$, and dbt of air leaving the air washer, (b) heating required in each coil per kg of dry air, (c) by pass factor of the air washer, and (d) efficiency of the air washer.

***Solution*** Figure 5.7 shows the block diagram of the air conditioning unit. Processes in its components can be shown on the psychrometric chart as follows

Since dpt = –3°C is given at point 1, find first intersection of –3°C dbt line with the saturation line, and then move horizontally till we come across the vertical line for dbt = 9°C. This locates point 1. From the chart, we read: $h_1$ = 16.5 kJ/kga, $\omega_1 = \omega_2$ = 0.0028 kgv/kga.

Heating is at constant $\omega$ process, so we again move horizontally to the right of point 1 and stop at point 2, which lies on the vertical line at dbt = 17°C. From the chart, we read: $h_2$ = 24.5 kJ/kga, $t_{wb,2}$ = 7.8°C.

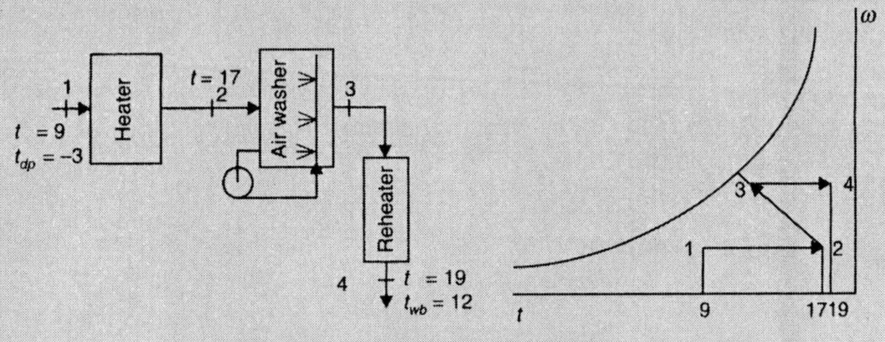


**Fig. 5.7** Air washer

Locate point 4 as dbt and wbt are given. From the chart, we read

$$h_4 = 34.2 \text{ kJ/kga}, \ \omega_4 = 0.0058 \text{ kgv/kga}.$$

Finally, draw a horizontal line through point 4, since the reheating process is a constant $\omega$ process, to the meet constant $h$ line through point 2 at point 3. We read the following values from the chart, $\phi_3$ = 0.86, $\omega_3$ = 0.0058, $h_3$ = 24.5 and $t_3$ = 10°C.

Heating load in the first heater = $h_2 - h_1$ = 24.5 – 16.5 = 8.0 kJ/kga

Heating load in the second heater = $h_4 - h_3$ = 34.2 – 24.5 = 9.7 kJ/kga

Efficiency of air washer = $(t_2 - t_3)/(t_2 - t_{wb,2}) = (17 - 10)/(17 - 7.8)$
$= 76.1\%$

By pass factor of the air washer = 100 – efficiency
$= 100 - 76.1 = 23.9\%$

#### 5.2.3.4 Chemical Dehumidification

Many salts and liquids are hygroscopic in nature and liberate heat during adsorption/absorption of moisture from the air. Such chemicals can be used for dehumidification and heating of air in an insulated chamber. If liquid chemical is used, then it is recirculated and sprayed like evaporative cooler, through a nozzle in the space through which air flows. The process is known as chemical dehumidification. For solid chemicals, air is passed through the bed of the chemical. Neglecting the amount of work done in recirculation, the process is essentially like evaporative cooling discussed above, an isenthalpic process. Hence, it is represented by a constant enthalpy line on a psychrometric chart.

**Example 5.9** A classroom for 60 students is to be air conditioned to 22°C dbt and 55% $\phi$ first by chemical dehumidification, and then cooling it by passing over a cooling coil. The outdoor conditions are 32°C dbt and 22°C dpt in September. The quantity of fresh air required is 0.5 m$^3$/min/student. Find (a) dbt of air after dehumidifier, (b) capacity of the dehumidifier, (c) capacity of the cooling coil, and (d) ADP of the cooling coil, assuming a bypass factor of 0.3.

***Solution***

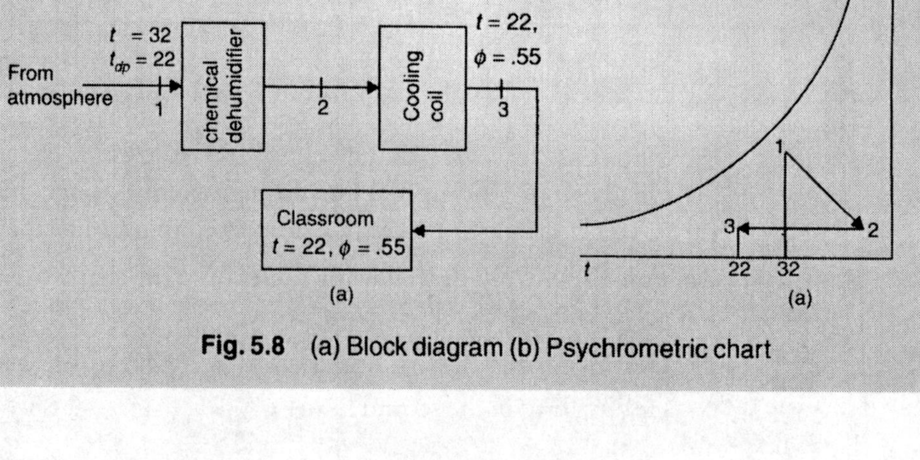


**Fig. 5.8** (a) Block diagram (b) Psychrometric chart

Given: Chemical dehumidifier and cooling chamber with mass flow rates and air conditions at different sections, as shown in Fig. 5.8 (a).

System: The chemical dehumidifier and cooling coil are open systems and are considered as control volume for analysis for which sfee can be used, since the flow is steady.

Assumptions: We assume that air conditioning is required for ventilation purpose only, and there is no heat load in the classroom. In such a case, the condition of air after cooling unit will be the same as that in the classroom. At first sight this assumption looks inappropriate but in the absence of heat load there is no other choice. Sometimes, such unrealistic assumptions have to be made; this example is chosen to illustrate this concept.

Analysis: Figure 5.8 (a) shows the block diagram of the system, and Fig. 5.8 (b) shows processes 12 and 23, respectively in the chemical dehumidifier and cooling coil on the psychrometric chart. Process 12 follows a constant $h$ line in the chemical dehumidifier, while process 23 in the cooling chamber moves along the horizontal line, as $\omega$ remains constant in the process. Locate points 1 and 3 on the psychrometric chart from the given data. Draw a line through point 1 along the constant $h$ line, and from point 3, a horizontal line. These two lines intersect at point 2. We read from the psychrometric chart the following.

Given: Volume of fresh air required for 60 students per minute = 0.5 × 60. $\omega_1 = 0.0168$, $\omega_2 = \omega_3 = 0.009$ kg/kga, $h_1 = h_2 = 75$, $h_3 = 45$ kJ/kga and $v_1 = 0.89$ m³/kga.

Mass of air supplied = 0.5 × 60/0.89 = 33.71 kg/min

Capacity of the dehumidifier = $m_a(\omega_1 - \omega_2)$

= 33.71 × (0.0168 – 0.009) × 60 = 13.13 kg/hr

Capacity of cooling coil = $m_a(h_2 - h_3)$ = 33.71 × (75 – 45)/(60 × 3.5) = 4.62 ton, and

dbt of air after dehumidifier from chart = $t_2$ = 50.8°C.

Had the whole air been in contact with the cooling coil, the temperature would have changed to ADP, $t_{adp}$. Therefore, bypass factor for the coil is

$$\text{B.F.} = (t_3 - t_{adp})/(t_2 - t_{adp}) \quad \text{or} \quad 0.3 = (22 - t_{adp})/(50.8\ t_{adp})$$

$$\rightarrow t_{adp} = 9.66°\text{C}$$

Discussions: In the actual situation there will be heating load in the space. Then, air from the cooling coil will be at a condition different from that in the room, so that it also takes care of heat addition, which may be in the form of sensible and latent heat. We shall discuss such a problem later.

### 5.2.3.5 Adiabatic Mixing of Streams

In large buildings, air is taken from different places and mixed before sending it to the air conditioning plant. Fresh air from the atmosphere is also added to control the pollutant level in the conditioned space. During mixing of air, there is no heat exchange with the surrounding. After mixing of air streams, state of the air will change and is obtained by making an energy and mass balance of the control volume, shown in Fig. 5.9. Mass balance of dry air and water vapour for the open system gives the following.

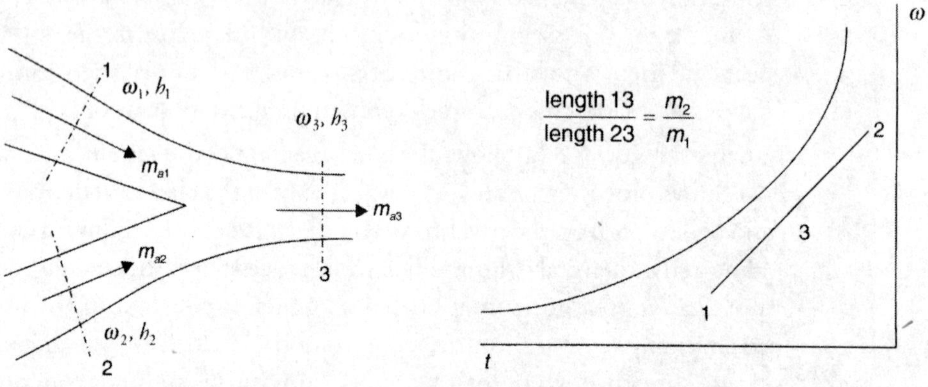


**Fig. 5.9** Adiabatic mixing

$$m_{a3} = m_{a1} + m_{a2} \quad m_{a3}\,\omega_3 = m_{a1}\,\omega_1 + m_{a2}\,\omega_2$$

$$\Rightarrow m_{a1}/m_{a2} = (\omega_2 - \omega_3)/(\omega_3 - \omega_1) = \text{length 23/length 31} \tag{5.22}$$

Similarly, energy balance gives $m_{a3}\,h_3 = m_{a1}\,h_1 + m_{a2}\,h_2$

$$m_{a1}/m_{a2} = (h_2 - h_3)/(h_3 - h_1) = \text{length 23/length 31} \tag{5.23}$$

These equations show that point 3 will lie on line 12, such that the segment of the line is proportional to the mass of dry air at the other state point. That is, the length of line 12 is proportional to the mass at point 3.

**Example 5.10** One stream of air at 55 m³/min at 15°C and 60% $\phi$ flows into another stream of air 35 m³/min at 25°C and 70% $\phi$. Calculate for the mixture (a) dbt, (b) wbt, (c) $\phi$ and (d) $\omega$.

***Solution*** Figure 5.9 shows schematic and the state points of air on the psychrometric chart. State points for two streams can be located on the chart as dbt and $\phi$ are known for both the streams. We read from the chart

$$v_1 = 0.821,\ \omega_1 = 0.00625,\ v_2 = 0.8582 \text{ and } \omega_2 = 0.014$$

$$\therefore\ m_{a1} = 55/0.821 = 67 \text{ kga/min} \quad\text{and}\quad m_{a2} = 35/0.8582 = 40.8 \text{ kga/min}$$

Join points 1 and 2 by a straight line and locate point 3 such that
Length 13/Length 12 = $m_{a2}/(m_{a1} + m_{a2})$ = 40.8/(67 + 40.8) = 0.3785
Point 3 can be located on the chart from where we read $t_d$ = 19°C, $t_{wb}$ = 15.2°C, = 67% and $\omega$ = 0.0094 kg/kga.

## 5.3 BINARY MIXTURE

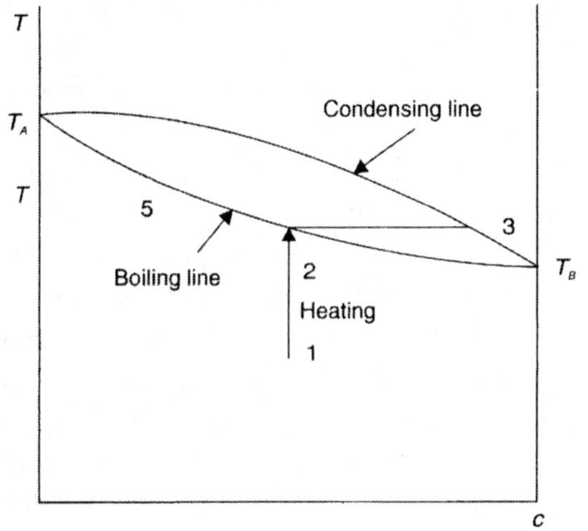


**Fig. 5.10** T-c chart

Binary means two, and therefore, any mixture of two pure substances in their any phase is a binary mixture, provided the constituents cannot be separated from each other by mechanical means. Separation of the components in such mixtures is possible only by heating. If the constituents of a binary mixture are gases, then such a system can be analysed by the methods discussed before. However, this analysis is not applicable to a mixture of a liquid and gas or a liquid and solid, which are commonly called solutions. Heating of the solution generates vapours of both the components, with more vapours of the component having the lower boiling point. Concentration of a substance, both in vapour and liquid phases, go on changing with continuous heating. Such mixtures have been used in refrigeration, in which the solution is changed to a mixture of vapours at one place, and the vapour mixture is converted into solution at some other place in the cycle, for example, ammonia–water solution, lithium–bromide solution. Binary vapours need, like psychrometric, a separate thermodynamic analysis, covered in the following paragraphs.

### 5.3.1 Temperature–Concentration Chart

Let us consider two pure substances $A$ and $B$ having boiling temperature $T_A$ and $T_B$, respectively with $T_A > T_B$ (assumed). When a solution of $A$ and $B$ is heated at constant pressure, its temperature increases, and at some temperature, liquid $B$ starts changing into vapour. Figure 5.10 shows the heating process with the initial condition of the solution at point 1. When vaporisation starts at point 2, the vapours at 3, occupying space above the liquid surface, will be in thermodynamic equilibrium with the liquid at state 2. Concentration of both liquid and vapours, as well as the temperature, may be measured. Such measurements repeated for different initial concentrations of the solution generate data of concentration of the liquid and vapour in equilibrium at different evaporation temperatures at a particular pressure; *concentration is defined as the ratio of mass of B to that of the solution.* The data is plotted on T-c chart, which is shown in Fig. 5.10, for some pressure. The experiment is repeated at different pressures to give a family of curves. We observe from the figure, that unlike a pure substance, there are two phase lines; these two are called *boiling and condensing lines,* respectively. We further observe that at the extreme end of concentration of $B$, the difference between the two temperatures decreases; the two lines will coincide when there is either component $A$ or $B$ in the solution.

The T-c diagram is useful, as it gives the concentration of $B$ in liquid and vapour mixtures during change of phase at a temperature $T$ at any pressure. As the saturation temperature depends on pressure too, there will be different boiling and condensing lines for different pressures. The two curves will shift upward with increasing pressure, since the boiling temperature of both the pure substances increases.

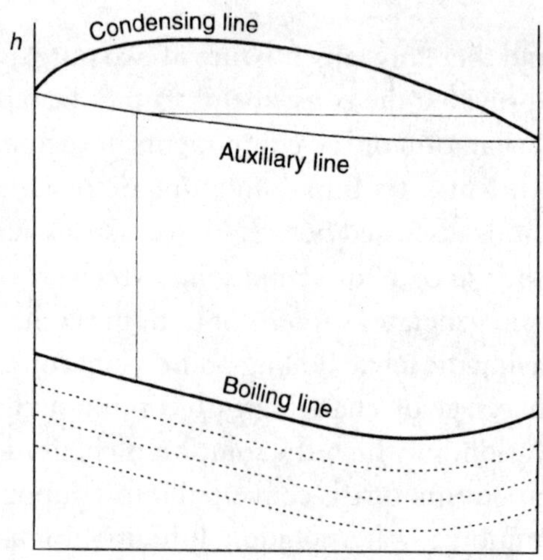


**Fig. 5.11** h–c chart

### 5.3.2 Enthalpy–Concentration Diagram

We learnt in Chapter 2 that enthalpy is an important property for first law analysis of any open system. For a binary solution, enthalpy will also depend on the concentration of the components, besides temperature and pressure. Enthalpy-concentration charts are therefore developed, so that enthalpy may easily be obtained once the process has been drawn on it. Figure 5.11 shows boiling and condensing lines at some pressure on the enthalpy–concentration chart. For different pressures, there will be different boiling and condensing lines, and the h–c chart has a family of such curves at selected pressures. Moreover, unlike the T-c diagram, the two ends of the boiling and condensing lines do not meet at $c$ equal to zero or 1; the vertical distance between the two curves is enthalpy of vaporisation of component $A$ and $B$, respectively.

Isotherms for different temperatures have been shown on the h-c chart by dotted lines in the liquid region, and are available in the vapour region also. However, there are no isotherms in the wet vapour region. Therefore, we cannot locate the state point on the condensing line, which is in thermal equilibrium with the point on the boiling line. The corresponding saturated vapour state is located on the h-c chart by using auxiliary or equilibrium lines at that pressure. (Auxiliary lines have been drawn for all those pressures for which condensing lines are drawn.) The procedure is to draw a vertical line through the point on the boiling line intersecting the auxiliary line at some point from where a horizontal line is drawn that meets condensing curve; this is the required point. Similarly, a point on the boiling line can be obtained from that on the condensing line adopting the reverse process. The equilibrium line has only this role and none else.

### 5.3.3 Steady Flow Processes with Binary Mixture

Processes to which a binary mixture is subjected in different applications may be classified as follows

- mixing of two streams of the binary mixture of the same components with heat transfer,
- separation of one component from the mixture by heating or cooling,
- rectification of the binary mixture.

In the following analysis, $c$ represents the concentration of component $B$, which is assumed to have lower boiling temperature, and subscripts $l$ and $v$ refer to the liquid and vapour state, respectively.

#### 5.3.3.1 Mixing of Two Streams

Figure 5.12 shows two binary mixtures at states 1 and 2 mixing with each other, resulting in a stream represented by the state 3. Let $Q$ be the rate of heat rejection during mixing, and the flow be a steady flow. We have to find the concentration and enthalpy of the mixture. There are two unknowns, and two equations are obtained from the principle of conservation of mass and energy. We get the following relations

$$m_1 + m_2 = m_3,\ m_1 h_1 + m_2 h_2 = m_3 h_3 + Q \text{ and } m_1 c_1 + m_2 c_2 = m_3 c_3 \tag{5.24}$$

Notice that we are assuming the solution as homogeneous while writing the above equations. This is possible when the two phases are mixed with each other thoroughly, like wet steam. This is done to eliminate the complexity of determining the concentration of a component in different phases, in which case the number of unknowns will be more. Eliminating $m_3$ from the above equations, we get

$$m_1/m_2 = (c_2 - c_3)/(c_3 - c_1) = \text{length } 23'/\text{length } 13' \tag{5.25}$$

$$h_3 = h_1 + m_2/m_3\,(h_2 - h_1) - Q/m_3 = h_{3'} - Q/m_3$$

and $$c_3 = c_1 + m_2/m_3\,(c_2 - c_1) = c_3 \tag{5.26}$$

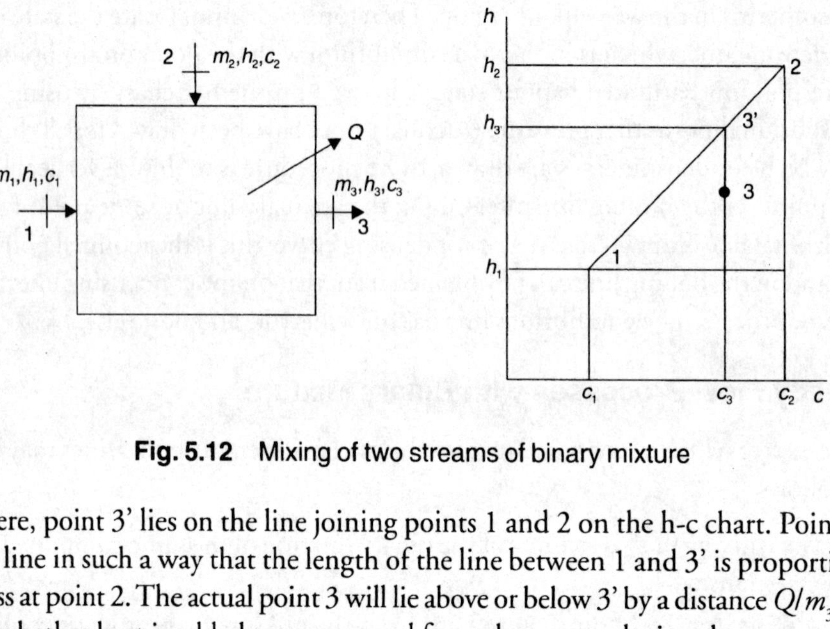


**Fig. 5.12** Mixing of two streams of binary mixture

where, point 3' lies on the line joining points 1 and 2 on the h-c chart. Point 3' divides the line in such a way that the length of the line between 1 and 3' is proportional to the mass at point 2. The actual point 3 will lie above or below 3' by a distance $Q/m_3$ depending on whether heat is added to or removed from the system during the process.

### 5.3.3.2 Heating or Cooling of Mixtures

A component of a binary mixture[7] can be removed from it only by heat addition. Heating will vaporise both the components, removing more of component *B*, which has lower

7. Here, the discussion is made with keeping in mind a solution in liquid phase. The reader should see for himself if the solution is in a different phase.

boiling point. This increases the concentration of $B$ in the vapour phase and decreases it in the liquid phase. Remember our earlier discussions in Section 5.3.1, that concentration of $B$ in vapour is maximum at a certain temperature, beyond which it starts decreasing. This point is to be kept in mind while deciding the temperature of the solution.

Now, imagine the reverse direction of flow in Fig. 5.12 That is, the binary mixture enters at Section 3 and leaves at Sections 1 and 2, while heat $Q$ is added per unit time. Because of heat addition, vapours of both the components are formed, which occupy the space above the liquid, which is collected in the bottom of the container. Therefore, vapour and the liquid always remain in equilibrium with each other in the container. We assume that the vapour mixture leaves at 2, and the liquid mixture collected at the bottom of the separator leaves at state 1. The states at points 1 and 2 will therefore, lie on the boiling and condensing line, respectively, on the h-c chart.

Energy and mass balance of the system under steady state gives the same equations, Equation. 5.24, and therefore, their solution is given by Eqs. 5.25 and 5.26. As heat is added in this case, the point 3 in Fig. 5.12 (b) will lie above point 3'. A little peep into the problem will tell us that it is not as easy as it looks, because it is not easy to locate two points on the boiling and condensing line, which are in thermal equilibrium with each other and also satisfy the above equations. In fact, the number of unknowns are four and not three, namely mass and concentration of the vapours and liquid leaving the chamber, assuming these are known for mixture entering into the chamber. There are three algebraic equations and conditions of thermal equilibrium. However, if temperature of the solution in the chamber is known, then from the T-c chart, it will be possible to find out the concentration of the component $B$ in the vapours and the liquid Then, the solution is possible.

Similarly, we shall get results if cooling of the solution is done. Equations. 5.25 and 5.26 will be applicable in this case, with $Q$ in the above equations replaced by $-Q$.

#### 5.3.3.3 Rectification of Binary Mixture

In many situations, we need to separate these two components completely, or to generate the pure component of at least one constituent, like in an absorption refrigeration system. This is known as *rectification*. The adopted procedure is to feed the solution of the binary mixture at some mid-point of a column, with heating at the bottom and cooling at the top end of the column. A heater at one end and a cooler at the other end of the column produces a temperature gradient along its height. Vapour formed at the bottom moves up, and the liquid from the cooler, as well as that fed to the column, flows down. The vapours and the liquid remain in equilibrium at all the sections of the column. At every section, heat transfer takes place from the vapours to the liquid, as the former is at a higher temperature.

Let the mixture consist of components *A* and *B*, with *B* having lower boiling point. This results in evaporation of more quantity of component *B* from the downcoming liquid, and condensing more of component A from the upgoing vapour. Therefore, as we go up the column, vapours at the top become rich in component *B*, and the liquid at the bottom in component *A*. In this way, it is possible to separate the components completely by proper design of the column, and control the heat added or removed. Component *B* comes out at the top, and the other one from the bottom.

Ammonia water solution has been used in the vapour absorption system. This system differs from the vapour compression system in the respect that the compressor of the latter is replaced by a generator where ammonia is separated from its solution in water using the concept of rectification. Ammonia is the refrigerant in this system, and water vapours have to be completely removed before the refrigerant enters the evaporator, as the temperature in it may be below zero, and water, if present, would change into ice choking the pipe. Therefore, thermodynamic study of binary mixtures is required to determine the performance of vapour absorption systems, which are gaining popularity day-by-day because of their being environment friendly and a means of utilising waste heat produced in industries, hospitals, commercial installations in a fruitful way.

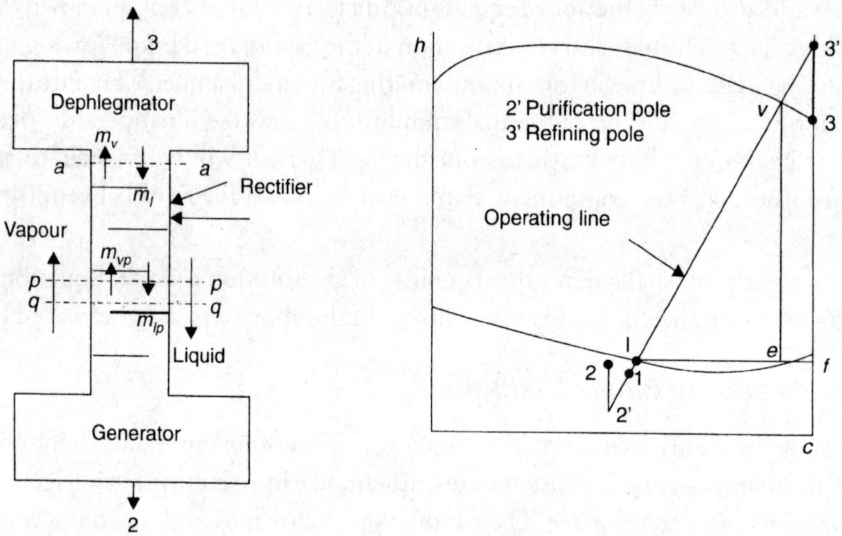


(a) Rectification column

(b) Rectification process on h-c chart.

**Fig. 5.13** Rectification

For any such system, we shall be required to find out the amount of heat transfer in the heating and cooling section, and also the variation of the state along the column. This is best done by showing the variation of the state of the mixture on the h-c chart. A commonly adopted method for control volume analysis is to consider two sections *pp* and $qq$, infinitesimally close to each other, as shown in Fig. 5.13 (a). Mass balance of the control volume formed between the sections $pp$ and $qq$ demands that mass crossing the section $pp$ should be equal to that crossing section $qq$ for a steady state. Taking mass flowing upward as positive and that flowing downward as negative, this gives,

$$m_{vp} - m_{lp} = m_{vq} - m_{lq} = \text{constant} \quad \Rightarrow \quad m_v - m_l = \text{constant} = m \text{ (say)} \tag{5.27}$$

Here, $m_v$ and $m_l$ are the mass of vapour and liquid respectively at any section, and the second subscript indicates section $pp$ or $qq$. We note from the equation that $m_v - m_l$ remains the same at all the sections of the rectifier, and represents net mass flow rate across the section. However, $m$ is still not known, since both $m_v$ and $m_l$ are not known. If $c_v$ and $c_l$ are the concentrations of $B$ in vapour and liquid, respectively, at section $pp$, then mass balance of component $B$ yields

$$\text{net mass of } B \text{ going upward} = m_v c_v - m_l c_l = m\, c \text{ (say)} \tag{5.28}$$

Here, net mass flow rate of $B$ across the section is written as $mc$, where $c$ is the concentration of $B$ for the combined liquid and vapour at that section. We may also consider mass balance of the total flow and flow of the component through the dephlegmator (control volume), using the earlier notations and subscript '$a$' for the entrance section to the dephlegmator, to give

$$\text{Total mass balance} \quad m_{va} = m_{la} + m_3 \quad \Rightarrow \quad m_3 = m_{va} - m_{la} = m \tag{5.29}$$

and mass balance of $B$

$$c_3 m_3 + c_{la} m_{la} = c_{va} m_{va} \quad \Rightarrow \quad c_3 m_3 = c_{va} m_{va} - c_{la} m_{la} = mc_a \quad \text{(say)} \tag{5.30}$$

Equations 5.29 and 5.30 give $c_a = c_3$ (5.31)

Mass flow and concentration of the fluid at the dephlegmator exit are known. Therefore, $m$ and $c_a$ are known from Eqs. 5.29 and 5.30. This point will lie on the vertical line, $c = c_a = c_3$. Let $h$ be enthalpy of the mass crossing the section $pp$. Enthalpy crossing the section in the upward and downward direction is $m_v h_v$, $m_l h_l$, respectively. This gives enthalpy of mass crossing section $pp$ in the upward direction as

$$m_v h_v - m_l h_l = m\, h \text{ (say)} \tag{5.32}$$

Substituting for $m$ from Eq. 5.27 in 5.31 or 5.28, we get respectively

$$\frac{m_v}{m_v - m_e} = \frac{m_v}{m} = \frac{h - h_e}{h_v - h_e} = \frac{c - c_e}{c_v - c_l} \tag{5.33}$$

Consider point 3' on the intersection of the vertical line $c = c_3$ with line joining points $l$ and $v$ on the boiling and condensation lines for the given mixture, see Fig. 5.13(b). That is, $c_{3'} = c_3$. (*It may be recalled that liquid and vapour at any section will be in thermal equilibrium. This implies that the liquid and vapour state will lie on the saturated liquid and saturated vapour line, respectively.*) This point 3' satisfies Eq. 5.32 as can be seen from the similar triangles lev and lf3' in Fig. 5.13. This point is known as *pole* because for all the sections, $l$ and $v$ vary in such a way that the line joining these points will meet at point 3'. Further, as the vapours rise, concentration of component $B$ increases. That is, the line 3'vl will swing to the right (anticlockwise around point 3') from section to section of the column in the upward direction. This line is called the operating line. Ultimately, the line coincides with the vertical line at $c = c_3$ at the top of the column.

Now, consider the unit consisting of a generator, rectifier and dephlegmator as control volume. The solution enters it at state 1, vapour leaves at 3, and a weak solution leaves at 2, during which heat $q_g$ is added in the generator and $q_d$ is rejected in the dephlegmator. Mass and energy balance of rectifier column results in the following equations.

$$m_1 = m_2 + m_3,\ m_1C_1 = m_2C_2 + m_3C_3 \text{ and } m_1h_1 + q_g = m_2h_2 + m_3h_3 + q_d$$

Solving these equations, we get

$$\frac{(h_3 + q_d/m_3) - h_1}{c_3 - c_1} = \frac{h_1 - (h_2 - q_g/m_2)}{c_1 - c_2} \tag{5.34}$$

Recalling $c_{3'} = c_3$, it can easily be seen from Fig. 5.13 (b), that point 3 lies below 3' by a distance $q_d/m_3$, and point 2 above 2' by distance $q_g/m_2$, where point 2' is obtained in the same way as point 3' was obtained. Points 3' and 2' are known as *refining and purification poles*, and the line 2'13' as the *operating line*.

## SUMMARY

Gas and gas–vapour mixtures form important working substances in many thermal systems. A gas mixture is normally assumed to be made from perfect gases, and such a mixture can be treated as a single gas for whom the various gas constants like specific heats, gas constant and molecular weight is determined using Eqs. 5.5, 5.7 and 5.8. These equations make use of either mass or mole fraction, defined by Eq. 5.1. If there are $n$ number of gases in a mixture, then there will be $(n-1)$ fractions, which, along with mixture pressure and temperature, describe state of the gas system. Individual gases occupy the same volume, the volume of the mixture, at the mixture temperature and pressure. However, their pressure is not the same; pressure of a constituent gas is called its partial pressure. The partial pressure of individual gases is such that their sum is the mixture pressure. In fact, all the temperature dependent properties are an algebraic sum of the corresponding property of the individual gases. This leads to Eq. 5.5, that gives the specific property of the mixture. But entropy change is also associated with change in pressure.

Mixture of air and water vapour forms another important working substance, as it represents the moist air present in the atmosphere, and whose temperature and humidity needs to be controlled for different situations. The study of behaviour of atmospheric air is called psychrometric. Psychrometric chart and equations were developed, that give the properties of air at any of its states. Air may be heated, cooled, humidified or dehumidified by adding/removing heat and water. Mass and energy balance results in equations that give the final state. However, all these processes can be shown on the psychrometric chart, and we need not solve the equations. We have defined various terms that are used in the topic. A special process is adiabatic saturation, as the name indicates, it is the process of saturating air with water adiabatically. The mechanism of air saturation is discussed, and it is found that the enthalpy difference is the thermodynamic potential for such a process.

A combination of these processes is used to achieve the desired temperature and relative humidity. This requires determination of sensible and latent heat transfer between the system and the surrounding, from which we know how much heat and moisture contents from air are to be removed/added in the air conditioning processes. Heat and moisture comes from outside, as well as from the occupants of the air-conditioned space. Air from the room is taken to the air conditioner, replacing with about 20–25% fresh air to maintain air pollution level in the room. It is then conditioned in the air conditioner to a state such that, after mixing with air in space, it will reach the desired temperature and relative humidity. Sensible heat factor scale and a circle on the chart are very handy to handle this kind of a problem, since the mixing process of air conditioned air with air in the room moves along a line parallel to the line joining a point on the SHF scale with the circle.

A mixture of air and petrol is another mixture, which is of importance in the internal combustion engine. This has not been discussed here.

The other mixture of importance is a binary mixture. Ammonia-water mixture is used in such absorption refrigeration system. This is a mixture of liquid and gas. In the analysis of such systems, h-c and T-c charts are very useful. Section 5.3 discusses these charts and auxiliary lines on the h-c chart. Analysis of a few processes used with binary mixtures is also discussed. Rectification is dealt in detail, as this is a process of separating the mixture constituents from the mixture. For this, a rectifier column is used, in which the mixture is supplied at about half the height of the column. The mixture is heated at the bottom and cooled at the top. Heating generates vapours of both the constituents, which travel upward and come in contact with down flowing cool liquid. It is shown that a pole plays an important role to determine the state (concentration and enthalpy) of the mixture at different sections of the rectifier column, and to determine heat added at the bottom or removed from the top. This analysis is a perquisite for the vapour absorption system. Binary mixtures may also be formed from solid crystals and liquid. Lithium-bromide water mixture used in such absorption system is an example.

# EXERCISES

## Review Problems

5.1 What is the meaning of partial pressure and partial volume? State the Dalton and Amagat's laws and explain their significance.

5.2 Define mass and mole fractions. Someone claims that mass and mole fractions for $CO_2$ and $N_2O$ are the same. Is it true and why?

5.3 Sum of the mole fractions for an ideal gas mixture is equal to unity. Is this true for real gases, too?

5.4 Consider a mixture of several gases of the same molar masses. Will all the mass fractions be identical? How about the mole fractions?

5.5 Are Dalton's law and Amagat's law the same for a mixture of perfect gases? Comment.

5.6 Can a mixture of ideal gases also be an ideal gas? Give an example.

5.7 In a mixture of several gases, which component will have the highest partial pressure—the one with the highest mole number or with the largest molar mass?

5.8 If one mol of monoatomic gas is mixed with one mol of diatomic gas, what will be the molar heat capacity at constant volume for the mixture?

5.9 A container divided into two portions contains a mixture of gases in one portion, while the other one has a vacuum. The partition wall is removed. Will the total pressure and the partial pressure of the individual gases change and why?

5.10 A rigid tank contains a mixture of two ideal gases. A valve is opened, and some gas escapes. Therefore, the pressure in the tank drops. Will the partial pressure of each component change? How about the pressure fraction of each component?

5.11 In determination of a property of enthalpy of a gas mixture, is it necessary to know that the datum for each of the component be the same?

5.12 Does entropy of a mixture always change from the sum of the constituent gases when mixed? Comment.

5.13 What is the difference between dry and moist air? Can we treat water vapour in moist air as a perfect gas? Justify your statement with reasoning.

5.14 What is the meaning of dew point, dry bulb, wet bulb and apparatus dew point temperature? Explain the difference between them with suitable diagrams.

5.15 Under what conditions will the wet bulb temperature of an air water vapour mixture equal its dry bulb temperature?

5.16 What is the difference between relative humidity and humidity ratio?

5.17 Is there any correlation between partial pressure of water vapour in air and RH of air?

5.18 What is the relation between the partial pressure of water vapour and the saturation pressure of water at the dew point temperature?

5.19 Explain the difference between thermodynamic wet bulb and wet bulb temperature. Is it possible that these two temperatures may be the same and when? Explain your answer in light of the difference of enthalpy principle.

5.20 Illustrate the typical process of psychrometric chart. Sketch one at standard atmospheric pressure and label each of the properties with customary units. Why is the term standard pressure used in this question?

5.21 Suggest an air conditioning method to cool and dehumidify the air and sketch it on psychrometric chart.

5.22 What is meant by the term adiabatic saturation? Is there any other kind of saturation? Make comments.

5.23 State the reason for the specific enthalpy to remain constant in adiabatic saturation. Are desert coolers effective in monsoon? Explain.

5.24 How can the adiabatic saturation process be used to cool another stream of air without change in its humidity ratio?

5.25 Why does adiabatic saturation find application in desert areas ? Name the applications and describe it.

5.26 What is the importance of potential difference in a process? Give a few examples. Which potential difference is responsible for moisture transfer between air and water surface?

5.27 How will you decide whether the air in contact with water (or steam) will be heated or cooled?

5.28 Why is evaporative cooling ineffective in humid tropical areas?

5.29 What happens to the chemical used in a chemical dehumidification process? Is it possible to re-use it and how?

5.30 What is the difference between sensible and latent heat? Illustrate the importance of the sensible heat factor scale provided in the psychrometric chart.

## Problems for Practice

5.1 A mixture of 50% propane and 50% oxygen by volume is compressed reversibly and adiabatically from 0.1 MPa and 300 K to 1 MPa. Determine the final temperature of the mixture and the work required to compress 100 mol of the gas mixture. The molar heat capacity Cp of the propane and oxygen are 74.0565 and 29.4912 kJ/kmol K, respectively whereas their *Cv* is 65.74442 and 21.1776 kJ/kmol K. [*Ans.* 433.3 K, – 579.3 kJ].

5.2 A 5 $m^3$ tank contains 30% $H_2$, 30% Argon and 40% methane by volume at 1 MPa and 300 K. It is necessary to change the composition of the gas mixture such that it contains 60% methane by volume. Determine the amount of methane to be added and the final pressure of the mixture in the tank. [*Ans.* 1.0023 kmol, 1.5 MPa].

5.3 A rigid tank and insulated tank of volume 2 $m^3$ is divided into two equal compartments by a partition. One compartment contains He at 500 kPa and 300 K, while the second compartment contains $CO_2$ at 1 MPa and 600 K. The partition is punctured and the gases are allowed to mix. Given: *Cp* for He and $CO_2$ = 27.786 and 37.144 kJ/kmol K, respectively. Determine (a) temperature and pressure of the mixture, (b) composition of the mixture by volume, (c) the entropy changes of He and $CO_2$, and (d) the net entropy change associated with the process. [*Ans.* 509.42 K, 849.18 kPa, $CO_2$ 50%, He 50%, 12.3652 kJ/kmol K, 1.0431 kJ/kmol K, 2.6884 kJ/kmol K].

5.4 It is proposed to use methane as a fuel in an internal combustion engine. During the suction stroke, 10 mol of air is taken for every mol of methane at 100 kPa and 300 K. If the compression ratio of the engine is 10, determine the pressure and temperature of the mixture at the end of the stroke. *Cp* for air and methane is 29.1062 and 36.1493 kJ/kmol K, respectively, and $C_v$ is 20.7889 and 27.8358 kJ/kmol K, respectively. [*Ans.* 2.443 MPa, 733 K].

5.5 A mixture of air and steam at 50°C, 1.013 bar has a gravimetric analysis of 4% moisture and 96% dry air. Calculate (a) the composition by volume of the mixture, (b) the partial pressure of the vapour and dry air, (c) the RH of the mixture. Also, calculate the enthalpy per kg of the mixture reckoned from °C and heat to be removed at a constant pressure of 1.013 bar for condensation to begin.
[*Ans.* 6.27%, 93.73%, 0.0636 bar, 0.949 bar, 51.5%, 151.8 kJ/kg, 13.3 kJ/kg].

5.6 The air in a cylinder fitted with a piston is saturated with water vapour. The volume is 0.3 $m^3$, the pressure is 3.5 bar and the temperature is 60.1°C. The mixture is compressed to 5.5 bar, the temperature remaining constant. Find (a) the masses of air and vapour present initially, and (b) the mass of vapour condensed on compression. [*Ans.* 1.035 kg, 0.0392 kg, 0.0148 kg].

5.7 A surface condenser is fitted with separate air and condensate outlets. A portion of the cooling surface is screened from the incoming steam and the air passed over these screened tubes to the air extraction and becomes cooled below the condensate temperature. The condenser receives 20000 kg/h of steam dry saturated at 36.2°C. At the condensate outlet, the temperature is 34.6°C, and at the air extraction, the temperature is 29°C. The volume of air plus vapour leaving the condenser is 3.8 $m^3$/min. Assuming constant pressure throughout the condenser, calculate (a) the mass of air removed per 10000 kg of steam, (b) the mass of steam condensed in the air cooler per minute, and (c) the heat to be removed per minute by the cooling water. Neglect the partial pressure of the air at the inlet to the condenser. [*Ans.* 2.63 kg, 0.492 kg, 807 MJ].

5.8 An air-conditioned room is maintained at a temperature of 21°C and an RH of 55% when the barometric pressure is 740 mm $H_g$. Calculate the humidity ratio of the air–water vapour mixture, and the temperature of the inside of the windows in the room if moisture is just beginning to form on them. What mass of water vapour per kg of dry air in the room must be removed from the mixture in order to prevent condensation on the windows when their temperature drops to 4°C? Calculate the initial RH to satisfy this condition if the temperature remains at 21 °C. The barometric pressure remains constant. [*Ans.* 0.00876, 11.62°C, 0.0036 kg, 32.7%].

5.9 In a bathroom, the water tap is accidentally left open and the room is flooded with water. The air inside the room becomes saturated with water vapour. If the air is at 40°C and 1 atm pressure, determine (a) the partial pressure of water vapour in air, (b) the humidity ratio of the air inside the room, and (c) dew point of the air–water vapour mixture in the room. [*Ans.* 7.374 kPa, 0.0489 kgv/kga, 40°C].

5.10 In an air-conditoning system, air at 28°C and 1.013 bar is drawn into a building with a percentage saturation of 50%. It is required to maintain the air in the building at 20°C and 40% RH. The total heat gains to the room (sensible and latent) are 15 kW, and the latent heat gain are 3 kW. Calculate the temperature to which the inlet air should be cooled at the cooling coil (assuming a coil bypass factor of 0.2), the refrigerating load, and the heat input to the heater. The system requires 5 $m^3$ of fresh air per second, and no air is re-circulated. [*Ans.* 6.5°C, 220 kW, 67.1 kW].

5.11 An auditorium of dimension 20 × 40 × 6 m has air with RH 60% at 1 atm pressure and 30°C. Determine (a) the partial pressure of water vapour in the air, (b) the specific humidity of the air, (c) the mass of water vapour in the auditorium, and (d) the dew point temperature. The saturation pressure of water at 20°C, 25°C and 30°C are 2.339, 3.169 and 4.246 kPa, respectively. [*Ans.* 2.5476 kPa, 0.016 kg/kga, 87.33 kg, 21.3°C.]

5.12 An air-conditioner delivers 2 $m^3$ of air at 5°C, which is completely saturated with water vapour. If this air is mixed with 5 $m^3$ air at 35°C and of 40% RH in a room, determine the temperature, specific humidity and the RH of the air after mixing. [*Ans.* 35°C, 0.0113 kgv/kga, 55%].

5.13 A small auditorium is required to be maintained at 22°C dry temperature and 70% RH. The ambient conditions are 30°C dbt and 75% RH. The amount of free air circulated is 200 $m^3$/min. The required conditions are obtained by first cooling and dehumidifying through a cooling coil having ADP of 14°C, and then by heating. Find: (a) cooling capacity of coil in tons of refrigeration and the bypass factor, (b) the amount of water vapour removed by the cooling coil in kg/h, and (c) assuming the bypass factor at 0.2, determine the capacity of the heating coil in kW and its surface temperature. [*Ans.* 37.88 T, 118.6 kg/h, 19.89 kW, 23.25 °C] (IES 1992).

5.14 In an air-conditioned space, 50 kga/s of fresh air at 45°C dbt and 30% RH is introduced. The room air at 25°C and 50% RH is recirculated at 450 kga/s. The mixed air flows over a cooling coil which has ADP temperature of 12°C and BF of 0.15. Determine the condition at the outlet of the cooling coil, RSH, RLH, the cooling load of the coil and the condensate rate. The saturation pressure of water at 12, 25 and 45°C is 0.014016, 0.03166 and 0.09584 bar, respectively. $h$ is given by: $h = 1.005\,t + \omega\,(2500 + 1.88\,t)$ where $t$ is in °C.

5.15 300 $m^3$/min of air at 10°C dbt and 90% RH is to be heated, and humidified at 35 °C dbt and 22.5°C wbt. The required condition is achieved by heating, humidifying and again by heating. The RH of air coming out of the humidifier is 90%. Find (a) the heating capacity of the first heater and the bypass factor if the surface temperature of the coil is 40°C, (b) the capacity of the humidifier in kg/h, (c) the heating capacity of the second heater and the coil surface temperature if the bypas. factor is 0.5, and (d) the humidifying efficiency of the air washer (humidifier). [*Ans.* 118.76 kW, 0.37, 103.04, 108.62 kW, 52.1°C, 91.7%] (IES 97).

5.16 On a particular day, a Sling psychrometer reads 30°C as the dbt and 20°C as wbt. Determine the humidity ratio, the RH, the partial pressure of water vapour and the enthalpy of the air water vapour mixture. [*Ans.* 0.0105 kgv/kga, 0.41, 1.55 kPa, 58 kJ/kga].

## Multiple-choice Questions

5.1 Standard barometric pressure is

(a) 1.03 bar, (b) 760 mmHg. (c) 1013.25 mbar. (d) all of the above.

5.2 Specific humidity is defined as
   (a) mass of water vapour contained in the air–vapour mixture per kg of dry air.
   (b) mass of water vapour contained per kg of the air–vapour mixture.
   (c) mass of dry air contained per kg of air–vapour mixture.
   (d) none of the above.

5.3 Specific humidity is also called
   (a) relative humidity. (b) humidity ratio.
   (c) absolute humidity. (d) saturation humidity.

5.4 Humidity ratio $W$ is expressed as ($P_v$ = partial pressure of vapour, $P_a$ = partial pressure of air, $P_b$ = barometric pressure of air)
   (a) $W = 0.622\, P_v/P_b$. (b) $W = 0.622\, P_v/P_b$.
   (c) $W = 0.622\, P_v/(P_b - P_a)$. (d) $W = 0.622\, P_v/(P_b + P_v)$.

5.5 Humidity ratio $W$ maybe expressed as ($ma$ = mass of dry air, $mv$ = mass of vapour in air vapour mixture)
   (a) $W = m_v/m_a$. (b) $W = m_v/(m_v + m_a)$.
   (c) $W = m_a/(m_v + m_a)$. (d) $W = m_v/(m_a - m_v)$.

5.6 Degree of saturation is defined as the ratio of
   (a) specific humidity of moist air to specific humidity of saturated air at the same dry bulb temperature.
   (b) moisture content contained per kg of dry air to the moisture content required to saturate 1 kg of dry air at the same dry bulb temperature.
   (c) both (a) and (b).
   (d) none of the above.

5.7 Relative humidity is defined as
   (a) the ratio of the mass of water vapour in a given volume at a given temperature to the mass of water vapour in the same volume at the same temperature, when air is saturated.
   (b) the ratio of actual partial pressure of water vapour in moist air to the saturation pressure of water vapour at the same dry bulb temperature.
   (c) both (a) and (b).
   (d) none of the above.

5.8 The barometric pressure = 1.03 bar the saturation pressure at the given dry bulb temperature = 0.06 bar, actual vapour pressure at the same dry bulb temperature is 0.03 bar. The relative humidity is
   (a) 33.33%. (b) 50%. (c) 48.5%. (d) none of the above.

5.9 The barometric pressure = 1.03 bar, the saturation pressure at the given dry bulb temperature = 0.06 bar, actual vapour pressure at the same dry bulb temperature = 0.02 bar. The degree of saturation is

(a) 0.3333. (b) 0.5. (c) 0.485. (d) none of the above.

5.10 The barometric pressure 750 mm Hg, the saturation pressure at the given dry bulb temperature is 40 mmHg, actual vapour pressure at the same dry bulb temperature is 20 mm Hg. The relative humidity would be

(a) 50%. (b) 33.33%. (c) 48.5%. (d) none of the above.

5.11 Wet bulb temperature of air whose relative humidity is 100% is

(a) less than the dew point temperature.
(b) equal to the dew point temperature.
(c) greater than the dew point temperature.
(d) may be more or less depending upon the dry bulb temperature.

5.12 Wet bulb temperature of air whose relative humidity is 100% is

(a) less than the dry bulb temperature.
(b) equal to the dry bulb temperature.
(c) greater than the dry bulb temperature.
(d) maybe more or less depending upon the humidity ratio.

5.13 Stream *A* has a dry bulb temperature of 35°C and wet bulb temperature of 25°C, stream *B* has a temperature of 40°C and wet bulb temperature of 25°C.

(a) Enthalpy per kg of dry air for stream *B* is higher than that of stream *A*.
(b) Enthalpy per kg of dry air for stream *B* is lower than that of stream *A*.
(c) Enthalpy per kg of dry air for stream *B* is equal to that of stream *A*.
(d) none of the above.

5.14 If the specific humidity of moist air remains the same, but its dry bulb temperature increases

(a) its dew point temperature increases.
(b) its dew point temperature decreases.
(c) its dew point temperature remains the same.
(d) its dew point temperature may increase or decrease, depending upon increase or decrease of humidity.

5.15 Stream *A* of moist air has specific humidity of 0.006 kg/kg of dry air and dry bulb temperature 25°C and stream *B* of moist air has specific humidity of 0.006 kg/kg of dry air and dry bulb temperature 30°C, then

(a) stream *A* has lower relative humidity and same dew point temperature.

(b) stream $A$ has higher relative humidity and same dew point temperature.
(c) stream $A$ has higher relative humidity and higher dew point temperature.
(d) stream $A$ has higher relative humidity and lower dew point temperature.

5.16 Stream $A$ of moist air has vapour pressure of 8 mmHg and relative humidity of 50%. The stream has vapour pressure of 8 mmHg and relative humidity of 100%.
(a) Stream $B$ has higher specific humidity than stream $A$.
(b) Stream $B$ has lower specific humidity than stream $A$.
(c) Stream $B$ has same specific humidity as stream $B$.
(d) none of the above.

5.17 If wet bulb depression is zero, then relative humidity is equal to
(a) 0%. (b) 50%. (c) 100%. (d) none of the above.

5.18 Room sensible heat factor is defined as (RSH = room sensible heat, RLH = room latent heat)
(a) RSH/(RSH + RLH). (b) RLH/(RSH + RLH).
(c) RSH/RLH. (d) none of the above.

5.19 Room sensible heat = 50 kW, Room Latent Heat = 50 kW. Then, room sensible heat factor is equal to
(a) 0.5. (b) 1. (c) 0.75. (d) none of the above.

5.20 During sensible heating process
(a) dry bulb temperature increases. (b) relative humidity decreases.
(c) wet bulb temperature increases. (d) all of the above.

5.21 During sensible heating process
(a) specific humidity remains constant.
(b) partial pressure of vapour remains constant.
(c) dew point temperature remains constant.
(d) all of the above.

5.22 5 kg of moist air stream $A$ has dry bulb temperature of 35°C and humidity ratio of 0.009 kg/kga. This stream mixes with 10 kg of another stream of moist air at dry bulb temperature of 20°C, humidity ratio of 0.003 kg/kg of dry air. The resultant mixture stream has
(a) dry bulb temperature of 25°C and humidity ratio of 0.005 kg/kg of dry air.
(b) dry bulb temperature of 30°C and humidity ratio of 0.007 kg/kg of dry air.
(c) dry bulb temperature of 30°C and humidity ratio of 0.005 kg/kg of dry air.
(d) insufficient data to find properties.

5.23 Air at dry bulb temperature of 35°C and dew point temperature of 20°C passes through a cooling coil, which is maintained at 25°C. The process would be

(a) cooling and dehumidification. (b) sensible cooling.
(c) cooling and humidification. (d) unpredictable.

5.24 Stream of moist air at dry bulb temperature of 35°C and dew point temperature of 25°C passes through a water shower whose temperature is maintained at 20°C. The stream will undergo a process of

(a) cooling and humidification. (b) cooling and dehumidification.
(c) sensible cooling. (d) evaporative cooling.

5.25 Room sensible heat loss = 30000 kJ/hr, room latent heat loss = 20000 kJ/hr. Then, sensible heat is equal to

(a) –0.6. (b) + 0.6. (c) + 0.3. (d) + 3.

CHAPTER SIX

# Reactive Mixtures



## 6.1 INTRODUCTION

In Chapter 4, we discussed mixtures of non-reactive gases. The special feature of such mixtures is that the chemical composition does not change during a process. But in many mixtures, the constituents react with each other, and therefore, the chemical composition is changed after the completion of the process. Some reactions take place in one direction only, and are called irreversible reactions, whereas the others take place in both the directions. That is, the products combine to form the reactants back. Such reactions are called reversible reactions, and under the equilibrium condition, all the reactants and products are present in the mixture; their number of moles depend on the degree of reaction in the forward direction. These concepts are discussed in detail in Sections 6.5. and 6.6.

In a reaction, the chemical substances called reactants come in contact and react to form the product. This is expressed in equation form written using the symbols $\rightarrow$ or $\Leftrightarrow$ between the reactants and products, like the following equations.

$$v_A A + v_B B \rightarrow v_C C + v_D D \quad \text{or} \quad v_A A + v_B B \Leftrightarrow v_C C + v_D D$$

in which $A$ and $B$ are the reactants and $C$ and $D$ the products. The first reaction an irreversible reaction as indicated by $\rightarrow$ whereas second reaction is reversible shown by $\Leftrightarrow$ symbol. While writing these equations, the law of conservation of mass is to be satisfied, which means the number of moles of a chemical element is conserved. That is, the number of each chemical element before and after the chemical reaction has to be the same.

A chemical reaction of importance to mechanical engineers is the combustion process, in which oxygen is one of the constituent of the reactants, which is obtained from atmospheric air. Air contains oxygen, nitrogen and traces of other gases. Atmospheric air, neglecting the presence of other gases, consists of 21% oxygen and 79% nitrogen on mole basis. That is, for each mole of oxygen, air contains 79/21 = 3.76 moles of nitrogen. Or, we need 4.76 moles of air to supply 1 mole of oxygen. However, if we consider the composition of air by mass, then it contains 23.1% $O_2$ and 76.9% $N_2$. That means, we will need $(1 + 76.9/23.1) \approx 4.32$ kg of air to provide 1 kg of $O_2$ for combustion. We may also find out the molecular weight of air using the methods discussed in Chapter 4; it is taken as $28.95 \approx 29$. Nitrogen does not participate in the reaction at low and normal temperatures but at high temperature above 1100°C, it does react with oxygen to form NOx. This is not desirable, and hence we need special techniques to control the quantity of NOx in the products of combustion. This has attracted special attention of designers due to increasing awareness about the air pollution, and stringent laws that are coming in force time to time. Section 6.2 describes the procedure to make a balance of the chemical elements in a chemical reaction.

A chemical reaction may be endothermic or exothermic. In an endothermic reaction, the mixture absorbs heat, whereas heat is given off in an exothermic reaction. One of the jobs for us is to determine this heat exchange. For example, in a combustion chamber in which fuel is burnt by oxygen in air, we would like to determine the heat liberated, so that the maximum temperature and pressure reached may be calculated. This is done by applying principle of the conservation of energy. However, in a chemical reaction, we face the problem of finding out the change in enthalpy or internal energy because of the different datum used for their measurement for different constituents. This is overcome by the introduction of chemical enthalpy, so that the reference may be the same. This is discussed in Section 6.3. Also, for first law analysis, it is convenient to define a reference state for chemical reactions, since the temperatures of products and reactants are different. The reference state is standard atmospheric conditions of 1 bar and 298 K. Section 6.4 explores the thermal analysis.

## 6.2 STOICHIOMETRIC REACTION

A combustion process is like any exothermic process. The procedure to write the chemical equation, described in this section, is applicable to any kind of chemical reaction, and is quite general. In no way is it connected to only a combustion process. But we shall confine our discussions to a combustion process because of its importance, as mentioned before. Normally, fuel is mixed with air in a combustion chamber, and is ignited to burn.

A fuel basically consists of atoms of carbon, hydrogen and a small percentage of sulphur and nitrogen, and can be represented by the chemical formula $CxHy$, neglecting other chemical components. That is, one kmole of the fuel contains $12x$ kg carbon and $y$ kg of hydrogen (recall, molecular weight of carbon and hydrogen is 12 and 1, respectively). Total mass of 1 kmol fuel = $(12x + y)$. The composition of fuel is expressed as mass fraction of the constituents, and is known as ultimate or gravimetric analysis. Thus, gravimetric analysis of this fuel is

$$C = 12x/(12x + y) \times 100\% \text{ and } H = y/(12x + y) \times 100\% \quad (6.1)$$

In case of gaseous fuel, rather, volumetric analysis is used. In this, the volume of each constituent at mixture pressure and temperature is measured, and then expressed as fraction of the total, or the mixture volume measured at mixture pressure and temperature. We learnt in Chapter 4 that volume fraction is also equal to the mole fraction, and hence both can be used interchangeably.

Yet another analysis, called proximate analysis, is also used for solid fuels in which per cent of volatile matter, ash, moisture and fixed carbon is determined. This information is regarding the quality of the fuel.

Mass or number of moles of air required for burning fuel can easily be obtained by writing the chemical reaction. For example, a chemical reaction for carbon, hydrogen and fuel $CxHy$ with just sufficient oxygen required for the reaction may be written as

$$C + O_2 \rightarrow CO_2$$

$$H_2 + \tfrac{1}{2}O_2 \rightarrow H_2O \text{ and}$$

$$CxHy + aO_2 \rightarrow bCO_2 + dH_2O \quad (6.2)$$

where $a$, $b$ and $d$ are the moles of the respective components. We have used the symbol $\rightarrow$ to indicate that the reaction is taking place in that direction. Chemical substances on the left side of this arrow are called reactants and those on right side the products. The reaction must satisfy the law of conservation of mass. This implies that the number of molecules or atoms of each chemical element should be the same on both the sides of the chemical equation. For example, in the second equation, the reactant consists of the mixture of 1 mole of $H_2$ and ½ mole of $O_2$, whereas the product also has the same number of moles of these elements in $H_2O$. This may be verified for other reactions, too. In general, balancing the number of moles of the elements give a number of algebraic equations from which the unknowns can be determined (this is explained later). Presence of $\rightarrow$ shows that we are taking them as irreversible reactions at this point of the discussion. In fact, we shall later see that the first two reactions are reversible at high temperatures beyond 1200 K.

In these equations, no reactant is present on the right hand side. That is, the number of moles of reactant components is just equal to that required for the reaction. This kind of reaction is known as theoretical, stoichiometric, or complete reaction, and coefficients of the chemical components as stoichiometric coefficient, represented by the symbol $\nu$ to differentiate it from the actual number of moles $n$ that may be present in the actual reaction, $\nu_1 \leq n_1$ of component $i$. In the actual combustion process, the number of moles of air is more than the stoichimetrically required value. Thus, in our analysis, both $n$ and $\nu$ will be used, depending on the application.

Stoichiometric coefficients are determined from the principle of conservation of number of moles of the chemical element like carbon, hydrogen, oxygen, nitrogen, etc. The balance of carbon, hydrogen and oxygen for the third reaction in Eq. 6.2 gives $b = x$, $d = y/2$ and $a = b + d/2$. We also observe from the first two reactions in Eq. 6.2, that one mole each of carbon and hydrogen requires 1 and ½ mole of oxygen, respectively. In terms of mass, one kg of carbon requires 1 × 32/12 = 8/3 kg, where 32 and 12 are the molecular weights of $O_2$ and C, respectively. Similarly, 1 kg of hydrogen consumes ½ × 32/2 = 8 kg of oxygen. Knowing the gravimetric analysis of the fuel, it is easy to determine the mass of air needed for its combustion. We may then determine the theoretical ratio of the mass of air and fuel, called the air fuel ratio.[1]

While writing down the chemical reaction for the fuel, we have assumed that the whole of the carbon burns to carbon dioxide. In actual application, the fuel–air mixture enters the combustion chamber, resides in it for some time, and then the products leave the chamber. The period of stay of the fuel in the chamber is known as residence time. The time required for combustion to occur may be more than the residence time, or the amount of oxygen supplied may be less than that required for the reaction to be completed. In both the cases, some carbon monooxide would also form, or some fuel may not burn and passes out of the combustion chamber in the burning state with the flame. Such combustion is called *incomplete combustion*, and the heat produced during the combustion process will naturally be less. In order to save this wastage, some excess air is supplied, so that the fuel burns completely. Excess air supplied for this purpose should be as small as possible. Normally, in most of the boilers, excess air supplied is around 5 to 10%. In some situations, on the contrary, excess air is supplied to control the temperature of the products of combustion. For example, in gas turbines excess air is supplied since the hot gases come in contact with the turbine blade, and may soften the material if the temperature is too high. Temperature of the gases is also kept less to reduce the formation of NOx, which are not environment friendly. In such a case, the chemical reaction can be written as

$$CxHy + aO_2 \rightarrow bCO_2 + dCO + eH_2O + fO_2 + 3.67aN_2 \tag{6.3}$$

The unknowns $a$, $b$, and $e$ are obtained by making mass balance of C, H and O atoms and requires knowledge of $d$ and $f$ from other sources. For example, excess air is

1. A/F ratio has importance because fuel burns *only and only* if this ratio lies within certain range, and temperature of the air fuel mixture is above self-ignition temperature of the fuel.

known from which $f$ can be obtained. Similarly, $d$ would be known, say from the flue gas analysis. Normally, the equation of type 6.3 is written for 100 moles of the dry product of combustion; the reason being that Orsat analysis gives all the coefficients on the RHS of the equation except that of $H_2O$, because it is removed from the sample. Once the equation is written, the actual A/F ratio can be determined. The excess air is given as percentage theoretical air or percentage excess air, defined as below.

Percentage theoretical air

= mass of actual air/mass of theoretical air (6.4)

Percentage excess air

= mass of actual air – mass of theoretical air/mass of theoretical air (6.5)

### 6.2.1 Orsat Analysis

The Orsat analysis is based on the principle that gases $CO_2$, CO and $O_2$ are absorbed in liquid solutions, which are prepared fresh at the time of use to give better accuracy. It gives the molar or volume fraction of these constituents at mixture pressure and temperature. The Orsat apparatus consists of three bulbs $A$, $B$ and $C$, connected as shown in Fig. 6.1, to a common pipe through two-way-cocks in individual line. Bulb $A$ contains 20% solution of potassium hydroxide, which absorbs $CO_2$ (but does not absorb $O_2$ and CO); bulb $B$ contains alkaline solution of pyrogallic acid that absorbs $O_2$, and bulb $C$ has acidic solution of cuprous oxide to absorb CO. The right end of the common pipe is connected to a 100 cc graduated burette, and the other pipe end is connected to the flue gas source through a three-way cock and water absorbent for collecting the sample. The third line of the three-way cock is open to the atmosphere. Water absorbent is used to remove water vapour from the gas sample, and is a hygroscopic crystal that absorbs water from the sample when it passes through it. The free end of the burette is connected to a bottle filled with water through a flexible rubber pipe, so that the water level can be raised or lowered in the burette by moving the bottle up and down. The burette is graduated from 100 at the top to 0 at the bottom.

The first step is to collect the flue gas sample inside the burette. For this, the three-way cock end of the main tube is connected to the sample which is to be analysed. In this position, the three-way cock is closed to the atmosphere. The water bottle that sucks charge into the burette is lowered. After the water has touched the bottom of the burette, the three-way cock is opened to the atmosphere, and the water bottle is moved up to reject the collected charge to the atmosphere. This process is repeated 7 to 10 times, to ensure that whole of the air, which was present initially in the apparatus, is removed and the gases inside the burette represent the true sample. In the last run, the water bottle is lowered so that the water level in the burette is at 0 division. This ensures that 100 ml of gas is taken in. The three-way cock is now closed to both the sample tube and the atmosphere.

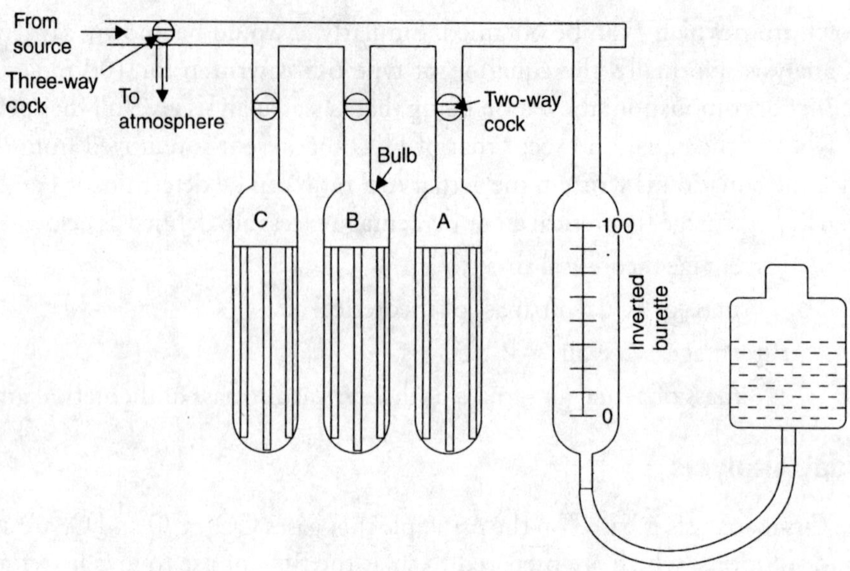


**Fig. 6.1** Orsat apparatus

Now, the cock for bulb *A* is opened (keeping cocks for bulb *B* and *C* closed), and the water bottle is raised and lowered a number of times. This procedure would pass the sample through the solution in *A*, where $CO_2$ would be absorbed. We have to keep the gas in contact with liquid for a long enough time, so that $CO_2$ is completely absorbed. For this, after a few cycles of upward and downward movement of the water bottle, we bring the water bottle to a level, such that the water level in the burette and the bottle are at the same level, and record the division value. If this value does not change in successive runs, then we may assume that the whole of the $CO_2$ has been absorbed. The reading of the burette gives directly the volumetric percentage of $CO_2$ in the flue gas at the atmospheric pressure and temperature. This procedure is then repeated with bulb B keeping other bulbs closed, so that the sample gas flows to it. The new burette reading minus the first reading gives the percentage of $O_2$. Lastly, we get the percentage of CO from bulb *C* in the same way, by passing the sample into bulb *C*. Since we are interested in percentage composition of the flue gas at mixture pressure and temperature, an inverted burette of 100 ml was taken. It is important to keep in mind that the order of the bulb is very important, because KOH solution absorbs only $CO_2$ and not $O_2$ or CO. If by mistake, we pass the gases first in bulb *B*, then not only $O_2$ but also $CO_2$ would be absorbed, as pyrogallic acid solution absorbs $CO_2$ as well.

Once the percentage of $CO_2$, $O_2$ and CO is obtained, then by difference, we determine the percentage of $N_2$ as well. This percentage is nothing but the volume fraction at the mixture pressure and temperature, and hence the mole fraction for dried flue gas. For this reason, this analysis is called the dry flue gas analysis. In industry, gas chromatograph is used for flue gas analysis and that also gives the unburnt fuel in exhaust.

**Example 6.1** Ethane is burnt using air. Write down the combustion equation for (a) Stoichiometric combustion, (b) 400% theoretical air with no CO in products of combustion, and (c) 80% theoretical air.

***Solution*** Stoichiometric reaction can be written as:

$$C_2H_6 + aO_2 + 3.76aN_2 \rightarrow bCO_2 + dH_2O + 3.76aN_2$$

Balance of number of hydrogen, carbon and oxygen atoms, respectively results in: $2d = 6 \rightarrow d = 3$, $b = 2$ and $2a = 2b + d$

$$\rightarrow a = 3.5$$

Therefore, complete combustion equation for ethane is

$$C_2H_6 + 3.5O_2 + 13.16N_2 \rightarrow 2CO_2 + 3H_2O + 13.16N_2$$

Nitrogen is present on both sides of the equation for the simple reason that we use air that contains it and it does not participate in the reaction, bringing no change in its number of atoms.

(b) The reaction with 400% theoretical air would mean that the number of moles of $O_2$ would be $4 \times 3.5 = 14$, and then that of $N_2 = 4 \times 13.16 = 52.64$. It is given that no CO is present, the combustion equation then is

$$C_2H_6 + 14O_2 + 3.76\,(14)N_2 \rightarrow 2CO_2 + 3H_2O + eO_2 + 52.64N_2$$

There would be no change in moles of $H_2O$ or $CO_2$, as the fuel is the same. Balance of $O_2$ atoms give $e = 10.5$. Thus, the reaction is

$$C_2H_6 + 14O_2 + 14\,(3.76)N_2 \rightarrow 2CO_2 + 3H_2O + 10.5O_2 + 52.64N_2$$

(c) We shall assume that there is no free oxygen in the products of combustion in this case. The equation may be written as

$$C_2H_6 + 0.8\,(3.5)O_2 + 0.8\,(3.5)\,(3.76)N_2 \rightarrow bCO_2 + dCO + 3H_2O + 10.528N_2$$

As above, making balance of carbon and oxygen atoms gives $b$ and $d$. The equation is

$$C_2H_6 + 0.8\,(3.5)O_2 + 0.8\,(3.5)\,(3.76)N_2 \rightarrow 0.6CO_2 + 1.4CO + 3H_2O + 10.528N_2$$

**Example 6.2** The dry products of combustion of a hydrocarbon fuel have the following Orsat analysis: 8% $CO_2$, 1% C0, 8.8% $O_2$ and the rest $N_2$. Determine the actual as well as theoretical air fuel ratio, percentage excess air and the composition of the fuel. Fuel contains only hydrogen and oxygen. Take molecular weight of air 28.95.

***Solution*** Let us assume the fuel is C*x*H*y*, and write the combustion equation based on 100 moles of dry products of combustion, since the Orsat analysis is given to us. We obtain percentage of $N_2$ in fuel gas by difference and it comes to 82.2%. Thus, the equation is

$$CxHy + aO_2 + 3.76\,aN_2$$
$$\rightarrow 8CO_2 + 1CO + bH_2O + 8.8O_2 + 82.2N_2$$

Nitrogen, oxygen, hydrogen and carbon balance gives $3.76\,a = 82.2 \rightarrow a = 21.85$, $a = 8 + ½ + b/2 + 8.8 \rightarrow b = 9.1$, $y/2 = b \rightarrow y = 18.2$ and $x = 8 + 1 = 9$. Knowing Orsat analysis, we can also determine the fuel composition as illustrated in this example. The fuel is $C_9H_{18.2}$. The fuel is from alkane series and it is probably $C_9H_{18}$. It seems that some error has been made in the Orsat analysis. However, we shall continue our analysis with $y = 18.2$ instead of 18. The chemical composition is

Mass of 1 mol of fuel = 9 × 12 + 1 × 18.2 = 126.2 = Molecular weight of fuel. Composition of the fuel is

$$C = 12 \times 9/126.2 = 85.6\% \text{ and } H = 18.2/126.2 = 14.4\%$$

Actual air fuel ratio = (4.76 × moles of $O_2$ × Mol wt. of air)/(moles of fuel × Mol wt. of fuel)

$$= (4.76 \times 21.85 \times 28.95)/(1 \times 126.2) = 23.85$$

Theoretical combustion equation can be written as follows

$$C_9H_{18.2} + aO_2 + 3.76aN_2 \rightarrow 9CO_2 + 9.1H_2O + 3.76\,aN_2$$

Oxygen balance gives, $a = 9 + 9.1/2 = 13.55$. Therefore, theoretical air fuel ratio is

Theoretical A/F ratio = (4.76 × 13.55 × 28.95)/(1 × 126.2) = 14.8

Percentage of excess air = 100 × (23.85 – 14.8)/14.8 = 61%

**Example 6.3** A certain coal has the following ultimate analysis: 88% C, 5.5% H, 4.5% O, 1% N and 1% S. Find the equivalent formula for the coal if the latter is to be represented by $C_aH_bO_dN_eS_f$. Also, write down the complete combustion equation.

***Solution*** Ultimate analysis is a way of expressing the composition of the solid fuel based on the basis of total fuel mass, and is obtained in the laboratory using a bomb calorimeter. This percentage is given for fuel weighing 100 kg. The fuel $C_aH_bO_dN_eS_f$ contains 12*a* kg C, *b* kg H, 16*d* kg O, 14*e* kg N and 32 kg S, respectively. Equating these to the percentage given, we get $a = 81/12 = 7.334$, $b = 5.5$, $d = 4.5/16 = 0.281$,

$e = 1/14 = 0.072$ and $f = 1/32 = 0.031$. This gives the chemical formula of the fuel as

$$C_{7.334}H_{5.5}O_{0.281}N_{0.072}S_{0.031}$$

Stoichiometric combustion reaction can be written as follows

$$C_{7.334}H_{5.5}O_{0.281}N_{0.072}S_{0.031} + aO_2 + 3.76a\ N_2 \rightarrow bCO_2 + dH_2O + eSO_2 + fN_2$$

Balance for C, H, N, O and S can be made as explained earlier. We get

$$C_{7.334}H_{5.5}O_{0.281}N_{0.072}S_{0.031} + 8.6O_2 + 3.76\ (8.6)N_2 \rightarrow 7.334CO_2 + 2.75H_2O + 0.331SO_2 + 32.45N_2$$

**Example 6.4** Determine the stoichiometric AF ratio for a hydrocarbon fuel $C_6H_{14}$. The fuel is burnt with 20% excess air. Determine wet volume analysis of the products (a) if all the vapour is present and (b) if the products are cooled to atmospheric temperature of 25°C. Also, determine the dry volumetric analysis and dew point temperature. Finally, estimate the chemical equation if only 80% theoretical air is supplied instead of excess air. Assume atmospheric pressure 1 bar.

***Solution*** Complete combustion equation for $C_6H_{14}$ is (verify it using the method discussed before)

$$C_6H_{14} + 9.5O_2 + 9.5\ (3.76)\ N_2 \rightarrow 6CO_2 + 7H_2O + 35.75N_2$$

AF ratio = (4.76 × 9.5 × 28.95)/(12 × 6 + 1 × 14) = 25.25

For 20% excess air, the combustion equation, assuming no CO in the products of combustion, becomes

$$C_6H_{14} + 1.2\ (9.5)O_2 + 1.2\ (9.5)\ (3.76)N_2 \rightarrow 6CO_2 + 7H_2O + 1.9O_2 + 42.89N_2$$

Total number of moles of the wet products is

$$6 + 7 + 1.9 + 42.89 = 57.79$$

(a) Wet volumetric analysis with all the moisture in products is obtained by expressing the mole of each constituent as mole fraction. This gives $CO_2$ = 6/57.79 × 100 = 10.38% and so on. The result is $CO_2$ 10.38%, $H_2O$ 12.11%, $O_2$ 3.29% and $N_2$ 74.22%. We may determine the dew point of the products if we know the partial pressure of water vapour in it, which is 0.1211 × 1 = 0.1211 bar. From the steam table, saturation temperature corresponding to this pressure is 89.78°C.

(b) From the steam table, saturation pressure of water at 25°C is 0.0174 bar, which is the partial pressure of water vapour in the products after it has been cooled to 25°C. Because of cooling of air, some but not all the water vapour will condense out. Let $y$ be the number of moles left of $H_2O$ vapour in the products. We may determine it using Eq. 5.3. The number of moles = $6 + y + 1.9 + 42.89 = 50.79 + y$, and the mole fraction of water vapour = $y/(50.79 + y)$. Using Eq. 4.3, we get

$$y/(50.79 + y) = 0.174/1 \rightarrow y = 0.9$$

Hence, the wet volumetric composition after cooling of the products is $CO_2$, 11.61%; $H_2O$, 1.74%; $O_2$, 3.68% and $N_2$, 82.97%. In the same way, we may determine the dry fuel gas analysis based by taking zero moles of water vapour. The result is $CO_2$, 11.81%; $O_2$, 3.74% and $N_2$, 84.45%.

When insufficient air is supplied then some of the carbon would burn to CO. It is also possible that some of the hydrogen may not burn at all, but the chances are less, as it has high affinity for oxygen. However, this does not rule out the presence of free hydrogen in the products, and therefore, we assume that no free hydrogen is present. The combustion equation may be written as

$$C_6H_{14} + 0.8\,(9.5)O_2 + 0.8\,(9.5)\,(3.76)N_2 \rightarrow aCO_2 + bCO + 7H_2O + 28.6N_2$$

Balance of carbon and oxygen gives $a + b = 6$ and $2a + b + 7 = 25.2$. Solving these two equations, we get, $a = 2.2$ and $b = 3.8$. Therefore, the equation is

$$C_6H_{14} + 0.8\,(9.5)O_2 + 0.8\,(9.5)\,(3.76)N_2 \rightarrow 2.2CO_2 + 3.8CO + 7H_2O + 28.6N_2$$

## 6.3 ENTHALPY OF FORMATION

We learnt earlier that the reference state for measurement of properties like enthalpy and entropy is different for different chemical compounds. In non-reactive mixtures, as mentioned in Chapter 5, it does not matter, because the composition of the mixture remains the same. However, in a reactive system, this matters a lot, as the components are changed during the reaction changing the chemical composition. Consider a combustion chamber in which a mixture of fuel and air is entering, and products are leaving; Eq. 6.3 represents the type of reaction. Applying sfee to the combustion chamber, we may write

$$H_P = H_R - Q \text{ or } Q = -(H_P - H_R) = -H_{RP} \tag{6.6}$$

in which $H_p$ and $H_R$ are the enthalpy of all the products and reactants, respectively. We may assume the reactants and products to be ideal gases, for which enthalpy is a function of temperature. Therefore, the enthalpy of the products will depend on its temperature, whereas that of reactants will depend on the reactants' temperature. If the gases are not ideal, which is normally so, then it would also depend on the pressure of the products and reactants. Lastly, the amount of heat produced during combustion would depend on (i) the temperature and pressure of the mixtures before and after the reaction, and (ii) the phases of the components of the mixture. For example, if the reactant is in liquid form, then some of the heat will be absorbed by it to vaporise, and hence heat liberated will be decreased. Similarly, the amount of heat produced will depend on whether $H_2O$ is in gaseous or liquid form. We will assume that the reaction is taking place at standard pressure of 1 bar (100 kPa) and temperature of 25°C. Then, $H_{RP}$ is known as *enthalpy of reaction*, and heat released during combustion as the calorific value of the fuel.

We may think of a number of chemical reactions to determine calorific value of the fuel using Eq. 6.6. The different methods would result in different values for calorific value, which is not true, as it depends only on the fuel and not the process. This happens because of different reference states of the same property for different chemical compounds. Therefore, there is need to develop a mechanism, so that the existing property tables for the chemical compounds may be used for such analysis.

This is achieved by defining a common reference state. This state is taken as 1 bar pressure and 25°C temperature, at which enthalpy of the element in its stable form is fixed at zero. Here, stable state is important. For example, oxygen exists in O and $O_2$ form. The stable form of oxygen is $O_2$, and we take enthalpy of $O_2$ and not that of O as zero. But it still does not absolve us from the difficulty we faced, as in combustion equations, we need enthalpy of compounds and not the elements. The solution comes in the form of first law analysis of a stoichiometric reaction, whose reactants are elements in their stable state, occurring at reference temperature and pressure. Say, we wish to find out enthalpy of $CO_2$. We consider a reaction between C and $O_2$, both at 1 bar and 25°C, to produce $CO_2$ (complete combustion) at 1 bar and 25°C. The reaction can be written as

$$C + O_2 \rightarrow CO_2 \tag{6.7}$$

The application of the first law to this reaction gives, $\Sigma v_i h_R = \Sigma v_i h_p + Q$, where summation on the left side is for all the reactants, and on the right for all the products, $v$ is the number of moles of the component $i$ in the reaction, and $Q$ is the heat liberated in this reaction. But for reactants, enthalpy is zero because of the adopted reference. Therefore, we get $-Q = \Sigma v_i h_p = H_p$. Heat produced in the reaction from the elements is called the *enthalpy of formation*, and is given in Table 6.1. The symbol $h_f^0$ is used to represent it; a superscripted 0 means that the reaction proceeds at 1 bar pressure.

The enthalpy for an ideal gas does not depend on pressure, but varies with temperature. It is obtained at any temperature by adding to the enthalpy of formation, the change in enthalpy from 25°C to the temperature, which can be obtained from any gas table (A1.5 and A1.6).

$$\begin{aligned} h^{\circ} &= \text{enthalpy at any temperature} \\ &= h_f^0 + \left(h_T^0 - h_{298}^0\right) \\ &= h_f^0 + \Delta h^0 \end{aligned} \tag{6.8}$$

and
$$H_p = \Sigma n_i \left(h_i^0\right)_p \text{ and } H_R = \Sigma n_i \left(h_i^0\right)_R \tag{6.9}$$

That is, knowing the enthalpy of formation and reading the enthalpy from the gas table for the component at 288 K and T, we can find out the enthalpy of the component at any temperature T. We have to also write down the chemical reaction to know the number of moles $n_i$ of the component $i$.

It is not always necessary to consider the chemical reaction with reactants as basic elements to determine the enthalpy of formation. We shall demonstrate through a solved example, that enthalpy of formation can also be determined by using the Hess law or the first law analysis of any reaction for which enthalpy of reaction is known from the experimental method or otherwise. Enthalpy of formation can also be determined by the application of statistical thermodynamics.

**Table 6.1** Enthalpy, entropy and Gibbs function of formation at 1 bar and 25°C

| Substance | Formula | State | $h_f^0$ kJ/kmole | $s_f^0$ kJ/kmole K | $g_f^0$ kJ/kmole |
|---|---|---|---|---|---|
| Carbon dioxide | $CO_2$ | gas | –393 522 | 213.795 | –394 392 |
| Carbon monoxide | CO | gas | –110 527 | 197.653 | –137 272 |
| Water | $H_2O$ | gas | –241 826 | 188.834 | –228 598 |
| | | liquid | –285 830 | 69.950 | –237 196 |
| Methane | $CH_4$ | gas | –74 873 | 186.251 | –50 794 |
| Ethane | $C_2H_6$ | gas | –84 740 | 229.597 | –32 886 |
| Propane | $C_3H_8$ | gas | –103 900 | 269.917 | –23 489 |
| Butane | $C_4H_{10}$ | gas | –126 200 | 306.647 | –17 255 |
| Octane | $C_8H_{18}$ | gas | –208 600 | 466.514 | 6 611 |
| | | liquid | –250 105 | 360.575 | –353 951 |
| Methanol | $CH_3OH$ | gas | –201 300 | 239.709 | –270 336 |

*(Contd)*

| Substance | Formula | State | $h_f^0$ kJ/kmole | $s_f^0$ kJ/kmole K | $g_f^0$ kJ/kmole |
|---|---|---|---|---|---|
| Ethanol | $C_2H_5OH$ | gas | –239 220 | 126.809 | –275 741 |
| Sulphur dioxide | $SO_2$ | gas | –296 842 | 248.212 | –368 327 |
| Sulphur trioxide | $SO_3$ | gas | –395 765 | 256.769 | –469 722 |
| Nitrogen oxide | $N_2O$ | gas | 82 050 | 171.800 | 32 572 |
| Benzene | $C_6H_6$ | gas | 82980 | 269.562 | 2 651 |
| ***Elements*** | | | | | |
| Carbon | C | solid | 0 | 5.740 | –1 711 |
| Oxygen | $O_2$ | gas | 0 | 205.147 | –61 134 |
| Sulphur | S | solid | 0 | 32.056 | –9 553 |
| Nitrogen | $N_2$ | gas | 0 | 191.610 | –57 097 |

**Example 6.5** Given the two reactions

$$H_2 + ½O_2 \rightarrow H_2O + 286189 \text{ and}$$
$$CO + ½O_2 \rightarrow CO_2 + 282217 \text{ kJ/kmole,}$$

show that the water gas reaction is endothermic requiring 3972 kJ/kmole heat.

***Solution*** Assuming that the reaction is occurring at standard temperature and pressure condition, we may write for the first reaction that enthalpy of formation of water, by definition of enthalpy of formation, is

$$(h_f^0)_{H_20} = -286189 \text{ kJ/kmol} \qquad (A)$$

Similarly, for the second equation, using Eq. 6.6, we get

$$-Q + H_R = H_P,$$

where $H_R = (h_f^0)_{CO} + ½\,(h_f^0)_{O_2} = (h_f^0)_{CO}$ and

$$H_P = (h_f^0)_{CO_2}$$

This gives after substituting the values

$$-286189 + (h_f^0)_{CO} = (h_f^0)_{CO_2}$$

$$\rightarrow (h_f^0)_{CO_2} - (h_f^0)_{CO} = -28217 \qquad (B)$$

Now, consider the water gas reaction, which is

$$CO + H_2O \rightarrow CO_2 + H_2$$

This reaction may be visualised as if $H_2O$ dissociates to its constituents as follows

$$H_2O = H_2 + ½\, O_2 - 286189 \text{ (from the given data)}$$

Oxygen so formed combines with CO to form $CO_2$, given by the chemical reaction

$$CO + ½\, O_2 \rightarrow CO_2 + 282217$$

That is, we may think of the given water gas process may be realised by combining the two intermediate processes. We may use, for determining enthalpy of this reaction, the Hess law, which states that change in enthalpy for any series of reactions between the same initial and final state is the same. This means that the enthalpy of the water gas reaction will be the same as that obtained by considering the two reactions given in the problem. After substituting from Eqs. *A* and *B*, the first law gives for the water gas reaction.

$$H_{RP} = H_P - H_R = [(h_f^0)_{CO_2} - (h_f^0)_{CO}] - (h_f^0)_{H_2O}$$

$$= -282217 - (-286189) = 3972 \text{ kJ/kmole}$$

That is, the water gas reaction is endothermic and heat required is 3972 kJ/kmol. It should be remembered that $H_{RP}$ is negative of heat transfer in the chemical reaction.

**Example 6.6** Determine enthalpy of formation of benzene (l) which gives off heat 41800 kJ/kg during the stoichiometric combustion process.

***Solution*** This is another way of determining enthalpy of formation. In this problem, we know enthalpy of reaction of the process (given). Molecular weight of benzene = 12 × 6 + 1 × 6 = 78. The reaction is

$$C_6H_6\text{ (l)} + 7.5O_2 \rightarrow 6CO_2 + 3H_2O\text{ (l)}$$

$$H_{RP} = H_P - H_R = 6 \times \left(h_f^0\right)_{CO_2} + 3 \times \left(h_f^0\right)_{H_2O(l)} - \left(h_f^0\right)_{C_6H_6(l)}$$

Reading values from Table 6.1, we get

$$= 6 \times (-393522) + 3\,(-285830) - \left(h_f^0\right)_{C_6H_6(l)}$$

$$= -3128622 - \left(h_f^0\right)_{C_6H_6(l)} = -41800 \times 78$$

or $$\left(h_f^0\right)_{C_6H_6(l)} = 41778 \text{ kJ/kmole}$$

## 6.4 FIRST LAW ANALYSIS

Let us assume the reactants are at pressure $p_1$ and temperature $T_1$, which are different from the standard temperature and pressure defined above. These reactants would react to form products at pressure $p_2$ and temperature $T_2$, since during the chemical reaction, the temperature and pressure of the products may change. We may, instead, think of a chain of processes as follows: (i) bring reactants through a non-reactive process to the standard conditions, (ii) control the reaction to proceed at standard pressure and temperature conditions, and (ii) have another non-reactive process, so that the products are at pressure $p_2$ and temperature $T_2$, as shown in Fig. 6.2.

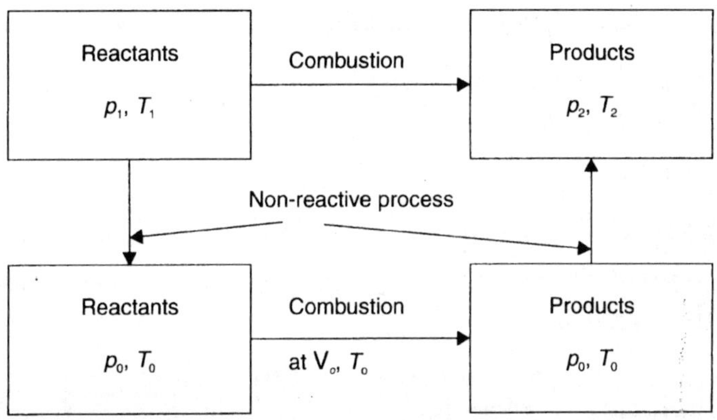


**Fig. 6.2** Combustion with reactants and products not at standard $p$ and $T$

By the Hess law, the change in enthalpy would be the same for the two different ways of performing chemical reactions—one direct from reactants to products, and the second just mentioned above. The reason for carrying out combustion at standard temperature and pressure condition is that *enthalpy of reaction* (A1.7), defined in the last section, is the amount of heat liberated when the reaction is assumed to occur at standard pressure and temperature. That is, the change in enthalpy during a reaction at standard temperature and pressure is known for the reaction, and is available from the tables. Let us use suffixes $R$, $P$ and $x$ to refer to initial, final and reference temperatures, and use $H_p$ and $H_R$ to mean enthalpy of all the components of products and reactants, respectively. That is, $H = \Sigma n_i h_i$ where $n_i$ is the number of moles of component $i$; summation is for all the components. For the alternate route suggested in the opening paragraph, we have

$$H_P - H_R = (H_P - H_{Px}) + (H_{Px} - H_{Rx}) + (H_{Rx} - H_R)$$
$$= (H_P - H_{Px}) + H_{RP} + (H_{Rx} - H_R)$$

Thus, knowing the enthalpy of reaction, we may find out change in enthalpy during the reaction, since the quantities in parentheses can be obtained from the gas tables if the temperature is known.

In exothermic reactions, enthalpy of combustion will be negative, since heat is given off. Enthalpy of combustion is also known as calorific value of the fuel. If $Q$ is the calorific value of the fuel, then

$$H_{RP} = -Q \tag{6.10}$$

In practice, heat of reaction of a fuel is referred to as the lower calorific value of the fuel when determined taking $H_2O$ in products as gas or to higher calorific value when fuel gas is cooled to standard temperature. Heat of reaction will also depend on the phase of the reactant, because if it is in liquid form, then some of the heat produced will be used up in evaporating the reactant to gaseous phase.

We may determine the internal energy of the reaction from the chemical reaction and the knowledge of enthalpy of the reaction. For any chemical reaction, we may write

$$H_{RP} = H_P - H_R \text{ and } U_{RP} = U_P - U_R$$

But $\quad H = U + pV \rightarrow H_P = U_P + (pV)_P$ and

$$H_R = U_R + (pV)_R$$

$$H_{RP} = (U_P - U_R) + [(pV)_P - (pV)_R] = U_{RP} + RT(n_P - n_R) \tag{6.11}$$

where $n_P$ is the total number of moles $= \Sigma n_{pi}$ summed over all the products and $n_R$ is the sum for all the reactants.

**Example 6.7** Calculate enthalpy of combustion of propane at 25°C both on kmole and kg basis under the following conditions. (a) liquid propane with liquid $H_2O$ in the products, (b) liquid propane with gaseous $H_2O$ in the products, (c) gaseous propane with liquid $H_2O$ in the products, and (d) gaseous propane with gaseous $H_2O$ in the products. The enthalpy of evaporation of propane is 370 kJ/kg. Given,

$$\left(h_f^0\right)_{C_3H_8(g)} = -\ 103847 \text{ kJ/kmole}$$

Recalculate the enthalpy of combustion for case (a) if the gas temperature is 200°C.

***Solution*** The combustion equation is

$$C_3H_8 + 5\ O_2 + 5\ (3.76)\ N_2 \rightarrow 3\ CO_2 + 4\ H_2O + 18.8\ N_2$$

Given: $\left(h_f^0\right)_{C_3H_8(g)} = -103847$ kJ/kmol M

$= $ Mol wt. of propane $= 44$

$$\rightarrow \left(h_f^0\right)_{C_3H_8(l)} = \left(h_f^0\right)_{C_3H_8(g)} - \text{Latent heat of propane}$$

$$= -103847 - 44 \times 370$$

$$= -120127 \text{ kJ/kmole}$$

(a) Liquid propane with liquid $H_2O$ in the products

$$H_P = 3\left(h_f^0\right)_{CO_2} + 4\left(h_f^0\right)_{H_2O(l)} + 18.8\left(h_f^0\right)_{N_2}$$

$$= 3\,(-393522) + 4(-285830) + 0 = -2323886$$

$$H_R = \left(h_f^0\right)_{C_3H_8(l)} + 5\left(h_f^0\right)_{O_2} + 18.8\left(h_f^0\right)_{N_2}$$

$$= -120127 + 0 + 0 = -120127$$

$$\therefore \quad H_{RP} = H_P - H_R = -2323886 - (-120127)$$

$$= -2203759 \text{ kJ/kmole}$$

$$= -2203759/44 = 50085.4 \text{ kJ/kg}$$

(b) Liquid propane with gaseous $H_2O$ in the products

$$H_P = 3\left(h_f^0\right)_{CO_2} + 4\left(h_f^0\right)_{H_2O(g)} + 18.8\left(h_f^0\right)_{N_2}$$

$$= 3\,(-393522) + 4\,(-241826) + 0 = -2147870$$

$$H_R = \left(h_f^0\right)_{C_3H_8(l)} + 5\left(h_f^0\right)_{O_2} + 18.8\left(h_f^0\right)_{N_2}$$

$$= -120127 + 0 + 0 = -120127$$

$$\therefore \quad H_{RP} = H_P - H_R = -2147870 - (-120127)$$

$$= -2027743 \text{ kJ/kmole}$$

$$= -2027743/44$$

$$= 46085.1 \text{ kJ/kg}$$

The above procedure may be repeated for cases (c) and (d). We have to use appropriate value of enthalpy of formation keeping in mind the phase of the chemical compound.

Products at 200°C

$$H_P = 3\left[\left(h_f^0\right) + \left(h_T^0 - h_{298}^0\right)\right]_{CO_2} + 4\left[\left(h_f^0\right) + \left(h_T^0 - h_{298}^0\right)\right]_{H_2O(l)} + 18.8\left[\left(h_f^0\right) + \left(h_T^0 - h_{298}^0\right)\right]_{N_2}$$

$$= 3\,[-393522 + 7154.64] + 4\,[-285830 + 5983] + 18.8\,[0 + 5117] = -2182290.5$$

$$H_R = \left(h_f^0\right)_{C_3H_8(l)} + 5\left(h_f^0\right)_{O_2} + 18.8\left(h_f^0\right)_{N_2}$$

$$= -120127 + 0 + 0 = -120127$$

$$\therefore \quad H_{RP} = H_P - H_R = -2182290.5 - (-120127)$$

$$= -2062163.5 \text{ kJ/kmole} = -2062163.5/44$$

$$= -46867.4 \text{ kJ/kg}$$

**Example 6.8** Calculate the internal energy of combustion of gaseous octane with liquid $H_2O$ in the products.

***Solution*** The stoichiometric reaction equation is

$$C_8H_{18}\ (g) + 12.5\ O_2 + 12.5\ (3.76)\ N_2$$
$$\rightarrow 8\ CO_2 + 9\ H_2O\ (l) + 47\ N_2$$

Internal energy of reaction is defined at standard temperature and pressure. Therefore, we assume the reaction to occur at these conditions. Reading enthalpy of formation from Table 6.1, we get

$$H_{RP} = 8\left(h_f^0\right)_{CO_2} + 9\left(h_f^0\right)_{H_2O(l)} - \left(h_f^0\right)_{C_8H_{18}(g)}$$
$$= 8 \times -393522 + 9 \times -285830 - (-208600)$$
$$= -5512046 \text{ kJ/kmole of fuel}$$

Also, $n_P - n_R = 8 + 47 - (1 + 12.5 + 47) = -5.5$ (number of moles of $H_2O$ is not counted as it is in liquid form) and molecular weight of the fuel = $12 \times 8 + 1 \times 18 = 114$

$$\therefore\ U_{RP} = H_{RP} - (n_P - n_R)RT$$
$$= -5512046 - (-5.5) \times 8.3143 \times 298$$
$$= -5498149 \text{ kJ/kmole}$$
$$= -5498149/114$$
$$= -48231.74 \text{ kJ/kg}$$

**Example 6.9** An IC engine burns liquid octane and uses 120% theoretical air. The air and fuel enter at 25°C and the products leave the engine at 900 K. In the engine, 80% carbon burns to $CO_2$, and the remainder burns to CO. The heat transfer from this engine is just equal to the work done. Determine (a) the power output of the engine if the engine burns 2 kg of fuel per hour, (b) the Orsat analysis of the products of combustion, and (c) dew point temperature.

***Solution*** The stoichiometric combustion equation can be obtained by conservation of moles as

$$C_8H_{18}\ (l) + 12.5O_2 + 12.5\ (3.76)N_2 \rightarrow 8CO_2 + 9H_2O\ (l) + 47N_2$$

Therefore, the actual combustion reaction with 120% theoretical air and 80% C burning to $CO_2$ is

$$C_8H_{18}\ (l) + 1.2 \times 12.5O_2 + 1.2 \times 12.5\ (3.76)N_2 \rightarrow 6.4CO_2$$
$$+ 1.6CO + 9H_2O\ (g) + 2.5O_2 + 56.4N_2$$

We may read enthalpy of formation from Table 6.1 and enthalpy of various chemical compounds at 288 and 900 K from the gas tables, and then determine enthalpy of products and reactants, as explained above. This gives

$$Hp = 6.4[-393522 + 28041] + 1.6[-110527 + 18397] + 9[-285830 + 21924] + 2.5 \times 19246 + 56.4 \times 18221$$
$$= -3770446$$

$$H_R = \left(h_f^0\right)_{C_8H_{18}(l)} = -250105$$

Applying the first law to the engine taking the contents inside the cylinder as the system, we get

$$Q + H_R = W + H_P \rightarrow 2W = H_R - H_P$$
$$= -250105 - (-3770446)$$

or $W = 1760171$ kJ/kmole of fuel

$= 1760171/114 = 15440$ kJ/kg

Power of the engine = $(2/3600) \times 15440 = 8.58$ kW.

Total number of moles or the dry products

$= 6.4 + 1.6 + 2.5 + 56.4 = 66.3$

Therefore, volumetric analysis is $CO_2$, 9.45% (6.4/66.3x100); CO, 2.36%; $O_2$, 4.88% and $N_2$, 83.31%. Also, partial pressure of water vapour = $9/(9 + 66.3) \times 1 = 0.1177$ bar. From the steam table, dew point temperature corresponding to this pressure is 48.68°C.

### 6.4.1 Adiabatic Flame Temperature

When a combustion reaction occurs, the temperature increase will depend on energy interaction of the system with its surrounding. If there were no such interaction then maximum temperature would be attained. This temperature is called the adiabatic flame temperature, and is an important parameter to be determined for design of the combustion chamber. Consider a combustion chamber in which the reactants enter from one side and the products leave from other side, and assume that it is insulated. Application of the first law gives $H_P = H_R$, where $H_P$ and $H_R$ can be obtained using Eq. 6.8 and 6.9. Since temperature is unknown, the solution of this equation is not straightforward, and is solved using the hit and trial method. We assume some temperature and read the enthalpy of all the components forming products from the gas table at the assumed temperature and 298 K. This gives $\Delta h^0$ for these compounds. This is, however, a time consuming process, and sometimes irritating, too. But there is no solution to it.

Adiabatic temperature obtained from the first law analysis is erroneous, because at high temperature, $CO_2$ dissociates into CO and $O_2$; nitrogen oxides also dissociate; even nascent oxygen and radicals are formed. Moreover, the dissociated components combine back, that is, the process of dissociation is reversible, and hence the number of moles will depend on the equilibrium condition of the chemical reaction. Equilibrium is reached when the rate of forward and backward reaction is equal. The number of moles can be determined by considering the equilibrium of the reaction (discussed later). Therefore, the trial and error technique, combined with the dissociation process, which is again a function of temperature, makes the whole process of determining the adiabatic flame temperature very involved and complex.

**Example 6.10** A certain fuel consisting of a mixture of various compounds has a chemical analysis of 86% C, 13% H and 1% N on mass basis and has a higher calorific value of 44350 kJ/kg. This fuel at 25°C is burned with 200% theoretical air at 25°C in a stable state, steady flow process. Determine the adiabatic flame temperature.

***Solution*** Since the ultimate analysis of the fuel is known, we may determine its chemical formula by equating the mass of C, H and N in 100 kg of fuel. This gives

$$12a = 86 \rightarrow a = 7.17,\ b = 13 \text{ and}$$
$$14d = 1 \rightarrow d = 0.071$$

Therefore, the chemical formula of the fuel is $C_{7.17}H_{13}N_{0.071}$ and its molecular weight is 100. The chemical reaction may be obtained by balancing the number of moles of C, H, and O to give

$$C_{7.17}H_{13}N_{0.071} + 10.49\ O_2 + 10.49(3.76)\ N_2$$
$$\rightarrow 7.17\ CO_2 + 6.5\ H_2O + 0.071\ NO2 + 39.4\ N_2 \quad \text{(A)}$$

Therefore, the actual reaction with 200% theoretical air, assuming C changing to $CO_2$ only, is

$$C_{7.17}H_{13}N_{0.071} + 2 \times 10.49\ O_2 + 2 \times 10.49(3.76)\ N_2$$
$$\rightarrow 7.17\ CO_2 + 6.5\ H_2O + 0.071\ NO2 + 10.49\ O_2 + 78.8\ N_2$$

Since the products are not at 25°C, we shall use Eqs. 6.8 and 6.9 to find the enthalpy of the products. We get

$$H_P = 7.17(-393522) + 6.5\ (-241826) + 0.071\ (33723) +$$
$$7.17\ (h_T^0 - h_{298}^0)_{CO_2} + 6.5\ (h_T^0 - h_{298}^0)_{H_2O} +$$
$$0.071\ (h_T^0 - h_{298}^0)_{NO_2} + 10.49\ (h_T^0 - h_{298}^0)_{O_2} +$$
$$78.8\ (h_T^0 - h_{298}^0)_{N_2}$$

$$= 7.17\,(h_T^0 - h_{298}^0)_{CO_2} + 6.5\,(h_T^0 - h_{298}^0)_{H_2O} +$$
$$0.071\,(h_T^0 - h_{298}^0)_{NO_2} + 10.49\,(h_T^0 - h_{298}^0)_{O_2} +$$
$$78.8\,(h_T^0 - h_{298}^0)_{N_2} - 4394987$$

We are neither given the higher calorific value, nor the enthalpy of formation of the fuel. By considering the stoichiometric combustion of the fuel at standard condition (Eq. A), it will be possible to find out the enthalpy of formation. We get

$$H_{RP} = H_P - H_R$$
$$= 7.17(-393522) + 6.5\,(-285830) +$$
$$0.071\,(33723) - (h_f^0)_{\text{fuel}} = 4677053 - (h_f^0)_{\text{fuel}}$$

Note that enthalpy of formation for $H_2O$ (l) has been taken in $H_{RP}$ relation because higher calorific value is given. Substituting the calorific value for $H_{RP}$ and simplifying we get

$$(h_f^0)_{\text{fuel}} = -244460 \text{ kJ/kmole}$$
$$= H_R \text{ for actual combustion reaction}$$

Since there is no work and heat transfer, $H_P = H_R$. Equating the two and simplifying, we get

$$= 7.17\,(h_T^0 - h_{298}^0)_{CO_2} + 6.5\,(h_T^0 - h_{298}^0)_{H_2O} +$$
$$0.071\,(h_T^0 - h_{298}^0)_{NO_2} + 10.49\,(h_T^0 - h_{298}^0)_{O_2} +$$
$$78.8\,(h_T^0 - h_{298}^0)_{N_2}$$
$$= 4150528 \qquad \text{(B)}$$

We may have some idea about the temperature that would reach by assuming that the product is all nitrogen for the reason that it has the largest number of moles. Thus, the total number of moles

$$= 7.17 + 6.5 + 0.071 + 10.49 + 78.8 \approx 103$$

Therefore, $H_P = H_R$ with only nitrogen as product gives

$$= 103\,(h_T^0 - h_{298}^0)_{N_2}$$
$$= 4150528 \text{ or } (h_T^0 - h_{298}^0)_{N_2}$$
$$= 40296.4 \qquad \text{(C)}$$

Now, from the gas table for nitrogen, we may find out the temperature that would satisfy Eq. C. It is more than 1500 K. However, the products also contain triatomic gases like $CO_2$, $H_2O$ and $NO_2$, which have higher specific heat as compared to diatomic gases. This results in decrease in temperature. Also, as discussed previously, dissociation of gases at high temperature would reduce the temperature. So, we assume two adiabatic flame temperatures, say 1400 and 1500 K, and read $(h_T^0 - h_{298}^0)$ for different components appearing in Eq. A. The calculation may be done in tabular form as shown in Table 6.2.

**Table 6.2** Determination of flame temperature

| Temperature | 1400 K | 1500 K |
|---|---|---|
| $7.17\ (h_T^0 - h_{298}^0)_{CO_2}$ | 400618 | 443504 |
| $6.5\ (h_T^0 - h_{298}^0)_{H_2O}$ | 282420 | 311708 |
| $0.071\ (h_T^0 - h_{298}^0)_{NO_2}$ | 3841 | 4226 |
| $10.49\ (h_T^0 - h_{298}^0)_{O_2}$ | 387020 | 426768 |
| $78.8\ (h_T^0 - h_{298}^0)_{N_2}$ | 2761440 | 3031308 |
| LHS | 3856259 | 4217514 |

Comparing the total given at the bottom with RHS of Eq. C, we observe that the temperature would lie between 1400 and 1500 K, and may be determined by linear interpolation. The adiabatic flame temperature is = 1400 + (40296.4 – 3856259)/(4217514 – 3856259) × 100 = 1481.5 K.

## 6.5 SECOND LAW ANALYSIS

Second law analysis revolves around determination of change of entropy of the system and universe, or of availability and irreversibility in a process. It was discussed in Chapter 3 and we would use the knowledge for analysis of a chemical reaction. For a flowing stream of a single component, the availability function (Darrius function) was defined as

$$b = h - T_o s \tag{6.12}$$

This can be extended to mixtures as well, in which h and s are replaced by $\Sigma n_i h_i$ and $\Sigma n_i s_i$, respectively, where summation is carried over all the components i in the mixture. It is assumed that all the constituents of the mixture are at reference temperature and pressure, since entropy in Table 6.1 is listed at this reference condition. That is, we assume that the fuel, both $O_2$ and $N_2$ in air, and each constituent of the products are at 1 bar

25°C. We may note that this does not represent the actual situation, as the individual constituents would be at their partial pressure and not at the mixture pressure, if the mixture of fuel and air is prepared first and then supplied to the combustion chamber. However, the partial pressure of each constituent may be calculated, as discussed in Chapter 5, and correction for effect of pressure on entropy be made. Correction for one mole of a component $i$ is given by

$$s_i - s^0 = -R\ln(p_i/p^0) = R\ln(1/y_i), \text{ where } y_i = \text{mole fraction} = p_i/p^0 \qquad (6.13)$$

where ° over the quantity represents its value at reference condition, and $p_i$ is the partial pressure of constituent $i$. The ratio of pressures in Eq. 6.13 is the mole fraction $y_i$ of the component $i$, which is known from the actual chemical reaction. Note the (–ve) sign is absorbed and the argument for function ln is $(1/y_i)$ instead of $y_i$. Figure 6.3 shows the difference in the two cases.

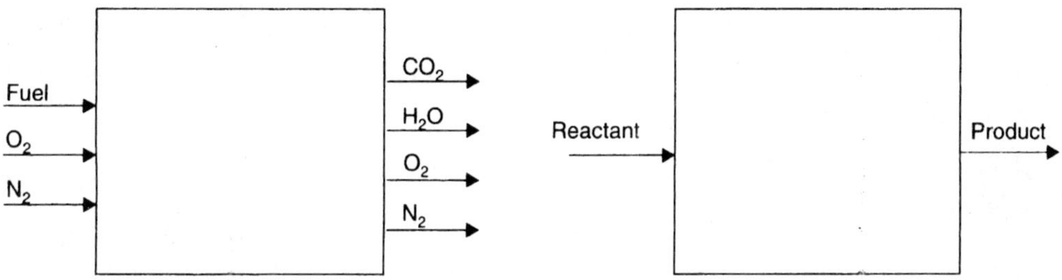


(a) Each constituent of reactant and product at 25°C, 1 bar

(b) Reactants and products at 1 bar and 25°C

**Fig. 6.3** Models for combustion of fuel

Maximum work is obtained if the products are brought to a dead state without any heat transfer to the surrounding, and may be obtained using Eq. 6.12. In absence of changes in kinetic and potential energy, this gives

$$W_{rev} = \Sigma n_R\left(h^0 - T_0 s^0\right)_R - \Sigma n_p\left(h^0 - T_0 s^0\right)_P$$

$$= -H_{RP} - T_0\left[\Sigma n_R\left(s^0\right)_R - \Sigma n_p\left(s^0\right)_P\right] \qquad (6.14)$$

Here, $h^\circ$ is obtained at any temperature using Eq. 6.8, and $s^\circ$ in the same way as for $h^\circ$, using the following equation

$$s^\circ = \text{enthalpy at any temperature} = s_f^0 + \left(s_T^0 - s_{298}^0\right) = s_f^0 + \Delta s^0 \qquad (6.15)$$

Equation 6.15 gives the entropy, assuming that the component $i$ is at standard pressure, and the correction in change in entropy due to change in pressure is subsequently made using Eq. 6.13.

Usually, the fuel and air in the reactants are at atmospheric pressure and temperature, which is close to the reference condition, and may be considered to be at reference pressure and temperature without significant error. Then, $(h^0 - Ts^0)$ for the fuel is its Gibbs function, which is tabulated in Table 6.1 for different compounds. If a chemical reaction takes place in such a manner that both the reactants and products are in temperature equilibrium with the surroundings, then '$g$' may be used in place of $h - Ts$. Equation 6.14 then becomes

$$W_{rev} = \Sigma(n_i\, g_i)_R - \Sigma(n_i\, g_i) = -\Delta G \tag{6.16}$$

where,

$$\Delta G = \Delta H - T_0\, \Delta S \tag{6.17}$$

*That is, when the final state of the products is in equilibrium with the surroundings, then the maximum reversible work that may be obtained from the fuel is availability of the fuel and air.*

Irreversibility in a combustion process is obtained by calculating the availability of the stream before and after the reaction. In case of adiabatic combustion, the difference will be the irreversibility due to combustion only. It is given by

$$I = T_0\left(\Sigma(n_i\, g_i)_R - \Sigma(n_i g_i)_p\right) = \psi_R - \psi_P \tag{6.18}$$

where, $\psi$ is availability. The combustion process is used in a gas turbine, boiler and internal combustion engine. The purpose of such a process is to burn fuel to raise the temperature of the products to the maximum temperature allowed by the blade material of the gas turbine, raising steam in the boiler and produce work in IC engines. In all the three applications, the input is the exergy of fuel. Therefore, any efficiency should be defined on the basis of the application.

**Example 6.11** Determine the maximum work that would be obtained in Example 6.9 if this combustion took place reversibly and in pressure and temperature equilibrium with the surroundings.

***Solution*** The combustion equation for Example 6.9 is

$$C_8H_{18}\ (l) + 15\ O_2 + 15\ (3.76)\ N_2 \rightarrow 6.4\ CO_2 + 1.6\ CO + 9\ H_2O\ (g) + 3.3\ O_2 + 56.4\ N_2$$

Since the products are at 900 K, there is increase in enthalpy and entropy due to temperature change from 298 K. Entropy also changes because the gaseous components are at their partial pressure, and this change is given by Eq. 6.13. Calculation in tabular form make determination of these quantities easy. The following tables list the entropy of the reactants and products, using Eqs. 6.13 and 6.15, with $R = 8.3143$ (column for $h$ is not added, since enthalpy of products and reactants were determined before, but it may be added too). Remember, the effect of change in pressure will be on the gaseous components.

**Table 6.3** (a) Entropy of the reactants with $R = 8.3143$

| Constituent | $n$ | $1/y$ | $R \ln(1/y)$ | $s_f^0$ | $s^0$ |
|---|---|---|---|---|---|
| $O_2$ | 15 | 4.76 | 12.972 | 205.147 | 218.119 |
| $N_2$ | 56.4 | 1.266 | 1.961 | 191.61 | 193.57 |
| $y = \Sigma n_i$ | 71.4 | | | | |

**Table 6.3** (b) Entropy of the products with $R = 8.3143$

| Constituent | $n$ | $1/y$ | $R \ln(1/y)$ | $s_f^0$ | $(\Delta s)_{900K}$ | $s^0$ | $(s^0)_P$ |
|---|---|---|---|---|---|---|---|
| $CO_2$ | 6.4 | 11.984 | 20.6494 | 213.795 | 49.852 | 284.296 | 1819.5 |
| CO | 1.6 | 47.938 | 32.175 | 197.653 | 33.423 | 263.258 | 421.228 |
| $H_2O$ (g) | 9.0 | 8.5222 | 17.8149 | 69.950 | 39.625 | 127.39 | 1146.51 |
| $O_2$ | 3.3 | 23.2424 | 26.1566 | 191.61 | 34.602 | 252.47 | 833.15 |
| $N_2$ | 56.4 | 1.3600 | 2.5561 | 191.500 | 32.968 | 227.026 | 12804.27 |
| $y = \Sigma n_i$ | 76.7 | | | | | | 20411.86 |

Now, the total enthalpy and entropy of the products and reactants can be obtained. We get

$Hp = -3770446$ and $H_R = (h_f^0)_{C_8H_{18}(l)} = -250105$ (obtained in Example 6.9)

$$(s^0)_P = 20411.86 \text{ (see table)}$$

$$(s^0)_R = 1 \times 360.575 + 15 \times 205.147 + 56.4 \times 191.61$$

$$= 14235.35$$

$$W_{rev} = (H_R - H_P) - T_0 \{(s^0)_R - (s^0)_P\}$$

$$= -250105 - (-3770446) - 298 \times (14235.35 - 20411.86)$$

$$= 5360949 \text{ kJ/kmole} = 47025.87 \text{ kJ/kg}$$

We obtained the actual work in example 6.4.3 = 15440 kJ/kg. The overall exergetic efficiency of the combustion process in the IC engine is

exergetic efficiency = 15440/47548.7 × 100 = 32.83%.

**Example 6.12** Let us consider an adiabatic combustion of ethene gas ($C_2H_4$) in 400% excess air, and assume that each constituent in the reactant and product is at 1 bar (to eliminate the effect of mole fraction on entropy), and products attain adiabatic flame temperature. The surrounding temperature is 25°C. Find out the irreversibility in the combustion process, and suggest how can it be reduced.

***Solution*** We assume that flow to the combustion chamber is steady. Following the method to write the reaction equation with 400% excess air, we get

$$C_2H_4\ (g) + 12\ O_2 + 12\ (3.76)\ N_2$$
$$\rightarrow 2\ CO_2 + 2\ H_2O\ (g) + 9\ O_2 + 45.1\ N_2$$

For adiabatic process, $H_P = H_R$. By the trial and error method, as discussed before, the temperature comes to 1016 K. At this temperature, entropy is determined from the gas table reading $s_f^0$ from Table 6.1, as it is required for calculation of irreversibility. The results for products are given in Table 6.4.

**Table 6.4** Entropy of the products with $R = 8.3143$

| Constituent | $n$ | $s_f^0$ | $s^0$ | $n \times s^0$ |
|---|---|---|---|---|
| $CO_2$ | 2 | 213.795 | 270.194 | 540.388 |
| $H_2O(g)$ | 2 | 69.950 | 233.355 | 466.71 |
| $O_2$ | 9 | 205.147 | 244.135 | 2197.215 |
| $N_2$ | 45.1 | 191.61 | 228.691 | 10313.96 |
| Total | | | | 13518.28 |

Also, $S_R = 219.33 + 12 \times 205 + 45.1 \times 191.61 = 1132.7$

$\therefore\ S_P - S_R = 13518.28 - 1132.7$

$= 2195.58$ kJ/kmole K and

$I = T_0 \times (S_P - S_R) = 654610$ kJ/kmole

$= 654610/28 = 23334$ kJ/kg fuel

where, 28 is the molecular weight of ethene. We may also find out the irreversibility by determining the availability at the inlet and outlet of the combustion chamber, and verify that it also gives the same result. However, the maximum reversible work that may be obtained from the reactant through a combustion process is obtained when the reaction occurs such that it is in equilibrium with the surroundings, and both the products and reactants are at 1 bar and 25°C. In this case, since the temperature is 25°C, no change in entropy or enthalpy with temperature is to be obtained. Since $O_2$ and $N_2$ are appearing on both the sides and their temperature is the same, these cancel out, except 3 moles of $O_2$ on the reactant side. Carrying out the procedure to find out reversible work, we get

$$\Delta H^0 = H_{RP} = H_P - H_R$$
$$= [2(-393522) + 2(-241826)] - (52467)$$
$$= -1323163 \text{ kJ/kmole}$$
$$\Delta S^0 = S_P^0 - S_R^0$$
$$= [2(213.795) + 2\,(188.843)] - [219.33 + 3(205.148)]$$
$$= -\,29.516 \text{ kJ/kmole K}$$
$$\Delta G^0 = \Delta H^0 - T_0 \times \Delta S^0$$
$$= -1314363 \text{ kJ/kmole fuel}$$
$$= -\,46851 \text{ kJ/kg fuel}$$
$$\therefore\ W_{rev} = -\Delta G^0 = 46851 \text{ kJ/kg}$$

and $\psi_P$ = availability of products = $W_{rev} - I$ = 23517 kJ/kg

We note that the availability of products is = $\psi_P/\Delta G^0$ 23517/46851 ≈ 50%. Since no work has been obtained during the combustion, 50% of available energy is lost in combustion itself, which is a major part. We also observe that in this case, $\Delta G^0$ and $\Delta H^0$ are of the same order of magnitude. However, this may not be so in all the fuels.

Since enthalpy increases with temperature linearly and entropy logarithmically, we note that if the temperature of products were increased, then irreversibility in the combustion process would decrease. Further, heat is available at high temperature; its availability would also be large. Therefore, it is advantageous to go for higher temperature in the combustion chamber, provided the material of the chamber permits it. The other restriction may be from the application. For example, in gas turbines, we have to also ensure that the life of the blade is not severely affected due to high temperature. A little insight into the chemical reaction reveals that temperature attained would be high if the number of moles is reduced.

> This singles out $N_2$ molecules, as their number is large, and are present because we get $O_2$ from air. Thus, using $O_2$ in place of air would decrease irreversibility in the combustion chamber.

## 6.6 CHEMICAL EQUILIBRIUM AND DISSOCIATION

So far, we have discussed irreversible reactions in which the reactants are completely consumed and the products are produced. Such a reaction was shown by the arrow $\rightarrow$. But there are many reactions in which the reaction proceeds in both directions. This is so because the products also combine to form reactants in which the reaction proceeds in the reverse direction. Such a reaction is shown by a double arrow $\Leftrightarrow$, and at any time, all the reactants and products will be present in the reaction vessel. Thus, a chemical reaction between reactants $A$ and $B$ to produce $C$ and $D$ under equilibrium condition is written as:

$$\nu_A A + \nu_B B \Leftrightarrow \nu_C C + \nu_D D \tag{6.19}$$

The reversible equation 6.19 under equilibrium condition, is called the equilibrium equation. Accordingly, the reactants and products are called the equilibrium reactant and product, respectively. It is also possible that some of the components of the reactants and products may dissociate under favourable conditions. For example, at high temperature, $CO_2$ and $H_2O$ dissociate. We shall consider dissociation later, and we assume no dissociation in our analysis at present.

We would like to know the equilibrium condition for any such reaction. We discussed many kinds of equilibrium in Chapter 1. Thermal equilibrium requires equality of temperature, whereas a mechanical equilibrium demands equality of pressure throughout the system and between the system and the surrounding. We also mentioned a few other intensive properties of the system that have to be the same for the system, and must be equal to that of the surrounding for equilibrium of the system. So, the question that arises is: what should be this property for chemical equilibrium? Let us consider a mixture of $n$ chemical components 1, 2, ..., $n$ having number of moles $n_1$, $n_2$, etc. at pressure $p$ and temperature T and occupy volume V; the number of moles may not necessarily be the stoichiometric coefficients required for an ideal reaction. For such a system, we may write

$$U = U(V, T, n_1, n_2, \ldots), H = H(p, T, n_1, n_2, \ldots)$$
$$F = F(V, T, n_1, n_2, \ldots) \text{ and}$$
$$G = G(p, T, n_1, n_2, \ldots)$$

or $dU = (\partial U/\partial V)_{T,n_i n_k}\, dV + (\partial U/\partial T)_{V,n_i n_k}\, dT + \Sigma(\partial U/\partial n_i)_{T,V n_k}\, dn_i$

(summation for all $k \neq i = 1, 2, \ldots, n$)

$$= (\partial U/\partial V)_{T,n_i n_k} dV + (\partial U/\partial T)_{V n_i, n_k} dT + \Sigma \mu_i dn_i$$

where $\mu_i = \Sigma(\partial U/\partial n_i)_{T, Vn_k}$

Similarly, we may write for change in Gibbs function, enthalpy and Helmholtz function. Relation for the Gibbs function is;

$$dG = (\partial G/\partial p)_{T,n_i n_k} dp + (\partial G/\partial T)_{P,n_i n_k} dT + \Sigma\mu_i dn_i$$

$$= \Sigma\mu_i dn_i \text{ when } dT = dp = 0$$

In the above equations, the notation $\mu_i$, called the *chemical potential*, was used and is defined by the following equations

$$\mu_i = (\partial U/\partial n_i)_{T,V n_k} = (\partial H/\partial n_i)_{T,P,n_k} = (\partial F/\partial n_i)_{T,V,n_k} = (\partial G/\partial n_i)_{T,P,n_k} \quad (6.20)$$

where, *i* represents one constituent and *k* stands for all other constituents. Examination of this equation shows that the chemical potential is a function of properties, and hence is itself a property, but its value does not depend on the type of the property. Further, its value does not depend on the size of the system, and hence it is an intensive property. Thus, we conclude that chemical property is an independent and intensive property of the system consisting of a mixture of pure substances. Lastly, we observe that $\mu_i$ is a property of the component *i* only, and depends on the temperature and pressure of the system, that is, $\mu = \mu(p, T)$. From Eq. 6.20 for a single component, we may write that

$$dG_i = \mu_i dn_i \text{ or } G_i = \mu_i n_i \rightarrow g_i = \mu_i$$

That is, chemical potential for a pure substance at constant pressure and temperature is equal to the specific Gibbs function of that substance. Replacing *g* by $\mu$ and noting from Chapter 4 (Eq. 4.18) for a pure substance that

$$s = -(\partial g/\partial T)_p = -(\partial \mu/\partial T)_p,$$

we get from $g = h - Ts$

$$\mu = h + T(\partial \mu/\partial T)_p \text{ or } h = \mu - T(\partial \mu/\partial T)_p$$

or

$$h = -\left[T(\partial \mu/\partial T)_p - \mu\right] = -T^2\left[\partial(\mu/T)/\partial T\right]_p \quad (6.21)$$

Now reverting back to the question of equilibrium of the chemical reaction, shown by Eq. 6.19, the chemical equation at any stage may be written as

$$\nu_A A + \nu_B B \rightarrow (1-\varepsilon)\nu_A A + (1-\varepsilon)\nu_B B + \varepsilon\, \nu_C C + \varepsilon\, \nu_D D \quad (6.22)$$

where, $\varepsilon$ is called the degree of reaction, and in general, it is less than 1. This equation shows that $\varepsilon\nu_A$ mole of *A* and $\varepsilon\nu_B$ mole of *B* react to form $\varepsilon\nu_C$ moles of *C* and $\varepsilon\nu_D$ moles of *D*, respectively, and components *A* and *B* are still present at that stage of the reaction. That is, *the right hand side of Eq. 6.22 represents the composition of the mixture at that time*. Thus degree of reaction gives change in number of moles of reactants as a fraction of their stoichiometric composition which combine at that instant. At any stage of the reaction, the number of moles of $A$ or $B$ will be

$$n_A = (1-\varepsilon)\nu_A + p, n_B = (1-\varepsilon)\nu_B + q, n_C$$
$$= \varepsilon\nu_C + r \text{ and } n_D = \varepsilon\nu_D + s \qquad (6.23)$$

where, $p$, $q$, $r$ and $s$ are constants greater than or equal to zero necessary for chemical balance. Equation 6.22 includes the effect of the change taking place continuously. Differentiating each equation in Eq. 6.23 with $\varepsilon$, since this is the independent variable, we get:

$$-dn_A/\nu_A = -dn_B/\nu_B = dn_C/\nu_C = dn_D/\nu_D = d\varepsilon \qquad (6.24)$$

This equation is called the *equation of constraint*. Here, it is worth repeating that in a reaction, it is not necessary to always have the number of moles of reactants equal to the stoichiometric coefficient. A mixture may be rich, in which case the number of moles of fuel present in the mixture of reactants is more than the stoichiometric value. On the contrary, the mixture may be lean, in which case the moles of air are more. However, Eq. 6.24 does not depend on the actual number of moles present.

Now, it is time to find the condition required for equilibrium of the system. This condition derives from the second law of thermodynamics, which states that for an isolated system, the entropy will never decrease during a process. That is, the entropy of a system in equilibrium is maximum. This condition can be modelled in different mathematical forms depending on the type of system. We can show that this principle of entropy leads to another principle, that the system will have minimum energy on the state of equilibrium. We will agree for it immediately, because we know that any effort to put a ball on the upper side of a slope brings it back to the level of minimum potential energy. We also learnt the temperature entropy equation

$$TdS = dU + pdV$$

for a pure substance subjected to a reversible process. However, in case of a irreversible process too, it takes the form

$$TdS \geq dU + pdV$$

This condition reduces for a constant volume ($V = c$) process occurring in an isolated system ($U = c$) to

$$(dS)_{U,V} \geq 0$$

The other criterion may be found out by writing in terms of $H$, $F$ and $G$. Change in Gibbs function for a small change $dT$ and $dp$ in temperature and pressure, respectively, for a reversible or irreversible process is

$$dG = Vdp - SdT \leq 0$$

$dG$ less than zero implies a spontaneous reaction generating entropy in the process, and zero means that the reaction is possible and will proceed till the equilibrium condition is reached. $dG > 0$ is not possible, because it will violate the second law. Thus, for equilibrium of a chemical reaction, the required condition is $dG]_{T,p} = 0$. But $G = \Sigma n_i \mu_i$. This gives, after substituting from Eq. 6.24, and simplifying

$$dG]_{T,p} = \Sigma\, dn_i \mu_i = \Sigma - \nu_i \mu_i d\varepsilon = 0$$

*Here, summation is for all the reactants and products,* with the (+ve) sign for the reactant and the (–ve) for the product. That is, for the reversible reaction, Eq. 6.19

$$dG]_{T,p} = (-\nu_A \mu_A - \nu_B \mu_B + \nu_C \mu_C + \nu_D \mu_D)\, d\varepsilon = 0 \tag{6.25}$$

But $d\varepsilon \neq 0$, therefore, the quantity in parenthesis in Eq. 6.25 has to be zero. This gives

$$\nu_A \mu_A + \nu_B \mu_B = \nu_C \mu_C + \nu_D \mu_D \tag{6.26}$$

This is an expression for conditions of equilibrium in terms of stoichiometric coefficients and chemical potential. Chemical potential, like any other property, has external reference and can be expressed at any temperature T and pressure $p$ as

$$\mu = \mu^0 + RT ln p \tag{6.27}$$

Substituting for $\mu$ in Eq. 6.26 and simplifying, we get

$$RT\, ln \frac{p_C^{\nu_c}\, p_D^{\nu_D}}{p_A^{\nu_A}\, p_B^{\nu_B}} = \nu_A \mu_A^0 + \nu_B \mu_B^0 - \nu_C \mu_C^0 - \nu_D \mu_D^0 = -\Delta G_T^0 \tag{6.28}$$

where, $\Delta G_T^0$ is the difference in standard chemical potentials at the reference state. We know that partial pressure of any component is related to its mole fraction. For example, if $p$ is the mixture pressure, then

$$p_A = x_A p,\ p_B = x_B p,\ p_C = x_C p,\ p_D = x_D p$$

We define the equilibrium constant, $Kp$ as follows, change partial pressure in terms of mole fraction and using Eq. 6.28, resulting in

$$Kp = \frac{p_C^{\nu_C}\, p_D^{\nu_D}}{p_A^{\nu_A}\, p_B^{\nu_B}} = \left[\frac{x_C^{\nu_C}\, x_D^{\nu_D}}{x_A^{\nu_A}\, x_B^{\nu_B}}\right] p^{\nu_C + \nu_D - \nu_A - \nu_B} = \exp\left(-\frac{\Delta G_T^0}{RT}\right) \tag{6.29}$$

Values of $Kp$ at 1 bar for different reactions is given in A1.8. Some authors have preferred to proceed with $G = H - TS$, instead of using the chemical potential. This leads to the same result. Students should try it themselves. It should be emphasised here, that for determining the equilibrium constant, we consider the ***equilibrium equation*** and not the actual chemical reaction in which the number of moles of any component may differ from its stoichiometric value, or some other gas, which does not participate in the chemical reaction, may be present, such as $N_2$ going with air.

The value of $Kp$ depends on the reaction, pressure and temperature. These values are tabulated at standard pressure for different reactions at different values of $T$. We may see again the formidable data to be tabulated for every reaction. We would prefer to express the values of $Kp$ for a chemical substance, rather than for the reaction. Like enthalpy of formation, we may consider a chemical reaction at standard pressure among the elements in their stable form to form the substance. Equilibrium constant for such a reaction is called equilibrium constant of formation (A1.9), and is represented as $K_f$. To establish the relation between $Kp$ and $K_f$, let us consider the following reaction to understand the procedure.

$$CO + \tfrac{1}{2}O_2 \rightarrow CO_2 \qquad \text{(A)}$$

Here, $O_2$ is the stable form. We may think of another process, as given below:

$$C + \tfrac{1}{2}O_2 \rightarrow CO \qquad \text{(B)}$$

and

$$C + O_2 \rightarrow CO_2 \qquad \text{(C)}$$

For each of these three reactions A, B and C, we can write the equilibrium constant as follows:

$$K_{p,A} = \frac{p_{CO_2}}{(p_{CO})(p_{O_2})^{1/2}},\ K_{f,B} = \frac{p_{CO_2}}{(p_{CO})(p_{O_2})^{1/2}} \text{ and}$$

$$K_{f,C} = \frac{p_{CO_2}}{(p_{CO})(p_{O_2})} \rightarrow K_{p,A} = \frac{K_{f,C}}{K_{f,B}} \qquad (6.30)$$

This equation can be generalised to the following forms

$$K_p = \frac{\Pi_{\text{products}}\,(K_{f,j})^{\nu_j}}{\Pi_{\text{reactants}}\,(K_{f,i})^{\nu_i}} \text{ or}$$

$$\ln Kp = \Sigma_{\text{products}}\,\nu_j \ln K_{f,j} - \Sigma_{\text{reactants}}\,\nu_i \ln K_{f,i} \qquad (6.31)$$

Mole concentration is also used to express the equilibrium constant. Mole concentration is defined as the number of moles per unit volume

$$[A] = n_A/V,\ [B] = n_B/V,\ [C] = n_C/V \text{ and } [D] = n_D/V \qquad (6.32)$$

[ ], the square brackets, are used to represent the concentration. From the gas law for perfect gas, $n/V = p/(RT)$. This gives

$$[A] = p_A/RT,\ [B] = p_B/RT,\ [C] = p_C/RT,\ [D] = p_D/RT$$

Substituting for partial pressures in terms of concentration in Eq. 6.29 for each component, defining equilibrium constant in terms of concentration as follows, and then simplifying, we get

$$Kc = \frac{[C]^{\nu_C}[D]^{\nu_D}}{[A]^{\nu_A}[B]^{\nu_B}} \text{ and } Kp = Kc\,(RT)^{\nu_C + \nu_D - \nu_A - \nu_B} \qquad (6.33)$$

Now, let us consider dissociation also. It is a phenomenon in which a fraction of some of the compounds like $CO_2$ and $H_2O$, dissociate into their constituents at high temperature. This fraction, which dissociates, is called the degree of dissociation. If $\alpha_1$

and $\alpha_2$ are the degree of dissociation of $CO_2$ and $H_2O$, then we may write the following dissociation equations

$$\alpha_1 CO_2 \Leftrightarrow \alpha_1 CO + \alpha_1/2O_2 \text{ and}$$
$$\alpha_2 H_2O \Leftrightarrow \alpha_2 H_2 + \alpha_2/2O_2 \tag{6.34}$$

To understand the effect of dissociation on the chemical composition of the mixture, let us consider the following stoichiometric reaction

$$C_6H_6 + 7.5\ O_2 \rightarrow 6\ CO_2 + 3\ H_2O$$

Writing from Eq. 6.34, the composition of products instead will be

$$= \left[6(1-\alpha_1)CO_2 + 6\alpha_1 CO + 3\alpha_1 O_2\right] +$$
$$\left[3(1-\alpha_2)H_2O + 3\alpha_2 H_2 + 1.5\alpha_2 O_2\right]$$
$$= 6(1-\alpha_1)CO_2 + 6\alpha_1 CO + 3(1-\alpha_2)H_2O +$$
$$3\alpha_2 H_2 + (3\alpha_1 + 1.5\alpha_2 O_2$$

A reaction that has been used later in solved examples, and is quite of importance in combustion, is $CO + ½\ O_2 \rightarrow CO_2$, which takes the shape

$$CO + \tfrac{1}{2}O_2 \rightarrow (1-\alpha)CO_2 + \alpha CO + \alpha/2O_2$$

After incorporating dissociation of $\alpha$ moles out of 1 mole of $CO_2$ into its constituents. Lastly, as equilibrium is reached, the degree of dissociation will reach $1-\varepsilon$, where $\varepsilon$ is the degree of reaction for the dissociation reaction.

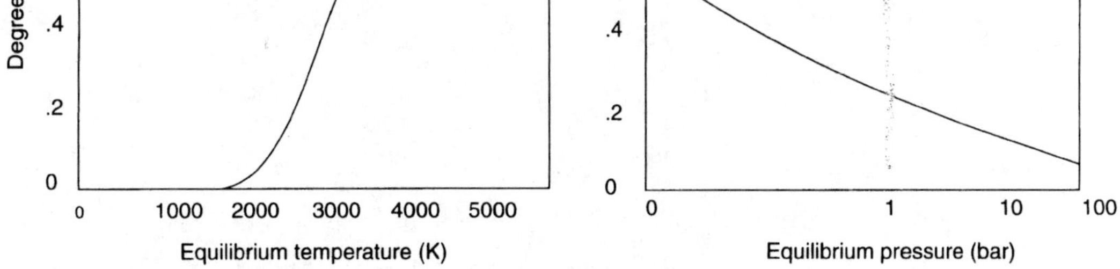


**Fig. 6.4** Variation of degree of reaction with *p* and *T* under equilibrium condition

Equilibrium constant for the dissociation reaction can also be determined considering this as a reaction, and Example 6.13 illustrates the procedure. Magnitude of the equilibrium constant is a good indicator of the degree to which a reaction would proceed to completion. If $Kc < 0.01$, then dissociation of $CO_2$ would not occur, and if greater than 100, then the dissociation reaction will be complete and no $CO_2$ would exist in the products. The value of $Kc$ depends on the pressure and temperature in the reaction vessel. Figure 6.4 shows the variation of the degree of dissociation with temperature and pressure. We observe that the equilibrium constant increases with increase in equilibrium temperature, and decreases with increase in pressure.

**Example 6.13** A correct mixture of CO and $O_2$ is exploded in a vessel in which the initial pressure and temperature are 1 bar and 60°C, respectively. Calculate the composition of the products of combustion and the pressure at 2500 K. Take equilibrium constant $Kp = 26.3$.

***Solution*** The combustion equation at any stage, as discussed above, is

$$CO + \tfrac{1}{2}O_2 \rightarrow (1-\alpha)CO_2 + \alpha CO + \alpha/2 O_2$$

where, $\alpha$ is the degree of dissociation of $CO_2$. The right hand side represents the composition during the reaction. From the reaction, total moles of reactants entered $= n_1 = 1 + \tfrac{1}{2} = \tfrac{3}{2}$, and that of products, $n_2 = (1-\alpha) + \alpha + \alpha/2 = 1 + \alpha/2$. Also, from the perfect gas equation, if $p_1$ and $p_2$ are respectively the pressure of the reactants and products, both occupying the same volume, then

$$p_2/n_2 = (p_1/n_1) \times (T_2/T_1) \rightarrow p_2/n_2 = (2500/333) \times (1/1.5) = 5 \qquad \text{(A)}$$

Further, for the products,

$$p_{CO2}/p_2 = (1-\alpha)/n_2,\ p_{CO}/p_2 = \alpha/n_2,\ p_{O2}/p_2 = \tfrac{1}{2}(\alpha/n_2)$$

$Kp$ for the stoichiometric reaction $CO + \tfrac{1}{2}O_2 \rightarrow CO_2$ is given as 26.3 in the problem. The stoichiometric equation gives the coefficients $\nu$ required in the equilibrium constant. Substituting for partial pressure in Eq. 6.29 and squaring the resultant equation, we get

$$\left(\frac{1-\alpha}{n_2}\right)^2 p_2^2 \left(\frac{n_2}{\alpha p_2}\right)^2 \left(\frac{2n_2}{\alpha p_2}\right) = (26.3)^2$$

Substituting for $p_2$ from Eq. A in it and solving the equation, we get

$$\alpha = 0.079,\ n_2 = 1.0395 \text{ and } p_2 = 5.2 \text{ bar}$$

Composition of the products. CO = (0.079/1.0395) = 7.6%

$CO_2 = (1 - .079)/1.0395 = 88.6\%$ and $O_2 = 3.8\%$

### 6.6.1 Equilibrium Constant and Heat of Reaction

Equation 6.29 can be rewritten as

$$\ln Kp = -\Delta G_T^0/(RT) = (1/RT)\left[\Sigma(\nu\mu^0)_R - \Sigma(\nu\mu^0/T)_p\right]$$
$$= (1/R)\left[\Sigma(\nu\mu^0/T)_R - \Sigma(\nu\mu^0/T)_p\right] \tag{6.35}$$

Noting that standard chemical potential $\mu^0$ depends only on temperature, and the stoichiometric coefficients are independent of temperature, differentiation of Eq. 6.35 w.r.t $T$ gives

$$\frac{d(\ln Kp)}{dT}\,\frac{1}{R}\left[\Sigma\left(\nu\frac{d(\mu^0/T)}{dT}\right)_R - \Sigma\left(\nu\frac{d(\mu^0/T)}{dT}\right)_p\right] \tag{6.36}$$

But from Eq. 6.27, $\mu/T = \mu^0/T + R\ln p$. Therefore, the derivative in Eq. 6.21 is transformed to

$$\left[\partial(\mu/T)/\partial T\right]_p = \left[\partial(\mu^0/T)/\partial T\right]_p = \left[d(\mu^0/T)/dT\right]_p$$

since the second term in ($\mu/T$) is a function of $p$, and hence its derivative w.r.t $T$ at constant pressure is zero. Moreover, $\mu^0$ is a function of temperature alone, and hence its partial derivative becomes the total derivative. Equation 6.21 becomes

$$h = -T^2\left[d(\mu^0/T)dT\right]_p \text{ or } \left[d(\mu^0/T)/dT\right]_p = -h/T^2$$

Substituting in Eq. 6.36, we get

$$d(\ln Kp)/dT = (1/RT^2)[\Sigma(\nu h)_p - \Sigma(\nu h)_R = H_{RP}/(RT^2) \tag{6.37}$$

This equation relates the equilibrium coefficient with the heat of reaction, and is known as the van't Hoff equation. Heat of reaction may be considered as constant for a small range of temperature, so that Eq. 6.37 can be integrated w.r.t $T$ from $T_1$ to $T_2$. This gives

$$\ln(Kp)_{T2} - \ln(Kp)_{T1} = (H_{RP}/R)\,[1/T_1 - 1/T_2] \tag{6.38}$$

In exothermic reactions $H_{RP}$ is negative, and $T_2 > T_1$. Equation 6.38 then concludes that $(Kp)_{T1} > (Kp)_{T2}$. That is, the equilibrium constant in an exothermic reaction decreases with increase in temperature. On the contrary, it would increase in endothermic relation, since the heat of reaction is positive.

Considering again the reaction

$$CO + \tfrac{1}{2}O_2 \rightarrow (1-\alpha)\,CO_2 + \alpha\,CO + \alpha/2_{O2}$$

we have shown in Example 6.13 that the equilibrium constant for this reaction, after substituting for $n_2 = 1 + \alpha/2$, is

$$K_p = \frac{p_{CO2}}{p_{CO}\sqrt{p_{O2}}} = \frac{1-\alpha}{\alpha}\left[\frac{2+\alpha}{\alpha}\frac{1}{p}\right]^{1/2}$$

$$= [1/\alpha - 1][2/\alpha + 1]^{1/2} p^{-1/2} \propto 1/\alpha^{3/2} p^{-1/2} \text{ for } \alpha >> 1$$

We note that as $Kp$ decreases, the degree of dissociation increases. Thus, there exists a relation between the equilibrium constant and the degree of dissociation with temperature.

### 6.6.2 Temperature Rise in Combustion with Dissociation

We shall now determine temperature rise in an exothermic chemical reaction. We have studied that dissociation changes the composition of the products, and is responsible for bringing down the temperature, as it is endothermic. Let us consider the following *adiabatic combustion occurring at constant volume in a closed system*

$$CO + 0.5\ O_2 + 0.5\ (3.76)\ N_2 \rightarrow CO_2 + 1.88\ N_2 \tag{6.39}$$

Let number of moles of CO be $1 + x$ (and not 1) mole, and 0.5 moles of $O_2$ in the reactant. The mixture is rich when $x > 0$, lean when $x < 0$ and stoichiometric when $x = 0$. Let degree of dissociation of $CO_2$ is $(\alpha - x)$. This will result in $(\alpha - x)$ moles of CO. However, $x$ extra moles of CO will be present. Therefore, the reaction with dissociation is given by

$$(1 + x)\ CO + 0.5\ O_2 + 1.88\ N_2$$
$$\rightarrow (1 + x - \alpha)\ CO_2 + CO + 0.5\ (\alpha - x)O_2 + 1.88\ N_2 \tag{6.40}$$

Since combustion is adiabatic in a closed system, $E_1 = E_2$. Neglecting kinetic and potential energies

$$E_1 = [(1 + x)\ (u_{CO})_1 + 0.5\ (u_{O_2})_1 + 1.88\ (u_{N_2})_1]_{T1} \text{ and}$$
$$E_2 = [(1 + x - \alpha)\ (u_{CO_2})_2 + \alpha\ (u_{CO})_2 + 0.5\ (\alpha - x)\ (u_{O_2})_2 + 1.88\ (u_{N_2})_2]_{T2} \tag{6.41}$$

Here, $T_1$ and $T_2$ are the temperatures of reactants and products respectively. For convenience, let us use Er to represent quantity $[(1 + x)\ (u_{CO}) + 0.5\ (u_{CO_2}) + 1.88\ (u_{N_2})]$. But $u = u° + u(T)$, where $u°$ is the internal energy at reference temperature; $u$ being a function of temperature alone for a perfect gas. This gives

$$Er = (1 + x)\ (u_{CO}) + 0.5\ (u_{CO_2}) + 1.88\ (u_{N_2}) + (1 + x)\ (u°_{CO}) +$$
$$0.5\ (u°_{CO_2}) + 1.88\ (u°_{N_2})$$
$$= E_R + E_{R0}, \text{ where } E_R = [(1 + x)\ (u_{CO}) + 0.5\ (u_{CO_2}) + 1.88\ (u_{N_2})]_T \tag{6.42}$$

Here, $E_R$ represents the internal energy of the reactant at the given temperature when measured from the reference state. Recall the definition of heat of reaction at constant volume,

$$Qv = u_{CO_2} - u_{CO} - 0.5\, u_{O_2} \tag{6.43}$$

Equation 6.41 can also be written in the form

$$E_2 = [(1 + x - \alpha)\,(u_{CO_2} - u_{CO} - 0.5\, u_{O_2})]_2 + (1 + x)\,(u_{CO})_2 + 0.5\,(u_{O_2})_2 + 1.88\,(u_{N_2})_2 \tag{6.44}$$

Substituting from Eq 6.42 and 6.43 in Eq. 6.44, we get

$$E_2 = [(1 + x - \alpha)Qv_2 + E_{R2} \tag{6.45}$$

Substituting in Eq. $E_1 = E_2$, we get after simplification

$$\alpha = (1 + x) - [E_{R1} - E_{R2}]/Qv \tag{6.46}$$

Equation 6.46 gives degree of dissociation. In order to calculate the final temperature $T_2$, the equilibrium constant is also required, which we have learnt how to find from the chemical reaction with dissociation. The following example illustrates this procedure.

**Example 6.14** Determine the equilibrium adiabatic combustion temperature for the steady flow constant pressure burning of a stoichiometric mixture of carbon monoxide and dry air, with inlet conditions of 1 bar and 25°C for the rich mixture having 10% more CO. Determine also the irreversibility of the process.

***Solution*** The stoichiometric combustion equation is

$$CO + 0.5\,O_2 + 1.88\,N_2 \rightarrow CO_2 + 1.88\,N_2 \tag{A}$$

But due to dissociation of $CO_2$, some CO and $O_2$ will also be present in the final products; the degree of dissociation of $CO_2$ depends on the temperature of the products, which is unknown. This reaction may be written as

$$CO_2 \rightarrow CO + ½\,O_2 \tag{B}$$

Let under the equilibrium conditions, degree of dissociation be $x$. Then, at that time the products will have the composition

$$(1 - x)\,CO_2 + (x + 0.1)\,CO + x/2\,O_2 + 1.88\,N_2$$

since 10% extra CO is supplied. The total number of moles of products is $(1 - x) + (x + .1) + x/2 + 1.88 = 2.98 + x/2$. Let $p_m$ is the mixture pressure, then the partial pressure of each constituent is

$$p_{CO_2} = \frac{1 - x}{2.98 + x/2}\,p_m$$

$$p_{CO} = \frac{x - 0.1}{2.98 + x/2}\,p_m$$

$$p_{O_2} = \frac{x/2}{2.98 + x/2} p_m$$

$$p_{N_2} = \frac{1.88}{2.98 + x/2} p_m \qquad \text{(C)}$$

For the stoichiometric reaction (B), the equilibrium constant (remember, Eq. B gives the power in Eq. 6.29 whereas Eq. C gives the partial pressure to be used), after simplification, is

$$K_p = \frac{(p_{CO}/p^0)(p_{O_2}/p^0)^{1/2}}{(p_{CO_2}/p^0)} = \frac{x + 0.1}{1 - x}\left(\frac{x}{x + 5.76}\right)^{1/2}\left(\frac{p_m}{p^0}\right)^{1/2}$$

$$= \frac{x + 0.1}{1 - x}\left(\frac{x}{x + 5.76}\right)^{1/2} \qquad \text{(D)}$$

$$H_p = (1 - x)\left[h^0_{CO_2} + \left(h^0_T - h^0_{298}\right)\right]_{CO_2} + x\left[h^0 + \left(h^0_T - h^0_{298}\right)\right]_{CO} +$$
$$x/2\left[h^0 + \left(h^0_T - h^0_{298}\right)\right]_{O_2} + 1.88\left[h^0 + \left(h^0_T - h^0_{298}\right)\right]_{N_2}$$
$$= (1 - x)\left(h^0_{T,CO_2} - 9364 - 393522\right) + x\left(h^0_{T,CO} - 8671 - 110530\right) +$$
$$x/2\left(h^0_{T,O_2} - 8683\right) + 1.88\left(h^0_{T,N_2} - 8670\right)$$

writing the values of enthalpy of formation and enthalpy at 298 K for different constituents from the tables. Similarly,

$$H_R = h^0_{CO} + \tfrac{1}{2}\, h^0_{O_2} + 1.88\, h^0_{N_2} = -110530 + 0 + 0$$
$$= -110530 \text{ kJ/kmole}$$

Since the process is adiabatic, the flow energy equation gives:

$$(1 - x)\left(h^0_{T,CO_2} - 402886\right) + x\left(h^0_{T,CO} - 119201\right) +$$

$$x/2\left(h^0_{T,O_2} - 8683\right) + 1.88\left(h^0_{T,N_2} - 8670\right) = -110530 \qquad \text{(E)}$$

This equation is to be solved by the hit and trial method, assuming some value of $T$, and evaluation of its left hand side. $x$ is obtained from Eq. D, since $Kp$ for the reaction at different temperatures can be obtained from Eq. 6.31, using equilibrium constant of formation at assumed temperature from the tables for the component. Coefficient of products and reactants in the equation will correspond to the reaction

$$CO + \tfrac{1}{2}\, O_2 \rightarrow CO_2$$

in which equilibrium constant of formation of $O_2$ is zero. Thus, from Eq. 6.31, we get

$$\ln Kp = \ln K_{f,CO2} - \ln K_{f,CO} \quad \text{(F)}$$

Once $Kp$ is known, then the value of $x$ is obtained from Eq. D. The hit and trial process continues till the difference of the left and right side of Eq. E is close to zero. The values of enthalpy and equilibrium constant of formation for different gases at temperature T are read from the tables. This solution is done in a tabular manner (Table 6.5).

**Table 6.5** Determination of $x$ and $T_2$

| $T$, K | ln $Kf$ | | – Ln $Kp$ | $Kp \times 10^3$ | 100 × | LHS – RHS |
|---|---|---|---|---|---|---|
| | CO | $CO_2$ | (Eq. F) | | (Eq. D) | (Eq. E) |
| 2200 | 7.190 | 9.398 | 2.208 | 6.17 | 5.828 | –44210 |
| 2300 | 7.072 | 8.988 | 1.916 | 12.13 | 8.938 | –22720 |
| 2400 | 6.961 | 8.613 | 1.652 | 22.29 | 13.03 | 1483 |
| 2500 | 6.856 | 8.268 | 1.412 | 38.73 | 18.14 | 28490 |
| 2600 | 6.756 | 7.949 | 1.193 | 64.10 | 24.2 | 58100 |

We may get the exact temperature and dissociation constant $x$ by plotting (LHS – RHS) against the temperature. The actual temperature is the temperature at which the curve meets the zero line. Similarly, the graph for $x$ can be plotted. But we observe that for $T = 2400$ K, the difference is small, and we may take the adiabatic temperature to be 2400 K at which the dissociation coefficient is 0.13.

Equations 6.13 and 6.15 combined give the change in entropy of any constituent at temperature $T$. We may use these equations to find out the change in entropy of the system during the combustion process, which in turn, will give the irreversibility of the process,

$$I = T_0\Big[(1-x)\left(s_f^0 + s_T^0 - s_{298}^0 - \bar{R}\ln y\right)_{CO_2} +$$

$$(0.1+x)\left(s_f^0 + s_T^0 - s_{298}^0 - \bar{R}\ln y\right)_{CO} +$$

$$x/2\left(s_f^0 + s_T^0 - s_{298}^0 - \bar{R}\ln y\right)_{O_2} +$$

$$1.88\left(s_f^0 + s_T^0 - s_{298}^0 - \bar{R}\ln y\right)_{N_2}\Big]_p$$

$$= T_0\left[(1.1 - x)\left(s_f^0 - \bar{R}\ln y\right)_{CO} - x/2\left(s_f^0 - \bar{R}\ln y\right)_{O_2}\right.$$

$$\left. - 1.88\left(s_f^0 - \bar{R}\ln y\right)_{N_2}\right]_R$$

The gas tables by Keenan, Chao and Kays give the entropy for different constituents, such that the values listed at 298 K correspond to the entropy of formation. Therefore, we may read the entropy at a temperature of 2400 K for products and 298 K for reactants. Also, the total number of moles of products for the given reaction under equilibrium condition and reactants is 2.98 + .13/2= 3.04 and 1 + 0.5 + 1.88 = 3.38, respectively. Substituting the various values, we get:

$$\begin{aligned} I = {} & 298[(1 - .13)\{320.387 - 8.314 \ln (.87/3.04)\} + .23\{265.366 \\ & - 8.314 \ln(.23/3.04)\} + 0.065\{275.706 - 8.314 \ln (.13/3.04)\} \\ & + 1.88\{258.683 - 8.314 \ln (1.88/3.04) - (197.653 \\ & - 8.314 \ln (1/3.38)\} - 0.5\{205.147 - 8.314 \ln (0.5/3.38)\} \\ & - 1.88\{191.609 - 8.314 \ln (1.88/3.38)\}] \\ = {} & 45000 \text{ kJ/kmole CO} \end{aligned}$$

## SUMMARY

Chemical reactions play an important role in chemical industry. Reactants combine with each other to form desired products under suitable conditions of pressure and temperature. These reactions are normally reversible, meaning the products re-combine to form the reactants. However, the rate of formation of reactants can be decreased by proper control of temperature and pressure. In chemistry, we learn that sometimes a catalyst is used to increase the rate of the forward reaction. Such substances do not participate in the reaction. This chapter does not study effect of a catalyst.

Heat may be released or absorbed by reactions and accordingly it is called endothermic or exothermic. Combustion is the exothermic reaction in which one of the constituents of the reactants is oxygen or air. We come across this reaction in many applications; a few significant applications are boiler, internal combustion engine, industrial furnaces, etc. Here the fuel composition may be given in terms of its chemical elements C, H, etc. (ultimate analysis) or ash, volatile matter, moisture, fixed carbon (proximate analysis) for solid fuels, which are based on mass. For gaseous fuel, composition is given on the volumetric or molar basis of the constituent gases. This chapter dealt with conversion of gravimetric to volumetric analysis and vice versa.

Air is used to give required oxygen and contains 21% or 23% of $O_2$ by volume or mass. This chapter dealt with writing chemical reaction with different quantities of air and also to determine the volumetric composition of the products of combustion, known as flue gas.

The chapter also explained first and second law analysis along with definition of enthalpy of formation. It is also required to determine equilibrium constant of any reaction. The discussion is exhausive including dissociation effects and deals with the analysis to determine the final state after combustion in closed space as well as steady flow under chemical equilibrium condition and to find out energy interaction with the surroundings.

# EXERCISES

## Review Questions

6.1 How many moles of air are needed to burn 1 kmole of carbon?

6.2 Why are products measured on a dry basis?

6.3 What is the dew point of hydrogen burnt with stoichiometric pure oxygen, or air?

6.4 In most cases, combustion products are exhausted above the dew point of $So_2$. Why?

6.5 Why does combustion contribute to global warming?

6.6 How is the fuel enthalpy of combustion connected to its enthalpy of formation?

6.7 Does it make a difference to the enthalpy of combustion, if it is burnt with pure oxygen or air? What about the adiabatic flame temperature?

6.8 What happens to the adiabatic flame temperature if it is burnt rich or lean?

6.9 Is the irreversibility in a combustion process significant? Why?

6.10 If the AF ratio is larger than the stoichiometric, is it more or less reversible?

6.11 How is a chemical reaction process different from the combustion process?

6.12 In a combustion process, is the adiabatic flame temperature affected by the reactions?

6.13 In equilibrium, the Gibbs function of products and reactants is the same; how about the energy?

6.14 Does a dissociation process require energy, or does it give out energy?

6.15 Does Kp depend on the type of reaction and pressure and the temperature of the products?

## Problems for Practice

6.1 A sample of pine bark has the following ultimate analysis on a dry basis, present by mass: 5.6% H, 53.4% C, 0.1% S, 0.1% N, 37.9% O and 2.9% ash. This bark will be used as a fuel by burning it with 100% theoretical air in a furnace. Determine the air fuel ratio on mass basis.

6.2 Liquid propane is burnt with dry air. A volumetric analysis on the products of combustion yields the following volume per cent composition on a dry basis: 8.6% $CO_2$, 0.6% CO, 7.2% $O_2$ and 83.6% $N_2$. Determine the per cent of theoretical air used in this combustion process.

6.3 A fuel, $c_x h_y$, is burned with dry air, and the product composition is measured on a dry mole basis to be: 9.6% $CO_2$, 7.3% $O_2$ and 83.1% $N_2$. Find the fuel composition ($x/y$), and the per cent theoretical air used.

6.4 Pentane is burnt with 120% theoretical air in a constant pressure process at 1 bar. The products are cooled to ambient temperature, 20°C. How much mass of water is condensed per kg of fuel? Repeat the answer, assuming that the air used in the combustion has a relative humidity of 90%.

6.5 The coal gasifier in an integrated gasification combined cycle power plant produces a gas mixture with the following volumetric per cent composition: $CH_4$ 0.3, $H_2$ 29.6, CO 41.0, $CO_2$ 10.0, $N_2$ 0.8, $H_2O$ 17.0, $H_2S$ 1.1 and $NH_3$ 0.2. This gas is cooled to 40°C, 30 bar, and the $H_2S$ and $NH_3$ are removed in water scrubbers. Assuming that the resulting mixture, which is sent to the combustors, is saturated with water, determine the mixture composition and the theoretical AF ratio in the combustors.

6.6 The hot exhaust gas from an internal combustion engine is analysed and found to have the following per cent composition on a volumetric basis at the engine exhaust manifold: 10% $CO_2$, 2% CO, 13% $H_2O$, 3% $O_2$ and 72% $N_2$. This gas is fed to an exhaust gas reactor and mixed with a certain amount of air to eliminate the carbon monoxide. It has been determined that a mole fraction of 10% $O_2$ in the mixture will ensure that no CO remains at the exhaust. What must be the ratio of flows entering the reactor?

6.7 A rigid vessel initially contains 2 kmole of carbon and 2 kmole of $O_2$ at 25°C, 2 bar. Combustion occurs, and the resulting products consist of 1 kmole of $CO_2$, 1 kmole of CO, and excess $O_2$ at a temperature of 1000 K. Determine the final pressure in the vessel, and the heat transferred from the vessel during the process. Calculate the irreversibility for the process.

6.8 In a test of rocket propellant performance, liquid hydrazine ($N_2H_4$) at 1 bar, 25°C, and oxygen gas at 1 bar and 25°C, are fed to a combustion chamber in the ratio of 0.5 kg $O_2/N_2H_4$ kg. The heat transfer from the chamber to the surroundings is estimated to be 100 kJ/kg $N_2H_4$. Determine the temperature of the products exiting the chamber. Assume that only $H_2O$, $H_2$, and $N_2$ are present. The enthalpy of formation of liquid hydrazine is + 50417 kJ/kmole. [*Ans.* 2854 K].

6.9 The combustion of heptane $C_7H_{16}$ takes place in a steady flow burner, where fuel and air are added as gases at $T_0$, $p_0$. The mixture has 125% theoretical air, and the products are passed through a heat exchanger, where they are cooled at 600 K. Find the heat transfer from the heat exchanger per kmole of heptane burned.

6.10 One alternative to using petroleum or natural gas as fuels is ethanol ($C_2H_5OH$), which is commonly produced from grain fermentation. Consider a combustion process in which liquid ethanol is burned with 120% theoretical air in a steady flow process. The reactants enter the combustion chamber at 25°C and the products exit at 60°C, 1 bar. Calculate the heat transfer per kmole of ethanol. Repeat it with fuel delivered as a vapour.

6.11 Another alternative to using petroleum or natural gas as fuel is methanol, ($CH_3OH$), which can be produced from coal. Both methanol and ethanol have been used in automotive engines. Repeat problem 6.10 using liquid methanol as fuel.

6.12 In a new high-efficiency furnace, natural gas, assumed to be 90% methane and 10% ethane (by volume) and 110% theoretical air each enter at 25°C, 1 bar, and the products (assumed to be 100% gaseous) exit the furnace at 40°C, 1 bar. What is the heat transfer for this process? Compare this with an older furnace where the products exit at 250°C, 1 bar.

6.13 A closed, insulated container is charged with a stoichiometric ratio of oxygen and hydrogen at 25°C and 150 kPa. After combustion, liquid water at 25°C is sprayed in such a way that the final temperature is 1200 K. What is the final pressure?

6.14 Wet biomass waste from a food processing plant is fed to a catalytic reactor, where, in a steady flow process, it is converted into a low energy fuel gas suitable for firing the processing plant boiler. The fuel gas has a composition of 50% methane, 45% $CO_2$ and 5% $H_2$ on a volumetric basis. Determine the lower heating value of this fuel gas mixture per unit volume.

6.15 Blast furnace gas in a steel mill is available at 250°C to be burned for the generation of steam. The percentage composition of this gas is, on a volumetric basis, $CH_4$ 0.1, $H_2$ 2.4, CO 23.3, $CO_2$ 14.4, $N_2$ 56.4, $H_2O$ 3.4. Find the lower heating value per unit volume of this gas at 250°C and ambient pressure.

[*Ans.* 1668 kJ/m³].

6.16 Acetylene gas at 25°C, 1 bar, is fed to the head of a cutting torch. Calculate the adiabatic flame temperature if acetylene is burned with: (a) 100% theoretical air at 25°C, and (b) 100% theoretical oxygen at 25°C. Explain the difference in the two cases, and justify the answer you get. [*Ans.* 2900, 7400 K].

6.17 Methane is burned with 80% of stoichiometric air in a steady flow process at 1 bar. Methane and air are both supplied at 25°C, and the products leave at 1700 K. Assuming that no $CH_4$, OH, NO, or free oxygen appear in the products, determine the amount of heat transferred per kg of methane. [*Ans.* 9127 kJ/kg methane].

6.18 The combustion chamber of a gas turbine uses a liquid hydrocarbon fuel and gas of an approximate composition of $C_8H_{18}$. During testing, the following data are obtained: $T_{air}$ = 400 K, $T_{prod}$ = 1100 K, $C_{air}$ = 100 m/s, $C_{prod}$ = 150 m/s, $T_{fuel}$ = 50°C and $FA_{actual}$ = 0.0211 kg fuel/kg air. Calculate the combustion efficiency for this process. Assume no heat loss during the combustion process. [*Ans.* 86.2%].

6.19 Methane supplied at 1 bar, 25°C is burned adiabatically in a steady flow burner with the stoichiometric amount of air supplied at the same conditions. Assuming that the products contain no $CH_4$ or NO, determine the temperature of the products. [*Ans.* 2240 K].

6.20 Methane is burned with enough excess air, both of which are supplied at the reference condition, to give a flame temperature of 1800 K. What are the per cent theoretical air and the irreversibility? [*Ans.* 150%, 287 MJ/kmole fuel].

6.21 Consider the combustion of methanol, $CH_3OH$, with 25% excess air. The combustion products are passed through a heat exchanger and exit at 2 bar, 400 K. Calculate the absolute entropy of the products exiting the heat exchanger, assuming that all the water is vapour. [*Ans.* 2150 kJ/kmole K].

6.22 2 moles of ammonia are burned in a steady flow process with x moles of $O_2$. The products, consisting of $H_2O$, $N_2$, and the excess $O_2$, exit at 200°C, 7 Mpa. (a) Calculate x if half of the water in the products is condensed, and (b) calculate the absolute entropy of the products at the exit condition. [*Ans.* 5.757, 1414.3].

## Multiple-choice Questions

6.1 Calorific value = (highest calorific value $-x \times h_{fg}$), where

(a) $x$ is the hydrogen fraction in fuel, and $h_{fg}$ is the enthalpy of evaporation at 25 °C saturation temperature.

(b) $x$ is the fraction of water vapour in products of combustion, and $h_{fg}$ is the enthalpy of evaporation at atmospheric pressure.

(c) $x$ is the fraction of water vapour in products of combustion, and $h_{fg}$ is the enthalpy of evaporation at standard temperature of 15°C.

(d) $x$ is the fraction of water vapour in products of combustion, and $h_{fg}$ is the enthalpy of evaporation at ambient temperature.

6.2 Calorific value of fuel determined at constant volume

(a) is equal to the calorific value of fuel at constant pressure.

(b) is more than the calorific value of fuel at constant pressure.

(c) is less than the calorific value of fuel at constant pressure.

(d) may be more or may be less, depending upon molecular contraction or expansion of the products of combustion.

6.3 The calorific value of hydrogen/kg when burnt at constant pressure in the atmosphere of air, as compared to that when burnt at constant volume is

(a) less (b) more (c) equal (d) may be more or less.

6.4 The lower calorific value of fuel/kg may be expressed as

(a) LCV = 339C + 1440 (H – O/8) + 92.95 8 – 222 H kJ/kg.

(b) LCV = 339C + 1440 (H + O/8) + 92.958 – 222 H kJ/kg.

(c) LCV = 339C + 1440 (H + O/8) + 92.95 8 + 222 H kJ/kg.

(d) LCV = 339C + 1440 (H – O/8) + 92.95 8 + 222 H kJ/kg.

6.5 The higher calorific value of the fuel whose composition is C = 86%, H = 6%, incombustible = 8%

(a) will be higher than its lower calorific value.

(b) will be lower than its lower calorific value.

(c) will be the same as its lower calorific value.

(d) will be lower or higher than the lower calorific value.

6.6 The oxygen percentage by volume in air is

(a) 21%. (b) 23%. (c) 30%. (d) 79%.

6.7 The oxygen percentage by weight in air is

(a) 21%. (b) 23%. (c) 30%. (d) 79%.

6.8 Ultimate analysis of fuel means determination of

(a) higher calorific value of fuel. (b) lower calorific value of fuel.

(c) percentages of carbon, hydrogen, nitrogen, sulphur and moisture.

(d) percentages of fixed carbon, ash, volatile matter and moisture.

6.9 Proximate analysis of fuel means, determination of

(a) higher calorific value.

(b) lower calorific value .

(c) percentages of carbon, hydrogen, nitrogen, sulphur and moisture.
(d) percentages of fixed carbon, ash, volatile matter and moisture.

6.10 The calorific value of carbon, when burnt at constant pressure in the atmosphere of air at NTP
(a) is equal to calorific value when burnt at constant volume at NTP.
(b) is more than the calorific value when burnt at constant volume at NTP.
(c) is less than the calorific value when burnt at constant volume at NTP.
(d) may be more or less at constant volume at NTP.

6.11 Orsat apparatus is used to determine
(a) products of all constituents of fuel combustion by weight.
(b) products of all constituents of combustion of fuel by volume.
(c) products of only dry constituents of combustion by weight.
(d) products of only dry constituents of combustion by volume.

6.12 Molecular volume of a perfect gas at given temperature and pressure
(a) increases with number of atoms in the molecule.
(b) decreases with the number of atoms in the molecule.
(c) remains the same irrespective of the atoms in the molecule.
(d) may increase or decrease, depending upon the structural set of atoms in the molecule.

CHAPTER SEVEN

# Vapour Power Cycles



## 7.1 INTRODUCTION

We know from the second law of thermodynamics, that an engine operating on a thermodynamic cycle delivers work only when it extracts heat from the source at higher temperature $T_1$ and rejects a part of it to a sink at low temperature $T_2$. An engine may use different working substances such as air, wet or superheated steam, mercury, liquid metals, or any other vapour.[1] The cycles are classified as gas or vapour power cycles accordingly. However, steam is the most commonly used vapour, and hence the steam power cycle is in common usage. Water was used as working substance, since it is easily available in nature, and is non-toxic, non-erosive, and non-harmful to living stock. However, it has a big limitation that its specific heat is high and requires a large quantity of heat to affect a necessary change in temperature. Its latent heat decreases as steam pressure increases and maximum temperature is limited; today, a steam power plant uses steam at a temperature around 630°C. This temperature is much lower than that used in gas power cycles, and therefore, efficiency of the steam cycle is expected to be comparatively less. The fuel is

1. Freon 11 has been used as another vapour in turbines in low pressure range.

burnt outside the engine and a large chunk of fuel energy—as high as 12% of the heat input—goes as waste with exhaust gas. Lastly, a major portion—around 45–50% of heat input—goes to cooling water in the condenser at a low temperature as waste. For these reasons, we have been able to increase the efficiency of a steam plant up to 45%. Steam vapour cycles and the methods used for improving efficiency are discussed in Sections 7.3–7.6.

Many reversible cycles were developed simultaneously using air as working substance—for example, Otto, Diesel, Dual, and the Brayton cycle. These cycles are called air standard cycles. The various air standard cycles, differing in the manner of heat addition to air and/or heat rejection, were developed to set standard cycle for using different kinds of fuels available and to meet varying operating conditions of the engine. Today, a gas turbine cycle operates around 1675°C that is much higher than that used in the steam cycle, and these cycles have been able to achieve an efficiency of around 60%. These cycles are discussed in Chapter 8.

In the beginning, reciprocating engines were used both in steam and air power cycles. The crank mechanism was required to convert the linear motion of a piston to the rotational motion of the crankshaft. The piston used to be very heavy, and its velocity would change during motion, producing varying inertia forces that were the main culprit for unbalanced forces. These devices were heavy, produced noise due to the presence of unbalanced forces, required heavy foundation and the speed would fluctuate. Therefore, attempts were made to change the momentum of the steam or air jet by moving it over curved vanes fitted on a drum to get rotational motion directly, and get rid of these problems. Steam and gas power turbines were developed with gas turbine systems used for flying aircraft because of their low weight. But the latter could not compete with the steam turbine cycle for the production of power, until the combined cycle was developed, mainly because the efficiency of a simple gas turbine cycle is less, since a very high portion of the input goes with the exhaust gas at a very high temperature. A combined cycle utilised this waste heat to raise steam for power production. Further, it could not attract much attention for aviation purposes too for the same reason, till the Second World War, when its importance was realised by the defence organisation. Since that day, the gas turbine system is used in passenger and defence planes with many significant improvements. Also, with the invention of the combined cycle, power production using both steam and gas turbines has become the scene of the day in the power production industry.

Before the invention of the combined cycle, binary and ternary cycles were also developed. These are discussed later in this chapter.

## 7.2 TERMS USED IN CYCLES

Besides efficiency, we come across many other terms like efficiency ratio, work ratio, specific steam consumption, and mechanical efficiency, while studying power cycles. These are explained first.

### 7.2.1 Ideal Cycle Efficiency and Efficiency Ratio

The performance of any thermodynamic cycle is measured in terms of ratio of desired output from it, and the input required for that output. In fact, this definition is not limited to these cycles only, as this is applicable in all walks of life. We always compare the result (output) with the input, for which a ratio is the best indicator, as it gives the percentage utilisation of the input. For example, the marks obtained in an examination are not enough to know the performance, and the maximum mark should also be known. But the percentage of marks obtained gives enough information about our performance. In power cycles, the desired output is the work produced per unit of time measured in kW (kilowatt) or MW (megawatt), and the required input is heat in the same units. This ratio is called the *efficiency* of the engine. Therefore, it is natural for us to think of a power cycle having high efficiency, and to find those factors that are responsible for it.

An ideal cycle efficiency refers to that of a reversible power cycle such as the Carnot cycle, and is an important parameter for comparison of performance of actual cycles. The actual cycle differs from the theoretical (also known as ideal cycle), because of the presence of energy dissipating agents such as friction, eddies, etc. in the system, and the finite potential difference between the system and the surrounding, resulting in the irreversibility of the process during energy transfer between the system and the surrounding. This results in less output for the same input, and therefore, a decrease in efficiency of the actual cycle. The efficiency of the actual cycle is termed *thermal efficiency*, and its ratio to that of its corresponding theoretical power cycle is called the *efficiency ratio*[2] *in a steam cycle.*

### 7.2.2 Work Ratio

A cycle involves an expansion as well as a compression process. While work is obtained by the surroundings during the expansion process, it does work on the system during the compression process (we assume there is no other form of work transfer in processes of the cycle). As discussed in Chapter 1, the work of compression is less than the work of expansion, and that is why there is net work output, which is equal to the difference of

2. Note that in gas power cycles, air is used as the working medium. Here, ideal cycle efficiency is called air standard efficiency, and the efficiency ratio as relative efficiency. Otherwise, the meaning of the terms remain the same; only the nomenclature is changed, and some day, someone may try to bridge it up, except for their historical importance.

the two values. The ratio of the net work output of the cycle to the positive work (produced during the expansion process) is called the *work ratio.* It is an important parameter for the comparison of performance of different cycles. Naturally, we would prefer a higher value of work ratio so that the cycle output is more.

If $w_e$ and $w_c$ are respectively the expansion and compression work in a given cycle per kg of working substance, then

$$\text{Work ratio or } WR = (w_e - w_c)/w_e \tag{7.1}$$

### 7.2.3 Specific Steam Consumption (ssc)

It is defined as the mass of steam required in kg per hour for producing unit power, measured in kW and abbreviated as ssc. If $m_s$ kg of steam is consumed in one second (or 3600 $m_s$ kg in an hour) to produce power $P$ kW and $w_{\text{net}}$ is the net work output of the cycle per kg of steam, then

$$P = m_s \times w_{\text{net}} \quad \text{and} \quad \text{ssc} = 3600\, m_s/P = 3600/w_{\text{net}} \tag{7.2}$$

In gas power cycles, however, we use specific fuel consumption (sfc), which is identical to ssc. If $m_f$ kg of fuel and $m_a$ kg of air are consumed in one second to produce power $P$, $f$ is the fuel air ratio, $w_{\text{net}}$ is the amount of work produced per kg of air, and CV is the calorific value of the fuel used, then we can write

$$f = m_f/m_a \text{ and sfc} = 3600\, m_f/P = (m_f \times 3600)/(m_a \times w_{\text{net}}) = 3600 f/w_{\text{net}} \tag{7.3}$$

and, $q$ = heat input per kg of air = $m_f \times \text{CV}/m_a = f \times \text{CV}$

$$\therefore\ \eta = \text{Cycle efficiency} = w_{\text{net}}/q = 3600/(\text{sfc} \times \text{CV}) \tag{7.4}$$

That is, if cycle efficiency increases, then sfc decreases, see Eq. 7.4. Therefore, the size of the plant for a given generation capacity also decreases. This is quite logical, as more efficiency means more conversion of input into desired output per kg of working substance, and therefore, for given amount of power to be developed, less amount of working substance required and hence less sfc or ssc. Thus, ssc and efficiency are related to each other, and a trend of one gives that of another.

### 7.2.4 Isentropic Efficiency

Some power cycles use the isentropic process, which is a reversible adiabatic[3] process, and therefore, the entropy of the system does not change during the process. However, due to irreversibility, entropy of the system increases and needs to be accounted. Consider a compressor or turbine through which air flows as the working fluid, and the flow is steady. Air is subjected to different processes as shown in Fig. 7.1 on the *T-s* diagram, namely, isentropic compression process 12, constant pressure heat addition and expansion

3. An adiabatic process is one in which there is no heat interaction between the system and the surrounding.

process 34. 12' and 34' are the actual adiabatic compression and expansion processes. We note that entropy of air increases in both the cases. Using sfee, we may write that work done in processes 12 and 12' is $(h_2 - h_1)$ and $(h_{2'} - h_1)$, respectively. Similarly, work done by the turbine in the expansion processes 34 and 34' is $(h_3 - h_4)$ and $(h_3 - h_{4'})$. We observe that more work is done in the compression and less work is obtained in the expansion process, due to irreversibility present in the system. Isentropic efficiency recognises this fact. It is defined as the ratio of actual to the isentropic work in the expansion process, and that of isentropic to actual work done in the compression process, so that its value is always less than 1. We may write the following expressions for isentropic efficiency of compressor $\eta_c$ and turbine $\eta_t$

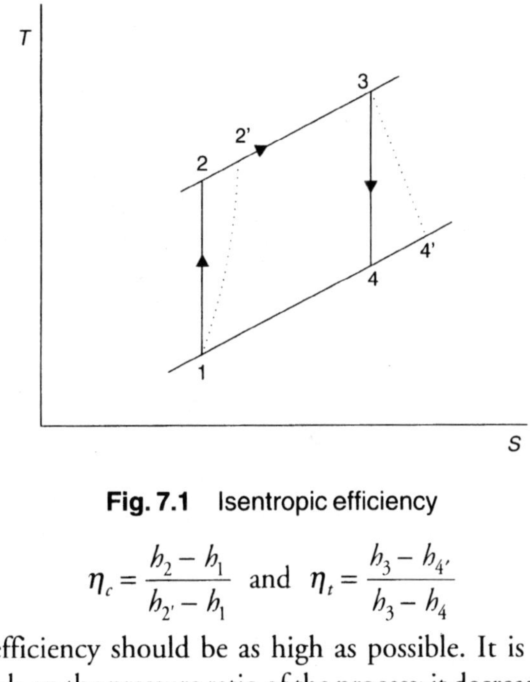


**Fig. 7.1** Isentropic efficiency

$$\eta_c = \frac{h_2 - h_1}{h_{2'} - h_1} \quad \text{and} \quad \eta_t = \frac{h_3 - h_{4'}}{h_3 - h_4} \tag{7.5}$$

Isentropic efficiency should be as high as possible. It is observed that isentropic efficiency depends on the pressure ratio of the process; it decreases for the compressor and increases for the turbine with increasing pressure ratio.

### 7.2.5 Swept Volume and Compression Ratio

A reciprocating engine was used for implementation of steam,[4] petrol or diesel cycles. In such a device, a piston reciprocates within the engine cylinder, because of the crank

4. Steam engines are now rarely used and diesel engines are used as captive power plants in industries for power requirements up to 10 MW. For more power, a steam plant is used, and if gas is available easily, then a gas turbine is used, as it can be started in no time. Steam power plants have a large start up period, and hence are mostly suitable for base load plants.

mechanism used. As a result, the piston has zero velocity at the extreme end of its journey, for which reason extreme ends are referred as dead centres. There are two dead centres—top and bottom dead centre (TDC and BDC) in vertical engines, or inner and outer dead centres (IDC and ODC) in horizontal engines. Figure 7.2 shows the position of IDC and ODC. We define displacement of the piston between the dead centres as stroke length $L$, diameter $D$ of the cylinder (not piston) as its bore and volume swept during piston movement between dead centres as swept volume, $V_s$. If $V_1$ and $V_2$ are the volumes of the gas inside the cylinder at ODC and IDC, respectively, then the ratio $V_2/V_1$ is known as *compression ratio* $r$ in case of the compression process, and *expansion ratio* in the expansion process. That is,

$$r = \text{compression ratio} = V_1/V_2 \tag{7.6}$$

and,

$$V_s = V_1 - V_2 = (\pi/4)\ D^2\ L \tag{7.7}$$

Moreover, $V_2$ is the volume occupied by the gas when the piston is at TDC or IDC, and is called clearance volume $V_c$ normally expressed as percentage $c$ of the swept volume (or stroke length).

$$\text{Since, } V_1 - V_2 = V_s,\ V_1/V_2 = r \text{ and } V_2/V_s = c \rightarrow r = (c + 1)/c \tag{7.8}$$

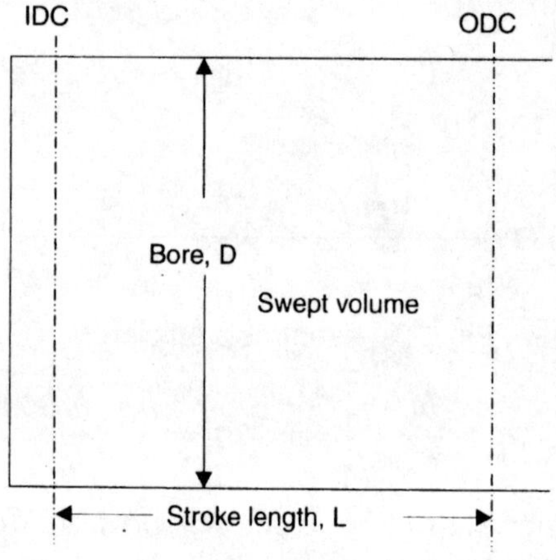


**Fig. 7.2** Dead centres, stroke length, and bore

### 7.2.6 Mechanical Efficiency[5]

The work output of the actual cycle in reciprocating engines is obtained from the *p-v* diagram, which is a plot of the variation of pressure with the movement of piston, and is

5. This should not be confused with mechanical efficiency used in power transmission from a gas turbine to a compressor. The turbine shaft is connected to the compressor shaft, which is supported on bearings. The engine and machine shaft may be directly connected, or through a gear box, in case the turbine and compressor revolve at different rotational speed. Some power is lost in shaft bearing, gearing, coupling, etc. Transmission efficiency is also used to represent this kind of loss.

obtained using an instrument called indicator. For this reason, a *p-v* diagram is also called an indicator diagram, and work calculated from the indicator diagram per unit of time, the indicated power. However, the power available at the engine shaft is less than this value due to losses within the system, such as friction loss between the cylinder and piston surface (reciprocating engine), eddies, and friction in bearings. These losses are not included in the calculation of indicated power. Power at the engine shaft is called brake power, which is obtained experimentally using a dynamometer. The ratio of brake to indicated power is titled *mechanical efficiency.*

$$\eta_{mech} = \text{brake power/indicated power} \tag{7.9}$$

There are two types of output, namely, indicated and brake power. It is possible to use any of these outputs for determining thermal efficiency of the engine, as defined earlier in this section. Accordingly, the two are named as indicated and brake thermal efficiency. Naturally, their ratio also gives mechanical efficiency.

### 7.2.7 Mean Effective Pressure

The mean effective pressure is the pressure of a hypothetical rectangular cycle, whose area is equal to that of the actual cycle. We will study its importance in the chapter on internal combustion engines.

## 7.3 VAPOUR POWER CYCLE

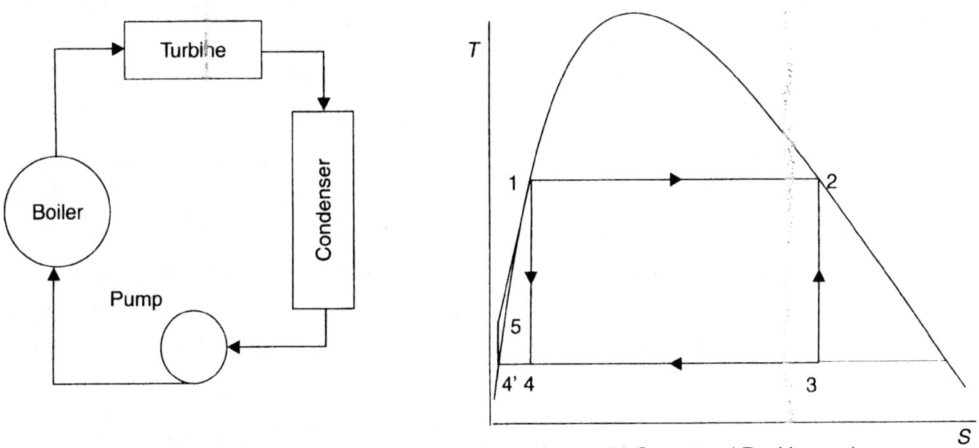


(a) Schematic diagram (b) Carnot and Rankine cycles

**Fig. 7.3** Components of a thermal plant and the Carnot cycle

The Carnot cycle is the first reversible power cycle that was implemented on a steam engine by James Watt in the eighteenth century. The Carnot cycle is drawn on the *T-s* diagram in Fig. 7.3 (b). It consists of isothermal heat addition process 12, isentropic expansion process 23, heat rejection process 34 and isentropic compression process 41, such that the cycle is complete. Note points 1 and 2 correspond to saturated liquid and saturated vapour states.

Since the dryness fraction of steam changes at constant temperature during heat addition and rejection, the Carnot cycle has the advantage, that these two processes can be made close to a reversible process during implementation. For implementation of this cycle, we need a boiler, a turbine, a condenser and a compressor.[6] When considered individually, these components form an open system. Moreover, the flow is steady, and therefore, each of these components can be analysed using the steady flow energy equation. Applying the steady flow energy equation for the processes 12 and 34, one can write the following equations

$$\text{heat added to steam in the boiler} = h_2 - h_1 = T_1\,\Delta S$$

$$\text{heat rejected by steam in the condenser} = h_3 - h_4 = T_2\,\Delta S$$

where, $T_1$ and $T_2$ are temperatures at which heat is added to and rejected from the cycle. The saturation pressure at these two temperatures are known as boiler and condenser pressure, respectively. Heat transfer in the processes 23 and 41 is 0, being isentropic processes. By Joule's law, the net work output of this cycle = sum of heat transfer in the four processes = $(T_1 - T_2)\,\Delta S$ .

$$\therefore \text{ cycle efficiency} = \text{net work output/heat added to steam} = 1 - T_2/T_1 \qquad (7.10)$$

Efficiency of this cycle depends upon the temperatures $T_1$ and $T_2$. Since this cycle consists of reversible processes, by the second law of thermodynamics, it has maximum efficiency for these two temperature limits.

If we apply the steady flow energy equation to the processes 23 and 41, we get the turbine work output and the work done in the compressor. It can easily be seen that their magnitude will be $(h_2 - h_3)$ and $(h_1 - h_4)$, respectively. Therefore, significant amount of work is consumed in the process 41 reducing the net work output of the cycle. For this reason, the Carnot cycle has work ratio of the order of 60–70%. Further, this cycle requires a strict control over the point 4, so that at the end of compression, the system is brought back to state 1. This, however, is not so easy. Lastly, compression involves a mixture of liquid and vapour, making compression difficult, as the liquid tends to separate out and affects the quality of water leaving the compressor.

It will be interesting to see whether decreasing $T_2$ or increasing $T_1$ by $\Delta T$ will be more effective in increasing the efficiency of the Carnot cycle. Say, the source and sink temperatures, respectively, are $T_1 + \Delta T$ and $T_2 - \Delta T$ . Then, efficiency of the engine is

---

6. While discussing reciprocating air compressors later, we shall show that such a system can be analysed using sfee.

$$\eta = 1 - \frac{T_2 - \Delta T}{T_1 + \Delta T} = 1 - \frac{T_2}{T_1} + \frac{T_2}{T_1}\left(\frac{\Delta T}{T_1} + \frac{\Delta T}{T_2}\right) + \text{higher order terms} \tag{7.11}$$

The binomial theorem has been used to change the denominator. Higher order terms may be neglected. We observe that efficiency is increased in both the cases. However, $\Delta T/T_2 > \Delta T/T_1$, since $T_1 > T_2$. Therefore, decrease in sink temperature is more effective. We may also get the same result if we consider the rate of change of efficiency with the source or sink temperature as we are asked to find it out. Differentiating Eq. 7.10 partially with respect to $T_1$ or $T_2$, we get $(\partial\eta/\partial T_1)_{T_2} = (T_2/T_1)(1/T_1)$ and $(\partial\eta/\partial T_2)_{T_1} = -(1/T_1)$; a –ve sign indicates that efficiency decreases with increase in sink temperature, and therefore, it has to be decreased for increase. This gives

$$|\{(\partial\eta/\partial T_2)_{T_1}/(\partial\eta/\partial T_1)_{T_2}\}| = T_1/T_2 > 1$$

### *Rankine Cycle*

Rankine suggested that instead of condensing steam to point 4, the steam may continue to condense up to point 4′, where it is converted into saturated water see Fig. 7.3(b). This way, raising the pressure of water from condenser pressure to boiler pressure from 4 to 5 is easy, because a single phase of the substance is compressed, and it does not require control of the point 4. Also, the amount of work required for raising liquid pressure is significantly small, to the extent that we neglect it in comparison to the turbine work output in actual calculations. This cycle is known as the Rankine cycle, for which the work ratio is high, of the order of 97%. Rogers and Mayhew(21) have reported that the efficiency ratio of the Rankine cycle is higher than that of the Carnot cycle, and this results in even higher actual efficiency for the Rankine cycle.

Therefore, the Rankine cycle is a more practical cycle compared to the Carnot cycle, though it is less efficient, since water fed to the boiler at state 5 is at low temperature, and hence the heat added during the process 51 in the boiler is irreversible and decreases the average temperature of heat addition. However, its suitability in implementation, and a better work ratio, has resulted in the Rankine cycle being adopted as the standard steam power cycle. The Rankine (cycle) efficiency can easily be obtained by applying sfee to the turbine and boiler, to determine the amount of turbine work and heat added to the cycle; these respectively are $(h_2 - h_3)$ and $(h_2 - h_{4'})$, neglecting work done in the pump that makes $h_5 = h_{4'}$.

$$\therefore \text{ Rankine efficiency} = (h_2 - h_3)/(h_2 - h_{4'}) \tag{7.12}$$

We can get $h_{4'}$, $h_2$ and $h_3$ from the steam table. At point 3, only pressure (or temperature) of wet steam is known. Therefore, its dryness fraction $x_3$ is determined by solving the process equation $s_3 = s_2$, as 23 is an isentropic process. Once $x_3$ is known, then $h_3$ can be determined.

This cycle has been used in power plants. Figure 7.3 (a) shows the block diagram of a steam power plant. Saturated water enters the boiler at state 5, and is heated to form saturated steam at state 2. Steam is sent through a big diameter insulated pipe (to reduce pressure drop and heat loss) to the turbine, where it expands isentropically to state 3. At this state, steam is wet, which is condensed in a condenser at constant temperature to the state 4′, where $H_2O$ is saturated liquid. Now, it is compressed in a pump to raise its pressure. Since the specific volume of water is very small, the amount of work done in raising the pressure is so small that in calculations, it is neglected.

The power cycle is normally shown on the *T-s* or *h-s* diagram, for the simple reason that the type of processes involved in the cycle can be shown easily on the *T-s* diagram, and the specific enthalpy required for the first and second law analysis can be easily read from the *h-s* diagram.

## 7.4 IMPROVEMENT IN CYCLE EFFICIENCY

In order that the Rankine cycle may operate within the wet dome region, the critical temperature makes the upper limit, which is 374.15°C or about 647 K for $H_2O$. This limits the source temperature. Similarly, the temperature of the atmosphere or water in ponds, which are natural sinks, limit the sink temperature to around 300 K. Thus, the maximum efficiency of the Carnot cycle, with these temperature limits, shall be around 46%. Efficiency of the Rankine cycle will be still less. That is, more than half of the heat available from natural fuels shall go as waste, affecting the environment to which heat is rejected. If cooling water from the condenser is left to the pond, then it will increase the temperature of the pond water that affects fisheries. Thus, not only are we paying for the waste heat, but we are also increasing thermal pollution, affecting our normal life in many ways. There are other problems too, which are associated with this cycle, as discussed below.

- Critical pressure of the steam is 221.2 bar, which is quite high and requires the components of the power plant to be heavy to withstand this high pressure safely para. Heat of vaporisation decreases sharply with increase in pressure, and therefore, it will reduce the work output per kg of the steam, too. In other words, for a given amount of power to be developed, we shall need a bigger sized plant.
- Expansion in the turbine makes steam wet on the other hand. If steam that flows over the turbine blades is too wet, then increased losses during expansion reduce the efficiency of the plant. Also, life of the blades is decreased, since water is separated out from wet steam, which hits the blade surface and erodes it. The dryness fraction 0.88 or more at the end of expansion has been observed to be safe for long life of the blades.

Therefore, improvisation in the cycle is required for better efficiency, low first cost of plant and life of the steam turbine. Many methods were thought of to increase efficiency of the power cycle, such as increasing boiler pressure, using superheated steam at the inlet to the turbine, lowering condensing temperature, regenerative feed heating, and reheating of steam. In all these processes, factors that count are how to (i) increase the average temperature of heat addition, (ii) decrease the average temperature of heat rejection, (iii) increase the power output of the cycle per kg of steam used to reduce steam requirement, and (iv) to ensure that the quality of steam is more than 0.88 at low pressure (LP) turbine exhaust for better blade life.

For large plants producing more than 50 MW, steam pressure at the inlet to the turbine is more than 100 bar,[7] and a very large amount of steam per hour (300T/hr +) is required. If such a large mass of steam at such high pressure is expanded in a single turbine, then the size of the turbine will be too large towards the low pressure region. This creates problems not only in transporting the turbine from the place of manufacture to the place of use, but its weight will also be quite high, as the whole unit has to be designed to sustain that high pressure. A better approach is to have a number of turbines, so that the high pressure turbine is only designed to sustain high steam pressure. An added advantage is that a designer may choose a different kind of steam turbine for a high pressure range, from that used for a low pressure range. In practice, an impulse turbine is used for a high pressure range, and reaction turbine for a low pressure range.

### 7.4.1 Superheating of Steam

One of the methods to meet our desired goals is to use superheated steam in place of saturated steam at reasonable boiler pressure, around 140 to 200 bar. Heat addition during superheating of steam makes the cycle irreversible,[8] but the increase in average temperature of heat addition compensates for it, and efficiency is actually increased. However, we need not calculate this average temperature, as the first law analysis gives the desired results.

Figure 7.4 (a) shows a power cycle with superheating on a *T-s* diagram. We observe that, superheating shifts the state point on the *T-s* diagram to the right, the output of the cycle increases by an area 2'233' and heat supplied by an area 2'2ba. Nevertheless, ratio of

---

7. Supercritical pressure cycles capable of producing 1000 MW power from a single unit are are also in operation. In these boilers, water is converted to steam, without passing through the 2-phase zone. The steam condition at the entrance to the turbine is 240 bar, 630°C.

8. Whenever sensible heat is added during heating water or superheating steam, a source of heat with increasing temperature is required for heat transfer to be irreversible, which is not a practical proposal and heat is provided by a source having highest temperature. That is why sensible heating process is irreversible one.

increase in work to heat is more than that of the original cycle. That is, a better cycle is superimposed on the cycle, and hence there is increase in efficiency of the cycle with superheat. Logically, by superheating, the average temperature of heat addition is increased, and hence the efficiency increases. Further, the vertical line 23 shifts to the right of 2'3' thereby improving the dryness fraction of steam at the end of the expansion process. This means that steam can be expanded further to lower pressure without damaging the turbine blades. Figure 7.4 (b) shows variation of efficiency and ssc with degree of superheat. It is observed that efficiency increases and ssc decreases. Though the cycle efficiency goes on increasing with temperatures, the metallurgical limits of the blade material put an upper limit on the maximum steam temperature, which today is around 630°C. Compared with modified 12% Cr rotor steel (used in plants up to 566°C), the new rotor steel uses less carbon, manganese, nickel, and molybdenum content, more tungsten content, and boron and cobalt have been added. Blade material having rupture strength of more than 120 N/mm$^2$ is suitable for application.

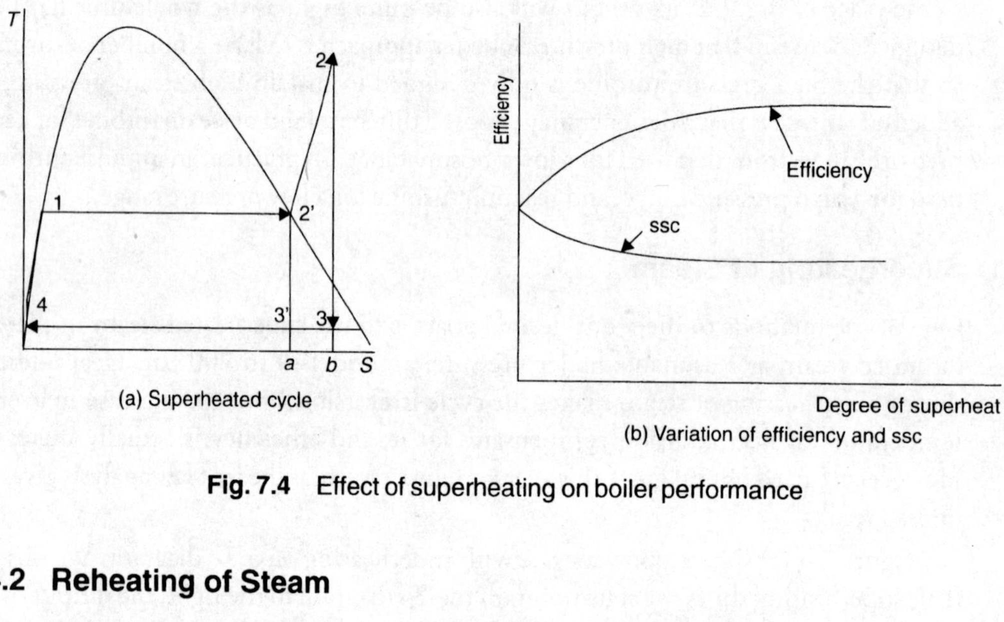


**Fig. 7.4** Effect of superheating on boiler performance

### 7.4.2 Reheating of Steam

With high pressure steam at inlet to turbine, it becomes wet during expansion, though steam pressure is still above atmospheric pressure. If allowed to expand further, it damages the blades. We can overcome this problem by expanding steam in a high pressure (HP) turbine up to a certain pressure, and then send it back to the boiler to reheat it to the initial steam temperature. It is then expanded in the next turbine. In this way, steam can be expanded through high pressure, intermediate and low pressure turbines, taken in that order. There can be more than one intermediate turbine. The number of turbines (commonly known as cylinders or modules) depend upon the size of the power plants, given in terms of MW capacity of the plant. A 200 MW plant with inlet steam pressure

140 bar, temperature 540°C has one stage of reheat after HP turbine and three modules, and a 500 MW plant with operating parameters at 200 bar, 600°C has five cylinders with single reheat after HP turbine. In supercritical plants, two stages of reheat have been used.

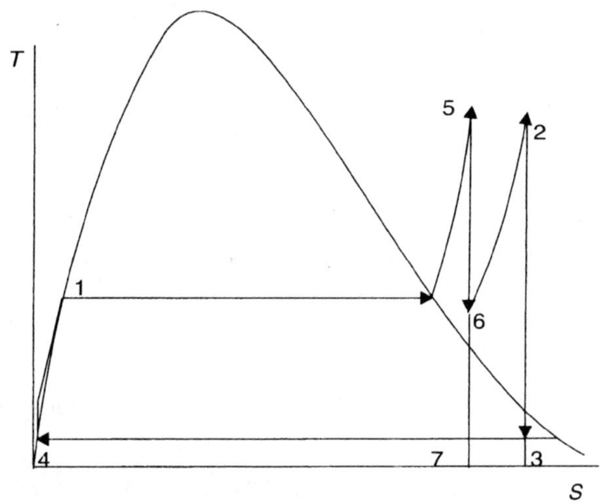


**Fig. 7.5** Reheat cycle

Figure 7.5 shows the reheat cycle on the *T-s* diagram. We observe that reheating moves the state point to the right, which is demanded because this improves the steam condition at the end of expansion in the turbine. The other reason is that constant pressure lines diverge which mean more work per kilogram of steam is produced during expansion of the steam between the same pressure limits. Reheating increases work output of the cycle by an area 6237, and therefore, reduces the size of the plant. We may think of reheat cycle as a cycle 6237 superimposed on Carnot cycle 41574. The efficiency of cycle 6237 is more compared to the original cycle, as heat addition in process 62 starts from temperature $T_6$, which is higher than $T_4$, at which heat addition starts in the main cycle. This increases the average temperature of heat addition in the superimposed cycle, while heat rejection is at the same temperature. Thus, efficiency of the cycle is increased. However, this gain is reduced when the regenerative cycle is also used, since water is then fed to the boiler close to the saturation temperature, and the original cycle efficiency is increased.

The average temperature of heat addition can be increased by multiple expansion through a small pressure drop in each expansion, followed with reheating. But this increases the complexities of the cycle, and is not worth implementation. On the contrary, the losses and the first cost are increased in taking steam from the turbine to the boiler and back. For these reasons, reheat is mainly used to ensure that the quality of steam at the exhaust of the LP turbine is above 0.88.

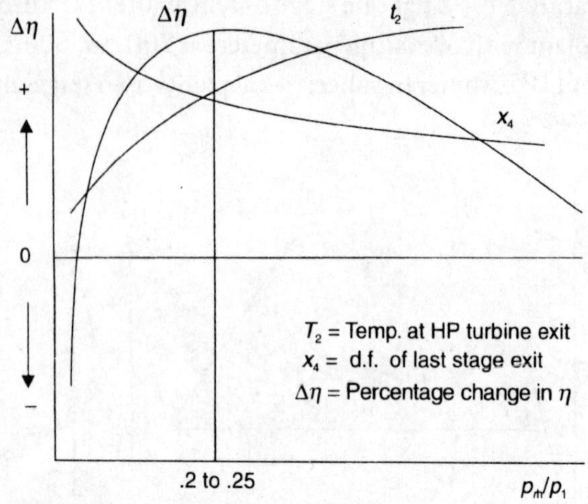


**Fig. 7.6** Effect of $p_{rh}/p_1$ on $\eta$, *d.f.* and $T_2$ for two-stage turbine

The pressure of reheat steam is decided by plotting the increase in efficiency and the dryness fraction of steam after expansion in the next stage for a different pressure ratio, $p_{rh}/p_1$, where $p_1$ and $p_{rh}$ are pressures of steam at the entrance and exit of the turbine from which steam is taken for reheating. That is, we choose some value for $p_{rh}/p_1$, and determine the efficiency of the cycle and the dryness fraction at the end of expansion in the last stage of the LP turbine. From this data, increase in efficiency from the cycle without reheat is calculated. This procedure is repeated for a different ratio of $p_{rh}/p_1$. Figure 7.6 shows the results of these calculations. It is observed that incremental efficiency is maximum around 0.2 to 0.25 pressure ratio.

### 7.4.3 Other Methods

**Increase in boiler pressure.** We know that the saturation temperature of steam increases with pressure. Figure 7.7 (a) compares the Rankine cycle at two different boiler pressures for the same sink temperature. The area of the cycle increases by area 1'1a2'1 and decreases by area 233'a2. However, these two areas are of the same order of magnitude, and therefore, the output of the cycle does not change significantly. On the contrary, heat addition is decreased by area 33'CB. Thus, there is gain in efficiency, which increases with increase in pressure, and after some pressure, the efficiency of the cycle starts falling, as can be seen from Fig. 7.7 (b). This behaviour can be explained by the fact that increase in boiler pressure decreases the latent heat addition and increases the sensible heat addition, resulting in lowering of average temperature of heat addition after a certain pressure.

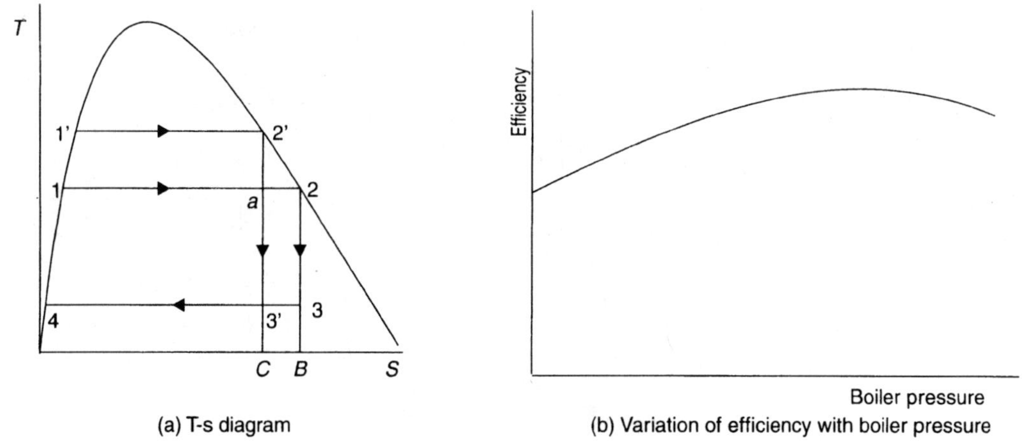


(a) T-s diagram

(b) Variation of efficiency with boiler pressure

**Fig. 7.7** Effect of boiler pressure

**Decrease in condenser pressure.** Lowering condenser pressure decreases $T_2$, and hence efficiency of the cycle increases. However, the temperature of the cooling media puts a limit on the condenser temperature, since temperature of steam in the condenser should be more than that of the cooling media by 8–10°C for effective heat transfer from steam to the cooling media. Also, as the condenser pressure decreases, air from the surroundings tends to leak through the various joints. An extraction pump is required to take out air leaking into the condenser and to pump out water (condensate) from it. Power is required to run this pump, and the advantage obtained by decrease in vacuum is to some extent nullified, to the extent that after a certain degree of vacuum, it is of no use. Condenser pressure is around 0.1 bar.

**Regenerative or bleed cycle.** So far, we have concentrated only on the right side of the *T-s* diagram to see ways and means for improving cycle efficiency. But in the Rankine cycle, the condensed water from the condenser is fed to the boiler at a much lower temperature, compared to the boiler temperature that makes heat transfer in the boiler irreversible. Therefore, for higher efficiency, we must make an attempt to eliminate this, too. It is done by heating feed water before supplying it to the boiler, for which steam is extracted from the turbine at different stages of expansion. Sufficient amount of steam is bled from the turbine at a stage, such that, after mixing it with water from the previous stage, water is saturated at the bled steam pressure. Such a cycle is called regenerative or bleed cycle. Figure 7.8 shows bleed cycle with steam taken from HP, IP and LP turbines to water heaters, and the analysis is based on mass and energy balance for each heater. (In this diagram, pumps are not shown, which will be required to raise water pressure from the condenser to the boiler pressure.) After the last stage of steam heating, feed water is heated further in an economiser, where the heat source is hot flue gas from the boiler. This way, we ensure that feed water is saturated at the boiler drum pressure.

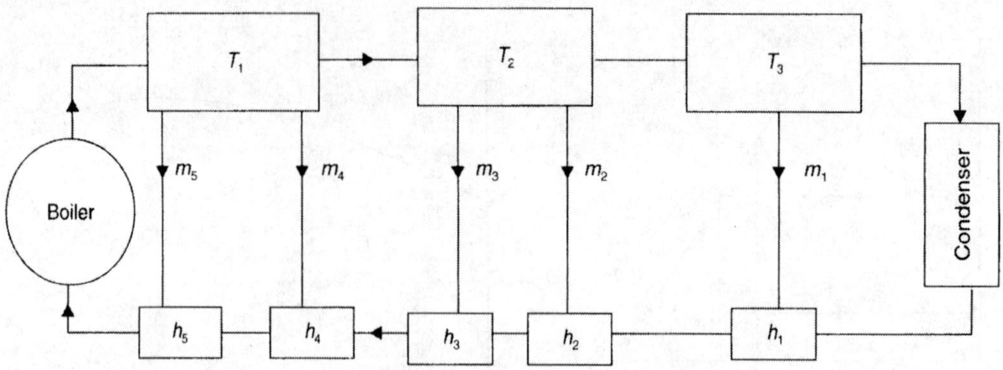


**Fig. 7.8** Bleeding of steam

The number of feed heaters used depends on the required temperature rise $\Delta T$, which is equal to the difference of saturation temperature, of steam at the boiler, and the condenser pressure, and is determined on the basis of the following considerations.

1. It can theoretically be proved that equal increase in enthalpy of feed water among all the heaters and the economiser results in maximum efficiency. This means the same rise in water temperature in all the heaters. If $n$ is the number of heaters, then rise in temperature in each feed heater and economiser will be $\Delta T/(n+1)$ and the total temperature rise in the feed heaters is $\Delta T_h = n\Delta T/(n+1)$. These results are tabulated in the following Table 7.1.

**Table 7.1** Effect of number of heaters on incremental gain

| $n$ | $\Delta Th$ | Incremental gain | $n$ | $\Delta Th$ | Incremental gain |
|---|---|---|---|---|---|
| 0 | 0 | – | 2 | $2\Delta T/3$ | $2\Delta T/3 - \Delta T/2 = \Delta T/6$ |
| 1 | $\Delta T/2$ | $\Delta T/2$ | 3 | $3\Delta T/4$ | $3\Delta T/4 - 2\Delta T/3 = \Delta T/12$ |

2. It is seen from the table that more is the number of heaters, more is the rise in feed water temperature, and hence more efficiency. But this also means less utilisation of waste heat from flue gas in the economiser, and hence less boiler efficiency. Moreover, incremental gain in temperature decreases with increase in number of heaters, resulting in diminishing returns, while the first cost increases. Therefore, one has to make a judicious choice for $n$.
3. We may determine the number of heaters in the same way, as we determined pressure at which steam should be taken for reheat. That is, for given boiler and

condenser conditions, we determine incremental gain in efficiency with one feed heater for different rise in feed water temperature. The results will be like the one shown by curve 1 in Fig. 7.9. It will have maximum efficiency at a temperature decided by the consideration mentioned at point 1. This process is repeated with 2, 3, 4, or more number of heaters. All of these, like curve 1, will have a maxima. The dotted line shows the locus of points of maximum efficiency. This decides $\Delta T_h$ for that value of $n$.

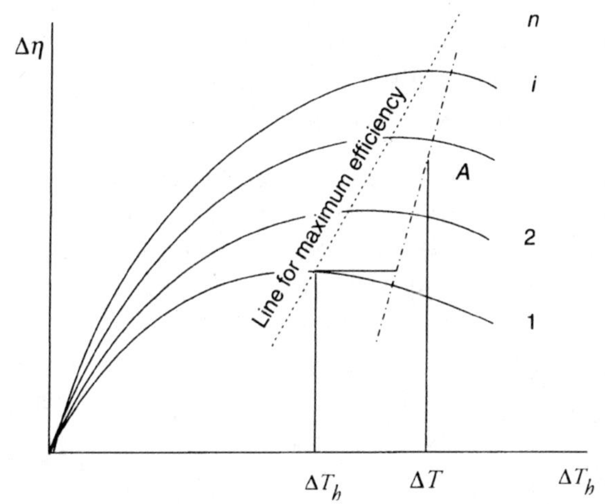


**Fig. 7.9** Incremental efficiency with heater numbers

4. We now determine $\Delta T$, the rise in temperature of water in heaters and the economiser, using equation $\Delta T = (n + 1)\,\Delta T_h/n$, where $\Delta T_h$ is the theoretical temperature rise at the point of maximum efficiency for given value of $n$. With this information, a point is obtained on the horizontal line drawn from the point of maxima for given $n$. It is shown in Fig. 7.9 by the centre line. The two lines appear to converge with $n$, reflecting that gain decreases with $n$, as mentioned at point 2.
5. Now, knowing the total rise in temperature for the given cycle, a vertical line is drawn to intersect the $\Delta T$ line, say, at A (Fig. 7.9). This gives the number of feed heaters required. The designer has the choice of choosing the upper or lower value of $n$.

   Different types of feed heaters are in use; these are shown in Fig. 7.10 for $m$th stage[9] ($1 < m < n$), where $n$ is the total number of feed heaters used in a bleed

9. Numbering of the feed heater stage is done from the condenser end. Thus, the heater at the last stage is the one which feeds water to the economiser.

cycle. Feed water may be heated directly or indirectly. Direct contact heater (Fig.7.10 (a)) is an example of direct heating, in which steam and water are mixed with each other. In indirect heaters, water does not come in contact with steam, and heat transfer is through a surface separating two streams. Hereafter, the condensate may be mixed ahead of the heater into the feed line in the 'Direct Pump Method' (Fig. 7.10 (b)). This is a more efficient method of heat utilisation of the drain. Both direct pump and direct contact methods however require a pump for each heater. The other indirect type heating is the 'Cascade Method', in which the drain is mixed with condensate from the previous stage, and finally fed to the condenser outlet from where it is pumped by a single feed pump. Example 7.2 illustrates the analysis procedure for the cascade method.

*Note*: One or other methods discussed above have been implemented in practice, considering the economic criterion of initial and running costs. Example 7.1 illustrates the analysis procedure and compares the effect of different parameters on cycle efficiency, quality of steam after expansion in the turbine and the work ratio.

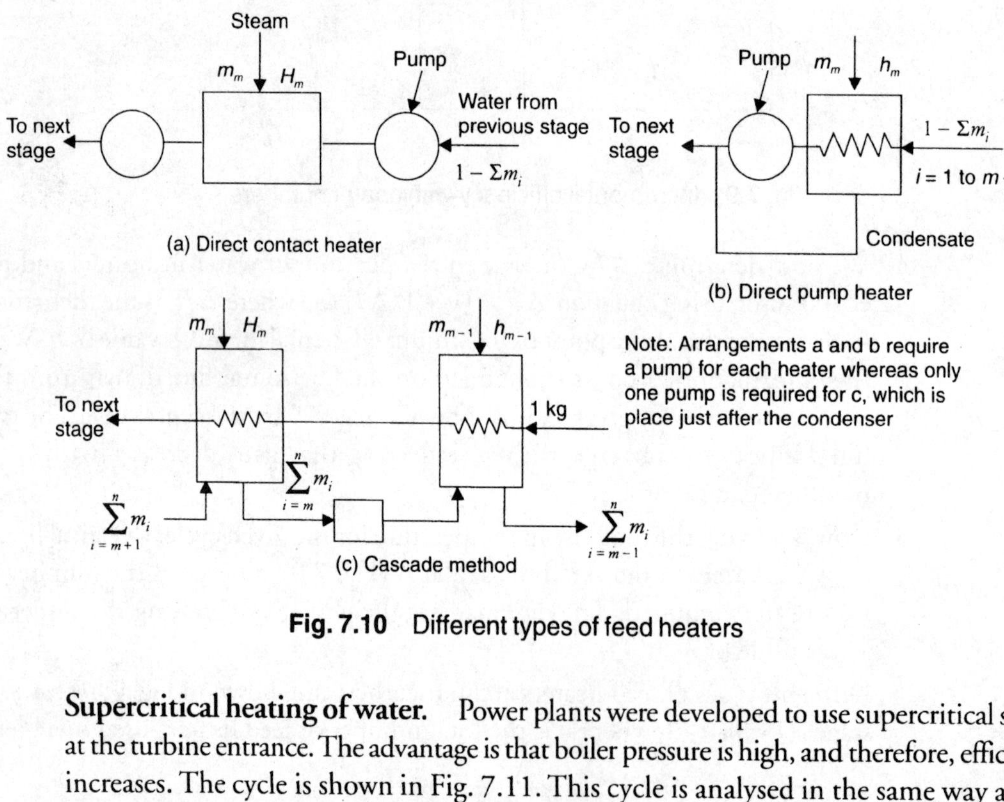


**Fig. 7.10** Different types of feed heaters

**Supercritical heating of water.** Power plants were developed to use supercritical steam at the turbine entrance. The advantage is that boiler pressure is high, and therefore, efficiency increases. The cycle is shown in Fig. 7.11. This cycle is analysed in the same way as any other steam power cycle, and it is left to the student.

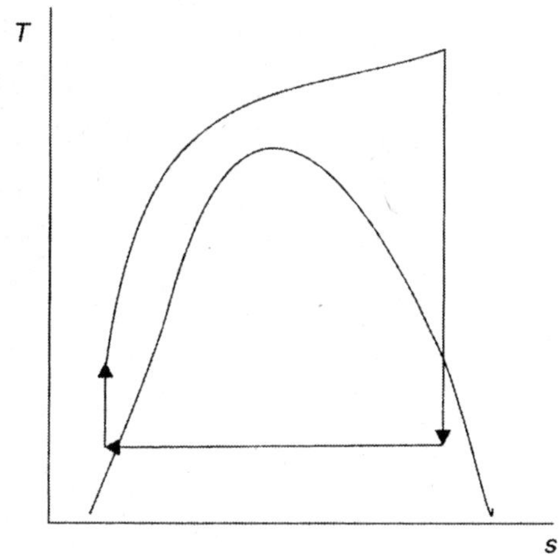


**Fig. 7.11** Supercritical pressure cycle

**Example 7.1** In a steam power plant, boiler pressure is 200 bar and the condenser temperature may be taken as 40°C. Find out the efficiency, work ratio and quality of steam in (A) Carnot cycle, (B) Rankine cycle (i) without superheat, (ii) with steam superheated to 500°C. How are these quantities modified if the steam is reheated at the stage when it is just dry during expansion?

***Solution*** The Rankine cycle is shown in Fig. 7.12.

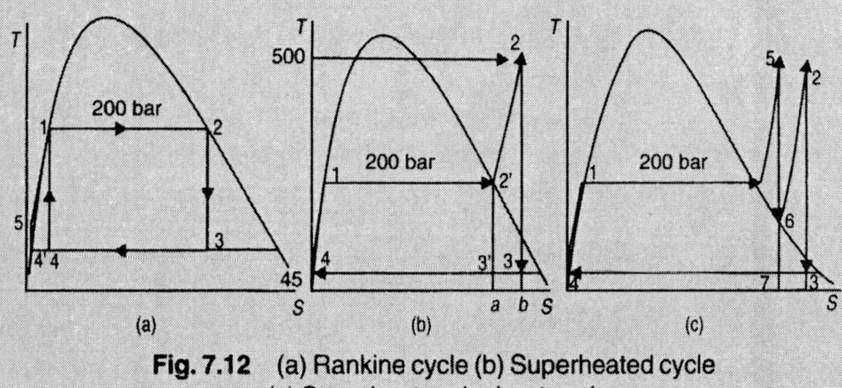


**Fig. 7.12** (a) Rankine cycle (b) Superheated cycle (c) Superheat and reheat cycle

Given: $p_1$ = 200 bar and $t_3$ = 40°C.

Read from steam table, at 200 bar pressure, $h_1 = h_f = 1826.3$,

$h_2 = h_g = 2409.7$ kJ/kg, $s_1 = 4.014$ and

$s_2 = 4.927$ kJ/kg K. At 40° C,

$$h_f = 167.6,\ h_{fg} = 2406.7 \text{ kJ/kg}$$
$$s_f = 0.573, \text{ and } s_{fg} = 7.685 \text{ kJ/kg K}$$

**Carnot cycle.** Enthalpy at points 3 and 4 is obtained by determining the dryness fraction of steam using the fact that $s_3$ $(= s_{f3} + x_3 \times s_{fg3}) = s_2$ and $s_4$ $(= s_{f4} + x_4 \times s_{fg4}) = s_1$ (since points 23 and 14 lie on isentropic lines), where $x_3$ and $x_4$ are the d. f. at these points. This gives $x_3 = 0.567$ and $x_4 = 0.448$.

$$\therefore\ h_3 = h_{f3} + x_3 \times h_{fg3} = 1531.14 \text{ and}$$
$$h_4 = h_{f4} + x_4 \times h_{fg4} = 1246.2$$

Applying sfee to processes 23 and 14, and noting that heat transfer is zero, we get

$$w_t = \text{turbine work} = h_2 - h_3$$
$$= 878.56 \text{ kJ/kg and}$$
$$w_c = \text{compression work}$$
$$= h_1 - h_4 = 581 \text{ kJ/kg}$$
$$w_{net} = \text{net work output of cycle}$$
$$= w_t - w_c = 878.56 - 581$$
$$= 297.56 \text{ kJ/kg and}$$
$$q = \text{Heat input to cycle in process 12}$$
$$= h_2 - h_1 = 586.4 \text{ kJ/kg}$$
$$\therefore\ \eta = \text{Cycle efficiency}$$
$$= w_{net}/q = 50.9\% \text{ and}$$
$$\text{Work ratio} = w_{net}/w_t = 297.56/878.56 \times 100$$
$$= 38.87\%$$

We observe that the Carnot cycle is not suitable for the given operating parameters. The dryness fraction is too poor after expansion and work ratio is 0.3887.

**Rankine cycle.** Refer Fig. 7.12 (a) $h_1$, $h_2$, $h_3$, $w_t$ and $h_{4'}$ (equal to $h_f$ at 40°C) were determined above. Work of compression of water from state 4' to 5 is obtained by noting that the density of water does not change, and the compression process is treated as constant volume compression. Further, the pump is an open system for which work of compression is given by $-\int v dp$, where $v$ is the specific volume of water = $1/\rho$, $\rho$ is the density of water that may be taken as 1000 kg/m$^3$ ($\therefore\ v = 0.001$ m$^3$/kg) and $dp$ is change in pressure in N/m$^2$. For the given problem, $p_{4'} = 0.007384$ bar from the steam table. Therefore,

$$dp = 200 - 007384 \approx 200 \text{ bar} = 2 x 10^7 \text{ N/m}^2. \text{ This gives}$$

$w_c$ = work done in compression is $0.001 \times 2 \times 10^7/10^3$
= 20 kJ/kg and

$w_{net}$ = net work output of cycle = $w_t - w_c$
= 878.56 – 20 = 858.56

Applying sfee respectively to pump and to boiler, we get

$h_5 = h_{4'} + w_c = 167.6 + 20$
= 187.6 kJ/kg

$q$ = Heat input to cycle
$= h_2 - h_5$
= 2222.1 kJ/kg

$\therefore \eta$ = Cycle efficiency
$= w_{net}/q$
= 38.6%

and work ratio = $w_{net}/w_t = 0.977$

If we neglect work of compression as this is only 20/878.56 = 2.28% of the turbine work, then we can find out efficiency = $w_t/(h_2 - h_{4'})$ = 39.18%. Normally, we neglect work of compression, as it is comparatively small.

**Rankine cycle with superheat.** Refer Fig. 7.12 (b). Steam enters the turbine at state 2 and leaves at 3. Steam is at 200 bar and 500°C at the entrance. From the steam table for superheated steam, we read

$h_2$ = 3238.2 kJ/kg and $s_2$ = 6.14 kJ/kg K. Following the procedure to determine $h_3$ for Carnot cycle, we get $x_3 = 0.7244$ and $h_3$ = 1911 kJ/kg. Neglecting pump work,

$w_t$ = Turbine work = $h_2 - h_3$
= 1327.2 kJ/kg = $w_{net}$

$q$ = Heat input to cycle = $h_2 - h_{4'}$
= 3238.2 – 167.6 = 3070.6 kJ/kg

$\therefore \eta$ = Cycle efficiency = $w_{net}/q$ = 46.22%

**Rankine cycle with superheat and reheat.** Refer Fig. 7.12 (c). There are two stages of expansion, that is, 56 and 23. At point 6, steam is saturated (given) and lies on the vertical line 56, that is, $s_5 = s_6$. From the steam table at point 5 ($p$ = 200 bar and $t$ = 500°C), $h_5$ = 3238.2 kJ/kg and $s_5$ = 6.14 kJ/kg K. We now look at the $s_g$ column in the table for wet steam, and locate pressures 33 and 34 bar at which specific entropy of saturated steam respectively, is 6.149 and 6.137. Pressure at point 6 therefore lies between

33 and 34 bar, and is obtained by linear interpolation to give $p_6$ = 36.75 bar and $h_6$ = 2806.75 kJ/kg. Now, for point 2, pressure is 36.75 bar and temperature 500°C, and therefore, again using linear interpolation on pressure and using the superheated steam table, we get $h_2$ = 3452.29 kJ/kg and $s_2$ = 7.1758 kJ/kg K. This means that for state 3, we know $t_6$ = 40°C and $s_6$ = 7.1758. We do not know whether steam is wet or superheated at this point. So, we see $s_g$ from the wet steam table at 40°C, which is 8.257 kJ/kg K and is more than 7.1758. Therefore, steam is wet after expansion in the second stage. We may calculate $x_3$ and $h_3$ in the usual way, and are equal to 0.86 and 237.36 kJ/kg, respectively.

$$w_t = \text{Turbine work} = (h_2 - h_3) + (h_5 - h_6)$$
$$= 1761.65 \text{ kJ/kg} = w_{net}$$
$$q = \text{Heat input to cycle} = (h_5 - h_{4'}) + (h_2 - h_6)$$
$$= 3719.14 \text{ kJ/kg}$$
$$\therefore \eta = \text{Cycle efficiency} = w_{net}/q = 47.37\%$$

**Comparison of different cycles.** Table 7.2 compares the cycles for efficiency and quality of steam at the outlet of the turbine.

**Table 7.2** Cycles for efficiency and quality of steam at the outlet

| Cycle → | Carnot | Rankine | Rankine with superheat | Rankine with superheat and reheat |
|---|---|---|---|---|
| Efficiency,% | 50.900 | 38.600 | 46.220 | 47.37 |
| $w_{net}$, kJ/kg | 297.560 | 858.560 | 1327.200 | 1761.65 |
| $w/w_{Carnot}$ | 1.000 | 2.890 | 4.460 | 6.92 |
| Steam d.f. | 0.567 | 0.567 | 0.7244 | 0.86 |

We see from the table that there is increase in efficiency, quality of steam and work done per kg of steam with superheat and reheat. The dryness fraction after expansion in the turbine has gone below the permissible limits in all these cases. In the actual cycle, irreversibility present in the system reheats the steam, increasing its dryness fraction, and we have to see this effect too before bothering for the dryness fraction of 0.86 as against the permissible value of 0.88. Lastly, work output has increased six times, and therefore, size of the plant decreases significantly.

**Example 7.2** Steam is supplied to a turbine at a pressure of 30 bar and superheated by 93°C. The turbine exhaust pressure is 0.96 bar. The main condensate is heated regeneratively in two stages by steam bled from the turbine at 4.7 bar and 1.8 bar, respectively. Calculate the mass of steam bled off at each pressure per kg of steam entering the turbine, and the theoretical efficiency of the cycle. The following assumptions are to be made.

1. That the expansion of the steam in the turbine is isentropic.
2. That the condensate is heated to the saturation temperature in each of the heaters, which are of indirect type.
3. That the drain water from the HP heater passes through a trap[10] without loss of heat into the steam space of the LP heater.
4. That the combined drains from the LP heater are cooled in a drain cooler[11] to the condenser temperature.

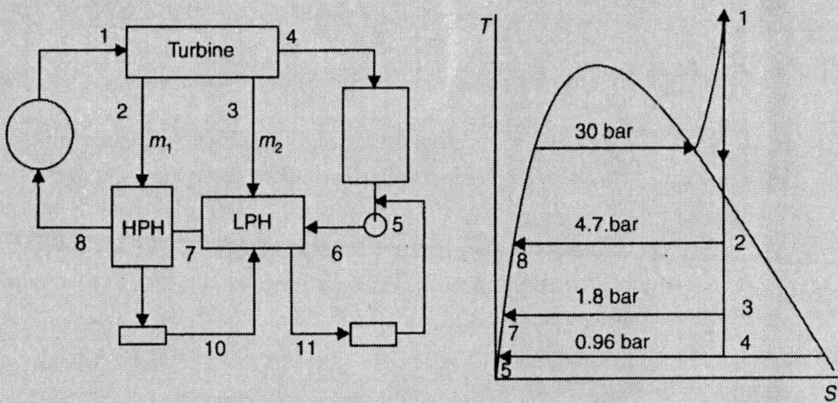


**Fig. 7.13** Analysis of bleed cycle

***Solution*** Figure 7.13 shows the schematic arrangement and the expansion of steam in the turbine on the *T-s* chart. Steam is bled off at points 2 and 3, which is used to heat feed water first in a low pressure and then in a high pressure heater. Pressure and the degree of superheat at point 1 is given, which determines all the required properties of steam either from the steam

10. A trap is a device which reduces pressure. Condensate from the HP heater is at a higher pressure than pressure of the steam in the LP heater and hence its pressure is required to be reduced by using a trap, before it is mixed with low pressure steam in the heater.
11. This example illustrates the cascade method. Condensate from the first stage heater is taken to a hot well through a drain cooler, which is nothing but an indirect type of heat exchanger. A trap reduces pressure.

table or the Mollier chart. Also, pressure at points 1, 2, 3 and 4 are known, and therefore, the enthalpy of saturated water and steam can be read from the steam table. Using the fact that expansion of steam in the turbine is isentropic, we write

$$s_4 = s_3 = s_2 = s_1 = 6.649 \text{ (from the steam table at state 1)}$$

or $$s_{f4} + x_4 \times s_{fg4} = s_{f3} + x_3 \times s_{fg3} = s_{f2} + x_2 \times s_{fg2} = 6.649$$

This yields $x_2 = 0.96$, $x_3 = 0.91$ and $x_4 = 0.88$. Finally, $h_2 = 2661.27$, $h_3 = 2502.89$ and $h_4 = 2402.33$ kJ/kg. Or else, we may show the process on the Mollier chart, and read the enthalpy values at different state points.

Mass balance for the turbine, high pressure heater, low pressure heater and the pump gives, respectively

$$m_4 = 1 - m_1 - m_2,\ m_{10} = m_1,\ m_{11} = m_1 + m_2, \text{ and } m_6 = 1 \text{ kg}$$

assuming that 1 kg of steam is entering the turbine. Water is pumped from the condensate well by a single pump, and therefore,

$$m_5 = m_6 = m_7 = m_8 = 1 \text{ kg}$$

Energy balance for the high pressure heater gives

$$m_1 h_2 + h_7 = m_1 h_{10} + h_8$$

We assume that the bled steam in high and low pressure heaters cool to saturated state. This and assumption 2 gives $h_8 = h_{10} = h_{f2} = 630.2$ and $h_7 = h_{11} = h_{f3} = 490.7$(from the steam table). Here all the unknowns, except $m_1$, are known, and this equation gives $m_1 = 0.069$ kg.

Similarly, applying energy balance to the low pressure heater, we get

$$m_1 h_{10} + m_2 h_3 + h_6 = (m_1 + m_2)\, h_{11} + h_7$$

This gives: $m_2 = 0.034$ kg. Finally, applying sfee to the turbine and boiler for 1 kg of steam at the entrance to the turbine results in

$$w = h_1 - \{m_1 h_2 + m_2 h_3 + (1 - m_1 - m_2)\, h_4\} \quad \text{and} \quad q = h_1 - h_8$$

or $w = 636.66$ and $q = 2429.07$ kJ/kg $\Rightarrow$ $\eta = w/q = 26.17\%$

## 7.5 IMPLEMENTATION OF STEAM POWER CYCLE

Figure 7.14 shows a block diagram for a large power plant. The major equipment needed is a boiler, turbine(s), condenser and feed pump. We also require other equipment called auxiliaries for the operation of the plant, without which the plant cannot function. These include air, ash, coal and cooling water handling equipment, pump to extract air and water from the condenser, de-aerator to remove dissolved gases from feed water, electrostatic precipitator to remove ashes from flue gases before discharging these into the atmosphere, and protective gear. Auxiliaries get power from the power generated, and is

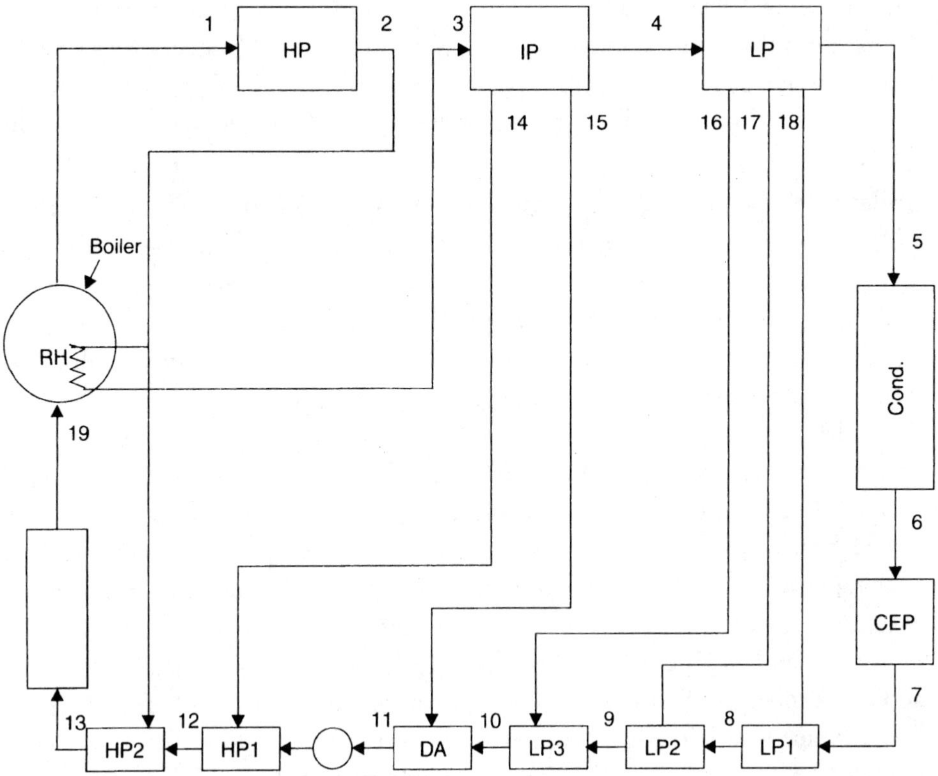


| Point | m | p | t |
|---|---|---|---|
| 1 | 564.0 | 143 | 533.9 |
| 2 | 564.0 | 35.6 | 346.0 |
| 3 | 530.8 | 33.3 | 529.0 |
| 4 | | | |
| 5 | | | |
| 6 | | – 0.918 | Liq. |
| 7 | | 23.9 | Liq. |
| 8 | | –0.74 | 53.0 |
| 9 | | –0.21 | 81.9 |
| 10 | | 1.43 | |
| 11 | 577.0 | 5.21 | 168 |
| 12 | | 13.70 | 189 |
| 13 | | 35.6 | 234 |
| 14 | | 13.7 | 415.0 |
| 15 | | 5.21 | 314.0 |
| 16 | | 1.43 | 229.0 |
| 17 | | – 0.21 | 84.9 |
| 18 | | – 0.74 | 66.5 |
| 19 | 598.0 | 143.0 | |
| 20 | 33.2 | 35.6 | 346.0 |

| | |
|---|---|
| Oil | 7.286 t/h |
| Feed water | 608 t/h |
| Air consumption | 669.9 t/h |
| RPM | 3000 |
| Steam consumption | 580 t/h |
| Pressure | kgf/cm$^2$ |
| Temperature | $^\circ$C |
| m, mass flow | t/h |

**Fig. 7.14** Power plant (200 MW)

of the order of 10%, which has been reduced significantly to a low level of 5%, so that the net output of the system is increased. Equipment like the economiser, super heater and air pre-heater are used to improve efficiency of the plant. A starter motor is required to start the system from cold for rolling of the turbine shaft, and to operate the lubrication system.

**Boiler.** Water is treated in a demineralisation plant to remove soluble impurities to the level of one part per billion, and is mixed with the condensate from the condenser to make up for loss of water, which is mainly because of leakage during its flow from unit to unit. Feed water is pumped to the boiler drum. The boiler is of massive size, and placed outside the steam turbine room. Distance between the boiler and the turbine should be kept as small as possible to reduce the length of the pipe carrying steam from the boiler to the turbine, and therefore, heat loss from the pipe to the surroundings. It also reduces the pressure drop due to friction. A boiler is fed a weighed amount of pulverised coal of proper size, and enough air (normally excess air of the order of 5%) is supplied by forced fans, so that complete combustion of coal occurs. Instruments control excess air, so that permissible level of carbon monoxide is present in flue gases. Primary air, mixed with pulverised coal, is fed centrally to burners, placed within the furnace at different points. Secondary air is supplied tangentially from the burner periphery, so that proper and quick mixing of fuel air occurs in the furnace for efficient and complete combustion. The furnace wall is made of copper tubes carrying cold water from the boiler drum through the down take pipe. Heat is transferred to water by radiation in the near furnace region, and by convection in the far region. A part of water is converted into steam, and a mixture of steam and water returns back through the uptake pipe to the drum, where steam is separated and collected in the space above the water level. Steam is then taken to the super heater to heat it to the desired temperature, and is fed to the HP turbine.

Fleu gases are left out in the atmosphere through a high chimney, at the base of which is placed an induced fan, that sucks the flue gases out of the furnace. For safety reason the forced and the induced fans maintain pressure in the furnace that is slightly lower than the atmospheric pressure. The super heater, economiser and air pre-heater are placed in this order in the flue gas chute from the boiler to the chimney to extract heat from it, so that the gases leave the chimney around 110°C. The limitation of having such high temperature at the exit from the chimney is due to the fact that at a lower temperature, steam and sulphur dioxide formed during combustion in the furnace, combine to form a toxic acid, which eats away the surfaces it comes in contact with. At the end of this passage, mechanical and electrostatic precipitators are placed to remove the fine ash particles coming with the flue gases before discharging these to the atmosphere. This is legally compulsory, otherwise penalty is levied. The heat loss in flue gas is around 12% of the input. The ash from the precipitator has been found to have good pozalinic quality, and is used by cement and fly ash brick industries.

Fuel contains ash, which may be as high as 45% in poor quality coal. The ash separates out in the furnace, is collected at the bottom of the furnace where it is cooled by mixing with water, and slurry is made and pumped to ash ponds using centrifugal pumps. The amount of ash is so large, that the ash ponds so formed are not only a source of pollution, but also occupy vast land. This ash, called pond ash, is used as filler material in civil works.

**Turbine.** A turbine is a device in which steam is expanded from boiler pressure to condenser pressure. Boiler pressure is quite high, and steam cannot be expanded in a single stage to condenser pressure[12]. That is, steam is expanded to some lower pressure in a nozzle or fixed blades (converging passages), changing pressure energy, to kinetic energy, such that flow in the turbine passage remains subsonic. High velocity steam then flows over a rotating or moving blade, to convert kinetic energy of steam into work, during which absolute velocity of the steam jet is reduced. This process is repeated in successive stages, till steam pressure is equal to condenser pressure. This means that steam flows alternately through fixed and moving blades. Fixed blades are required between two rows of moving blades to ensure smooth entry of steam to moving blades and also to expand steam. We may, therefore, imagine a turbine to be made up of a rotor carrying rows of rotating blades along its length, interspersed with fixed blade rows mounted on the casing in which the rotor is housed. Such machines are axial flow machines, as flow is along the rotor shaft axis.

The number of blades is thus fixed around the casing and rotor at alternate cross-sections at different axial locations along the turbine shaft. These blade rows shall be referred to as fixed and moving blades, respectively. It essentially implies that the first row in the direction of flow is made up from nozzles or fixed blade row, then rotating (moving), and so on. A pair of fixed and moving blade is called a stage of turbine. As one moves down the stages, pressure decreases and specific volume increases. Therefore, size of the turbine increases. Stator blades are designed so that the area of flow decreases as steam flows over them, resulting in expansion of steam, which increases relative velocity of steam with respect to the blade surface. But in an impulse turbine, area of flow remains constant, and therefore, the relative velocity of steam w.r.t. the blade remains constant. Turbines are designed so that kinetic energy of steam after the last stage is minimum, which is of the order of 10% of its value at the inlet. This much loss is to be accepted, as, if the velocity of the jet becomes zero, then no mass flow can take place.

---

12. If a large pressure drop in a nozzle is even possible, then velocity of the steam jet will be too high. We know that when a fluid flows through a passage, losses occur due to viscosity of the fluid, and these losses are proportional to the square of velocity. Therefore, losses will be immense if a very high velocity is used over the turbine blades. Steam being compressible, shock losses would also occur.

When a high speed steam jet flows over the moving blade, its momentum changes at least due to change in direction of the flow, and a force acts on the blades (Newton's second and third law of motion). Change in momentum is more if the velocity of steam is also increasing. This force generates torque, rotating the drum in its direction. Decrease in area increases the velocity of steam with respect to the rotor blade. However, note that the absolute velocity of steam decreases, since part of the kinetic energy of steam at the entrance to the turbine blades is converted into shaft work, resulting in decrease of kinetic energy at the exit of the blade. Care has to be taken so that steam enters the moving blade smoothly or without shock, for which the direction of the jet's relative velocity vector with respect to the blade itself at its inlet should be along the tangent to the blade at the inlet. This is possible at the designed operating conditions, but not at other load conditions. That is why, efficiency at operating conditions other than designed, decreases.

**Condenser.** The third major component of a power plant is the condenser. It is used to reduce pressure at the exit of the turbine below atmospheric pressure. Many types of condensers are in use, but a surface condenser is mostly used In brief, it consists of a round shell having end plates. Tubes carrying cooling water are fitted between the end plates, and these tubes fill the space inside the shell. Steam enters from the top of the condenser, and the condensate is collected at the bottom of the shell from where it is removed by an extraction pump. Hot steam comes in contact with the cold tubes, and is condensed into water due to heat transfer to cooling water. The condensed water is called condensate, and is usually undercooled up to 5°C below the saturation temperature at the condenser pressure. Air extraction pump removes air present in the condenser, or gases liberated from feed water at low pressure to maintain the desired vacuum. The condenser is housed below the ground level of the steam power plant, and is the lowest floor in a power plant. Cooling water requirement is quite high, and therefore, the same cooling water is reused for the purpose. Cooling water coming out of the condenser is cooled in a cooling tower or spray ponds, where hot water is sprayed in the form of jet, the fine particles of water come in contact with cold air and get cooled. In spray pond cooling, loss of water is significant, as the air currents carry water vapours away.

**Feed pump.** The fourth major component is the feed water pump. This pump supplies water to the boiler at high pressure, and therefore, a positive displacement type pump, like the electric driven reciprocating pump, is most suited. This pump is placed at the ground level.

## 7.6 BINARY AND TERTIARY CYCLES

We studied different methods used for improving the performance of the steam cycle. However, Carnot cycle efficiency could still not be attained. This is mainly due to the fact that it is not possible to implement various reversible processes of the cycle. But in addition, there is a problem with the working fluid as well. The working fluid too affects the efficiency, and puts, a limit on temperature range under which a cycle can operate. The degree of the effect depends upon its thermodynamic properties.

An ideal fluid should have the following thermodynamic properties.

- It should have small specific heat (to reduce sensible heat) and large enthalpy of evaporation (latent heat) at boiling temperature, to increase average temperature of heat addition.
- It should have high critical temperature and pressure; high enough so that working pressure is moderate even at the metallurgical temperature limit. This will mean more latent heat can be added even at reasonable pressure.
- It should have saturation pressure above atmospheric pressure at the ambient temperature, so that no vacuum is required in the condenser.
- It should have a very steep saturation vapour curve, so that the dryness fraction at the end of expansion in the turbine is high enough to avoid reheating.
- It should not freeze at low temperature at which heat is rejected.
- The other factors that affect choice of a working fluid are its availability, cost, non-poisonous nature, environment friendliness, wetting of the tube through which the fluid flows and corrosive action on the surface. The wetting property is essential, so that the fluid remains in contact with the tube surface to increase heat transfer.

Unfortunately, no single fluid meets the above mentioned requirements. If one is suitable in a high temperature range, then it is not suitable in a low temperature range. For example, mercury has saturation pressure of 12 bar at 560°C and $2.7 \times 10^{-4}$ cm Hg at 30°C, respectively and critical pressure and temperature of 1080 bar, 1460°C. So, it is very suitable in the high temperature range and not in the low temperature range (requires a high degree of vacuum). However, mercury does not wet the surface, which is overcome by adding 0.002% solution of magnesium and titanium to mercury. Moreover, it is poisonous and costly, and therefore, use of this as a working medium does not find commercial favour. Table A.10 lists properties of saturated mercury.

Other fluids in the high temperature range notably are diphenyl ether, aluminium bromide and liquid metals like sodium or potassium. Saturation temperature for $Al_2Br_6$ at $p$ = 12 bar is 482.5°C, which is comparable with that for mercury. Liquid metals are other suitable candidates for this range.

For a medium temperature range, water is the most suitable candidate amongst all available fluids. In the low temperature range, refrigerants like ammonia, sulphur-dioxide, and Freon 11, may be considered. For example, saturation pressure of F-11 is 8.2 bar at 100 and 1.74 bar at 40°C, which is a very comfortable range. Heat may be rejected to cooling media without creating vacuum. Also, water can reject heat to the F-11 cycle at

100°C or a little higher temperature at which saturation pressure of water is around 1 bar only. Therefore, F-11 is most suitable for low temperature range.

In fact, there exists no such fluid that satisfies all the mentioned requirements. But there is no sanctity to use only steam as the working fluid, or only a single vapour cycle. Alternatively, we may look for other vapours having some or more required properties in different operating ranges and run two or more power cycles coupled with each other. Here, coupling of two cycles is done by using a condenser of a high temperature range cycle as the boiler for the next stage cycle. That is, heat rejected in the first cycle is used for heating vapour in next cycle, and so on. Or, in other words, a heat exchanger[13] is used. This would therefore, require only one boiler for heating fluid in the first cycle, and subsequently, heat exchangers only. If only two fluids are used, then the cycle is termed binary vapour cycle with the high temperature range cycle as the topping cycle and the second one as bottoming cycle.

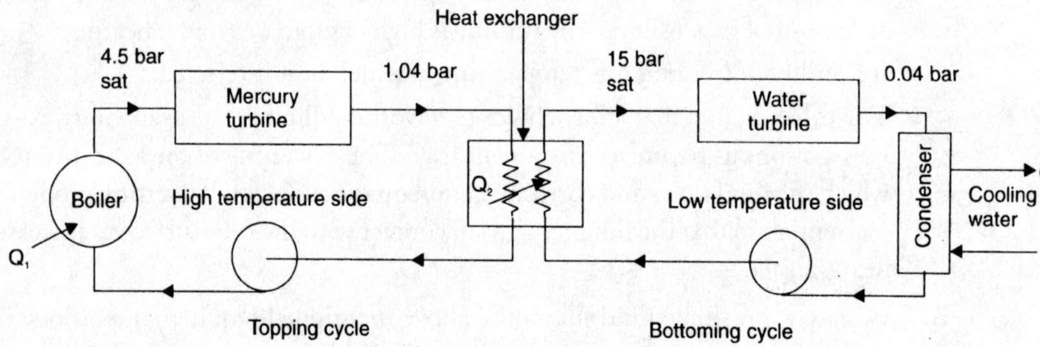


**Fig. 7.15** Schematic arrangement of a binary cycle

This series coupling may further be extended for three or more fluids. But it is limited to three fluids giving a tertiary cycle, which serves our purpose. The three fluids may be mercury for topping, $H_2O$ for intermediate and F-11 for bottoming cycles. However, we have to also see the cost and complexity involved in increasing this number. An important factor that goes in favour of increase in number, is that efficiency of the cycle is increased in the process. To demonstrate it, consider Fig. 7.15 (a) showing the schematic arrangement of a binary cycle. The amount of heat and work transfer at different points is also shown. Let $\eta_1$, $\eta_2$, and $\eta$ are efficiencies of the topping, bottoming and

13. A heat exchanger is a device in which two fluids, one at high temperature and another at low temperature, flow through adjacent passages, separated by a wall. Normally, one tube is placed in another bigger tube to make the passages. Further, one fluid enters from one end of tube and the second fluid enters another tube from the other end, from which side the first fluid comes out to reduce irreversibility in heat transfer.

whole cycle respectively, $Q_1$, $Q_2$ and $Q_3$ are heat added in boiler, exchanged in heat exchanger and rejected in condenser, respectively. Using the definition of efficiency, we may write the following equations,

$$Q_2 = Q_1(1 - \eta_1) \quad \text{and} \quad Q_3 = Q_2(1 - \eta_2)$$

$$\eta = 1 - Q_3/Q_1 = 1 - \{Q_2(1 - \eta_1)\}/Q_1 = 1 - (1 - \eta_1)(1 - \eta_2)$$

$$= \eta_1 + \eta_2 - \eta_1\eta_2 \tag{7.13}$$

This shows that efficiency of the combined cycle is less than the sum of the efficiencies of the individual cycles by the product of these efficiencies. It can be seen assuming any values for two efficiencies, that the total efficiency is more than that of any of these. For example, if $\eta_1 = 0.32$ and $\eta_2 = 0.3$, then $\eta = 0.32 + 0.3 - 0.32 \times 0.3 = 0.524$, which is greater than both efficiencies. We further note that we have added a less efficient cycle, to a better cycle but even then, the results are encouraging.

That is why, this concept has been utilised in gas power cycles, where it is known as a combined cycle. This concept somehow could not become popular with the vapour cycle though there are some power plants in USA based on mercury–water fluids. This could be because mercury is not environment friendly, is costly and not available in large quantity. A significant improvement in efficiency in steams cycle is also responsible for this situation. The requirement of mercury in such a cycle is much more that those of water—a fact that can be verified by making energy balance of the coupling heat exchanger—which may be another reason that limited its application. However, with our quest to find more and more liquid metals, some day in the future, these cycles also become popular, like combined cycles.

In case of a cycle made from $n$ individual cycle cascaded in series, we may similarly write that the overall efficiency of such cycle will be

$$\eta = 1 - \Pi(1 - \eta_i), \text{ where product is over i} = 1, 2, \ldots, n \tag{7.14}$$

We may also execute three Carnot cycles simultaneously using mercury in the high temperature range, water in the moderate temperature range and Freon 11 in the low temperature range, and couple them in series to give a tertiary cycle. These cycles can be analysed for calculation of specific work output, specific fuel input and efficiency, in the same way as done for a steam cycle, using appropriate property tables or charts.

**Example 7.3** A binary vapour cycles uses mercury and water with the mercury cycle as topping cycle. Mercury enters the turbine as saturated vapour at 10 bar, expands to 0.2 bar, and then gives heat in the heat exchanger to water to produce steam at 40 bar, 400°C. Steam expands isentropically in the steam turbine to a condenser temperature of 40°C, condensed in the condenser and then sent to the mercury–water heat exchanger. Assuming expansion in the mercury turbine as isentropic, determine (a) the mass of mercury required per kg of steam produced, (b) the total output, and (c) the thermal efficiency of the individual cycle as well as the overall efficiency. Data for the mercury cycle is given in Table 7.3.

**Table 7.3** Data for saturated mercury

| *p*, bar | *t*, °C | *h*, kJ/kg | | s, kJ/kg K | | *v*, m³/kg | |
|---|---|---|---|---|---|---|---|
| | | $h_f$ | $h_g$ | $s_f$ | $s_g$ | $v_f$ | $v_g$ |
| 10 | 515.5 | 72.23 | 363.0 | 0.1478 | 0.5167 | $80.9 \times 10^{-6}$ | 0.0333 |
| 0.2 | 277.3 | 38.35 | 336.55 | 0.0967 | 0.6385 | $77.4 \times 10^{-6}$ | 1.163 |

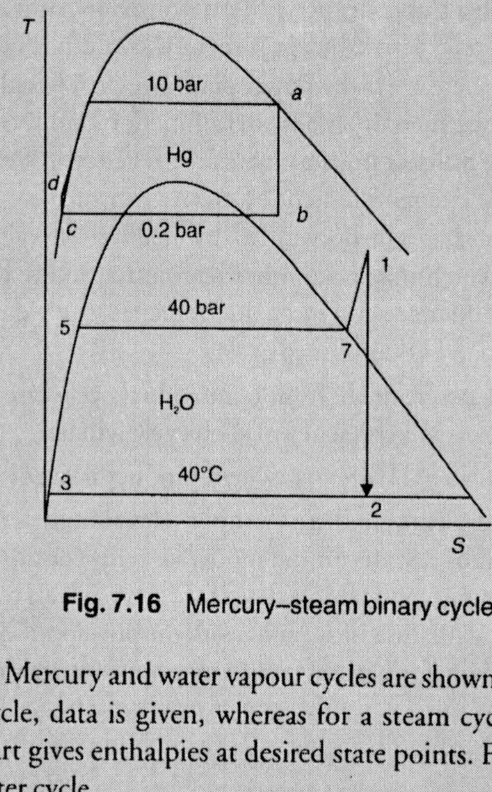


**Fig. 7.16** Mercury–steam binary cycle

***Solution*** Mercury and water vapour cycles are shown in Fig. 7.16. For a mercury cycle, data is given, whereas for a steam cycle, steam table or Mollier chart gives enthalpies at desired state points. From the chart, we read for water cycle

$$h_1 = 3214,\ h_2 = 2075,\ h_3 = 167.57,\ h_5 = 1087.31, \text{ and}$$
$$h_7 = 2801.4 \text{ kJ/kg}$$

From the mercury table, we can determine enthalpy at various state points. These are

$$h_a = 363,\ h_c = h_d = 38.35 \text{ kJ/kg}$$

To determine $h_b$, we use the process equation, $s_a = s_b \rightarrow x_b = 0.7751$ and $h_b = 269.48$ kJ/kg.

Let mass of mercury required per kg of steam produced in the heat exchange is mkg , then making its heat balance, we get

$$m\,(h_b - h_c) = 1 \times (h_1 - h_3)$$

$$\rightarrow \quad m = 13.175$$

Heat supplied in and work output of mercury cycle per kg of water in the water cycle are obtained by making energy balance of the mercury boiler and the turbine. This gives

$$Q_m = m\,(h_a - h_c) = 4276.5 \text{ kJ and}$$

$$W_m = m\,(h_a - h_b) = 1225.25 \text{ kJ}$$

This implies efficiency of mercury cycle = $\eta_m$

$$= 1225.25/4276.5 \times 100 = 28.65\%$$

Similarly, for water cycle, we get

$$Q_w = h_1 - h_3 = 3046.4, \quad W_w = h_1 - h_2 = 1139 \quad \text{and}$$

efficiency = $\eta_w$ = 37.39%

Total output of the combined cycle per kg of water

$$= 1225.25 + 1139 = 2364.25$$

$\therefore$ Overall efficiency = $2364.25/Q_m \times 100 = 55.28$.

From Eq. 7.14, we get overall efficiency

$$= 1 - (1 - \eta_m)\,(1 - \eta_w) = 55.3\%$$

which is same as obtained above.

This verifies the equation 7.14.

## 7.7 COGENERATION

In a 200 MW plant with steam pressure of 140 bar and temperature 540°C, about 48% of heat is lost in the condenser, which is not only waste, but also a source of thermal pollution. In industries, heat is required in many processes at a temperature around 130–200°C, which is obtained from wet steam. The advantage of wet steam is that temperature remains constant during heat transfer from it. We may expand steam to a pressure, such that steam temperature at the turbine exit is that required for process heat. Steam is then passed through the process heater, which acts as condenser for the power cycle and heat source for industrial process. Thus, heat loss from the power cycle, is usefully utilised, while simultaneously satisfying the second law requirement for a power cycle. This is the central theme of cogeneration, that has been usefully explored. Steam is cooled in the process, so that it is just condensed to saturated liquid, and is then fed to the boiler. However, there are chances of contamination of feed water due to leakage during the 'process heat' transfer operation that needs to be controlled through suitable maintenance programs. This way, up to 80–90% of heat generated may be used in combined electricity generation and process heat.

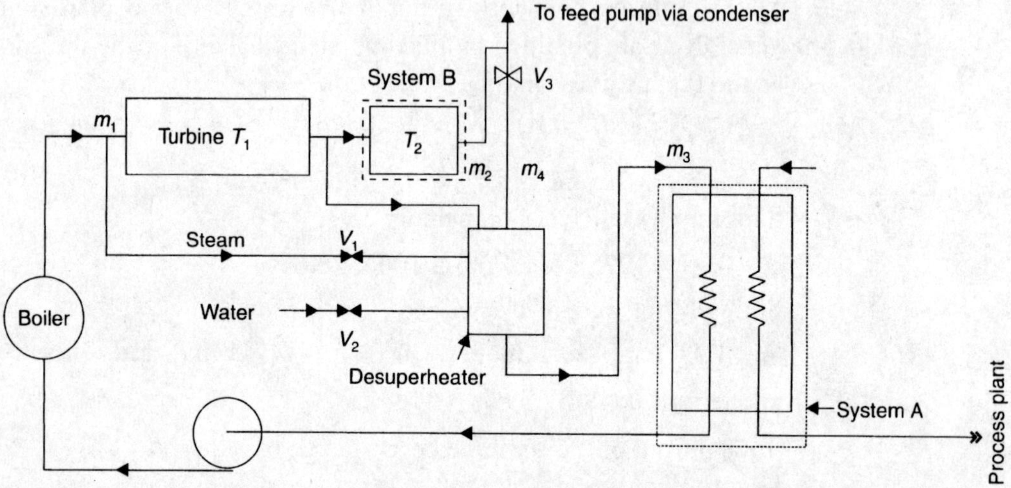


**Fig. 7.17** Block diagram of a cogeneration plant

Figure 7.17 shows a schematic view of such a plant. Wet steam is ensured at the inlet of the process heater by taking steam from turbine $T_1$ to a unit called desuperheater, where regulated quantity of superheated steam or water, or both, may be added to it by means of control valves $V_1$ and $V_2$ such that required mass of wet saturated steam is available at the desuperheater outlet. If the mass of steam from the LP turbine is more than required, then valve $V_3$ is opened. Steam coming out of the valve $V_3$ is condensed in the condenser, and reused as feed water. In addition the figure also shows another turbine $T_2$, which is used if power requirement is more than process heat requirement. In that case, the amount of steam needed for process heat is taken to the process equipment after turbine $T_1$, and the rest is expanded further in the second turbine to generate electricity.

From the above discussions, it is clear that steam pressure at the LP turbine exit is more than the atmospheric pressure. The mass flow rate of steam is determined on the basis of the power and process heat requirement for that industry, so that the industry is self-sufficient and power need not be purchased from the power grid. There may be the following requirements.

- Power requirement is more than what can be produced from the amount of steam needed for process heat. In such a case, extra power requirement is met from power supplying units or grids. The turbine is known as *back pressure turbine.* Alternately, more steam may be generated and expanded in turbine $T_1$. At the end of expansion, the amount of steam required for process heat is extracted, and the rest expands over turbine $T_2$. This can be properly designed to meet the factory requirements. In such a situation, the main turbine is called a *pass out turbine.*

- More steam is required for the process. If whole of the steam is expanded over the turbine, then power developed is more than required. In such case, the industry has the option to expand enough amount of steam over the turbine to produce power required and bypass extra amount of steam to the desuperheater, or to produce electricity for sale.

Thermal analysis. Let $m_1$, $m_2$, $m_3$ and $m_4$ kg be the steam flow rate per second through turbine $T_1$, $T_2$, process heater (heat exchanger) and rejected to condenser through valve $V_3$, respectively. $w_1$ and $w_2$ kJis work produced per kg of steam by turbines $T_1$ and $T_2$, $Q$ kJ of heat is required per hour for the process, and heat given by one kg steam in a heat exchanger is $\Delta h$, which can be determined knowing the condition of the steam at entrance, and water at the exit of the process heater.

Then, by applying conservation of mass and energy balance to systems $A$, $B$ and turbine $T_1$ (see Fig. 7.17), we can write the following equations.

$$Q = 3600 \times m_3 \times \Delta h \tag{7.15}$$

$$\text{Power produced by turbine } T_2 = P_2 = m_2 \times w_2 \tag{7.16}$$

$$\text{Power produced by } T_1 = P_1 = m_1 \times w_1 \text{ where } m_1 = m_2 + m_3 + m_4 \text{ and} \tag{7.17}$$

$$\text{Total power produced} = P = P_1 + P_2 \tag{7.18}$$

The solution of these equations gives the desired results. Remember $m_2 = 0$ for back pressure turbine.

Attempts have also been made to utilise the waste heat through flue gases, which are of the order of 10–12% by using this heat at a temperature of 105–110°C for air-conditioning, using the vapour absorption method. We shall not be able to go into further details of all these techniques at this stage.

**Example 7.4** In a cogeneration plant, the power load is 5.6 MW, and the heating load is 1.163 MW. Steam is generated at 40 bar, 500°C. It is expanded isentropically in a turbine to a condenser pressure of 0.06 bar. The heating load is supplied by extracting steam from the turbine at 2 bar, which is condensed in the process heater to saturated liquid at 2 bar, and then pumped back to the boiler. Compute (a) the steam generation capacity of the boiler in *t/h*, (b) the heat input to the boiler in kW, (c) the fuel burning rate of the boiler in *t/h* if coal of calorific value 25 MJ/kg is burned and the boiler efficiency is 88%, (d) the heat rejected to the condenser, (e) the rate of flow of cooling water in the condenser if the temperature rise of water is 6°C. Neglect pump work.

***Solution*** Figure 7.18 shows the schematic arrangement of the plant. This is a general scheme for any back pressure or pass-out turbine. In different applications, the inlet, outlet and the steam extraction condition may change depending upon the requirement of the plant. Or, the expansion may not be isentropic, and the isentropic efficiency of the turbine may be known. This only means that care has to be taken for these different variations. But the procedure outlined below remains the same for analysis of such plants. There is also a possibility that steam from other sources is available in the plant, and may be used purposefully by mixing it with main steam after expansion in the turbine, for which an energy balance will be required to find out the condition of the steam after mixing. That is, in a nutshell, the solution procedure will be a combination of the various methods we have learnt so far. You must realise this and try to develop solution skills accordingly, instead of treating every problem as a different kind of problem.

Given: $P = 5.6$ MW $= 5600$ kW, $Q = 1163$ kW, $p_1 = 40$ bar,
$T_1 = 500°C$, $p_2 = 2$ bar, $p_3 = 0.06$ bar

We may show the steam expansion process in the turbine on a Mollier diagram, and read from there

$$h_1 = 3445 \text{ kJ/kg}, h_2 = 2200, h_3 = 2180 \text{ kJ/kg}$$

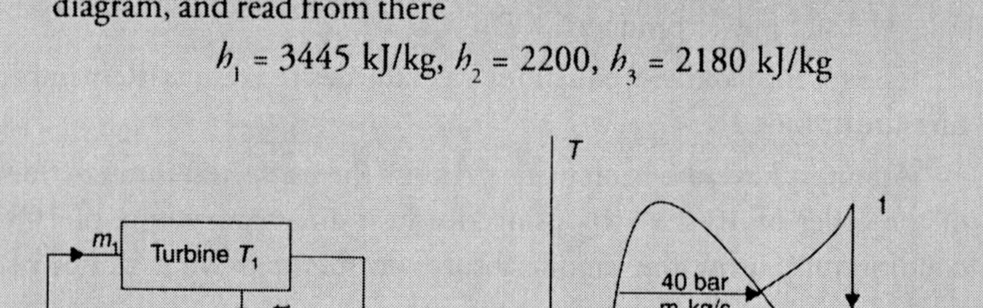


**Fig. 7.18** Cogeneration plant

$h_4$ = 151.5 and $h_7$ = 504.7 kJ/kg, which corresponds to $h_f$ at the corresponding steam pressure.

Let $m_1$ and $m_2$ kg be the masses of seam entering the steam turbine and the process plant, respectively, per second. Since $h_2$, $h_7$ and $Q$ are known, we can find out $m_2$ by making energy balance of the process plant, which is a steady flow open system. This gives

$$Q = m_2(h_2 - h_7) \quad \rightarrow \quad m_2 = 1901.4 \text{ kg}$$

$P$ =Power developed by the turbine is

$$= m_1(h_1 - h_2) + (m_1 - m_2) \times (h_2 - h_3)$$

since $m_1$ kg of steam flows in the Section 1–2 and $(m_1 - m_2)$ kg in the Section 2–3 of the turbine. Here everything is known in this equation, except $m_1$, we solve this equation for $m_1$ to give[14]

$$m_1 = 16730 \text{ kg/h} = 16.73 \text{ t/h}$$

It is given that pump work is negligible, and hence enthalpy of water after the pump will be the same as that before it. Two streams of water, one $(m_1 - m_2)$ kg having enthalpy $h_4$ and another $m_2$ kg having enthalpy $h_7$, are added to the boiler from which steam $m_1$ kg having enthalpy $h_1$ comes out. Flow to the boiler and from the boiler is steady, and therefore, we may apply sfee to it, assuming no heat loss from it. This gives, if $Q_b$ is the amount of heat added per second in the boiler,

$$(m_1 - m_2)h_4 + m_2h_7 + Q_b = m_1 h_1$$

$$\rightarrow Q_b = 15110 \text{ kJ/s} = 15.11 \text{ MW}$$

Let $m_f$ of mass of coal used per second, then taking into account the boiler efficiency, which also includes combustion efficiency, we get

$$\eta_b = Q_b/(m_f \times \text{C.V.})$$

$$\rightarrow m_f = 15110/(0.88 \times 25000) = 0.687 \text{ kg/s}$$

$$= 0.687 \times 10^{-3} \text{ (ton)}/(1/3600) \text{ (hour)} = 2.473 \text{ ton/hour.}$$

$(m_1 - m_2)$ kg/s of the steam is condensed in condenser from state 3 to 4, at which enthalpies were obtained above. From this data, we can determine the heat given by condensing steam. This heat is taken by cooling water. If $m_c$ is mass of cooling water in kg/s, then by making this energy balance, we get

$$(m_1 - m_2)(h_3 - h_4) = m_c c_p \Delta t \quad \rightarrow \quad m_c = 333 \text{ kg/s}$$

---

14. Ton is unit of mass = 1000 kg and, is represented by the symbol *t*.

## SUMMARY

This chapter discusses length and breadth of the various aspects of the vapour cycles used in reciprocating and rotating machines. The terms used and methods of improving the efficiency of any such cycle are discussed. A thumb rule to increase efficiency is to increase the average temperature of heat addition, and to decrease for heat rejection. Ideally, our efforts are to reach Carnot efficiency. However, we have to also ensure simultaneously that work output per cycle is more, so that the size of the plant is reduced. While using vapours, we have to also see that dryness fraction, pressure and temperature are within reasonable limits, study shows that the Rankine cycle improves work ratio of the cycle though efficiency in decreased. However superheating, reheating and feed heating improves efficiency, the work ratio of the Rankine cycle and the quality of the vapour at the exit of the last stage of expansion. A look at the Mollier diagram reveals that the constant pressure lines diverge to its right hand side. This means that if the state of steam at the entrance is shifted to the right of the chart, then expansion work is more and steam is more dry. That explains why reheat is desirable. Not only this, the average temperature of heat addition is more in the reheat cycle portion, and therefore, efficiency of the cycle is increased. The other method of improving efficiency is increase in boiler pressure. However, rate of gain in efficiency with boiler pressure would decrease, as latent heat becomes less at high pressure. The supercritical pressure cycle has higher efficiency.

We have also discussed how to use the criterion of maximising efficiency to determine the pressure at which steam should be taken for reheat. It is seen that ratio of reheat pressure to the steam pressure at the turbine inlet is around 0.2 to 0.26. Similarly, the procedure for determining the number of feed heaters is discussed. To determine this number, it can be proved that efficiency of the regenerative cycle becomes maximum when rise in temperature is equal in all the feed heaters and the economiser.

Water is not a suitable working substance for high or low pressures. But it is most desirable for moderate pressures ranges. On the other hand, there are a number of working substances which are most suited either for higher pressure or lower pressure. This led to an important concept of binary and tertiary cycles. In these cycles, heat is supplied to the fluid in topping cycle (high pressure cycle) only. Heat rejected by this cycle is used for heating the working substance of the next power cycle, and so on. That is, a number of cycles are cascaded, so that the condenser of the upper cycle is the boiler of the following cycle. If $\eta_i$ is efficiency of the $i$th cycle, then overall efficiency is given by $\eta_0 = 1 - \Sigma(1 - \eta_i)$, where summation is over the cycles. It has been shown that the efficiency of an individual cycle may not be very good, but the overall efficiency is considerably improved. Moreover, improving the efficiency of an individual cycle does improve the overall efficiency, but the rate of efficiency gain with increase in efficiency of individual cycle decreases.

Earlier, mercury was used in the topping cycle and $SO_2$, Freon, or ammonia in the bottoming cycle, with water for the intermediate cycle. This way, highest pressures were brought to reasonable values while achieving the heat addition to the cycle at high temperature ($\approx$600°C) and rejecting heat in the condenser of the bottoming cycle at atmospheric temperature but pressure above the atmospheric pressure to avoid leakage. The concept of binary cycle has been extended and does not require only the vapours in all the cycles. These days, gas turbine cycle, thermionic, thermoelectric, or MHD power cycles are used for the topping cycle. Combined cycle is the name given to such cycle, which is discussed in detail in Chapter 8.

In our efforts to utilise every kJ of heat produced by the fuel, concept of cogeneration evolved in which condenser of the power cycle is replaced by process heater adding a few controls. Therefore back pressure and pass-out turbines were designed. Lastly, using absorption refrigeration system, the energy wasted through flue gases from boiler could be tapped to produce air-conditioned space for the control room.

# EXERCISES

## Review Questions

7.1 Why is it not possible to operate a power plant on the Carnot cycle?

7.2 Is it possible that the steam from a LP turbine be compressed and fed to the boiler? Explain.

7.3 Why is the Rankine cycle considered as the standard vapour cycle, though it is irreversible? Discuss the philosophy that is utilised to improve its efficiency and the specific work output.

7.4 Are we justified in neglecting pump work while making the calculation of efficiency of the Rankine or the modified Rankine cycle? Will this be true if steam from the turbine outlet is directly pumped back to the boiler?

7.5 Explain the increase in efficiency due to the use of superheated steam in reciprocating steam engines. A steam engine is supplied with steam at 10 bar. The exhaust to the atmosphere takes place at 1.1 bar. Determine the Rankine efficiency: (a) if the supply steam is 90% dry, (b) if the steam temperature is 200°C, (c) what is the percentage increase in efficiency?

7.6 Dry saturated steam is supplied to a turbine at 15 bar. The isentropic expansion continues to 1.1 bar. Determine: (a) The Rankine engine efficiency, and the equivalent mean effective pressure. (b) How is the Rankine engine efficiency affected if the exhaust is sent to the condenser, where pressure is maintained at 0.2 bar.

7.7 How does reheating and superheating help in improving the dryness fraction of steam at the end of expansion, and improving the efficiency of the cycle?

7.8 Do you agree with the statement that efficiency of a power plant would increase with boiler pressure? Discuss the positive and negative effects of increase in boiler pressure.

7.9 What condition is required to be satisfied for deciding the pressure of steam at which it should be taken to the boiler for reheat? What are the practical values used in a modern plant? Would you recommend a number of reheating stages? Discuss the criteria used.

7.10 Why does the regenerative cycle have higher efficiency? How do we decide the number of feed heaters?

7.11 Devise a regenerative cycle that has the same thermal efficiency as the Carnot engine. Show the cycle on a *T-s* diagram.

7.12 Sketch the power cycle for a supercritical pressure system on the *T-s* and *h-s* diagrams. Is it different from the normal steam plants?

7.13 Is decreasing the sink temperature more effective than increasing the source temperature by the same temperature difference? If yes, then why do we not go below a certain condenser vacuum?

7.14 Write down the considerations for the design of a power plant and how these are met in practice.

7.15 Explain the importance of combined cycles, and discuss how this concept helps in combining different cycles even with low efficiency to result in high efficiency. What are the other advantages? May we also consider the binary cycle as a combined cycle?

7.16 Why is mercury a suitable substance of topping cycle, of a binary vapour cycle but not for a bottoming cycle? Why was the binary cycle using mercury discontinued?

## Problems for Practice

7.1 A thermal power plant operates on a Rankine cycle in which the condenser is maintained at 0.1 bar. It is desired to study the effect of pressure on the cycle efficiency for which steam is assumed to be available at 2.5 and 3.5 MPa. Assume that steam temperature in both the cases is 250°C. Make comments and verify whether your result agrees with that available in literature. [*Ans.* 0.316, 0.33].

7.2 In a steam power plant, superheated steam at 30 bar, 300°C enters a turbine and leaves at 0.1 bar with a moisture of 0.1. Determine the isentropic efficiency of the turbine. [*Ans.* 0.70].

7.3 Saturated water at 10 kPa is to be pumped to a boiler, which is maintained at 3 MPa in a power plant. If the isentropic efficiency of the pump is 0.85, determine the work done per kg in pumping the water. [*Ans.* 3.55 kJ/kg].

7.4 It is desired to study the effect of variation of degree of superheat on the efficiency of a steam power cycle. For this, steam at the inlet to the turbine may be taken at 2.5 Mpa, but at 250 and 350°C. Assume condenser pressure in both the cases to be 0.1 bar. Make your comments. [*Ans.* 0.316, 0.3264].

7.5 In solved Example 7.1, we took steam for reheating when it is saturated at the end of last stage expansion. We discussed in the text that reheat pressure is taken equal to 0.2 to 0.25 times the pressure at the inlet. Find out the difference we get in efficiency, and the d.f. of steam, choosing pressure on this criterion; take it as 0.2, 0.21, 0.212, 0.23,0.24 and 0.25 times the steam pressure at the turbine inlet, and see how the two quantities to be calculated vary with it.

7.6 A thermal power plant operates on a regenerative Rankine cycle. Superheated steam at 25 bar, 250°C is fed to the turbine, and a fraction of the steam is extracted at 0.5 Mpa, and the rest of the steam is expanded to the condenser pressure of 10 kPa. The condensate, after compressing to 0.5 Mpa, is mixed with the steam extracted from the turbine at 0.5 MPa in an open feed water heater, and the resulting mixture is saturated liquid at 0.5 MPa. This saturated liquid is then pumped back to the boiler. Calculate the fraction of steam extracted and the thermal efficiency of the power plant. [*Ans.* 0.1881, 0.3334].

7.7 The condition of steam in a steam plant employing one stage of regenerative feed heating is at the entrance to the HP 100 bar, 600°C and LP 20 bar, 400°C. The correct amount of steam is bled for feed heating at the exit from the HP stage. Calculate the mass of steam bled per kg of steam passing through the HP stage, and the amount of heat supplied in the boiler per second for an output of 10 MW. Neglect pump work. [*Ans.* 0.235, 25445 kJ/s] (IES 1993).

7.8 Steam enters the first stage of a turbine at 100 bar, 500°C, and expands isentropically to 10 bar. It is then reheated to 500°C, and expanded in the second stage to the condenser pressure of 0.1 bar. Steam is bled from the first stage at 20 bar and fed to a closed feed water heater. Feed water leaves the closed heater at 100 bar, 200°C ($h$ = 856.8 kJ/kg), while the condensate is supplied to the open heater into which steam is bled at 4 bar. Saturated liquid at 4 bar exits from the open heater and enters the closed heater. The net output of the turbine is 50 MW. Assuming the turbine and pump process to be isentropic, determine the mass of steam bled at each feed water heater per kg of steam entering the first stage, the mass of steam entering the first stage per second, and the thermal efficiency. [*Ans.* 0.125, 0.1082, 27.36 Ton/s, 57.96%] (IES 95).

7.9 A binary vapour cycle operates on mercury and steam with saturated mercury vapour at 4.5 bar at the inlet to the mercury turbine, from which it exhausts at 0.04 bar. The mercury condenser generates 50,000 kg/h of saturated steam at 15 bar, which is expanded in a steam turbine to 0.04 bar. (a) Find the overall efficiency of the cycle. (b) What is the rate of flow of mercury through the turbine? (c) Assuming that all processes are reversible, what is the useful work done in the binary vapour cycle? (d) If a superheater is placed in the mercury boiler to heat steam to 300°C, and if the internal efficiencies of the mercury and steam turbine are 0.85 and 0.87, respectively, calculate the overall efficiency of the cycle. The properties of saturated mercury are given in Table 7.4.

**Table 7.4** Properties of saturated mercury

| *P*, bar | *t*, °C | *h* | | *S* | | *v* | |
|---|---|---|---|---|---|---|---|
| | | $h_f$ | $h_g$ | $s_f$ | $s_g$ | $v_f$ | $v_g$ |
| 4.5 | 450 | 62.93 | 356.98 | 0.1352 | 0.5397 | | 0.068 |
| 0.04 | 216.9 | 29.98 | 329.85 | 0.0808 | 0.695 | | 6.178 |

[*Ans.* 53%, 59.35 × 104 kg/h, 28.5 MW, 46.2%]

7.10 A sodium–steam cycle operates between 1000°C and 40°C. Sodium rejects heat at 670°C to mercury. Mercury boils at 24.6 bar and rejects heat at 0.141 bar. Both sodium and mercury cycles are saturated cycles. Steam is formed at 30 bar and is superheated in the sodium boiler to 350°C. It rejects heat at 0.08 bar. Assume isentropic expansion; no heat losses, no regeneration and neglect pumping work. Find (a) the amount of sodium and mercury used per kg of steam, (b) the heat added and rejected in the composite cycle per kg steam, (c) the total work done per kg steam, (d) the efficiency of the individual and composite cycle, (e) the efficiency of the corresponding Carnot cycle, and (f) the work, heat added and efficiency of a supercritical steam (single fluid) cycle operating at 250 bar, and between the same temperature limits. Given, for mercury at 24.6 bar, $h_g$ = 366.78, $s_g$ = 0.48; at 0.0141 bar, $s_f$ = 0.09, $sg$ = 0.64, $h_f$ = 36.01 and $h_g$ = 330.77. For sodium at 1000°C, $h_g$ = 4982.53; at the turbine exhaust,

$h = 3914.85$ kJ/kg. At 670°C, $h_f = 746.29$ kJ/kg. For a supercritical steam cycle, the specific enthalpy and entropy at the turbine inlet may be computed by linear extrapolation from the steam table. All properties above are given in their standard units.

7.11 A 50 MW steam plant built in 1935 operates with steam at the inlet at 60 bar, 450°C and exhausts at 0.1 bar, with 80% turbine efficiency. It is proposed to scrap the old boiler and put in a new boiler and a topping turbine of efficiency 85%, operating with the inlet steam at 180 bar, 500°C. The exhaust from the topping turbine at 60 bar is reheated to 450°C and admitted to the old turbine. The flow rate is just sufficient to produce the rated output from the old turbine. Find the improvement in efficiency with the new set up. What is the additional power developed?

7.12 In a nuclear power plant, saturated steam at 30 bar enters a HP turbine and expands isentropically to a pressure at which its quality is 0.841. At this pressure, the steam is passed through a moisture separator, which removes all the liquid. Saturated vapour leaves the separator and is expanded isentropically to 0.04 bar in LP turbine, while the saturated liquid leaving the separator is returned via a feed pump to the boiler. The condensate leaving the condenser at 0.04 bar is also returned to the boiler via a second feed pump. Calculate the cycle efficiency and turbine outlet quality taking into account the feed pump term. Recalculate the same quantity for a cycle with the same boiler and condenser pressure, but without moisture separation. Give your comments on the results.

[*Ans.* 34.5%, 0.824, 35%, 0.716].

7.13 Geothermal energy from a natural geyser can be obtained as a continuous supply of steam, 0.87 dry, at 2 bar and at a flow rate of 2700 kg/h. This is utilised in a mixed pressure cycle to augment the superheated exhaust from a high pressure turbine of 83% internal efficiency, which is supplied with 5500 kg/h of steam at 40 bar and 500°C. The mixing process is adiabatic, and the mixture is expanded to a condenser pressure of 0.10 bar in a LP turbine of 78% internal efficiency. Determine the power output and the thermal efficiency of the plant.

[*Ans.* 1745 kW, 35%].

7.14 In a cogeneration plant, steam enters the HP stage of a two-stage turbine at 10 bar, 200°C, and leaves it at 3 bar. At this point, some of the steam is bled off and passed through a heat exchanger, which it leaves as saturated liquid at 3 bar. The remaining steam expands in the LP stage of the turbine to 0.4 bar. The turbine is required to produce a total power of 1 MW, and the heat exchanger to provide a heating rate of 500 kW. Calculate the required mass flow rate of steam into the HP stage of the turbine. Assume (a) steady conditions throughout the plant, (b) the velocity and gravity terms to be negligible, (c) both turbine stages are adiabatic with isentropic efficiency of 0.80. [*Ans.* 2457 kg/s].

7.15 In a cogeneration plant, the steam generator provides $10^6$ kg/h of steam at 80 bar, 480°C, of which $4 \times 10^5$ kg/h is extracted between the first and second turbine stages at 10 bar, and diverted to a process heating load. The condensate returns from the process heating load at 9.5 bar, 120°C, and is mixed with liquid exiting the lower pressure pump at 9.5 bar. The entire flow is then pumped to the steam generator pressure. Saturated liquid at 0.08 bar leaves the condenser. The turbine stages and the pumps operate with isentropic efficiency at 68% and 80%, respectively. Determine (a) the heating load, in kJ/kg/h, (b) the power developed by the turbine, in kW, (c) the rate of heat transfer in the steam generator, in kJ/kg/h. [*Ans.* 9.529 × 108, 236500, 3.032 × 109].

## Multiple-choice Problems

7.1 The Rankine cycle operating on steam low pressure limit of $p_1$ and high pressure limit of $p_2$

(a) has higher thermal efficiency than the Carnot cycle operating between same pressure limits.

(b) has lower thermal efficiency than Carnot cycle operating between same pressure limits.

(c) has same thermal efficiency as Carnot cycle operating between same pressure limits.

(d) may be more or less depending upon the magnitudes of $p_1$ and $p_2$.

7.2 Rankine cycle efficiency of a good steam power plant maybe in the range of

(a) 10% to 15%. (b) 35% to 45%. (c) 70% to 80%. (d) 90% to 95%.

7.3 Rankine cycle for a steam power plant has $T_1$ = highest temperature in the cycle $p_1$ = highest pressure in the cycle and $p_2$ = lowest pressure in the cycle. If $p_1$, *is* raised, keeping $T_1$ and $p_2$ the same, the Rankine efficiency

(a) decreases. (b) increases. (c) remains the same. (d) may increase or decrease.

7.4 If $t_1$ is raised keeping $p_1$ and $p_2$ the same, the Rankine efficiency

(a) decreases. (b) increases. (c) remains the same. (d) may increase or decrease.

7.5 Between the two options of either increasing $p_1$ or $T_1$

(a) increasing $p_1$ increases Rankine cycle efficiency more than increasing $T_1$.

(b) increasing $p_1$ increases Rankine cycle efficiency less than increasing $T_1$.

(c) both options are equally effective for increasing Rankine efficiency.
(d) none of the options give any increase in Rankine efficiency.

7.6 If $p_2$ is increased keeping $T_1$ and $p_1$ fixed, the Rankine efficiency
(a) increases. (b) decreases. (c) remains the same. (d) may increase or decrease.

7.7 The Rankine efficiency of a steam power plant
(a) improves in summer as compared to that in winter.
(b) improves in winter as compared to that in summer.
(c) is unaffected by climatic conditions. (d) None of the above.

7.8 The reheat cycle has been developed for a steam power plant
(a) to take advantage of increased efficiency at high pressures.
(b) to avoid excessive moisture in the last stages of the turbine.
(c) to obtain (a) and (b) both.
(d) None of the above.

7.9 Reheating of steam at the intermediate stage in a steam turbine plant
(a) increases the thermal efficiency of the cycle and improves the quality of exhaust steam.
(b) decreases the thermal efficiency of the cycle, but increases the dryness fraction of exhaust steam.
(c) increases the thermal efficiency of the cycle, but decreases the dryness fraction of exhaust steam.
(d) does none of the above.

7.10 The increase in thermal efficiency of the cycle due to reheating
(a) depends upon the ratio of reheat pressure to initial pressure of steam.
(b) depends upon the ratio of reheat pressure to exhaust pressure of steam.
(c) depends only on reheat pressure which should be as high as possible.
(d) depends upon none of the above.

7.11 With fixed initial pressure and fixed exhaust pressure, the increase in thermal efficiency of the reheat cycle
(a) increases with increase in temperature.
(b) decreases with increase in temperature.
(c) remains the same, unaffected by increase in temperature.
(d) is not predictable with increase in temperature.

7.12 Regenerative heating that is, bleeding steam to reheat feed water to boiler
(a) decreases the thermal efficiency of the cycle.
(b) increases the thermal efficiency of the cycle.

(c) does not affect the thermal efficiency of the cycle.

(d) may increase or decrease the thermal efficiency of the cycle, depending upon the point of extraction *of* steam.

7.13 Regenerative cycle thermal efficiency

(a) is always greater than simple Rankine cycle thermal efficiency.

(b) is greater than simple Rankine cycle thermal efficiency only when steam is bled at a particular pressure.

(c) is same as a simple Rankine cycle thermal efficiency.

(d) is always less than simple Rankine cycle thermal efficiency.

7.14 Depending upon the ratio of the reheat pressure to the initial pressure for the reheater

(a) the efficiency of the reheat cycle may be higher or equal to that of the Rankine cycle.

(b) the efficiency of the reheat cycle is always higher than that of the Rankine cycle, whatever the ratio.

(c) the efficiency of the reheat cycle is the same as that of the Rankine cycle, whatever the ratio.

(d) none of the above.

7.15 In a regenerative feed heating cycle, the optimum value of the fraction of steam extracted for feed heating

(a) decreases with increase in Rankine cycle efficiency.

(b) increases with increase in Rankine cycle efficiency.

(c) is unaffected by increase in Rankine cycle efficiency.

(d) None of the above.

7.16 In a regenerative feed heating cycle, the greatest economy is affected

(a) when steam is extracted from only one suitable point of the steam turbine.

(b) when steam is extracted from several places in different stages of the steam turbine.

(c) when steam is extracted only from the last stage of the steam turbine.

(d) when steam is extracted only from the first stage of the steam turbine.

7.17 The maximum percentage gain in a regenerative feed heating cycle thermal efficiency

(a) increases with number of feed heaters increasing.

(b) decreases with number of feed heaters increasing.

(c) remains the same, unaffected by number of feed heaters.

(d) none of the above.

CHAPTER EIGHT

# Gas Power Cycles



## 8.1 INTRODUCTION

In steam power plants, combustion of fuel takes place outside the engine and considerable amount of heat is lost in the boiler. It was thought that if an engine is developed, in which fuel burns inside the engine itself, then this loss may be reduced to increase the efficiency of power generation. A few air standard cycles, employing different ways of heat addition or rejection for a reciprocating system, called internal combustion engine, or IC. Engine in short, were developed; notably Otto, Diesel and Dual cycles. These differ only in the manner in which heat is added to air from the surroundings; otherwise all the three remaining processes are the same, namely, isentropic compression, isentropic expansion and constant volume heat rejection. Heat is added at constant volume in the Otto cycle, at constant pressure in the Diesel cycle, and partly at constant volume and then at constant pressure in the Dual cycle. Here, the system is made up of air confined in a cylinder subjected to these processes, and can be analysed using the non-flow energy equation.

Brayton thought of a cycle for an open system (gas turbine) in which air is isentropically compressed in a rotary compressor, heat added at constant pressure in a combustion chamber, isentropic expansion in a turbine, and finally, heat rejected to the atmosphere at constant pressure. The advantage of such a cycle is that the flow is continuous unlike the power cycles of a reciprocating engine, and a large amount of mass flow is possible, thereby increasing the output of the plant. The Brayton cycle is therefore most suited for a large power plant.

The importance of power cycles was discussed in Chapter 7, where it was also mentioned that air is one of the working substances that has been used in ideal gas power cycles around which a number of reversible cycles were developed. We shall discuss in detail these cycles in this chapter, and compare these with the actual internal combustion engine cycle later in the chapter on IC engines.

## 8.2 ANALYSIS OF RECIPROCATING CYCLES

Many air standard cycles were developed using different processes for heat addition to or rejection from the cycle, while retaining isentropic compression and expansion processes used in the Carnot cycle. These cycles were developed at a time when it was already established that the Carnot cycle has maximum efficiency of $1- T_2/T_1$, where $T_1$ and $T_2$ are temperatures of the source and the sink. In these cycles, like the Carnot cycle, air was assumed to be confined in a cylinder, and hence formed the basis for comparison of performance of internal combustion engines using different kinds of fuels as the heat source.

These cycles are executed using reversible processes, and therefore have the same efficiency as that of the Carnot cycle. Since the temperature of air changes during heat addition or rejection processes, we have to use the average temperature of heat addition and heat rejection in the expression for Carnot efficiency to verify the statement. This is illustrated later for the Otto cycle.

We shall now discuss air standard cycles, used in reciprocating engines, in the following section. For analysis of these cycles, we note that the system is air, which is assumed to behave as perfect gas at all temperatures, and that the system is a closed system, and hence nfee shall be used.

### 8.2.1 Otto Cycle

The cycle is shown in Fig. 8.1 (a) on the *p-v* diagram. We may obtain the cycle output by determining the work transfer during processes 12 and 43 (work transfer in the processes 23 and 41 is zero, being constant volume processes. Or, an easier way is to find out heat transfer in the processes 23 and 41, as heat exchange in the other two processes, namely, 12 and 34, is zero (isentropic process). We may write the following equations for the processes 23 and 41

$$\text{Heat added to air in process 23} = C_v(T_3 - T_2)$$
$$\text{Heat rejected by air in process 41} = C_v(T_4 - T_1)$$

$\therefore$ Heat exchange in cycle $= \oint \delta q = C_v [(T_3 - T_2) - (T_4 - T_1)] = \oint \delta w$

$= $ Work output of cycle by Joule's law

$\therefore$ $\eta_0$ = Efficiency of the Otto cycle = work output/heat input

$= 1 - (T_4 - T_1)/(T_3 - T_2)$

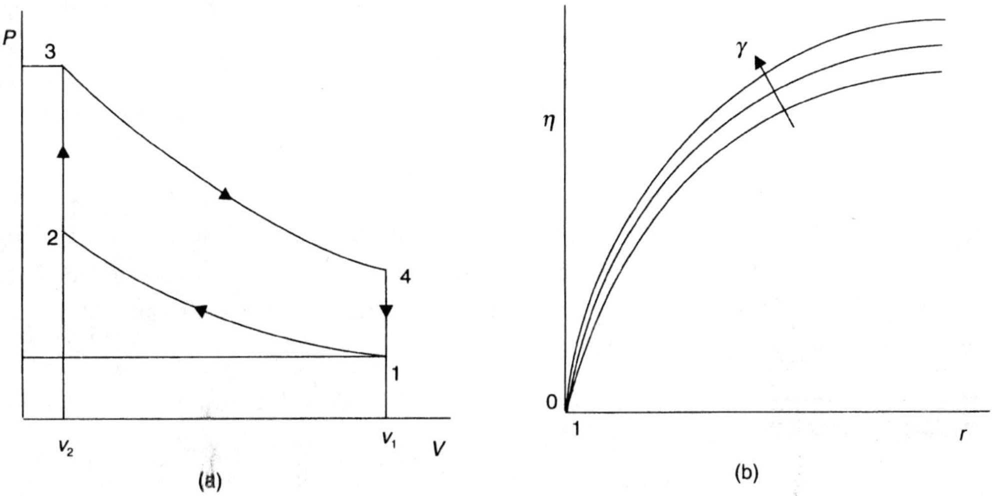


**Fig. 8.1** Otto cycle (a) on *p-v* diagram (b) $\eta$ *versus* *r* for different g

Applying the perfect gas law $TV^{(\gamma - 1)}$ = constant for isentropic process 12 and 34, and using the symbol *r* for the compression ratio $V_1/V_2$ or expansion ratio $V_4/V_3$, we get after simplification (note $V_1 = V_4$ and $V_2 = V_3$ in this cycle)

$$T_2/T_1 = T_3/T_4 = r^{(\gamma - 1)}, \text{ or } T_4/T_1 = T_3/T_2$$

and $$\eta_0 = 1 - T_1/T_2 = 1 - 1/r^{(\gamma - 1)} \tag{8.1}$$

It is seen that efficiency $\eta$ of the Otto cycle is a function of two variables *r* and $\gamma$. To show variation of $\eta$ on a 2-D plot, $\gamma$ is taken as a parameter, and a family of curves is drawn for different values of the constant $\gamma$. Figure 8.1 (b) shows one such plot. Since *r* cannot be less than 1, the scale of *r* starts with the value 1 in the figure. We may easily verify from Eq. 8.1 that efficiency increases with *r* for a given value of $\gamma$, and with $\gamma$ for a given value of *r*. However, the rate of increase of efficiency with *r* decreases with increase in *r*. That is, the efficiency curve for the Otto cycle tends to 1 with *r* asymptotically.

We may determine the Carnot efficiency of this cycle. The average temperature of heat addition and rejection are $(T_2 + T_3)/2$ and $(T_4 + T_1)/2$, since specific heat at constant volume is not a function of temperature for perfect gas and is the same during heat rejection or rejection process. Noting $T_4/T_1 = T_3/T_2$ for the cycle,

Carnot efficiency = 1 – average temperature of heat rejection/average temperature of heat addition

$$= 1 - (T_4 + T_1)/(T_2 + T_3) = 1 - T_1/T_2 = 1 - T_4/T_3$$

which is the same as that of the Otto cycle. This, however, does not undermine the importance of the Carnot cycle, since the average temperature of heat addition in the Otto cycle will always be less than the maximum temperature at which heat could be added in the Carnot cycle. Similarly, the average temperature of heat rejection in the Otto cycle will be more than that of the sink. This difference is due to variation in temperature during the heat addition or rejection processes. For these reasons, the Carnot cycle is still more efficient compared to air standard cycles, except for Stirling and Ericsson cycles.

**Example 8.1** In an Otto cycle, the minimum temperature and pressure are 310 K and 1 bar, respectively. The compression ratio and the maximum cycle temperature are 8 and 1500 K. Calculate the thermal efficiency, net work output and the mean effective pressure.

***Solution*** Refer Fig. 8.1 (a) to see the state points used in the solution.

*Given*: $T_1$ = 310 K $p_1$ = 1 bar $T_3$ = 1500 K and $r = 8$

Assume $\gamma$ = 1.4, $c_v$ = 0.71 kJ/kg K

$$\eta_{th} = 1 - 1/r^{(\gamma-1)}$$

$$= 1 - 1/(8)^{0.4} = 57.47\%$$

Using the process equation for isentropic compression 12 and expansion 34, respectively, we get:

$$T_2 = T_1\, r^{(\gamma-1)} = 712.2 \text{ K, and}$$

$$T_4 = 1500/r^{(\gamma-1)} = 652.92 \text{ K}$$

Heat added in the cycle = 0.71 × (1500 – 712.2)

= 559.34 kJ/kg

Heat rejected in the cycle = 0.71 × (652.92 – 310)

= 243.67 kJ/kg

Specific work output of the cycle = 559.34 – 243.67

= 317.67 kJ/kg

Perfect gas equation gives $V_1 = (0.287 \times 310)/(1 \times 10^2)$

= 0.8897 m$^3$/kg and $V_1/V_2$

= 8 (given)

This gives $V_1 - V_2 = 0.77849$

mep = work output/stroke volume
$= 317.67/(.77849 \times 10^2)$
$= 4.06$ bar

### 8.2.2 Diesel Cycle

Figure 8.2 shows the diesel cycle on the *p-v* and *T-s* diagrams. We follow the above procedure for analysis of the Otto cycle, and write the following equations:

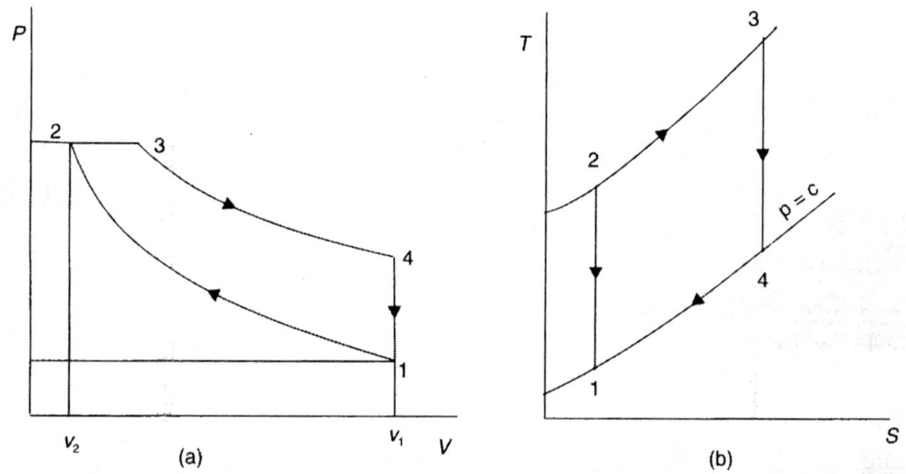


**Fig. 8.2** Diesel cycle (a) *p-v* (b) *T-s* diagram

Heat added to air in process 23 = $C_p(T_3 - T_2)$, and
heat rejected by air in process 41 = $C_v(T_4 - T_1)$

$\therefore$ Heat exchange in cycle $= \oint \delta q = C_p(T_3 - T_2) - C_v(T_4 - T_1) = \oint \delta w$

= work output of cycle by Joule's law.

and $\eta_d$ = Efficiency of Diesel cycle = work output/heat input

$= 1 - (T_4 - T_1)/(T_3 - T_2)\ (1/\gamma)$

To change this in terms of *r*, we can invoke perfect gas relations for processes 12, 23 and 34, and express temperatures at points 2, 3 and 4 in terms of temperature $T_1$. Thus,

$$T_2/T_1 = r^{(\gamma - 1)}, \quad T_3/T_2 = V_3/V_2 = \rho \text{ and}$$
$$T_4/T_3 = (V_3/V_4)^{(\gamma - 1)} = (\rho/r)^{(\gamma - 1)}$$

where, $\rho$ is defined as ratio of volume of air in cylinder at the end of heat supply to clearance volume of the cylinder. It is known as the *cut-off ratio*. We get on simplification,

$$\eta_d = 1 - \frac{1}{\gamma}\frac{\rho^\gamma - 1}{\rho - 1}\frac{1}{r^{\gamma-1}} \tag{8.2}$$

We may compare the efficiency of the Otto and Diesel cycles with that of Otto cycle by taking $\rho = 1 + \delta$, where $\delta$ is a small quantity, so that $\rho^\gamma$ may be expanded by the Binomial theorem resulting in

$$\rho^\gamma = (1+\delta)^r = 1 + \gamma^\delta + \gamma(\gamma-1)\delta^2/2 + ...$$

or $$\left(\rho^r - 1\right)/r = \delta + (r-1)\,\delta^2/2 + ...$$

and $(\rho^\gamma - 1)/\{\gamma(\rho - 1)\} = 1 + (\gamma - 1)\delta/2 + ...$

Therefore, the coefficient of $1/r^{(\gamma-1)}$ in Eq. 8.2 is more than 1, since both $\delta$ and $\gamma$ are positive quantities. Therefore, efficiency of the Diesel cycle is less than that of the Otto cycle for given value of compression ratio; the difference in two efficiencies increases as $\delta$ or $\rho$ increase. We also note that this efficiency limits to that of Otto cycle as $\rho \to 1$. But this will mean addition of less heat and hence less cycle output, and in the limit $\rho \to 1$, both will be zero. In fact, the value of $\rho$ will be decided by the size of the engine as well.

Efficiency depends on three variables, namely $r$, $\rho$ and $\gamma$. Since air is used in the cycles, $\gamma$ is taken as 1.4. So, efficiency is plotted against $r$ for different values of $\rho$ as a parameter (see Fig. 8.3); the plot for the Otto cycle is also shown for comparison. Note that efficiency of the Diesel cycle approaches that of the Otto cycle as the value of $\rho$ tends to 1. In fact, the two cycles should be compared on the basis of their application.

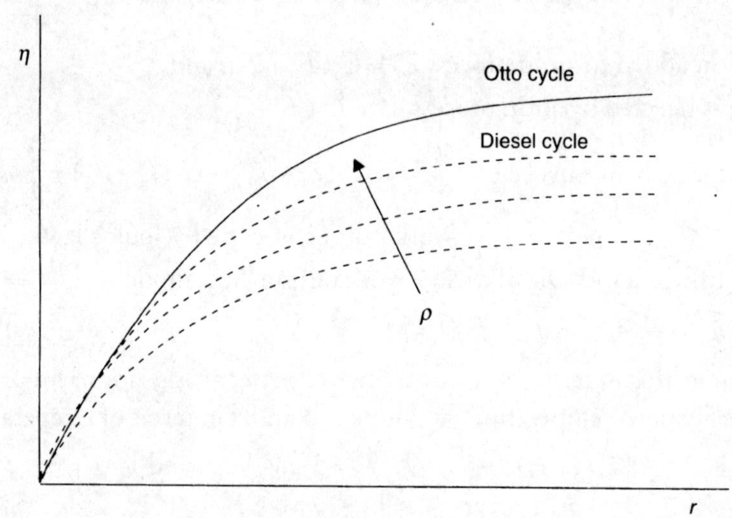


**Fig. 8.3** Efficiency versus $r$ for the diesel cycle

**Example 8.2** In an air standard Diesel cycle, the temperature at the start and at the end of the compression stroke are 310 K and 930 K, respectively. The energy added in the form of heat is 600 kJ/kg. Determine the compression ratio, the cut-off ratio, the maximum cycle temperature and the thermal efficiency. If pressure at the beginning of the compression stroke is 1 bar, then find out the mean effective pressure.

***Solution*** Refer Fig. 8.2(a).

Given: $T_1 = 310$ K $T_2 = 930$ K $q_{23} = 600$ kJ/kg

For air: $\gamma = 1.4$, $c_p = 1.005$ and $c_v = 0.71$ kJ/kg K

Since temperatures are known before and after compression, we get

$$r = (T_2/T_1)^{1/(\gamma-1)} = 17.588$$

Heat added is given. Therefore, $q_{23} = c_p(T_3 - T_2) = 1.005 \times (T_3 - 930)$

$$\rightarrow T_3 = 1527 \text{ K}$$

Using the perfect gas equation, we get $V_3/V_2 = \rho = T_3/T_2 = 1.642$

Using Eq. 8.2, we get,

$$\eta_{th} = 1 - (1/17.588)^{0.4} \times (1/1.4) \times (1.642^{1.4} - 1)/(1.642 - 1)$$
$$= 62.83\%$$

Using equation of state, we get: $V_1 = 0.287 \times 310/(1 \times 100) = .8897$ and $V_2 = V_1/r = 0.0571$ m³/kg.

Work output of the cycle $= \eta_{th} \times q_{23} = 377 = \text{mep} \times (V_1 - V_2)/100$

$$\rightarrow \text{mep} = 4.528 \text{ bar}$$

### 8.2.3 Dual Combustion Cycle

Whether heat is added at constant pressure or constant volume in actual applications depends upon the rate of heat added and the rate at which pressure decreases due to movement of the piston. In an ideal cycle, the piston is assumed stationary during heat addition at constant volume, but in practice this does not happen. As the speed of the engine increases, the rate of decrease of pressure due to movement of the piston increases, and heat addition cannot be assumed at either constant volume or constant pressure. In such a case, a better approach for modelling the situation will be to assume heat addition initially at constant volume, and then at constant pressure. That is why a dual combustion cycle was conceived. Figure 8.4 shows the Dual cycle on the *p-v* and *T-s* diagram. Heat addition during 23 is at constant volume and in 34 at constant pressure. We can write the following relations, following procedure used for the Otto cycle.

Heat added to air = $C_p(T_3 - T_2) + C_v(T_4 - T_3)$, and
Heat rejected by air = $C_v(T_5 - T_1)$

Heat exchange in cycle = $\oint \delta q = C_p(T_3 - T_2) + C_v(T_4 - T_3) - C_v(T_5 - T_1) = \oint \delta w$

Temperatures at all the points may be expressed in term of temperature $T_1$ at point 1, like the Diesel cycle. We get on simplification,

$$\eta_{dc} = 1 - \frac{\alpha_1 \rho^{\gamma} - 1}{(\alpha_1 - 1) + \gamma \alpha_1 (\rho - 1)} \frac{1}{r^{\gamma - 1}} \tag{8.3}$$

where, $\alpha_1$ is pressure ratio $p_3/p_2$, and $\rho$ is the cut-off ratio, defined before.

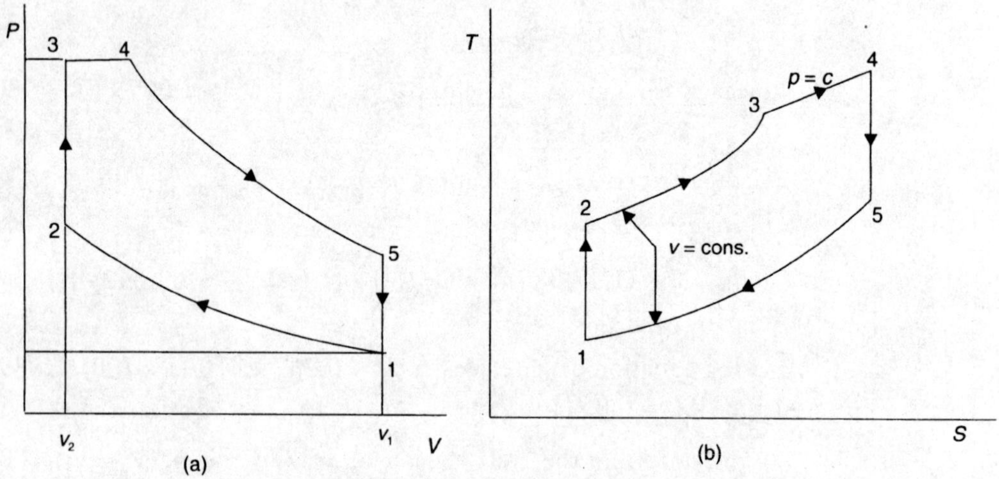


**Fig. 8.4** Dual cycle (a) *p-v* diagram (b) *T-s* diagram

Besides efficiency, there are other factors too that have to be considered for selecting the Otto, Diesel or Dual combustion cycle. For example, the maximum temperature of air reached in a cycle will depend on the amount of heat added during the heating process, and the temperature of air at which the addition of heat starts (which depends on the compression ratio). Not only the temperature but maximum pressure reached also depends on these two factors. Therefore, the amount of heat added, the compression ratio, the output and the maximum pressure and temperature of the cycle make basis of comparison. The results for the efficiency of the three cycles on the following comparison bases are

Same compression ratio and heat input $\eta_{\text{Otto}} > \eta_{\text{Dual}} > \eta_{\text{Diesel}}$
Same heat input and maximum pressure $\eta_{\text{Otto}} < \eta_{\text{Dual}} < \eta_{\text{Diesel}}$
Constant maximum pressure and same output $\eta_{\text{Otto}} < \eta_{\text{Dual}} < \eta_{\text{Diesel}}$
Constant maximum pressure and same heat input $\eta_{\text{Otto}} < \eta_{\text{Dual}} < \eta_{\text{Diesel}}$

We shall demonstrate the procedure for the first criterion only, as the similar process may be used by the reader to verify the results for the other criteria. Figure 8.5 shows three cycles on the *T-s* diagram for air; cycles 12341, 123″4″1 and 123′4′5′1 are the Otto, Diesel and Dual cycles, respectively. While drawing various processes, remember that constant volume curves for air on the T-s diagram are steeper than the constant pressure process. Since *r* is the same for all the three cycles, point 2 shall be the same for the same point 1. We know that the area under the curve 23 up to *s* axis is the amount of heat supplied in the process. We draw cycles 12341 (Otto), 123″4″1 (Diesel) and 123′4′5′ (Dual) such that the area under heat addition processes 23, 23″ or 23′4′ is the same. Now the isentropic expansion can easily be drawn, as shown. Area a14b, a15′c or a14″d is the amount of heat rejected by the corresponding cycle. That is, heat rejected in the Otto cycle is the least, and that in the Diesel cycle the most. Or, we may write, $\eta_{Otto} > \eta_{Dual} > \eta_{Diesel}$.

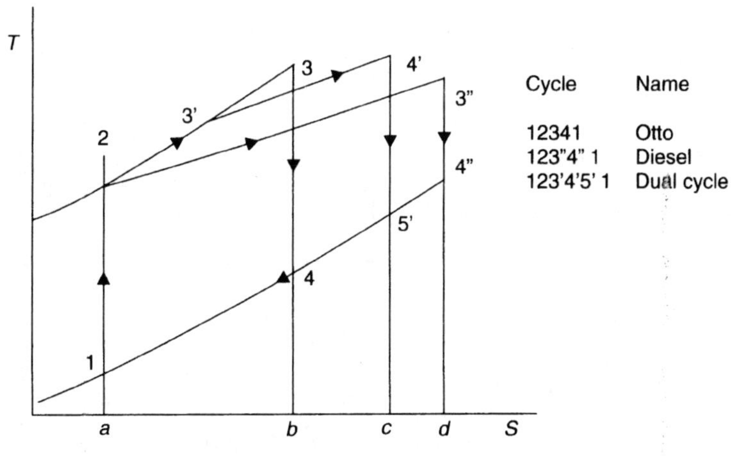


**Fig. 8.5** Comparison on the basis of same *r* and heat input

**Paradox.** We use petrol and diesel fuel to run engines for which Otto and Diesel cycles are considered to be ideal cycles because of the manner in which volume or pressure change during burning of fuel. Because of practical problems, petrol engines cannot work for a compression ratio of more than 10, whereas a diesel engine operates at a compression ratio of more than 16. Since these operate at different compression ratios, it turns out that the efficiency of a diesel engine is more. Hence, we quite often make a remark that diesel engines are more efficient than petrol engines. Otherwise, the Otto cycle is more efficient compared to the Diesel cycle, when compared at the same compression ratio.

### 8.2.4 Stirling and Ericsson Cycles

The Carnot cycle will have maximum efficiency since the whole of the heat is supplied at the highest temperature and rejected at constant sink temperature. But its implementation is difficult, especially of isentropic process that requires movement of piston with infinite speed to allow any time for heat transfer to take place during the process. Stirling and Ericsson replaced isentropic processes by constant volume and constant pressure process, respectively. Since heat addition and rejection in both the cycles is at constant temperature, both these cycles are contemporary to the Carnot cycle. The Ford Motor Company, General Motors Corporation and Philips Research Laboratories have developed the Stirling engine used in trucks and buses. Their use is limited because of use of a regenerator that is not 100% efficient and having many complexities in implementation. More research is needed to make vehicles based on this cycle acceptable to a common user.

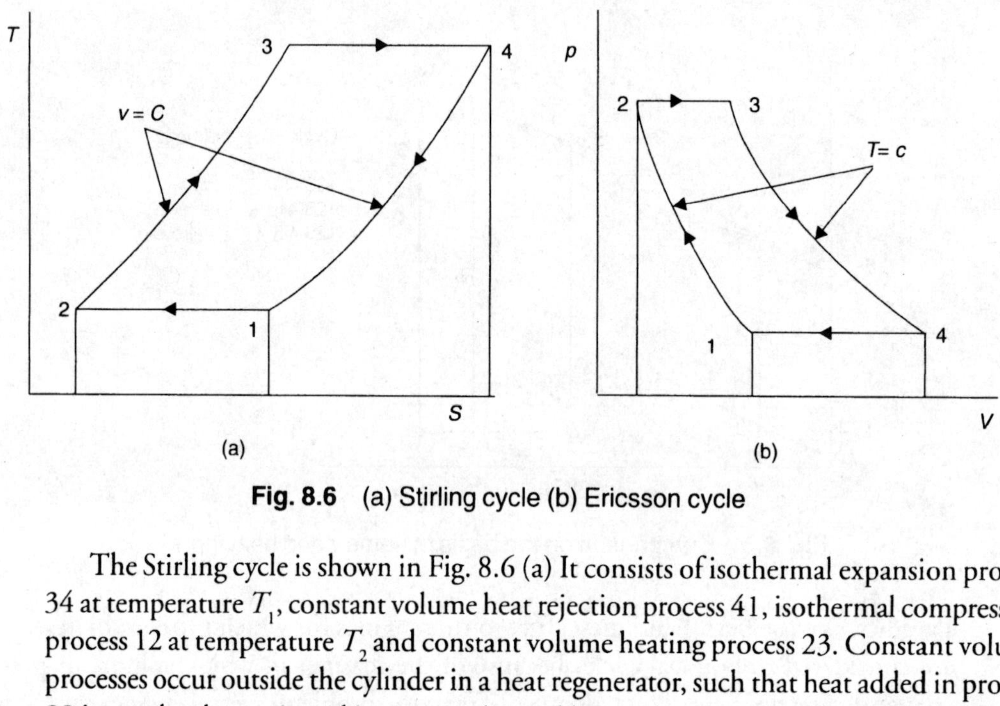


**Fig. 8.6** (a) Stirling cycle (b) Ericsson cycle

The Stirling cycle is shown in Fig. 8.6 (a) It consists of isothermal expansion process 34 at temperature $T_1$, constant volume heat rejection process 41, isothermal compression process 12 at temperature $T_2$ and constant volume heating process 23. Constant volume processes occur outside the cylinder in a heat regenerator, such that heat added in process 23 is equal to heat rejected in process 41, and hence the two cancel out while summing up heat transfer in the cycle. For 1 kg of air,

$$q_{12} = w_{12} = -R\,T_2 \ln (v_1/v_2) \qquad q_{23} = c_v\,(T_2 - T_1);\; w_{23} = 0$$

$$q_{34} = w_{34} = R\,T_1 \ln (v_4/v_3) \qquad q_{41} = -c_v\,(T_2 - T_1);\; w_{41} = 0$$

Since $q_{23} = q_{41}$, the cycle efficiency is given by

$$\eta = (q_{12} + q_{34})/q_{12} = (T_1 - T_2)/T_1$$

This is Carnot efficiency. We may similarly show that the Ericsson cycle has Carnot cycle efficiency.

Figure 8.6 (b) shows the Ericsson cycle on a *p-v* diagram. It consists of four processes, namely, isothermal compression 12, constant pressure heat addition 23, isothermal expansion 34 and isobaric heat rejection 41. In the Ericsson cycle, like the Stirling cycle, processes 2, 3 and 41 occur in a regenerator, which is made up of a wire matrix as shown in Fig. 8.7. Air after completion of process 34 enters the regenerator at state 4, and leaves at state 1 from its other end. Air rejects heat to the matrix, which heats it in such a manner that a temperature gradient exists along its length with temperature $T_1$ at one end and $T_2$ at the other end of the regenerator, as shown in the figure. Similarly, air at state 2 is passed through the regenerator in the reverse direction, that is, from the cold end, and leaves it at state 3 from the hot end to realise isobaric heating process 23. Thus alternately, streams of hot and cold air are passed through the regenerator matrix. If we assume that the regenerator is 100% efficient, then heat rejected to the matrix during process 41 is equal to heat taken by air during process 23.

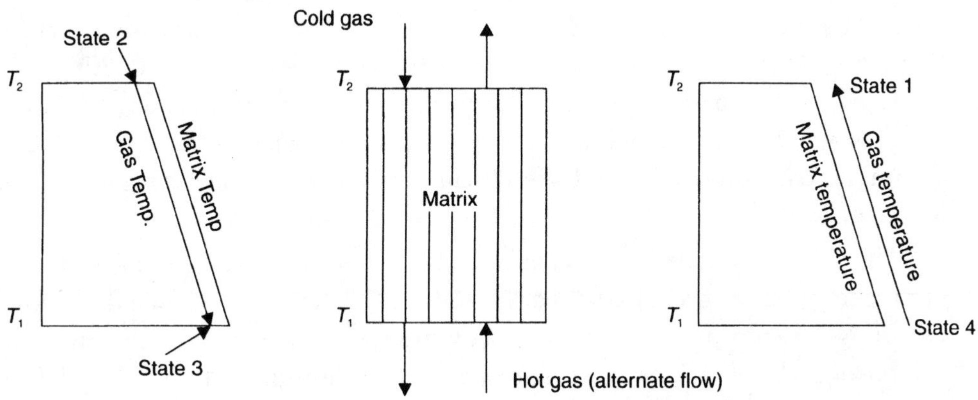


**Fig. 8.7** Temperature distribution in matrix

We note that both in the Stirling and Ericsson cycles, heat is added or rejected, like the Carnot cycle, during isothermal processes only. Therefore, both these cycles have Carnot efficiency = $1 - T_2/T_1$. Another advantage is that fuel burns externally in both cycles, and hence the quality of fuel does not matter. Or, in other words, engines based on these cycles are not internal combustion engines like petrol or diesel engines.

## 8.3 GAS POWER CYCLES FOR OPEN SYSTEMS

Gas power cycles were developed for open systems, too, for the reason that such systems can handle flow of large amount of fluid, and hence more power can be produced by a relatively small size plant, compared to the reciprocating system. The Crank mechanism used in reciprocating systems and heavy weight pistons introduces inertia and vibration problems, too.

### 8.3.1 Brayton or Open Simple GT Cycle

While the implementation of a constant temperature expansion or compression process is easy for wet steam, it is not that simple for gases. The Ericsson cycle faced this problem. Brayton replaced the isothermal compression or expansion process with isentropic—that is again to square one following a circular root. Stirling replaced the isentropic processes of the Otto cycle with isothermal, because of difficulties faced in implementation of the isentropic process and then Ericsson replaced the isometric process of the Stirling cycle with isobaric process, and Brayton replaced isothermal process back with isentropic. We mentioned earlier that many cycles were developed depending on the application.

The Brayton cycle has been used in gas turbines, and is also known as a simple gas turbine cycle. This cycle found importance in gas turbine plants because of its capacity to handle large volume flow rate, and it has the efficiency of the Otto cycle (shown later) besides easy implementation for an open system. Comparing this cycle with the Rankine cycle, the other power cycle, we can logistically arrive at the conclusion that for the Brayton cycle, work ratio is poor, since a sizeable portion (about 70–80%) of work produced in the expansion process is consumed in the compression of gas, unlike the compression of water. Though the average temperature of heat addition is high but so is for heat rejection process too and it is difficult to say if its efficiency is higher than that of Rankine cycle.

The Brayton cycle has been implemented in two ways in a gas turbine plant, resulting in an open and closed cycle. In the open gas turbine cycle, air is sucked from the atmosphere, compressed isentropically in a rotary compressor, heated at constant pressure in a combustion chamber, expanded isentropically in a turbine, and is finally left to the atmosphere. Figure 8.8 (a) shows the schematic arrangement of the Brayton cycle, and 8.8 (b) shows it on a *T-s* diagram. The turbine runs the compressor, and the excess work produced by the turbine is the net output of the cycle. *It may be noted that constant pressure lines are diverging as we move from the left to the right on the T-s diagram.* That is, the vertical distance between constant pressure lines increases from left to right, resulting in net power output. Had the two lines been parallel, then work of expansion would be equal to work of compression and net work output would be zero.

Energy analysis of a simple gas turbine cycle is done by applying sfee to its different components and assuming.[1]

- the gas mixture, formed after burning of fuel in a combustion chamber may be treated as air,

---

1. These assumptions were made only to simplify the solution, but in practice, many of them cannot be justified, especially the negligible change in KE. Hence, it is common to work with stagnation properties rather than static ones to overcome this problem. Other assumptions have also been taken care of.

- air may be treated as perfect gas,
- the mass flow rate is the same throughout the cycle,
- change of kinetic energy of the working fluid between the entrance and the exit is negligible,
- there is no pressure or heat loss in ducts and the combustion chamber.

This gives, turbine work $w_t = h_3 - h_4 = c_p\,(T_3 - T_4)$, compression work $w_c = h_2 - h_1 = c_p\,(T_2 - T_1)$, net work output $w_{net} = w_t - w_c$ and heat supplied in combustion chamber $q_{cc} = h_3 - h_2 = c_p\,(T_3 - T_2)$. Both the compressor and the turbine have the same pressure ratio $p_2/p_1 = p_3/p_4 = r$. Making use of the isentropic relation, $T \propto p^{(\gamma-1)/r}$, we get

$$T_2/T_1 = (p_2/p_1)^{(\gamma-1)/\gamma} = r^{(\gamma-1)} \quad \text{and}$$

$$T_3/T_4 = (p_2/p_1)^{(\gamma-1)/\gamma} = r^{(\gamma-1)}$$

$$\rightarrow\ T_2/T_1 = T_3/T_4 = r^{(\gamma-1)/\gamma} = c \tag{8.4}$$

where $c$ is used for $r^{(r-1)/r}$, since it would appear often.

$\therefore\ \ \eta$ = net work output/heat input = $1 - 1/c$ (8.5)

It is seen that efficiency of the Brayton cycle is the same as that of the Otto cycle, discussed before. We may also express the net work output in terms of pressure ratio using isentropic relations to give

$$w/(c_p\,T_1) = t(1 - 1/c) - (c - 1) = (c - 1)\,(t/c - 1) \tag{8.6}$$

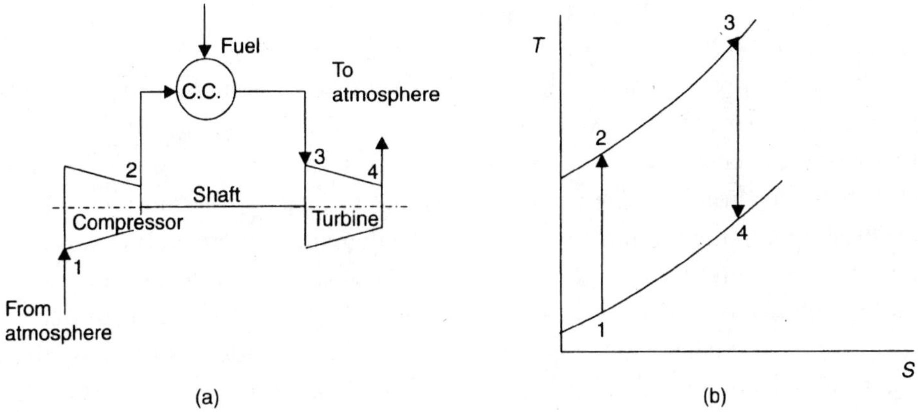


**Fig. 8.8** (a) Block diagram (b) Brayton cycle

where, $t = T_3/T_1$, which plays an important role in improving the performance of the cycle. However, $T_3$ is limited by metallurgical limits of the material used for gas turbine blades; efforts are going on to develop better blade material to withstand high temperature and methods to cool the blade. Temperatures of 1675 K at the turbine inlet has been achieved with the development of better material and cooling techniques for the turbine blade.

Figure 8.9 shows variation of specific work output against pressure ratio for different values of t. Work output is zero when c is either 1 or $t$ (see Eq. 8.6). Therefore, the plot shall have at least one maximum or minimum, somewhere between these two limits. Differentiation with respect to $r$ gives optimum pressure ratio $c_{opt} = \sqrt{t}$ for maximum specific work output. This condition may be written as $(T_2/T_1) \times (T_3/T_4) = t$ or $T_2 = T_4$, using Eq. 8.4. *That is, the pressure ratio for maximum specific work output should be such that temperatures of gas at the exit from the compressor and the turbine be equal.*

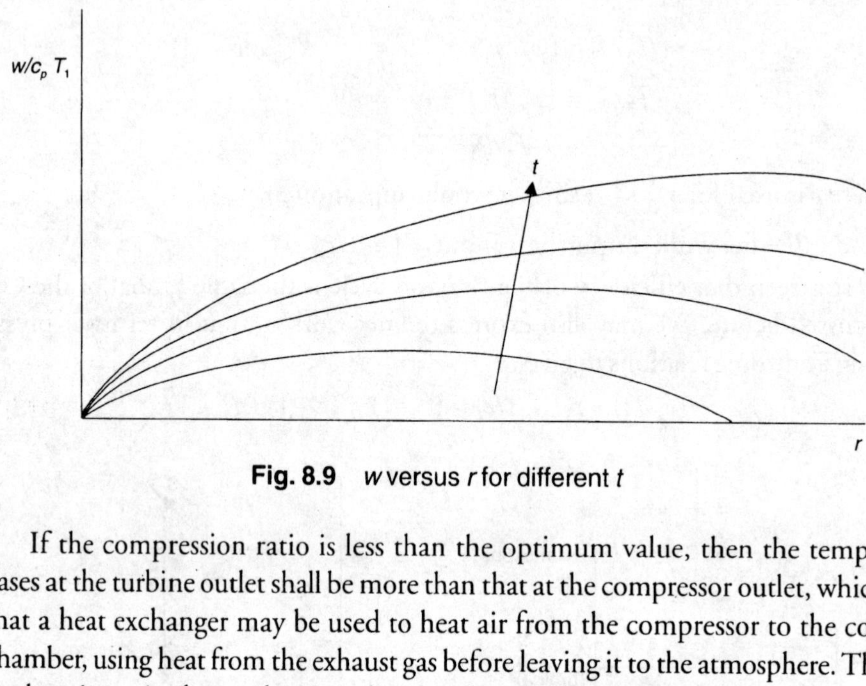


**Fig. 8.9** *w* versus *r* for different *t*

If the compression ratio is less than the optimum value, then the temperature of gases at the turbine outlet shall be more than that at the compressor outlet, which suggests that a heat exchanger may be used to heat air from the compressor to the combustion chamber, using heat from the exhaust gas before leaving it to the atmosphere. This reduces the heat input in the combustion chamber, and therefore, increases efficiency of the cycle. But the work output does not change. Analysis of the cycle is left to the reader, as one more application of sfee to the gas-to-air heat exchange is required to determine air temperature at the heat exchanger outlet. However, this adds weight to the equipment, and is not recommended for use in the aviation industry, where a gas turbine has found the most use and because of which significant development took place after the second world war. Therefore, a heat exchanger cycle finds use with land and marine applications, especially for power generation.

It is obvious that if the compression ratio is more than the optimum value, then the heat exchanger, if employed, will cool air from the compressor instead of heating it.

We note that the gases are exhausted to the atmosphere after expansion in the turbine.

Therefore, such a cycle is an open cycle. There is a closed cycle too for gas turbines. In a closed gas cycle, the turbine exhaust is cooled and reused. Therefore, for better performance, it is possible to use fluids other than air in these cycles with heat exchanger replacing a combustion chamber, and any kind of fuel can be used.

### 8.3.2 Improvement of Cycle Efficiency and/or Work Output

We mentioned above that constant pressure lines for air, on the *T-s* diagram, diverge towards the right. This suggests that as far as possible point 3 (entrance of the gas turbine) should be shifted to the right and point 1 (compressor inlet) to the left. The key to the solution is to increase the turbine output and reduce compressor work. This is achieved by introducing multi-stage expansion with reheating gases to initial temperature $T_3$ before the next stage expansion, and multi-stage compression with intercooling to cool air to initial temperature $T_1$ before the next stage compression. Minimising work input for the compression process, or maximising work output for the expansion process determines the pressure for intermediate compression or expansion. *It turns out that the pressure ratio for different stages of compression or expansion should be the same.* We shall demonstrate it for a two-stage expansion process with reheat cycle, assuming that expansion in turbines is isentropic or their isentropic efficiency is 100%. Figure 8.10 (a) shows a two-stage expansion with reheat. Assume reheating of gas (process 45) occurs at pressure $p$, and pressures at points 3 and 6 be $p_2$ and $p_1$, respectively, then specific work output of two turbines, $w$, is

$$w = c_p\,[(T_3 - T_4) + [(T_3 - T_6)], \text{ since } T_5 = T_3$$

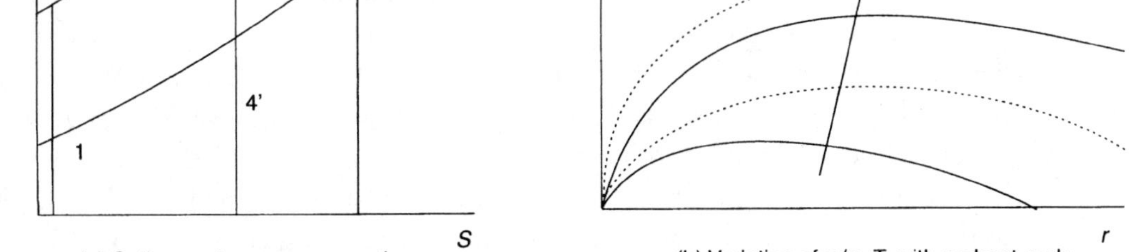


(a) Optimum stage pressure ratio

(b) Variation of $w/c_p\,T_1$ with $r$ reheat cycle

**Fig. 8.10**

But, $T_3/T_4 = (p_2/p)^{(\gamma-1)/\gamma}$ and $T_3/T_6 = (p/p_1)^{(\gamma-1)/\gamma}$, using isentropic relations. Substitution for $T_3$ and $T_6$ in the expression for specific work, and differentiating it with respect to $p$ results, after simplification, in $p = (p_1 p_2)^{1/2}$, that is reheat pressure is geometric mean. This relation may also be written as $p_1/p = p/p_2$. That is, the pressure ratio for the two stages should be the same. If more number of stages are used, we may show similarly that the pressure ratio for all the stages is the same for maximum work output. The result is, however, affected by poor isentropic efficiency of the compressor and the turbine.

We may proceed in the same manner for multi-stage compression, and show that the pressure ratio for each stage has to be equal for minimum work input in ideal case in which air is perfectly cooled to the initial temperature. In ideal situation, the pressure ratio for the compressor and the turbine shall be the same. But if pressure losses are considered, then pressure at the inlet to the turbine is less than that at the compressor outlet. Similarly, pressure at the turbine outlet will be more than the atmospheric pressure. Thus, expansion pressure ratio will be less than compression pressure ratio. Therefore, pressure ratio for different stages in the turbine will be less than that for compressor stages.

As for a simple gas turbine cycle, we may obtain the following equations for single-stage compression and two-stage expansion cycle, with reheat of gas to the initial temperature:

$$\frac{w}{c_p T_1} = 2t\, c - 1 - \frac{2t}{\sqrt{c}} \quad \text{and} \quad \eta = \frac{2t - c + 1 - 2t/\sqrt{c}}{2t - c - t/\sqrt{c}} \tag{8.7}$$

Variation of specific work output of the cycle with pressure ratio $r$ for different values of the parameter '$t$', given by Eqs. 8.6 and 8.7, is shown in Fig. 8.10 (b). Comparison of work output with that for simple cycle for the same $t$ reveals that reheating results in more specific work output for any given $r$. Also, that the operating pressure ratio for maximum work output shifts to the right. This means that we will require less number of compressor stages for the same rise in pressure. However, efficiency decreases, which may be explained from the fact that the reheat cycle may be thought of as superimposition of cycle 4564′4 on the original cycle 1234′1 (Fig. 8.10 (a)). Temperature limits under which the two cycles operate are $T_3$, $T_{4'}$ and $T_3$, $T_1$, respectively. Since $T_{4'} > T_1$, efficiency of reheat cycle is less than that of original cycle, and the net effect is to reduce cycle efficiency. This effect decreases as $T_3$ or t increases. These can be verified by plotting Eq. 8.7 on a $\eta$ versus $c$ graph with $t$ as a parameter. We further observe that temperature at the exit of the last turbine is more than that would be if no heating is employed. That means that using a heat exchanger will be advantageous to compensate for low efficiency of the reheat cycle.

**Example 8.3** Air at 300 K, 1 bar is compressed through a pressure ratio of 5, and after heating it in the combustion chamber, its temperature is raised to 1000 K. Air then expands to 1 bar. Find out the efficiency of the cycle for (a) Brayton cycle, (b) for the cycle when isentropic efficiency of compression is 0.85 and that for expansion 0.87, and (c) a regenerator with effectiveness 0.9 is used with the system in *b* for heating air from the compressor.

***Solution*** Given $r = 5$, $T_1 = 300$, $T_3 = 1000$, $p_1 = p_4 = 1$. Also $\gamma = 1.4$ $\Rightarrow$ $(\gamma - 1)/\gamma = 1/3.5$.

(a) Efficiency of Brayton cycle, by Eq. 8.4 gives

$$\eta = 1 - 1/(r)^{(\gamma-1)/\gamma} = 1 - 1/(5)^{1/3.5} = 37.26\%$$

(b) Figure 8.11 shows the Brayton and actual cycle with 12 compression, 23 combustion and 34 expansion processes; 12′ and 34′ correspond to isentropic compression and expansion processes. When a heat exchanger is used, then air is heated to point 5, as shown on the cycle. For air, temperature may be used in place of enthalpy for isentropic efficiency in the compression and expansion process, since $h = c_p T$.

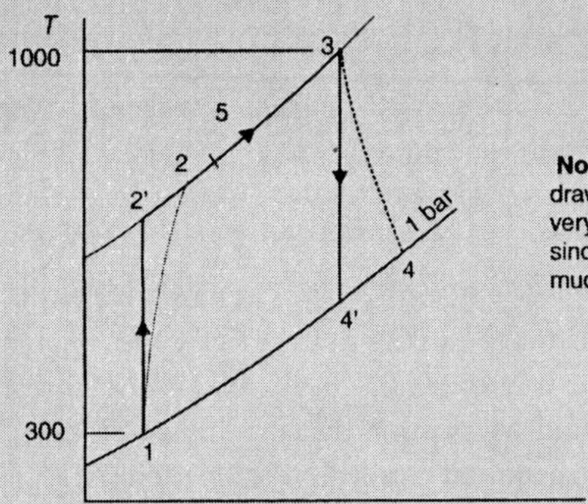


**Note**: Point 5 is above point 4 in the figure drawn, which is not possible. This error is very common when we draw this cycle since slope of constant pressure curves is much more than used in diagram

**Fig. 8.11** Actual gas turbine cycle with regenerator

Given: $T_1 = 300$ K, $r = 5$ and $T_3 = 1000$K. Also for air, $\gamma = 1.4$
$\Rightarrow$ $\gamma/(\gamma - 1) = 3.5$
Using relation for isentropic process $T \propto p^{(\gamma-1)/\gamma}$, we get
$T_{2'} = (5)^{1/3.5} \times 300 = 477.17$. and $T_{4'} = 1000/(5)^{1/3.5} = 631.39$ K

Relation for isentropic efficiency gives $T_2$, and similarly, $T_4$ is obtained after expansion in the turbine.

$$0.85 = (477.15 - 300)/(T_2 - 300) \Rightarrow T_2 = 507.06, \text{ and}$$
$$0.87 = (1000 - T_4)/(1000 - 631.39) \Rightarrow T_4 = 679.30 \text{ K}$$
$$w_c = 1.005\,(T_2 - T_1) = 204.56,\ w_t = 1.005\,(T_3 - T_4)$$
$$= 317.30,\ w = w_t - w_c = 112.74 \text{ kJ/kg}$$

and $q = 1.005\,(T_3 - T_2) = 491.49$ kJ/kg

$\therefore$ $\eta = (w_t - w_c)/q = 22.93\%$

In this example, effect of irreversible compression and expansion were considered, which has resulted in decrease in efficiency by (37.26 – 22.93)/37.26 = 36%.

(c) Effect of heat exchanger (HE) is to raise temperature before the combustion chamber to point 5. In an ideal situation, air temperature can be raised to $T_4$. When HE is not 100% efficient (ideal situation), then actual temperature rise will depend upon the HE effectiveness, which is a ratio of actual temperature rise to that for the ideal situation. This gives:

$$T_5 - T_2 = 0.9\,(T_4 - T_2) \rightarrow T_5 = 620.72 \text{ K},$$

and $q = 1.005\,(1000 - 620.72) = 381.18$

$$\therefore \eta = w/q = 112.74/381.18 = 29.58$$

It is observed that the heat exchanger before the combustion chamber has improved the efficiency.

### 8.3.3 Stagnation Plane

One of the assumptions made for the analysis of the GT cycle, that change in kinetic energy from the inlet to the exit of the GT is negligible, is not correct, and has to be taken care of. We learn in Fluid Mechanics, that if a stream of air is brought to rest isentropically, then the state so acquired is called the stagnation state, and is represented by a subscript '0' placed with the property. Also, Bernoulli's and steady flow energy equations give (referring Section 8.3.1 for state points)

$$h_0 = h + C^2/2 \rightarrow T_0 = T + C^2/2c_p \tag{8.8a}$$
$$-w_c = (h_2 - h_1) + (C_2^2 - C_1^2)/2 = c_p(T_{02} - T_{01}) \tag{8.8b}$$
$$w_t = (h_3 - h_4) + (C_3^2 - C_4^2) = c_p(T_{03} - T_{04}) \tag{8.8c}$$

and
$$q = c_p(T_{03} - T_{02}) \tag{8.8d}$$

Here, all the symbols have the usual meaning. Since air is brought to rest isentropically, we may write

$$p_0/p = (T_0/T)^{\gamma(\gamma-1)}$$

$$\frac{p_{02}}{p_{01}} = \frac{p_{02}}{p_2} \times \frac{p_1}{p_{01}} \times \frac{p_2}{p_1} = \left(\frac{T_{02}}{T_2} \times \frac{T_1}{T_{01}} \times \frac{T_2}{T_1}\right)^{\gamma/(\gamma-1)} = \left(\frac{T_{02}}{T_{01}}\right)^{\gamma/(\gamma-1)} \tag{8.9a}$$

and $$\frac{p_{02}}{p_1} = \left(\frac{T_{02}}{T_1}\right)^{\gamma/(\gamma-1)} \tag{8.9b}$$

We have now established that the relations between static properties are applicable for stagnation values, too. Thus, we can work as well with stagnation properties, as we do with static ones. However, this eliminates the assumption of negligible change in kinetic energy. Therefore, it is usual practice to work in the stagnation plane rather than the static plane, when the change in kinetic energy between the exit and the inlet of an open system is not negligible.

The enthalpy–entropy diagram, usually used in analysis, can easily be changed to the stagnation plane. Note that the difference arises because of the inclusion of kinetic energy, which does not affect entropy. Hence, there is no change in abscissa. Change in the ordinate is obtained by adding the term $C^2/2$ to the enthalpy. This is shown for the compression process 12 in Fig. 8.12. 12 is the actual compression process from pressure $p_1$ to $p_2$. Adding $C_2^2/2c_p$ to $h_2$ gives $h_{02}$. Similarly, we get $h_{01}$ by adding $C_1^2/2\,c_p$ to $h_1$. It is shown by vertical lines drawn from 1 and 2 of lengths $C_1^2/2\,c_p$ and $C_2^2/2\,c_p$, respectively the new points are shown as 01 and 02. Constant pressure lines passing through these new points have values $p_{01}$ and $p_{02}$, respectively. We may now express isentropic efficiency of the compressor and turbine in terms of stagnation temperature in the same way defined using static enthalpy or temperature. Thus,

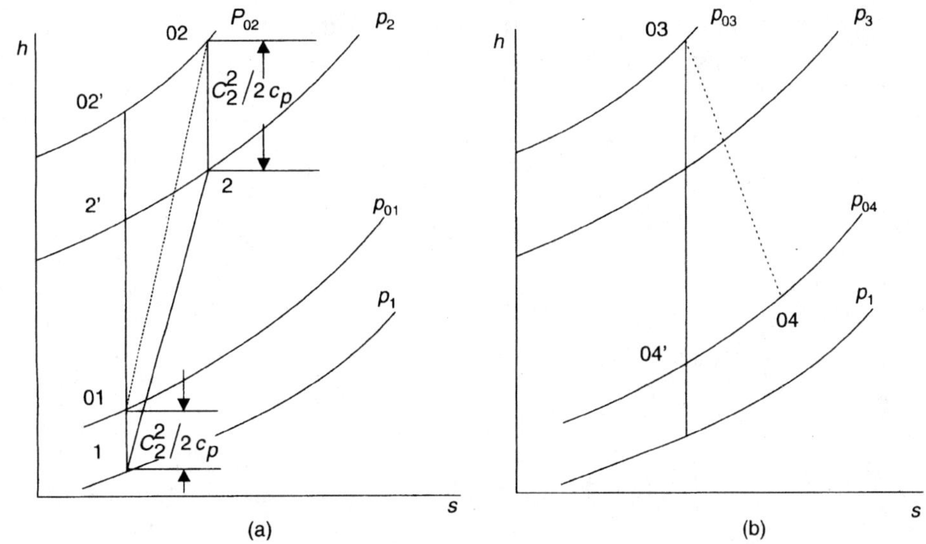


**Fig. 8.12** Stagnation states (a) compression (b) expansion

$$\eta_c = \frac{T'_{02} - T_{01}}{T_{02} - T_{01}} \quad \text{and} \quad \eta_t = \frac{T_{03} - T_{04}}{T_{03} - T'_{04}} \tag{8.10}$$

or
$$T_{02} - T_{01} = \frac{T_{01}}{\eta_c}\left[(p_{02}/p_{01})^{\gamma/(\gamma-1)} - 1\right]$$

and
$$T_{03} - T_{04} = \eta_t T_{03}\left[1 - \left(\frac{1}{p_{03}/p_{04}}\right)^{\gamma/(\gamma-1)}\right]$$

Here, 12 and 34 are actual compression and expansion processes, whereas 12′ and 34′ are corresponding isentropic processes. A comment will help in deciding stagnation values, since air is taken from the atmosphere, and burnt gases are left to it from the turbine. Thus, how the air is taken from or left to the atmosphere will affect the values of $p_{01}$ and $p_{04}$, respectively. In a stationary system, air is taken from the stagnant atmosphere (it is usual to neglect any velocity of flow of air). Thus, the atmospheric state is taken as the stagnation state. But normally, a filter is used, and hence stagnation pressure will decrease by the pressure loss in the filter. In some applications, it may be required to use a long pipe from the inlet section of air to the compressor eye, or a converging passage to develop some velocity at the inlet and vice versa. In such cases, pressure loss has to be subtracted from atmospheric pressure to get $p_{01}$. On the other hand, if the gas turbine is part of an aeroplane, then air is moving with respect to the plane, and hence its pressure and temperature are static values and not stagnation values. Then, stagnation values at the inlet to the compressor are to be determined.

Similarly, if gases are left to the atmosphere from the turbine through an exhaust diffuse (used to recover kinetic energy of exhaust), then atmospheric pressure $p_a$ is the corresponding stagnation pressure $p_{04}$, since the kinetic energy of the gas at its exit is negligible. Normally, $p_{04}$ is taken as $p_a$, the atmospheric pressure.

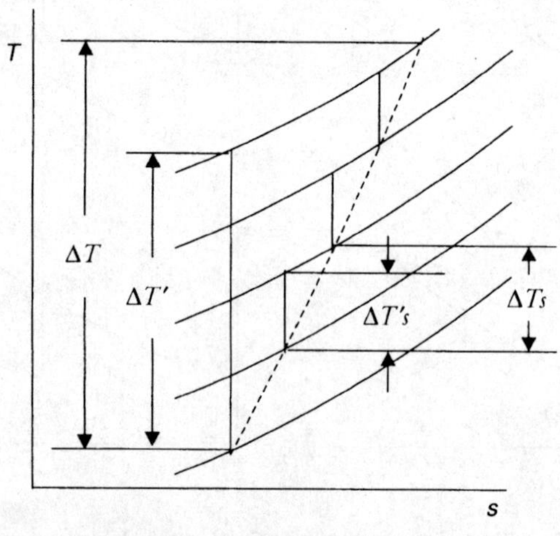


**Fig. 8.13** Polytropic efficiency

Isentropic efficiencies are seen to depend on the pressure ratio (see Eq. 8.10). This led to the concept of *small-stage* or *polytropic efficiency*, which does not depend on the pressure ratio. We assume that compression or expansion occurs in infinite stages, so that the isentropic efficiency for all the stages is the same. This efficiency is called polytropic efficiency, $\eta_{\infty c}$ for compression and $\eta_{\infty t}$ for expansion. A relation for $\eta_c$ and $\eta_{\infty c}$ can be obtained by considering compression[2], as shown in Fig. 8.13. Air is assumed to be compressed in stages; that is, output of one stage is fed as input to the next stage. Then, for the nomenclature shown in the figure, we may write:

$$\Delta T = \Sigma \Delta T_s = \Sigma \frac{\Delta T's}{\eta_s} = \frac{1}{\eta_s} \Sigma \Delta T_s' \Rightarrow \frac{\eta_s}{\eta_c} = \frac{\Sigma \Delta T_s'}{\Delta T'}, \text{ where,}$$

$$\Delta T = \Delta T'/\eta_c$$

Here, $\eta_s$ is the stage efficiency, which approaches $\eta_{\infty c}$ when the number of stages tend to infinity, $\Delta T$ and $\Delta T'$ are the actual and isentropic rise in stagnation temperatures, and $\eta_c$ is isentropic efficiency. Since constant pressure lines are diverging, the numerator $\Sigma \Delta T_s'$ will be greater than the denominator $\Delta T'$. This implies that polytropic efficiency of compression is more than the actual isentropic efficiency. (For expansion, it is the other way.) For an isentropic compression process,

$$T = C\, p^{(\gamma - 1)/\gamma}, \Rightarrow dT'/T = (\gamma - 1)/\gamma \; dp/p$$

Here, $dT'$ and not $dT$ is written to indicate that the process is isentropic. Using the definition of isentropic efficiency, and substituting for $dT'/T$, we get on simplification and then integrating from 1 to 2,

$$\eta_{\infty c} \frac{dT}{T} = \frac{\gamma - 1}{\gamma} \frac{dp}{p} \quad \Rightarrow \quad \eta_{\infty c} = \frac{\ln (p_2/p_1)^{(\gamma - 1)/\gamma}}{\ln (T_2/T_1)} \tag{8.11a}$$

or,

$$\frac{T_2}{T_1} = \left(\frac{p_2}{p_1}\right)^{\frac{\gamma - 1}{\gamma \eta_{\infty c}}}$$

But

$$\frac{T_2}{T_1} = \left(\frac{p_2}{p_1}\right)^{\frac{n - 1}{n}}$$

$$\therefore \quad (n - 1)/n = (\gamma - 1)/\gamma \eta_{\infty c}$$

and similarly for turbine $(n - 1)/n = \eta_{\infty t} (\gamma - 1)/\gamma$ (8.11b)

Figure 8.14 shows variation of isentropic efficiency for the compressor and turbine with $r$ for different values of. It is seen that isentropic efficiency increases for the same $r$

2. The same procedure can be followed for expansion.

with increase in polytropic efficiency, but decreases with $r$ for the compression process, while it increases for the expansion process.

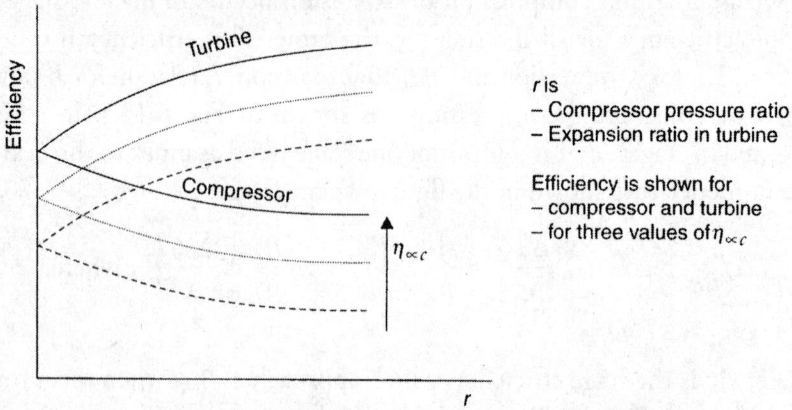


**Fig. 8.14** Effect of pressure ratio and polytropic efficiency on stage efficiency of the compressor and turbine

**Example 8.4** A gas turbine cycle is developed with a separate LP power turbine, in which gas leaving the HP turbine is passed through the heat exchanger before it enters the LP turbine. Draw this cycle on a *T-s* diagram and find out its efficiency. Show that for all values of $c$ and $t$, efficiency is higher than that of the normal scheme.

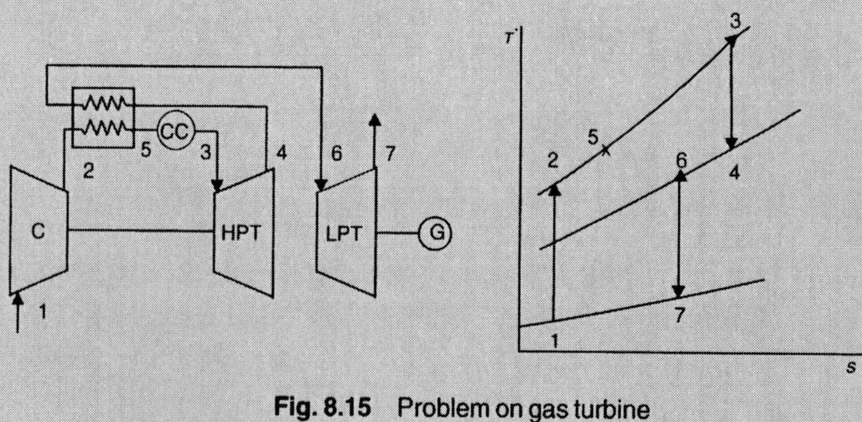


**Fig. 8.15** Problem on gas turbine

***Solution*** Figure 8.15 shows the schematic arrangement and the cycle on the *T-s* diagram. It may be noted that entropy in the process 46 is decreasing. This is due to heat lost by gases to air. Since HPT is deriving compressor, its output should be equal to that required for compression. This gives $p_4$ and $T_4$. We may now write following equations

$$T_2/T_1 = (p_2/p_1)^{(\gamma-1)/\gamma} = c \text{ or } T_2 = c\,T_1 \quad (1)$$

$$w_c = w_{HPT} \rightarrow (T_2 - T_1) = (T_3 - T_4)$$

$$T_4 = T_1(t - c + 1) = T_5 \quad (2)$$

or $$r = p_2/p_1 = p_3/p_4 \times p_6/p_7 = (T_3/T_4 \times T_6/T_2)^{(\gamma-1)/\gamma}$$

$$\rightarrow T_7/T_6 = t/\{c(t - c + 1)\} \quad (3)$$

But $$T_6 = T_2 \rightarrow T_7 = tT_1/(t - c + 1) \quad (4)$$

Efficiency, $$\eta = (T_6 - T_7)/(T_3 - T_6)$$

$$= \{(c-1)(t-c)(t+1)\}/\{c(1+t-c)\,(t-1)\}$$

Efficiency of conventional cycle with heat exchanger = $1 - t/c$. Thus the limits of c are 1 and $t$, that is, $1 \le c \le t$. This means $(c-1)\,(t-c) > 0$

or $$c/t > 1/(t - c + 1)$$

or $$1 - c/t < 1 - 1/(t - c + 1) = (t - c)/(t - c + 1)$$

That is, thermal efficiency of the proposed cycle is always more than that of a conventional heat exchanger cycle. However, as seen from the diagram, output of the cycle is decreased and therefore a large sized unit will be required for the same output. In other words, such a cycle is the most suited for stationary power plants.

$$\therefore \text{ ratio} = -\{(c-1)(t+1)\}/\{(1+t-c)(t-1)\}$$

**Example 8.5** A peak load generator is to be powered by a simple gas turbine with free power turbine delivering 20 MW of shaft power. Data given: compressor pressure ratio 11.0, compressor isentropic efficiency = 0.82, combustion pressure loss 0.4 bar, combustion efficiency 0.99, turbine inlet temperature 1150 K, gas generator turbine isentropic efficiency 0.87, power turbine isentropic efficiency 0.89, mechanical efficiency (each shaft) 0.98 and ambient conditions 1 bar and 288 K. Calculate the air mass flow required and the *sfc*.

***Solution*** Refer gas turbine cycle drawn in Fig. 8.16 without heat exchanger. The above two examples have shown how to tackle the problem of irreversibility and free power turbine. However, this problem is solved, as it considers heat release in the combustion chamber and calculation of *sfc*. We shall carry out calculations for 1 kg of air, and shall assume that temperature and pressure are given in the stagnation plane.

Thus the limits of c are 1 and $t$ i.e., $\rightarrow$. This means

$$(c-1)\,(t-c) > r \text{ or } c/t > 1/(t - c + 1)$$

or $$1 - c/t < 1 - 1/(t - c + 1) = (t - c)/(t - c + 1)$$

That is, thermal efficiency of the proposed cycle is always more that that of conventional heat exchanger cycle. However, as seen from the diagram, output of the cycle is decreased and therefore a large sized unit will be required for the same output. In other words, such a cycle is the most suited for stationary power plants.

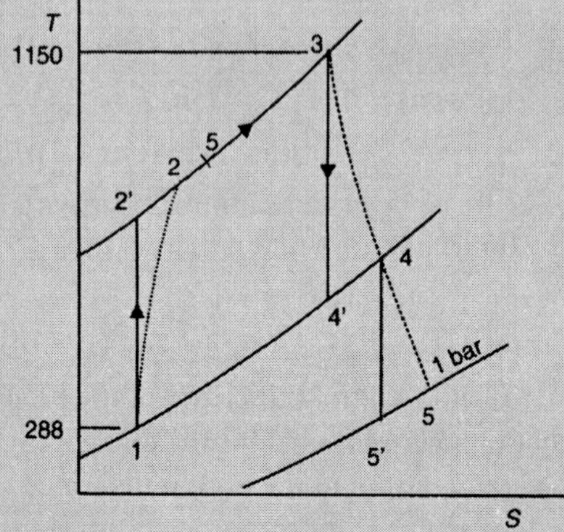


**Fig. 8.16** Free power turbine

Given: $p_{01} = 1$, $T_{01} = 288$, $r = 11$, $\eta_c = 0.82$, $\eta_{tg} = 0.87$ $\eta_{tp} = 0.89$, $\eta_b = 0.99$, $\eta_m = 0.98$, $\Delta p_b = 0.4$, $T_{03} = 1150$ K, $P = 20$ MW

Assumption: $\gamma = 1.4$, $c_p = 1.005$, $\eta_g = 1$(generator efficiency not given)

Using relation $T_{02'} = (p_{02}/p_{01})^{\gamma/(\gamma-1)}$ and Eq. 8.10, we get

$T_{02} = 347.5$ K and then, $wc = c_p(T_{02} - T_{01}) = 357.33$ kJ/kg.

$T_{04}$ is determined form equation $w_c/\eta_m = c_p(T_{03} - T_{04})$

$\rightarrow T_{04} = 859.90$ K

$T'_{04}$ is then obtained using Eq. 8.10 and pressure $p_{04}$ from the isentropic relation between $T_0$ and $p_0$. This gives: $T'_{04} = 807.2$ K and $p_{04} = 2.54$ bar. Since $p_{05} = p_{01}$, we may continue in the same way to obtain $T_{05} = 700$ K. Having obtained temperatures at the entrance and exit of the free turbine, specific work produced by this turbine is

$$w_t = c_p(T_{04} - T_{05}) = 1.005 \times (859.9 - 700) = 160.7 \text{ kW/kg/s}$$

Some power is lost in transmission from the free turbine to the generator, which is accounted by mechanical efficiency.

$$\therefore\ \eta_m = 20 \times 10^3/c\, w_t \ \rightarrow m = 127.0 \text{ kg/s}$$

For *sfc*, we should know the fuel air ratio. Fuel air ratio is function of temperature $T_{02}$ at inlet to combustion chamber and rise in temperature $\Delta T_0 = T_{03} - T_{02}$ in it. Graph has been developed between FA and $\Delta T$ for constant $T_{02}$. To determine fuel air ratio from the graph, we have $T_{02} = 347.5$ and $\Delta T_0 = T_{03} - T_{02} = 1150 - 347.5 = 803.7$. From the graph, we read $f = 0.0217$.

$\therefore$ actual fuel air ratio $f = .0215/0.99 = 0.02172$

and $\quad sfc = 3600f/w_t = 3600 \times 0.0217/160.7 = 0.487$

## 8.4 COMBINED CYCLE

In a gas cycle power plant, exhaust gases at the exit are at about 650–700°C temperature, with a lot of excess air that is supplied for the safety of turbine blades to keep gas temperature within safe limits at the entrance to the turbine. Thus, if these gases are left to the atmosphere, it shall be a colossal waste of energy. The temperature of the gases is so high, that energy of the waste gases can be recovered to produce superheated steam for running a steam turbine plant. Such a plant in which a steam power plant is cascaded with a gas turbine plant is called a combined cycle plant. New power plants are based on this concept, because their efficiency is higher than that of individual plants (Chapter 7).

Natural gas is available in abundance, which is fuel for gas plants. Even coal, or any other waste solid fuel may be converted into gas in a gas producer (gasifier), and used in a power plant. This preposition has found good response, because this shall also help in eliminating municipal, industrial and agricultural waste disposal problem while producing electricity. Research is going on to develop methods for implementation, as this shall require a great control over different parameters that are bound to vary. The use of gas produced from coal will encourage generating gas at the collieries and transporting it through pipes to distant places. This shall, in addition, reduce pressure on the rail and road transport, and use the ash to fill the cavities formed at excavation sites. It is expected that by 2010, the plants based on poor-quality fuel will run at an efficiency of 50%.

A combined cycle basically is a binary cycle in which the topping cycle runs on air and the bottoming cycle on vapour. The two cycles are coupled through the waste heat recovery unit. Such cycles were shown to have higher efficiency compared to the efficiency of individual cycles and hence it has great future. Figure 8.17 shows a gas steam combined cycle. If $\eta_1$ and $\eta_2$ are the efficiencies of the two cycles, then efficiency of the combined cycle is $\eta = \eta_1 + \eta_2 - \eta_1 \eta_2$.

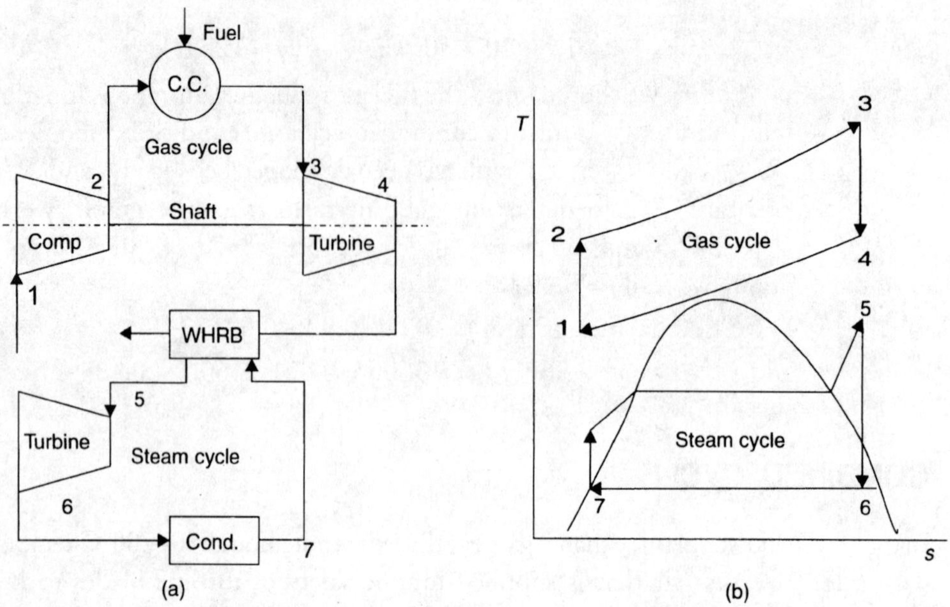


**Fig. 8.17** Combined cycle: (a) Block (b) *T-s* diagram

We may even increase the temperature of exhaust gases from the turbine by burning fuel, as turbine exhaust has free air in it; we have to add only burners and a suitable control for the same to meet load requirements. This, however, is decided by factors like pinch point and approach point temperature, stack temperature, steam temperature and pressure. The minimum temperature difference required for reasonable heat transfer from the gas for formation of steam plays an important role. This temperature difference is known as the pinch point temperature. For effective heat transfer between the two fluids, the pinch point temperature should be at least 20°C, but in practice, it may even be 30°C. The other temperature difference that matters is the difference of saturation temperature and the temperature at the exit of the economiser, that is required to avoid formation of steam before the boiler. This temperature difference is called the approach point temperature. Lowering this temperature will increase the probability of formation of steam in the economiser. The stack temperature is decided by the dew point temperature of acidic gases present in the flue gas. The stack temperature of 107°C is normally acceptable, as it is above the dew point temperature of sulphur dioxide.

Steam temperature and pressure are other very crucial parameters. Pressure is obtained on the basis that irreversibility of heat transfer is minimum. Figure 8.18 shows variation of temperature of water and the hot gases in the Waste Heat Recovery Boiler (WHRB) from their entrance to the exit, for a combined gas/steam (COGAS) plant for both single and dual-pressure systems; the names derive from the fact whether steam is produced at single or two pressures. This will depend on temperature and heat available from exhaust gases, and irreversibility considerations. We know from the second law of thermodynamics that more the temperature difference, more is the irreversibility. That is, the vertical height between the two temperature profiles is a measure of the irreversibility. We shall compare the two pressure situations as shown in Fig. 8.18 (a). We observe that for a high pressure steam cycle, irreversibility will be less in the high temperature zone, and more in a low temperature region, compared to a low pressure cycle. This suggests that there should be some pressure when the total irreversibility in the heat transfer process in WHRB will be minimum. This is the desired pressure.

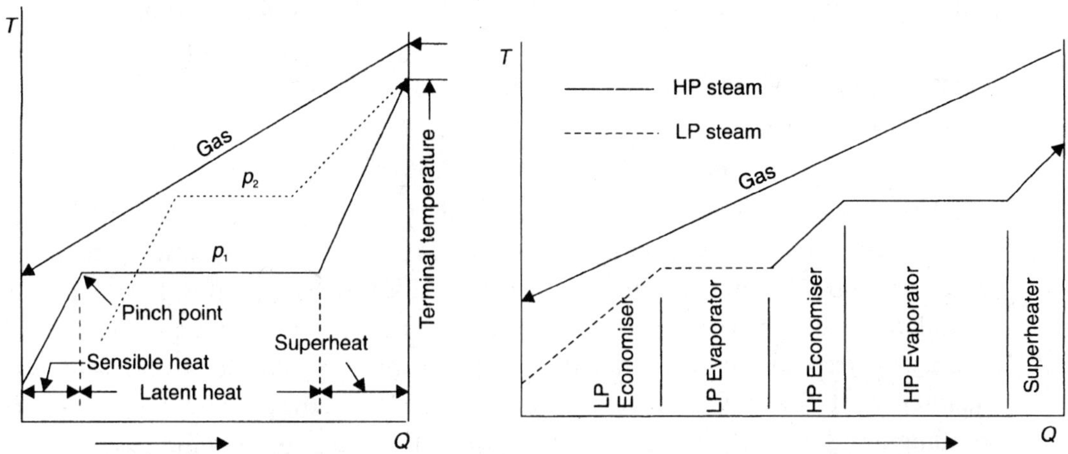


(a) Single pressure cycle (b) Dual pressure cycle

**Fig. 8.18** *T-Q* diagram for single and dual pressure COGAS scheme

Whether a single, dual or more pressure cycle will be needed, is determined on the basis of minimising irreversibility in the heat transfer process, for which similar treatment can be done for a dual pressure cycle. This will also depend on the temperature of the gases at the entrance, since the terminal temperature, defined as the difference in temperature of gas and steam, will fix the maximum temperature of the steam. This temperature difference is about the same as the pinch point temperature difference. Therefore, the three temperatures are to be optimised for maximum efficiency and the mass of steam generated. The study of such plants reveals the following information:

- The mass flow rate of steam decreases with increase in steam pressure (also called drum pressure), but output per kg of steam expanded increases. Therefore, the overall output and efficiency of the combined cycle increases up to a certain drum pressure. This is the pressure when irreversibility associated with heat transfer will be minimum.
- Approach, pinch and terminal point temperatures are important design variables. These should be small, as a large value of these reduces the combined cycle

output. But this requires a large surface area of the heat transfer equipment, which increases the cost and pressure drop on the gas side.

- Like the steam vapour cycle, increase in steam temperature and decrease in condenser pressure will increase the efficiency of the steam cycle, and hence that of the combined cycle.

A comment on adding a supplementary burner before the WHRB will be in order. This would increase gas temperature at the entrance, and hence we can think of a more efficient steam cycle that would increase the overall efficiency of the combined cycle. To find the effect of $\eta_1$ or $\eta_2$, the efficiencies of topping and bottoming cycle, on the overall efficiency, we may find out the rate of change of $\eta$ with either of them. From the expression of $\eta$, we get

$$\partial\eta/\partial\eta_1 = 1 - \eta_2 \quad \text{and} \quad \partial\eta/\partial\eta_2 = 1 - \eta_1$$

This shows that the rate of increase in overall cycle efficiency decreases with increase in efficiency of either cycle, but efficiency increases. That is, increase in efficiency diminishes with more supplementary burning of the fuel. Besides, we have to consider other factors too while deciding how much increase in efficiency is desirable, since any such attempt will also increase the complexity, and hence the cost of the plant. It may also add to the maintenance bill later. Whereas when we talk of efficiency, the purpose is to decrease the running cost. However, the overall cost consists of the initial and running cost. Economy demands that the electricity bill should be the least. Therefore, a combination of all the parameters affecting economy decides how much gain in efficiency will be desirable. For these reasons, supplementary firing is becoming less and less attractive.

Lastly, we studied that the compression ratio of the cycle is different for maximum output and efficiency. Nowadays, gas turbines are manufactured for maximum output because of their big market. Incidentally, this also corresponds to the optimum pressure ratio of the combined cycle. Hence, a gas turbine for the combined cycle is designed for maximum output.

### Anta Gas Power Plant[3]

A Dual pressure cycle has been used at the NTPC power plant of Anta in Rajasthan in their phase 1 plant. Figure 8.19 is its schematic diagram. Gases from the GT plant are supplied to the Waste Heat Recovery Boiler (WHRB), also called the waste heat steam generator, through a combustion chamber, where some fuel may be burnt to raise the temperature of the gases entering the WHRB, since oxygen is present in the flue gases.

3. A study of Exergy Analysis of Anta Gas-based Power Plant was done by a group of students of B.E (Final) Mechanical Engineering, Engineering College Kota in 1991 under the guidance of the author. In this analysis, many simplifications have been made to avoid rigorous treatment.

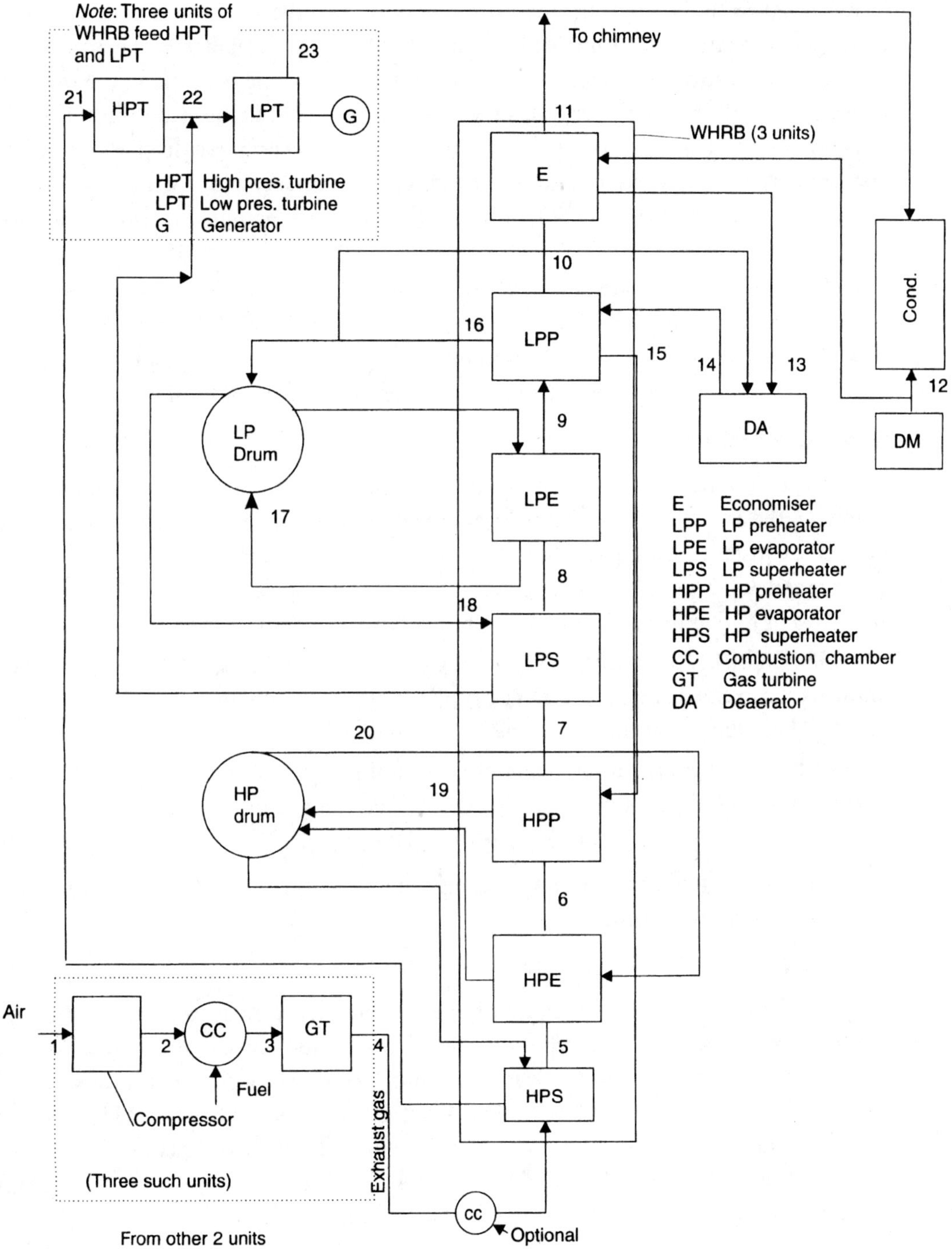


**Fig. 8.19** Layout of Anta gas (NTPC) power plant—phase 1

Water is fed to the WHRB through the economiser to the evaporator, where water is converted into saturated steam. Steam is superheated in the super heater. These elements are arranged in this order, so that flow of gases and water or steam is in the opposite direction to make heat transfer reversible. Gases are then left to the atmosphere through a chimney, and steam is used in the steam cycle. There is pressure drop as the various components in the gas path offer resistance and hence pressure at the turbine exit is more than the atmospheric pressure. Not only this, the cost of the equipment also increases the cost.

There are three gas turbines each of a 88 MW capacity, and one steam turbine of 149 MW capacity. Properties like pressure, temperature, enthalpy, entropy, as well as the mass flow rate of the fluid at different points, during operation of the plant under normal conditions, are listed in Table 8.1 in the gas cycle, and in Table 8.2 for the steam cycle.

### *Analysis of Power Plants*

Two type of analysis, namely energy and exergy, are required, which can be done easily following the methods discussed before. However, some sample calculations are done here. The constant C in Eq. (3.15) is determined for air by taking exergy zero at 1.01 bar pressure and 300 K temperature. For air, enthalpy and entropy at $p_a = 1$ bar and $T_a = 300$ K are 299.05 kJ/kg and 2.45796 kJ/kg K, respectively. Eq. (3.15) gives,
$C = 438.338$ kJ/kg. Similarly for $H_2O$, we get $C = 7.02416$ kJ/kg. Once the constant is known, then exergy per kg of air '*e*' is obtained at different sections of the plant using Eq. (3.13). The value of e at these sections is tabulated in Tables 8.1 and 8.2.

Further, since gas is compressible, the effect of pressure change has also been taken into account, and the change in entropy is obtained using the equation,

$$s_2 - s_1 = \phi(T_2) - \phi(T_1) - R \ln (p_2/p_1) \tag{8.12}$$

where, $\phi(T)$ is function of temperature,[4] which is available from gas tables. Steam table was used for properties of steam.

### *Analysis of the Combustion Chamber*

We shall demonstrate the procedure of energy and exergy analysis for a combustion chamber, and the same is followed for the rest of the components. Figure 8.18 shows the combustion chamber with air at the inlet Section 2, products of combustion at section 3 and fuel inlet at section g. From Table 8.1, we get: $h_2 = 663.90$, $h_3 = 1371.66$ kJ/kg; $s_2 = 2.544$ and $s_3 = 3.306$ kJ/kg K; mass flow rate in kg/hr of air and gas, $m_2 = 1330236$, $m_3 = 1354680$ and $m_g = 24444$; and $e_2 = 338.94$, $e_3 = 818.18$ and $e_g = 0.00$. The energy and exergy balance criterion for a combustion chamber results in,

4. In actual conditions, $\phi$ also depends on temperature.

Total increase in enthalpy = $m_3 \times h_3 - m_2 \times h_2 - m_g \times h_g$

$= (13.716592 \times 13.7166 - 13.30236 \times 7.639 - 0.24444 \times 3.3451) \times 10^7$

$= 99 \times 10^7$ kJ/hr

Increase in available energy = $m_3 \times e_3 - m_2 \times e_2 - m_g \times e_g$

$= (13.716592 \times 8.1818 - 13.30236 \times 3.3894) \times 10^7 = 67.14 \times 10^7$ kJ/hr

Heat input from gas = $24444 \times 42349.482 = 103.52 \times 10^7$ kJ/hr. $= 2.876 \times 10^5$ kJ/s,

Exergy supplied to c.c. = Exergy of fuel + exergy of air from compressor

$= (1.033 \times 103.52 \times 10^7 + 1330236 \times 338.94)/3600 = 4.23 \times 10^5$ kJ/s

$\therefore$ first law efficiency of c.c. = Increase in enthalpy of air/heat input from gas × 100

$= 99 \times 10^7/103.52 \times 10^7 \times 100 = 97.63$ %

Exergetic efficiency = increase in exergy/exergy supplied by the fuel

$= 67.14/107.935 \times 100 = 62.79$ %

Similarly, all the other components of the combined cycle may be analysed. However, loss in exergy of individual components is of more interest, since this gives an idea of irreversibility present in the system and the ongoing process. This is shown in Table 8.1 for gas; a negative sign means decrease in exergy of the gas in that component. The reader may verify that exergy of steam is increased during the process.

**Table 8.1** Properties of gas at different sections of WHRB

| Section | $p$ kgf/cm² | $T$ K | $h$ kJ/kg | $s$ kJ/kg K | $m$ kg/hr | $e$ kJ/kg | $\Delta e$ kJ/kg |
|---|---|---|---|---|---|---|---|
| 1 | 1.003 | 300.0 | 299.05 | 2.458 | 1330236 | 0.00 | 0.00 |
| 2 | 12.020 | 653.2 | 663.90 | 2.544 | 1330236 | 338.94 | 338.94 |
| 3 | 11.680 | 1278.0 | 1371.66 | 3.306 | 1354680 | 818.18 | |
| g | | | 334.51 | | 24444 | 0.00 | |
| 4 | 1.026 | 780.8 | 801.93 | 3.443 | 1354680 | 207.31 | |
| 5 | 1.026 | 718.0 | 733.56 | 3.352 | 1354680 | 167.29 | –41.02 |
| 6 | 1.026 | 562.1 | 567.07 | 3.092 | 1354680 | 72.77 | –93.52 |
| 7 | 1.026 | 500.9 | 503.38 | 2.972 | 1354680 | 50.04 | –22.73 |
| 8 | 1.026 | 497.2 | 500.70 | 2.966 | 1354680 | 49.28 | –00.76 |
| 9 | 1.026 | 441.5 | 442.27 | 2.842 | 1354680 | 27.29 | –21.39 |
| 10 | 1.026 | 407.2 | 407.24 | 2.756 | 1354680 | 17.28 | –10.61 |
| 11 | 1.026 | 377.1 | 374.70 | 2.683 | 1354680 | 8.13 | –9.15 |

**Table 8.2** Properties of steam and water at different sections

| Section | $p$ kgf/cm² | $T$ K | $h$ kJ/kg | $s$ kJ/kg K | $m$ kg/hr | $e$ kJ/kg |
|---|---|---|---|---|---|---|
| 12 | – | 318.9 | 192.17 | 0.652 | 197259 | 1.70 |
| 13 | – | 371.4 | 412.40 | 1.289 | 197259 | 30.76 |
| 14 | 1.53 | 384.4 | 467.25 | 1.434 | 197259 | 42.08 |
| 15 | 87.70 | 433.8 | 679.10 | 1.948 | 152800 | 99.65 |
| 16 | 17.90 | 433.9 | 682.03 | 1.950 | 38500 | 99.54 |
| 17 | 7.22 | 434.9 | 2761.74 | 7.735 | 38500 | 747.17 |
| 18 | 7.61 | 480.0 | 2867.00 | 7.040 | 38500 | 758.97 |
| 19 | 84.30 | 559.4 | 1270.55 | 3.280 | 152800 | 291.56 |
| 20 | 69.70 | 557.6 | 2774.03 | 7.223 | 152800 | 1032.11 |
| 21 | 63.90 | 757.9 | 3381.17 | 7.207 | 152800 | 1344.02 |
| 22 | 7.36 | 454.5 | 2812.24 | 7.944 | 152800 | 734.10 |
| 23 | 0.10 | 318.9 | 2300.23 | 7.190 | 191 | 143.19 |

### First and Second Law Efficiency of the Cycle

The main purpose of this study was to carry out the exergy analysis, and therefore, data for only one gas turbine was considered. But the steam turbine gets flue gases from all the three gas turbines. It was therefore assumed that it is getting equal feed from these turbines. Therefore, power developed by the generator connected to the individual gas turbine and steam turbine for the operating parameters is equal to (88.71+ 153/3) $10^3$ kJ/s.

$\therefore$ overall thermal efficiency of the cycle = (88 .71+ 153/3) $10^3/2.876 \times 10^5$

= 48.56%

Exergetic efficiency = (88 .71+ 153/3) $10^3/2.97 \times 10^5$ = 47.00%

It is also possible to find out isentropic efficiency of the gas turbine and the compressor from these data, since actual temperatures are known, and theoretical temperatures may be obtained by using isentropic expansion/compression with $\gamma$ = 1.4 for both air and products of combustion. However, the actual value of $\gamma$ depends on the temperature and the fuel-air ratio. $\gamma$ at different fuel-air ratio and air temperature may be obtained from the charts. In most of the operating conditions of gas turbine applications, the recommended values of $\gamma$ for compression and expansion processes are 1.4 and 1.333, or $\gamma/(\gamma - 1)$ = 3.7. and 4.0, respectively. Representing the state of the fluid after compression and expansion by the points as 2′ and 4′, and applying isentropic relations, we get

$$T_{2'} = (12.02/1.003)^{1/3.5} \times 300 = 609.95 \text{ K},$$
$$T_{4'} = (1.026/11.68)^{1/4.0} \times 1278 = 697.76 \text{ K},$$
$$\eta_c = (609.95 - 300)/(653.2 - 300) = 87.76\%$$

and,
$$\eta_t = (1278 - 780)/(1278 - 697.76) = 87.53\%$$

Isentropic efficiency of a steam turbine can also be found out, as the actual and isenthalpic enthalpy drop can be read from the Mollier diagram by showing the processes on it.

### Results and Discussions

Change in exergy in going from one section to another can be determined by taking the difference of exergy '*e*' at the two sections, tabulated in Tables 8.1 and 8.2. Also, for a component, the inflow and outflow of exergies may be calculated, from which gain or loss of exergy in that component is evaluated. Finally, the exergy balance sheet can be prepared, as shown in Table 8.3. It is observed that

- Major loss of exergy occurs in the gas combustion chamber, which is about 38.5%. That is, the exergetic efficiency of the combustion chamber is 62.79%. However, the first law analysis yields a combustion chamber efficiency of 97.63%, and is quite misleading.
- Exergetic and thermal efficiency of combined gas turbine and steam turbine are 47% and 48.56%, respectively.
- Exergy loss in the boiler unit is only 4.36% of the exergy supplied to the plant through gas. This is due to the fact that the boiler is basically a heat exchanger, and has been designed efficiently. Further break up of this loss is given in Table 8.4.
- In a steam boiler, major losses occur in the HP evaporator and superheater, which is due to the large temperature difference between water (and steam) and flue gas. The other source for loss of exergy is flue gas left to the atmosphere, with which significant amount of exergy is led out.

**Table 8.3** Exergy balance sheet

| Exergy supplied/utilised | Percentage |
|---|---|
| Exergy supplied by combustion of fuel | 100.0 |
| Exergy utilised in producing electricity | |
| GT generator | 29.86 |
| ST generator | 17.15 |
| Exergy losses | |
| In combustion chamber | 38.51 |

*(Contd)*

| Exergy supplied/utilised | Percentage |
|---|---|
| In gas turbine | 7.21 |
| In compressor | 0.15 |
| In boiler | 4.36 |
| In HP steam turbine | 0.69 |
| In LP steam turbine | 1.27 |
| In condenser | 2.53 |
| Miscellaneous | 0.27 |

**Table 8.4** Losses in Boiler

| Component | % | % |
|---|---|---|
| Losses in boiler as a whole | 4.36 | 100.00 |
| HP superheater | 0.74 | 17.98 |
| HP steam evaporator | 1.27 | 29.06 |
| HP water preheater # 2 | 0.14 | 3.15 |
| LP superheater | 0.05 | 1.15 |
| LP evaporator | 0.39 | 8.74 |
| HP heater # 1 and LP heater # 1 | 0.11 | 2.43 |
| Economiser | 0.63 | 14.30 |
| To atmosphere | 1.03 | 23.64 |

### *Conclusion*

It is concluded from the above results, that while energy analysis accounts for all energy transfers between the system and the surrounding, exergy analysis takes into account the capacity of getting work from given energy exchange. It is exergy analysis, and not energy analysis, which reveals that majority of losses occur in the combustion chamber. It may also be mentioned that the Anta Gas Power Plant is quite energy efficient, though there is still some scope in the improvement of the design of the combustion chamber, and heat exchangers like the HP steam evaporator and superheater. The second plant is more efficient and will have an efficiency of 54%.

It will be interesting to see the temperature variation of water and gas along the WHRB, for which the temperature of the two fluids at different points along the WHRB is reproduced in Table 8.5, and then drawn on the $T$ versus $Q$ chart as shown in Fig. 8.20. It is seen that this chart resembles Fig. 8.17 (b), as expected.

**Table 8.5** Temperature of fluids at different points along the WHRB

| St. No. | *ha* | *q* | *Ta* | *Ts* |
|---|---|---|---|---|
| 4 | 801.93 | 0 | 780.8 | 757.9 |
| 5 | 733.56 | 25727.631 | 718 | 552 |
| 6 | 567.07 | 88377.218 | 562.1 | 552 |
| 7 | 503.38 | 112344.365 | 500.9 | 433.8 |
| 8 | 500.7 | 113352.849 | 497.2 | 432 |
| 9 | 442.27 | 135340.058 | 441.5 | 432 |
| 10 | 407.24 | 149048.667 | 407.2 | 371.4 |
| 11 | 374.69 | 160770.412 | 377.1 | 318.9 |

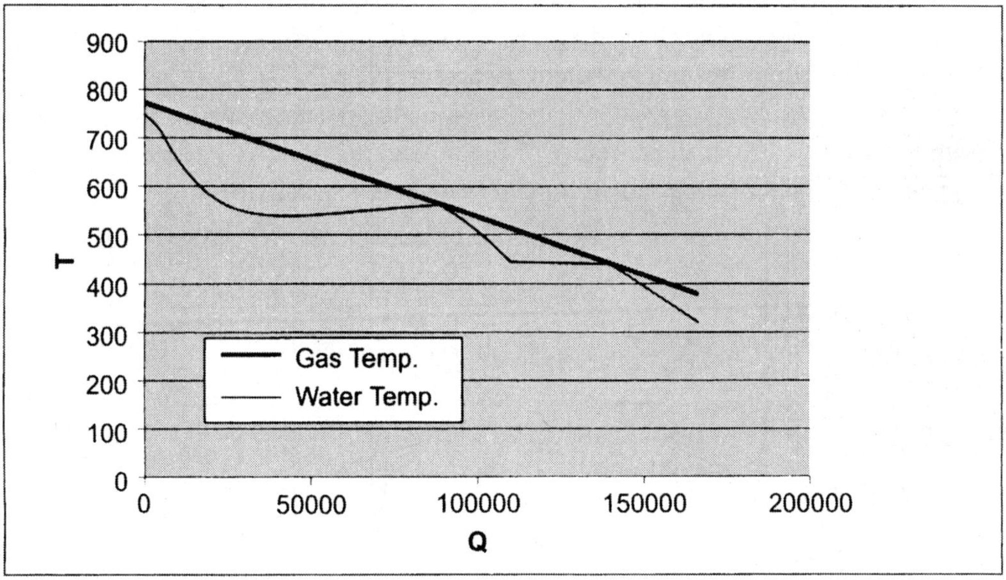


**Fig. 8.20** *T* versus *Q* chart for temperature variation of water and gas along the WHRB

## 8.5 JET PROPULSION

The first application of a gas turbine as power unit was in aircraft propulsion, where the main consideration remained to lower down the weight to power ratio by running the machine at the maximum output condition. The first aircraft were propeller aircraft, in which a propeller was rotated by a turbine shaft that developed forward thrust. Since then, many developments have taken place, and today, even a turbine is not required at very high flight speed, corresponding to Mach number of more than 7 or so. NASA successfully launched a scramjet in NN 2004 with Mach number 10. In between lie the different implementations using the property of jet action, which states that a jet coming out from a device would produce a thrust on the device in the direction opposite to the direction of the jet. Momentum in this process is conserved. Using this fact, the thrust can be determined. The thrust moves the plane in the forward direction. The amount of force would depend on the change in the momentum of the fluid.

Figure 8.21 (a) shows a schematic diagram to explain the working of a jet plane. The plane is moving with speed *Ca* from right to left. We may fix the origin of the coordinates with the plane. Then, the plane is stationary, and air is moving from left to right; it enters the plane at section '*a*' with velocity *Ca*. Air velocity is increased to *Cj* at the exit of the device due to energy supplied by the fuel. A power unit, consisting of a compressor, combustion chamber and the turbine, is placed inside the duct. Flue gases are expanded in the turbine to a pressure, so that power generated by the turbine is enough to drive the compressor. Gases come out of the power unit at high pressure, and are expanded further in a nozzle to some pressure $p_j$, thereby gas acquires velocity *Cj*. Pressure at the nozzle exit section *j* may be equal or greater than atmospheric pressure $p_a$, depending on the Mach number of the flow at the entrance of the nozzle; it can never be less. Gases leave the plane at section *j*. The various sections are shown in Fig. 8.21(b).

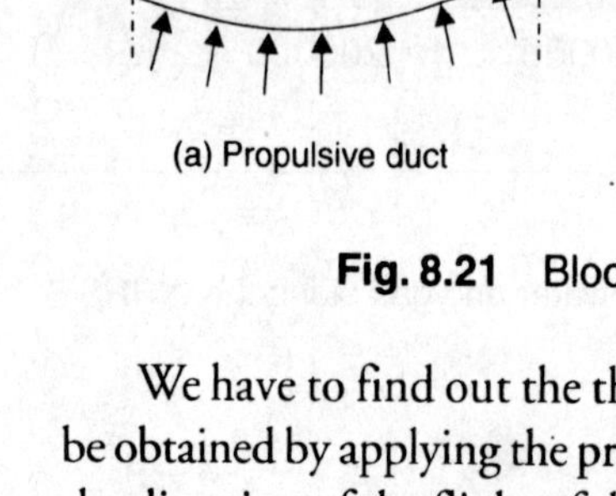

(a) Propulsive duct

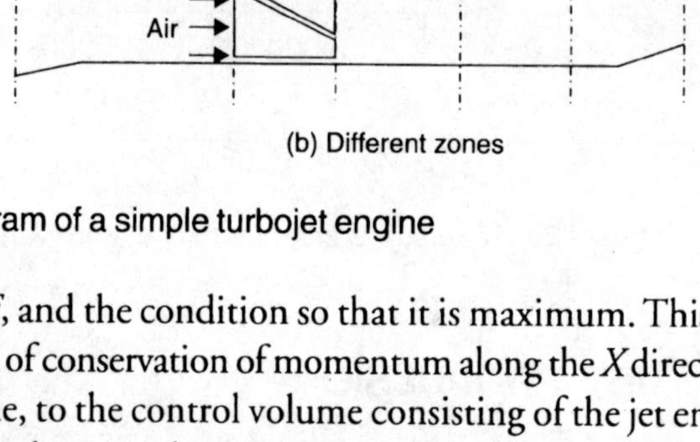

(b) Different zones

**Fig. 8.21** Block diagram of a simple turbojet engine

We have to find out the thrust *F*, and the condition so that it is maximum. This can be obtained by applying the principle of conservation of momentum along the *X* direction, the direction of the flight of the plane, to the control volume consisting of the jet engine duct. For this purpose, the pressure on the control volume may be measured with reference to atmospheric pressure, and say, $p_j$ represents the gauge pressure at the nozzle exit section. We get:

$$F_x = \int_s U(\rho U . ds) + \partial/\partial t \int_v \rho U dV = m_j C_j - m_a C_a$$

Here, we have assumed a steady state, and therefore, the $\partial/\partial t$ term vanishes from the right side, and the surface integral changes as shown in the equation. If $m_f$ kg of fuel is supplied, so that the fuel air ratio is $f = m_f/m_a$, then $m_j = m_a + m_f = m_a(1+f)$ by the mass conservation principle. Also, the external force $Fx = F - p_j A_j$, where $A_j$ is the cross-sectional area of the nozzle exit section, and the negative sign with $p_j$ is because it is acting in the negative $X$ direction (the positive direction is assumed in the direction of $Cj$). Writing these and simplifying, we get,

$$F = m_a[(1+f)\,Cj - Ca] + A_j p_j \approx m_a(Cj - Ca) + A_j p_j \tag{8.13}$$

neglecting the mass of fuel as $f << 1$. The forward thrust may be increased by increasing the mass of air $m_a$ or change in velocity $(Cj - Ca)$ from the entrance to the exit or both. Let us develop some criterion to decide which method would be more effective. The total energy at the exit of the turbine is the sum of the thrust power $(F\,Ca)$, also called propulsive energy, and the energy going as waste with the flue gases. The absolute velocity of the flue gas is $(Cj - Ca)$ (remember $Ca$ and $Cj$ are relative velocity, as these are measured with reference to a moving plane) and therefore, the kinetic energy with $m_a$ kg of gas is equal to $\frac{1}{2} m_a (Cj - Ca)^2$, and the thrust power by Eq. 8.13 is equal to $[m_a(Cj - Ca) + A_j p_j] \approx m_a Ca(Cj - Ca)$, assuming that the gases leave the nozzle at atmospheric pressure. If this is not true, then the second term may also be considered, which we shall discuss later. We define propulsive efficiency $\eta_p$ as the ratio of the thrust power $(F\,Ca)$ to the total energy at the exit of the turbine. Thus, it expresses the effectiveness with which the propulsive duct is being used for propelling the aircraft. Propulsive efficiency therefore is

$$\eta_p = \frac{m_a Ca(Cj - Ca)}{m_a[Ca(Cj - Ca) + (Cj - Ca)^2/2]} = \frac{2}{1 + Cj/Ca} \tag{8.14}$$

Propulsive efficiency, the way it is defined, should be as high as possible, so that the kinetic energy going as waste with the exhaust gas is minimum. From Eq. 8.14, we observe that this is so when $Cj$ and $Ca$ are of the same order of magnitude. That is, $(Cj - Ca)$ is almost zero. Thus, thrust can be increased by increasing the air mass flow rate and not $(Cj - Ca)$.

We used the energy at the turbine exit to define the propulsive efficiency. However, the input to the system is the energy of the fuel. Therefore, we define another term, the efficiency of energy conversion $\eta_e$, for the purpose. For the control volume, kinetic energy of air is increased by the amount $\frac{1}{2} m_a (Cj - Ca)^2$ as a result of the energy supplied to the system, which is $m_f \times CV$, where $CV$ is the calorific value of the fuel per kg of fuel. The energy of conversion is therefore, defined as the ratio of increase in kinetic energy of air to the energy input. Thus

$$\eta_e = \tfrac{1}{2} m_a (Cj - Ca)^2/(m_f \times CV) \tag{8.15}$$

Finally, we may also define overall efficiency $\eta_o$, which is based on the useful output, the thrust power, and the input to the system. Thus,

$$\eta_o = m_a\, Ca\,(Cj - Ca)/(m_f \times CV) = F\,Ca/(m_f \times CV) \tag{8.16}$$

It can be easily verified that $\eta_o = \eta_p \times \eta_e$

We infer from this analysis, that both the thrust produced and the overall efficiency are linked with the aircraft velocity, and the more the velocity, the more is the value of the two variables. That is, we can make a choice of $m_a$ and *Ca*. With higher speeds of craft, *Ca* is high, and therefore, the mass flow rate for the same thrust is less. This led to the development of a family of aircrafts in which the air mass flow decreases as the aircraft speed increases. These are (arranged in order of decreasing mass flow rate) (i) turboprop engine, (ii) turbofan engine, (iii) turbojet engine, and (iv) ramjet engine; all of them, except the first, work on the principle of jet. In a turboprop engine, a propeller is mounted on the front end of the aircraft, which produces thrust by its rotation. We shall discuss specially the turbofan engine later, as these are used by almost all the airlines in the world for passenger planes. However, the choice of the family of jets depends on the Mach number of the flight. Not only this, but the height at which the plane has to fly also depends on the Mach number. Figure 8.22 shows the altitude at which an aircraft operating fitted with any such engine would normally fly, and the range of operating Mach number.

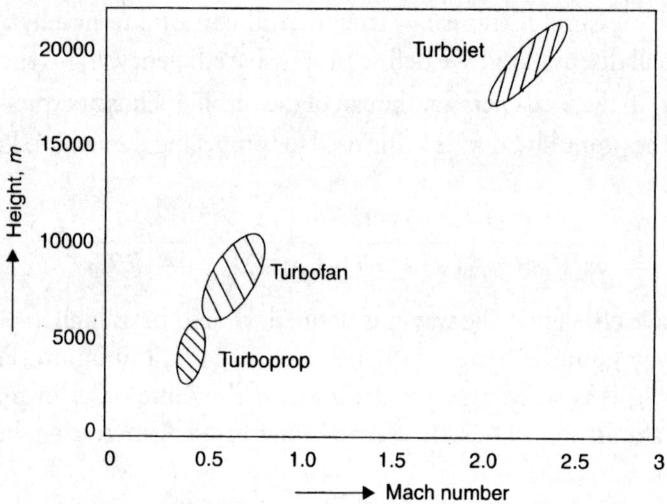


**Fig. 8.22** Regimes of different types of propulsion engines

Thrust and fuel consumption of any power unit depends on the aircraft cruising speed and the altitude of flight, since air density decreases with increasing height. Therefore, choice of the power unit will depend on these factors, and in addition, the range of the aircraft (it decides the amount of fuel to be stored and hence weight of the power unit) as well as the rate of climb (fuel consumption is more). We note that the turbofan is most suitable for passenger flight, as for this type, the Mach number is close to 1, the normal Mach number used in the aviation industry.

Other terms used for performance of aircraft engines are specific fuel consumption and specific thrust. Specific fuel consumption is defined as the fuel consumed per hour per unit thrust. From Eq. 8.16, the overall efficiency can be written in the form:

$$sf_c = \frac{m}{F} \text{ and } \eta_0 = \frac{C_a}{(m/F) \times C_v} = \frac{C_a}{sf_c \times C_v} \tag{8.17}$$

The other performance parameter, specific thrust, is defined as the thrust per unit mass flow rate of air; its SI unit is N s/kg, and is represented by the symbol *Fs*. This indicates the mass flow rate of air for the required thrust to be produced by the engine, and therefore of its relative size, which is an important parameter in the design of the engine. This is important not only to decrease the total mass, but the frontal area is also quite important to reduce the drag force on the craft. From the definition of the two parameters, we find the following relation between the two

$$Sfc = m_f/F = (m_f/m_a)/(F/m_a) = f/Fs \tag{8.18}$$

## 8.5.1 Intake and Propelling Nozzle Efficiencies

Air enters a simple turbojet engine through a duct, so that the velocity at the compressor inlet is reasonable. If the aircraft speed is low, then air is accelerated. In supersonic crafts, velocity is decreased by formation of shock waves. The inlet section is shown in Fig. 8.21(b). A diverging flare at the inlet of the pipe only assists in the smooth entrance of air into the duct. The flow in the propulsive duct is divided into five zones, as indicated in Fig. 8.21 (b). We have discussed the three zones that leaves the two zones, namely, the inlet and propelling nozzle sections, for discussions. Flow through the nozzle is taught in the compressible flow theory. We shall quickly see some of the important results, so that a turbojet cycle may be evaluated for its performance.

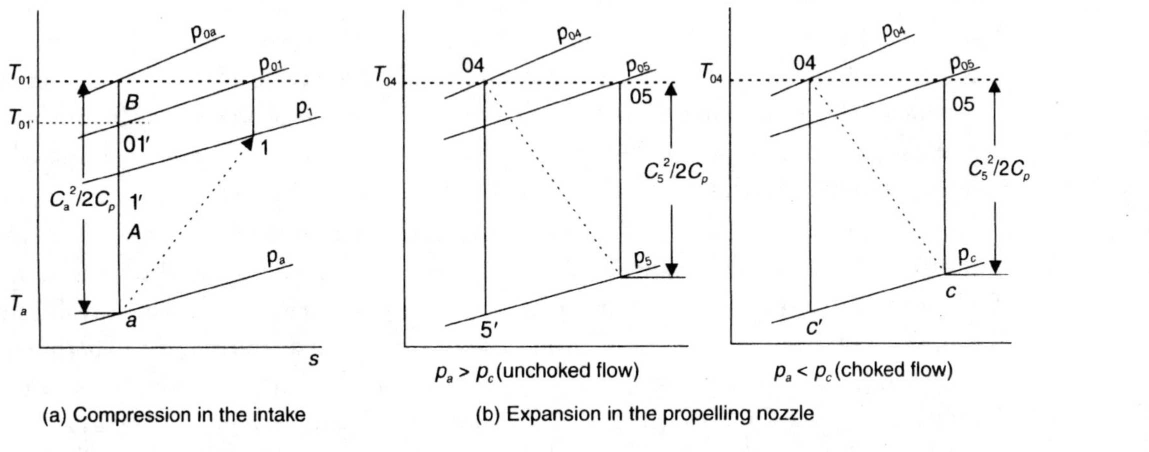


**Fig. 8.23** Flow in the intake and propelling nozzles

Figure 8.23 (a) shows the compression process in the intake nozzle. We have learnt before how to draw the process on the *h-s* chart. The static pressure rises from $p_a$ to $p_1$ in the duct. However, the stagnation pressure would decrease from $p_{0a}$ to $p_{01}$ because of various losses from the inlet of the nozzle to the compressor suction end. The isentropic process $a_1'$ will be a vertical line, which is shifted to a because of the losses in the nozzle. We may define a ratio of $(p_{01} - p_a)$, called the ram pressure rise, with $(p_{0a} - p_a)$, to indicate the loss in stagnation pressure. This ratio is named ram efficiency $\eta_r$.

$$\eta_r = (p_{01} - p_a)/(p_{0a} - p_a) \tag{8.19}$$

During this compression process, the stagnation temperature would not change, as there is no energy transfer to the duct. We may even use the term isentropic efficiency to represent irreversible compression in the intake nozzle. The isentropic efficiency $\eta_i$ is;

$$\eta_i = (T_{01'} - T_a)/(T_{01} - T_a) \tag{8.20}$$

where, $T_{01'}$ is the stagnation temperature in case of reversible compression. Writing $T_{01} - T_a = C_a^2/2c_p$ in Eq. 8.20, we get from Eqs. 8.19 and 8.20,

$$\frac{p_{01}}{p_a} = \left[\frac{T_{01'}}{T_a}\right]^{\gamma/(\gamma-1)} = \left[1 + \frac{T_{01'} - T_a}{T_a}\right]^{\gamma/(\gamma-1)} = \left[1 + \eta_i \frac{C_a^2}{2c_p T_a}\right]^{\gamma/(\gamma-1)} \tag{8.21}$$

Knowing isentropic efficiency, we can determine $p_{01}$. Isentropic efficiency and ram efficiency are two methods out of several used for expressing irreversibility in the compression process. We may visualise that the ratio $p_{01}/p_{0a}$ will be a function of the velocity or the Mach number at the inlet to the duct. This ratio is called the pressure recovery factor. If the pressure recovery factor is known, then $p_{01}$ can be obtained.

We study in fluid mechanics that a subsonic flow through a nozzle accelerates and it is likely possible that the velocity at the nozzle throat is sonic. Once this condition is reached, then mass flow through the nozzle cannot be increased, and the flow condition is known as choked condition, and pressure is called the critical pressure $p_C$, which in general will be more than $p_a$. The expansion process in the propelling nozzle for unchoked and choked flows are shown in Figs 8.23 (b) and (c), respectively. In case of unchoked flow, pressure at Section 5 will always be $p_a$. That is, $p_5$ can be taken as $p_a$.

Let us now consider a choked flow situation. If a convergent nozzle is used, then pressure at its exit will be more than $p_a$ and we may use a convergent divergent nozzle. However, though a convergent divergent nozzle will produce more thrust, in reality, it is not so for pressure ratios $p_{04}/p_a$ of around 3. At off-design operation, the experimental results indicated that a convergent nozzle is even more efficient. However, pressure ratio may be as high as 10–20 at a flight Mach number 2–3, a convergent divergent nozzle is used in such a situation. A convergent nozzle is also preferred because of ease in using a variable area nozzle. This is required for (i) reducing the jet noise caused due to mixing of hot stream with outside cold air, (ii) to improve starting and part-load performance, and (iii) to reduce the length of runway during landing.

We would first like to calculate the critical pressure at the nozzle exit. There is no energy transfer between Sections 4 and 5, and therefore, $T_{04} = T_{05}$. The isentropic efficiency may be defined as usual by the equation,

$$\eta_i = (T_{04} - T_5)/(T_{04} - T_{5'})$$

$$\rightarrow T_{04} - T_5 = \eta_i \; T_{04} \, [1 - \{1/(p_{04}/p_5)\}^{(\gamma-1)/\gamma}] \qquad (8.22)$$

But $T_{05} = T_5 + ½ \, C_5^2/c_p$ and $M_5$ = Mach number at section $5 = C_5/\sqrt{(\gamma RT_5)}$. This gives,

$$T_{04}/T_5 = T_{05}/T_5 = 1 + C_5^2/(2c_pT_5) = 1 + \{(\gamma - 1)/2\}\, M_5^2$$

Mach number at Section 5 will be 1 under choked condition. This gives, replacing subscript 5 by $c$ to indicate critical section at 5, see Fig. 8.23 (c)

$$T_{04}/T_c = (\gamma + 1)/2 \qquad (8.23)$$

Equation 8.23 gives the static temperature at the critical section. Under a choked condition, the static pressure and temperature at Section 5 will be $p_c$ and $t_c$, respectively. However, we have to first ascertain the nature of the flow through the nozzle. Let us say the flow is choked, then pressure at the nozzle exit will be $p_C$, more than $p_a$ and hence the ratio $p_{04}/p_c < p_{04}/p_a$. That is, if $p_{04}/p_a$ is less than $p_{04}/p_C$ for given flow, then it is an unchoked flow. We, therefore, determine the value of $p_{04}/p_a$, and compare it with $p_{04}/p_C$. Then, we take $p_5$ as $p_a$ for an unchoked flow, and $p_C$ for choked flow.

Knowing the isentropic efficiency, we can now find out, using Eq. 8.22, the static temperature $T_5$ for isentropic expansion, and then using the isentropic process equation, we get static critical pressure.

$$p_c = p_{04}\left(\frac{T_{c'}}{T_{04}}\right)^{\gamma/(\gamma-1)}$$

$$= p_{04}\left[1 - \frac{1}{\eta_i}\left(1 - \frac{T_c}{T_{04}}\right)\right]^{\gamma/(\gamma-1)} \rightarrow \frac{p_{04}}{p_c} = \left[1 - \frac{1}{\eta_i}\left(\frac{\gamma-1}{\gamma+1}\right)\right]^{-\gamma/(\gamma-1)} \qquad (8.24)$$

after substituting for $T_{04}/T_c$ from Eq. 8.23. Velocity and density at the exit section can be obtained using the following equation.

$$\rho_5 = p_5/RT_5$$

$$C_5 = [2c_p\,(T_{04} - T_5)]^{½}$$

and $Aj$ is then obtained from the continuity equation, which gives $Aj = m/(\rho_5 C_5)$. As $p_j$ is known, the second term on the right hand side of Eq. 8.13 can now be determined, which would be either $(p_C - p_a)$ for choked flow or zero for unchoked flow.

## 8.5.2 Turbofan

Turbofan was introduced because the simple turbojet engines produced a lot of noise, thus making them unsuitable for commercial flights. The noise is produced because of high jet velocity, *Cj*. In order to reduce the value of *Cj*, the mass of the jet was increased by taking more air from the atmosphere, and compressing its small quantity to produce enough power for running the fan and the compressor. The rest of the air, compressed in the air inlet duct, is passed to the propelling nozzle section, bypassing the compressor route, and exhausted through another nozzle. That is, there are two nozzles, one each for hot and cold streams. If $m_h$ and $m_c$ are masses of hot and cold air respectively, then the bypass ratio $B$ is defined as

$$B = m_c / m_h \tag{8.25}$$

As mentioned before, turbofan is used as passenger carrier. Figure 8.24 shows the arrangement of the fan, compressor and turbine in a turbofan engine. The LP turbine is used to run the fan and the LP compressor, whereas the HP turbine for the HP compressor. Therefore, the work output of these turbines has to be equal to that required by the corresponding compressor. These conditions will give the temperature at the exit of HP and LP turbines. Once the temperature is known, then using the definition of isentropic efficiency, the temperature that would have reached in the isentropic process is obtained, from which the actual pressure at the exit of the turbine can be obtained. However, if polytropic efficiency is known, then the value of $n$, the index of compression or expansion, is obtained, and then there is no need of finding out the isentropic temperature. The procedure is illustrated in the solved examples.

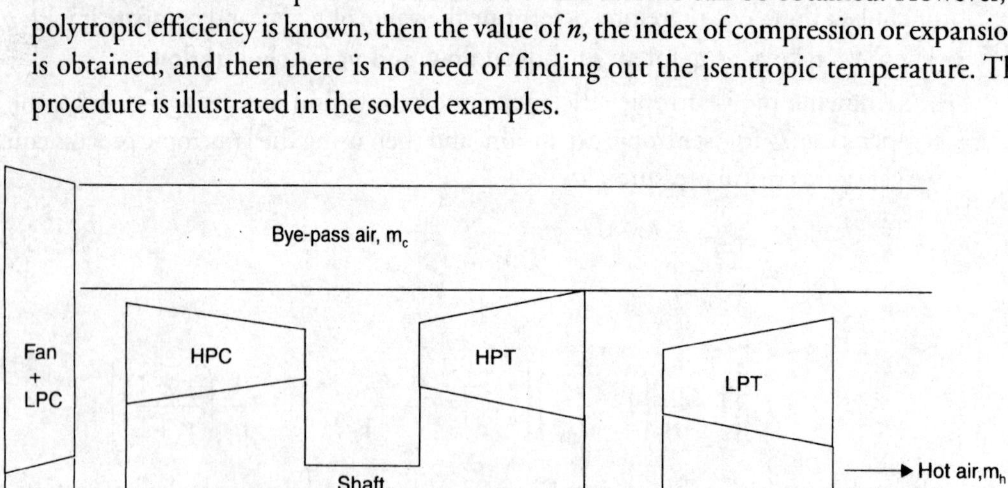


**Fig. 8.24** Twin-spool bypass engine

**Example 8.6** A simple turbojet is operating with a compressor pressure ratio of 8.0, a turbine inlet temperature of 1200 K and a mass flow of 15 kg/s, when the aircraft is flying at 260 m/s at an altitude of 7000 m. Assuming the following component efficiencies, and ISA conditions, calculate the propelling nozzle area required, the net thrust developed, and the *sfc*. Assume the fuel–air ratio for this condition as 0.01785.

Polytropic efficiencies of compressor and turbine 0.87

Isentropic efficiency of intake 0.95

Isentropic efficiency of propelling nozzle 0.95

Mechanical efficiency 0.99

Combustion chamber pressure loss 6% compressor delivery pressure

Combustion efficiency 0.97.

***Solution*** Given data may be written here; we are skipping it here. Take $\gamma$ = 1.4 and 1.333 for air and gases, respectively. For air $c_p$ = 1.005, and for gas 1.148 kJ/kg K. The subscripts a, 1,2,3 and 4 used in the solution refer to the usual convention, which we have adopted in our discussions.

From Eq. 8.11 (b) $(n-1)/n = (1/0.87)\{1.4/(1.4-1)\} = 0.3284$ for compression, and $(0.87)\{1333/(1.333-1)\} = 0.2175$, or

$$n/(n-1) = 4.5977 \text{ for expansion.}$$

From the ISA table, at a height of 7000 *m*, $p_a$ = 0.411 bar and $T_a$ = 242.7 K.

$$½\, Ca^2/c_p = 2602/(2 \times 1.005 \times 1000) = 33.63 \text{ K}$$

$$\rightarrow T_{01} = 242.7 + 33.63 = 276.33 \text{ K}$$

$$p_{01} = p_a\,[1 + \eta_i\ Ca^2/(2c_pT_a)]^{\gamma/(\gamma-1)}$$

$$= 0.441\,[1 + 0.95 \times 33.63/242.7]^{3.5} = 0.633 \text{ bar}$$

$$p_{02} = 8\,p_{01} = 5.068 \text{ bar}$$

$$\rightarrow T_{02} - T_{01} = T_{01}\,[(p_{02}/p_{01})^{(n-1)/n}$$

$$= 276.33\,[8.0^{0.3284} - 1] = 270.7 \text{ K}$$

Since power generated by the turbine is just sufficient to run the compressor,

$$T_{03} - T_{04} = c_{pa}\,(T_{02} - T_{01})/(c_{pg}\,\eta_m)$$

$$= 1.005 \times 270.7/(1.147 \times 0.99) = 239.6 \text{ K}$$

$$\rightarrow T_{04} = 960.4 \text{ K}$$

$$p_{03} = p_{02}\,(1 - 0.06) = 5.068\,(1 - 0.06) = 4.763 \text{ bar}$$

$\therefore\ p_{04} = p_{03}\,(T_{04}/T_{03})^{n/(n-1)} = 1.710$ bar

$\rightarrow$ Nozzle pressure ratio $= p_{04}/p_a = 4.160$

From the course on compressible fluid flow, $p_{04}/p_C = 1.918$ for $\gamma = 1.333$. Therefore, the nozzle is choked and hence

$p_5/p_C = 1.710/1.918 = 0.891$. Also

$$T_5 = T_c = \{2/(\gamma + 1)\}\, T_{04}$$
$$= 2 \times 960.4/2.333 = 823.3 \text{ K}$$

From equation of state,

$$\rho_5 = 100 \times 0.891/(0.287 \times 823.3) = 0.377 \text{ kg/m}^3$$

$$C_5 = (\gamma RTc)^{1/2} = (1.333 \times 0.287 \times 1000 \times 823.3)^{1/2} = 561.22$$

From the continuity equation, $A_5 = m/(\rho_5\, C_5)$

$$= 15/(0.377 \times 561.22) = 0.0708 \text{ m}^2$$

From Eq. 8.13,

$$F = 15\,(561.22 - 260) + 0.0708\,(0.891 - 0.411) \times 10^5$$
$$= 7916.7 \text{ N}$$

We may determine fuel air ratio, since $T_{02}$ and $(T_{03} - T_{02})$ are known. However, theoretical f is given.

$\therefore$ actual fuel air ratio $= 0.01785/0.97 = 0.0184$

$$sfc = 3600\, f\, m/F = 0.0184 \times 3600 \times 15/7916.7$$
$$= 0.01255 \text{ kg/h kN}$$

**Example 8.7** Under take-off conditions, when the ambient pressure and temperature are 1.01 bar and 288 K, the stagnation pressure and temperature in the jet pipe of a turbojet engine are 2.4 bar and 1000 K, and the mass flow is 23 kg/s. Assuming that the expansion in the converging propelling nozzle is isentropic, calculate the exit area required and the thrust produced.

For a new version of the engine, the thrust is to be increased by the addition of an aft fan, which provides a separate cold exhaust stream. The fan has a bye-pass ratio of 2.0, and a pressure ratio of 1.75, the isentropic efficiencies of the fan and fan-turbine sections being 0.88 and 0.90, respectively. Calculate the take-off thrust, assuming that the expansion in the cold nozzle is also isentropic, and that the hot nozzle area is adjusted so that the hot mass flow remains the same.

***Solution*** We take the usual notations. However, for the cold air stream bypassing the compressor route, we shall use subscripts 7 and 8 to represent the cold nozzle inlet and outlet sections, respectively. We are given $p_{04}$ = 2.4 and $p_a$ = 1.01 bar. This gives $p_{04}/p_a$ = 2.376. But,

$$p_{04}/p_C = \{(\gamma + 1)/2\}^{\gamma/(\gamma - 1)} = (2.333/2)^4 = 1.851$$

Therefore, the flow is choked.

So $p_5/p_C = 2.4/1.851 = 1.296$ bar

We may proceed as in Example 8.6 to get

$$T_5 = T_c = 857.26 \text{ K}, \quad C_5 = 572.68 \text{ m/s},$$

$$\rho_5 = 0.526 \text{ kg/m}^3 \text{ and } A_5 = 0.0763 \text{ m}^2$$

$$F = 23 \times 572.68 + 0.0763\,(1.296 - 1.01) \times 10^5$$

$$= 15353.8 \text{ N}$$

**New version**

Given, $p_{07}/p_a = 1.75 \;\rightarrow\; p_{07} = 1.768$ bar

$$T_{07} - T_a = (T_a/\eta_c)\,\{(p_{07}/p_a)^{(\gamma - 1)/\gamma} - 1\}$$

$$= (288/0.88)\,[(1.75)^{1/3.5} - 1] = 56.74 \text{ K and}$$

$$T_{07} = 344.74 \text{ K}$$

(Note that here, $\gamma$ has been taken as 1.4, since it is air and not the gas.) Since the LP turbine runs the fan, and the LP compressor, its output has to be equal to that required for running these. This condition gives

$$m_h C_{pg} (T_{04} - T_{05}) = m_c C_{pa} (T_{07} - Ta)$$

$$\rightarrow T_{04} - T_{05} = 99.43 \text{ K, since } m_h/m_c = 2.0 \text{ (given)}$$

or, $T_{05} = 900.57$ K. From Eq. 8.3.7, $p_{04}/p_{05} = 1.597$

$$\rightarrow p_{05} = 1.503 \text{ bar}$$

$P_{05}/p_a = 1.503/1.01 = 1.488$, and hence the nozzle is not choked. So, $p_5 = p_a = 1.01$ bar.

$$T_{05}/T_5 = (p_{05}/p_5)^{(\gamma - 1)/\gamma}$$

$$\rightarrow T_5 = 815.36 \text{ K and } T_{05} - T_5 = 85.2 \text{ K}$$

$$\therefore C_5 = [2(T_{05} - T_5)]^{1/2}$$

$$= (2 \times 1.147 \times 85.21 \times 1000)^{1/2}$$

$$= 442.12 \text{ m/s}$$

Hot nozzle thrust = 23 × 442.12 = 10168.8 N

Cold nozzle pressure ratio = $p_{07}/p_a = 1.75$

The critical pressure ratio is

$$p_{07}/p_C = [(\gamma + 1)/2\}\gamma/(\gamma - 1) = (2.4/2)^{3.5} = 1.893$$

This nozzle is also unchoked. Since the expansion is given to be isentropic,

$$T_{07}/T_8 = (p_{07}/p_a)^{(\gamma-1)/\gamma} = 1.1733$$

$$\rightarrow T_8 = 293.8 \text{ K, or } T_{07} - T_8 = 50.94 \text{ K}$$

$$\therefore\ C_8 = [2c_p(T_{07} - T_8]^{1/2} = 320 \text{ m/s}$$

Cold nozzle thrust = 46 × 320 = 14720 N

Total thrust = 14720 + 10168.8 = 24888 N

### 8.5.2.1 Performance of Turbojet Engine

The performance of a turbojet is measured in terms of specific fuel consumption and specific thrust, which depend on the variables such as gas temperature $T_{03}$ at the inlet of the turbine, compressor pressure ratio, altitude, and cruising speed. Figure 8.25 shows some performance curves for subsonic cruise condition. For all these curves, polytropic efficiency for the compressor and turbine is kept fixed. Figure a shows the effect of $T_{03}$ and compressor pressure ratio $R$ at a particular altitude and cruising speed, whereas b shows the effect of altitude at different flight Mach numbers.

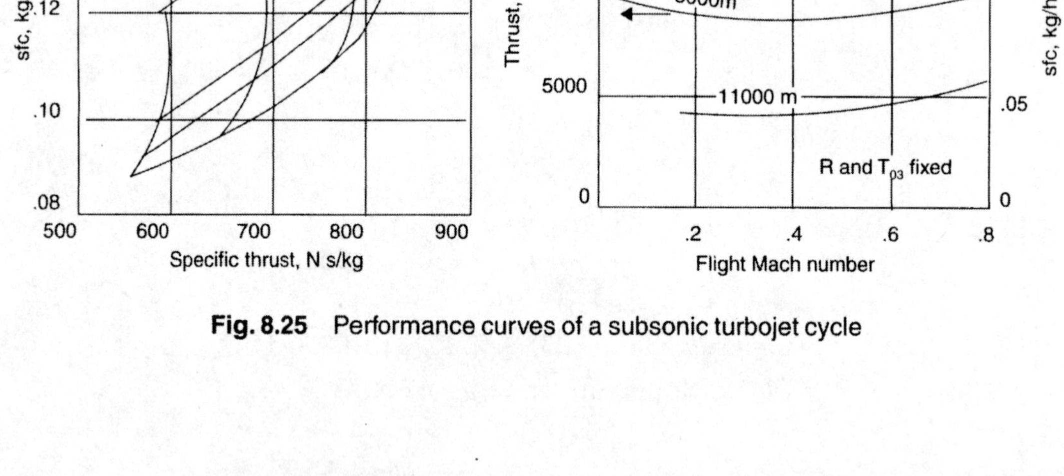


**Fig. 8.25** Performance curves of a subsonic turbojet cycle

We observe that

- *sfc* decreases with increase in compression ratio. The specific thrust increases up to some *R*, but after that it decreases. These observations hold true at all $T_{03}$. Thus, there is an optimum compressor pressure ratio, like shaft engines, which decreases with increasing $T_{03}$.
- At constant *R*, specific thrust increases fast with $T_{03}$ compared to increase in *sfc*. Though the latter is not desirable from the consideration of fuel economy, yet it is acceptable, since specific thrust increases more. This decreases the engine size and hence the cost, weight and drag force. However, costly alloy will be needed for the turbine blade to withstand high temperature flue gases.
- At a fixed compressor pressure ratio and $T_{03}$, *sfc* increases and thrust decreases with flight Mach number at any altitude. This is explained because increasing craft speed increases the drag and the temperature at the compressor inlet, that increases the work requirement of the compressor.
- Specific thrust increases and *sfc* decreases with increase in altitude at all flight Mach numbers. This is because the ambient temperature decreases with altitude that reduces temperature at the compressor inlet, and hence work required for compression.
- With increasing cruising speed, the thrust initially decreases due to increasing drag force, and then increases due to the beneficial effects of the ram pressure ratio.

Besides the above factors, other factors like engine life, initial cost and fuel consumption also influence the choice of operating variable. Engines operating at high turbine inlet temperature have low life and need special alloys. Moreover, for such engines the pressure ratio is likely to be high, and therefore, the number of stages required for the compressor and turbine will be large. This adds to the first cost, weight and complexity required for making the arrangement for cooling of turbine blades, and even multi-spooling may be preferred. In the aviation industry and military applications, normally, the thrust to weight ratio is the main consideration, and not the life of the engine. That is, the guiding factors to decide the operating parameters that influence the *sfc* and specific thrust are: (i) pressure ratio, (ii) turbine inlet temperature, (iii) life, and (iv) weight and cost. Figure 8.26 gives some comparative idea and the application of turbojet.

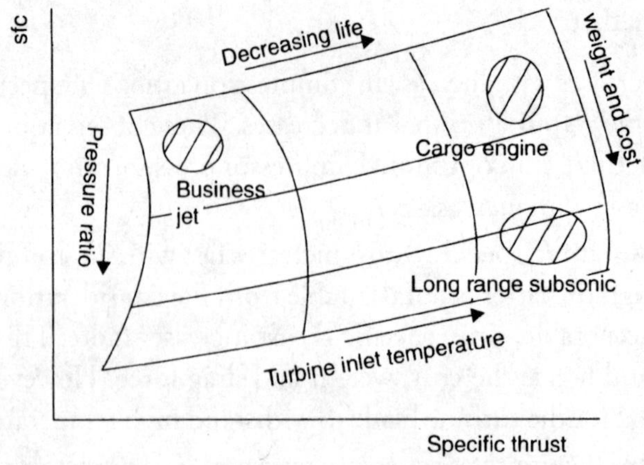


**Fig. 8.26** Turbojet application

## SUMMARY

This chapter discusses the length and breadth of various aspects of the gas and vapour cycles used in reciprocating and rotating machines. First law analysis of Otto cycle, diesel cycle and dual cycle has been done to find their thermal efficiencies. These cycles are used in a reciprocating system. Importance of stirling and Ericsson cycles is also highlighted, besides the determination of their thermal efficiencies. At these gas power cycles are reversible cycles. However, in the Otto, diesel and dual cycles, averages heat is neither added at the maximum temperature of the cycle nor is it rejected at the minimum temperature. Rather, heat is added at some average temperature lower than the maximum temperature and rejected at an average temperature higher than the minimum cycle temperature. It results in an efficiency lower that that of Carnot cycle even though all these cycles are reversible; the difference is the result of different temperatures of the source and the sink. This difficulty does not exist in Stirling and Ericsson cycles, which are the cycles of the future.

The thermodynamic cycle used in gas turbines is the Brayton cycle. This chapter dealt with determination of its efficiency, determination of optimal pressure ratio and different methods used for improving the specific work output or improvement in efficiency. This cycle is adopted in gas turbine power plants in cascading with steam cycle. Such a cycle is called a combined cycle. Complete first and second law analysis of Anta power plant is given such that it explains not only the analysis procedure but also points out the processes which need improvement for better utilisation of available energy.

Gas turbines have found important applications in aviation industry where weight to power ratio is kept to the minimum. Therefore, specific power output and not the efficiency of cycle is the important criterion. In modern air crafts, the gas turbine is used as gas producer so that gases may be expanded through a propelling nozzle to produce the forward thrust. This chapter discusses in depth the analysis procedure and various efficiencies required in the procedure.

# EXERCISES

## Review Questions

8.1 How does a gas power cycle differ from the vapour power cycle?

8.2 The gas and vapour power cycles are reversible cycles. Will they have the same efficiency as that of the Carnot cycle? Explain.

8.3 Do internal combustion engines operate on open or closed cycle?

8.4 Define compression ratio and mean effective pressure. Will these two change as the engine becomes old?

8.5 How do the efficiencies of an ideal Otto cycle and the Carnot cycle compare for the same temperature limits?

8.6 An ideal Otto cycle with a given compression ratio is executed with (a) air, (b) argon, (c) ethane as the working fluid. For which case will the thermal efficiency be the highest and why?

8.7 List the four events in the internal combustion engine. Is it essential that an IC engine has to complete four strokes of its piston to complete these events?

8.8 What is the significance of the term air standard cycle? Sketch the Otto, Diesel and the Dual cycle on a *T-s* diagram.

8.9 On what single factor does the efficiency of the Otto cycle depend? Is there any upper limit on it? Give comments.

8.10 How does the modern IC engine achieve higher power outputs without the use of higher compression ratio?

8.11 Define the cut-off ratio in the Diesel cycle. Write down the expression for its efficiency when the cut-off ratio is expressed as percentage of the stroke volume.

8.12 How does the Diesel engine avoid the problem of preignition or detonation?

8.13 Why is combustion in petrol and diesel engines considered at constant volume or pressure, respectively? What are the factors that decide this?

8.14 Find out the efficiency of the Stirling and Ericsson cycles, assuming a perfect regenerator.

8.15 The Stirling and Ericsson cycles use the common concept in their operation. Describe this concept and its principle.

8.16 Find out the efficiency of the Brayton cycle, assuming isentropic efficiencies of compression and expansion as $\eta_c$ and $\eta_t$, respectively.

8.17 Someone claims that at very high pressure ratios, the use of regeneration actually decreases the thermal efficiency of a gas turbine engine. Is there any truth in this claim?

8.18 In large compressors, the gas is frequently cooled while being compressed to reduce power consumption. Explain how this reduces power consumption. Will this also hold true for a pump?

8.19 On a *T-s* diagram, does the actual exit state of an adiabatic turbine have to be on the right hand side of the isentropic exit state?

8.20 What is the disadvantage of the cycle shown in Fig. 8.15.

8.21 Show that the efficiency of the conventional Brayton cycle with the heat exchanger connected to the exhaust from the LPT and not from the HPT is $1 - t/c$, where $t$ and $c$ have usual meaning.

8.22 Prove Equation 8.7.

## Problems for Practice

8.1 An engine operating on an air standard Otto cycle has a compression ratio of 8. The conditions at the start of the compression stroke are 1 bar, 300 K. Determine the thermal efficiency of the cycle and the net work done per kg air, if $C_V$ for air is 0.71 kJ/kg K. Also, determine the mean effective pressure for the cycle. [*Ans.* 0.565, 182.57 kJ, 2.35 bar].

8.2 In an air standard Otto cycle, the ambient conditions determine the minimum temperature, while the design conditions and metallurgical limits determine the maximum temperature. For fixed values of minimum and maximum temperatures $T_1$ and $T_3$, respectively, find the optimum compression ratio for maximum work output.

8.3 In an air standard Diesel cycle, the conditions at the beginning of the compression stroke are 300 K and 1 bar. The air is compressed to a pressure of 5 Mpa, and then the fuel is injected such that 20 kJ of energy is added per mole of air. Determine the compression ratio, the cut-off ratio and the thermal efficiency of the cycle. Also, determine its mean effective pressure. Take molar specific heat $Cp = 3.5$ R. [*Ans.* 17.35, 1.75, 0.3].

8.4 An air standard dual combustion cycle has a mean effective pressure of 10 bar. The minimum pressure and temperature are 1 bar and 17°C, respectively, and the compression ratio is 16:1. Calculate the maximum cycle temperature when the thermal efficiency is 60%. The maximum cycle pressure is 60 bar. [*Ans.* 1959°C].

8.5 In an IC engine operating on dual cycle, the temperature of air at the beginning of compression is 27°C. The ratio of the maximum and minimum pressures of the cycle is 70, and the compression ratio is 17. The amounts of heat added at constant volume and at constant pressure are equal. Compute the air standard thermal efficiency of the cycle. State three main reasons why the actual thermal efficiency is different from the theoretical value. [*Ans.* 67.4%] (IES1993).

8.6 An air standard limited pressure cycle has a compression ratio of 15, and compression begins at 0.1 MPa, 40°C. The maximum pressure is limited to 6 Mpa, and the heat added is 1.675 MJ/kg. Compute (a) the heat supplied at constant volume per kg of air, (b) the heat supplied at constant pressure per kg of air, (c) the work done per kg of air, (d) the cycle efficiency, (e) the temperature at the end of the constant volume heating process, (f) the cut-off ratio, and (g) the m.e.p. of the cycle.
[*Ans.* 235 kJ/kg, 1440 kJ/kg, 1014 kJ/kg, 60.5%, 1252 K, 2.144, 1.21 MPa].

8.7 A gas turbine unit receives air at 1 bar and 300 K, and compresses it adiabatically to 7.2 bar. The compressor efficiency is 0.8. The fuel has heating value of 44186 kJ/kg, and the fuel air ratio is 0.017 kgf/kga. The turbine internal efficiency is 0.9. Calculate the turbine and compressor work per kg of air compressed, and the thermal efficiency. For products of combustion, $c_p$ = 1.147 kJ/kg K and $\gamma$ = 1.333. [*Ans.* 471.25 kJ, 234.72 kJ, 31.49%] (IES 1992).

8.8 Determine efficiency and the optimum pressure ratio for maximum work output for the Brayton cycle, assuming that the compressor and the turbine have some isentropic efficiency.

8.9 In a Brayton gas turbine power plant, the minimum and maximum temperature of the cycle are 300 and 1200 K. The compression is carried out in two stages of equal pressure ratio, with intercooling of the working fluid to the minimum temperature of the cycle after the first stage of compression. The entire expansion is carried out in one stage only. The isentropic efficiency of both compressors is 0.85, and that of the turbine is 0.9. Determine the overall pressure ratio that would give the maximum net work per kg of fluid. Derive the expression you use. Take $\gamma$ = 1.4. [*Ans.* 13.6] (IES 95).

8.10 A gas turbine plant draws in air at 1.013 bar, 10°C and has a pressure ratio of 7.7. The maximum temperature in the cycle is limited to 750°C. Compression is conducted in an uncooled rotary compressor having an isentropic efficiency of 82%, and expansion takes place in the turbine with an isentropic efficiency of

85%. A heat exchanger with an efficiency of 70% is fitted between the compressor outlet and combustion chamber. For airflow of 40 kg/s, find (a) the overall cycle efficiency, (b) the turbine output, and (c) the air fuel ratio if the C.V. of the fuel is 47.22 MJ/kg. [*Ans.* 30.4%, 4272 kW, 115].

8.11 A combined power plant consisting of a closed GT cycle unit using air as the working fluid and a ST unit is to be designed such that the heat rejected at the GT unit is utilised to produce steam at the generator for the ST. The air leaving the generator is at 200°C, and it is cooled to the compressor inlet by a second cooler, which rejects the heat to waste.

Draw the flow and *T–s* diagram of the plant, and on the basis of ideal cycle, neglecting the feed pump work of the Rankine cycle, calculate: (a) the mass flow rate of steam per kg/s of air flow, (b) the total power output per kg/s of air flow, and (c) the overall efficiency of the plant. Take $c_p$ for air as 1.005 kJ/kg K and $\gamma$ = 1.4. The particulars of each cycle are as follows.

| | GT cycle | | ST cycle | |
|---|---|---|---|---|
| | Minimum | Maximum | Minimum | Maximum |
| Pressure, bar | 1 | 5 | 0.07 | 30 |
| Temperature, °C | 20 | 830 | – | 300 |

[*Ans.* 0.0762, 314.78, 49%].

8.12 A 50 MW steam plant built in 1935, operates steam at the inlet at 60 bar, 450°C and exhausts at 0.1 bar, with 80% turbine efficiency. It is proposed to scrap the old boiler and put in a new boiler, and a topping turbine of efficiency 85% operating with inlet steam at 180 bar, 500°C. The exhaust from the topping turbine at 60 bar is reheated to 450°C, and admitted to the old turbine. The flow rate is just sufficient to produce the rated output from the old turbine. Find the improvement in efficiency with the new set up. What is the additional power developed?

8.13 In a nuclear power plant, saturated steam at 30 bar enters a HP turbine, and expands isentropically to a pressure at which its quality is 0.841. At this pressure, the steam is passed through a moisture separator, which removes all the liquid. Saturated vapour leaves the separator and is expanded isentropically to 0.04 bar a in LP turbine, while the saturated liquid leaving the separator is returned via a feed pump to the boiler. The condensate leaving the condenser at 0.04 bar is also returned to the boiler via a second feed pump. Calculate the cycle efficiency and turbine outlet quality, taking into account the feed pump term. Recalculate the same quantity for a cycle with the same boiler and condenser pressure, but without moisture separation. Give your comments on the results. [*Ans.* 34.5%, 0.824, 35%, 0.716].

8.14 Geothermal energy from a natural geyser can be obtained as a continuous supply of steam 0.87 dry at 2 bar and at a flow rate of 2700 kg/h. This is utilised in a mixed pressure cycle to augment the superheated exhaust from a high pressure turbine of 83% internal efficiency, which is supplied with 5500 kg/h of steam at 40 bar and 500°C. The mixing process is adiabatic and the mixture is expanded to a condenser pressure of 0.10 bar in a LP turbine of 78% internal efficiency. Determine the power output and the thermal efficiency of the plant. [*Ans.* 1745 kW, 35%].

8.15 In a cogeneration plant, steam enters the HP stage of a two-stage turbine at 10 bar, 200°C, and leaves it at 3 bar. At this point, some of the steam is bled off and passed through a heat exchanger, which it leaves as saturated liquid at 3 bar. The remaining steam expands in the LP. stage of the turbine to 0.4 bar. The turbine is required to produce a total power of 1 MW, and the heat exchanger to provide a heating rate of 500 kW. Calculate the required mass flow rate of steam into the HP stage of the turbine. Assume (a) steady conditions throughout the plant, (b) the velocity and gravity terms to be negligible, (c) both turbine stages are adiabatic with isentropic efficiency of 0.80. [*Ans.* 2457 kg/s].

## Multiple-choice Problems

8.1 The air standard Otto cycle comprises

(a) two constant volume processes and two constant entropy processes.
(b) two constant pressure processes and two constant entropy processes.
(c) two constant pressure processes and two constant volume processes.
(d) none of the above.

8.2 The thermal efficiency of theoretical Otto cycle

(a) increases with increase in compression ratio.
(b) increases with increase in isentropic index $\gamma$.
(c) is independent of the pressure ratio.
(d) follows all the above.

8.3 The work output of the theoretical Otto cycle

(a) increases with increase in compression ratio.
(b) increases with increase in pressure ratio.

(c) increases with increase in adiabatic index $y$.
(d) all of the above.

8.4 For a fixed compression ratio, the thermal efficiency of the Otto cycle, when operating on standard air
(a) is more than that when operating on $CO_2$.
(b) is less than the when operating on $CO_2$.
(c) is same as that when operating on $CO_2$.
(d) none of the above.

8.5 For the same compression ratio
(a) thermal efficiency of the Otto cycle is greater than that of the diesel cycle.
(b) thermal efficiency of the Otto cycle is less than that of the diesel cycle.
(c) thermal efficiency of the Otto cycle is the same as that for the diesel cycle.
(d) thermal efficiency of the Otto cycle may be more or less and cannot be predicted.

8.6 In air standard diesel cycle, at fixed compression ratio and fixed value of adiabatic index ($y$)
(a) thermal efficiency increases with increase in heat addition cut-off ratio.
(b) thermal efficiency decreases with increase in heat addition cut-off ratio.
(c) thermal efficiency remains the same with increase in heat addition cut-off ratio.
(d) none of the above.

8.7 In a dual cycle of limited pressure, the heat added at constant volume is $x$ and at constant pressure is $y$. Also $x + y$ = constant
(a) if $x$ is increased, the thermal efficiency is increased.
(b) if $y$ is increased, the thermal efficiency is increased.
(c) if $x = y$, the thermal efficiency is the maximum.
(d) none of the above.

8.8 For dual combustion, air standard cycle, for fixed value of heat supplied and fixed value of compression ratio,
(a) mean effective pressure will be greater with increase in $rp$ and decrease in $rc$.
(b) mean effective pressure will be greater with decrease in $rp$ and increase in $rc$.
(c) mean effective pressure will remain the same with increase or decrease *of* $rp$ and $re$.
(d) none of the above.

8.9 For the same temperature of source and sink
  (a) the thermal efficiency of the Carnot cycle is greater than that of the Stirling cycle.
  (b) the thermal efficiency of the Carnot cycle is less than that of the Stirling cycle.
  (c) the thermal efficiency of the Carnot cycle is equal to that of the Stirling cycle.
  (d) the thermal efficiency of the Carnot cycle may be greater or less than that of the Stirling cycle, depending upon the working substance.

8.10 Basic closed cycle for gas turbine is
  (a) Carnot cycle. (b) Rankine cycle.
  (c) Brayton cycle. (d) Stirling cycle.

8.11 The air standard efficiency of a closed gas turbine cycle is given by ($r$ = pressure ratio for the compressor and turbine)
  (a) $\eta = 1 - (1/r)^{\gamma - 1}$. (b) $\eta = 1 - 1/r^{\gamma - 1}$.
  (c) $\eta = 1 - (1/r)^{(\gamma - 1)/\gamma}$. (d) $\eta = r^{(\gamma - 1)/\gamma} - 1$.

8.12 For a closed cycle gas turbine plant, if $T_1$ = minimum temperature of cycle, $T_3$ = maximum temperature of cycle, then $r = (T_3/T_1)^{\gamma/(\gamma - 1)}$ gives
  (a) the maximum work output pressure ratio.
  (b) the maximum thermal efficiency pressure ratio.
  (c) the pressure ratio for no work output.
  (d) the pressure ratio for best work ratio.

8.13 For close cycle gas turbine plant, if $T_1$ = minimum temperature of cycle, $T_3$ = maximum temperature of cycle, the pressure ratio that will give maximum work output is given by
  (a) $r = (T_3/T_1)^{\gamma/2(\gamma - 1)}$. (b) $r = (T_3/T_1)^{\gamma/(\gamma - 1)}$.
  (c) $r = (T_3/T_1)^{2\gamma/2(\gamma - 1)}$. (d) $r = (T_3/T_1)^{\gamma}$.

8.14 The thermal efficiency of a gas turbine plant, as compared to a diesel engine plant, is
  (a) lower. (b) higher. (c) same. (d) maybe lower or higher.

8.15 The mechanical efficiency of a gas turbine, as compared to an internal combustion reciprocating engine is
  (a) lower. (b) higher. (c) same. (d) unpredictable.

8.16 The pressure ratio for a gas turbine may be in the range *I*
  (a) 2 to 3. (b) 3 to 5. (c) 16 to 18. (d) 18 to 22.

8.17 The work ratio of a gas turbine plant, that is, the ratio of net work output to turbine work is equal to ($T_1$ = minimum temperature, $T_3$ = maximum temperature)

(a) $1 - (r)^{(\gamma - 1)/\gamma}$. (b) $1 - (T_1/T_3)(r)^{(\gamma - 1)/\gamma}$.

(c) $1 - (T_3/T_1)(r)^{(\gamma - 1)/\gamma}$. (d) $1 - (T_1/T_3)(r)^{(\gamma - 1)/2\gamma}$.

8.18 The work ratio of a closed cycle gas turbine plant depends upon

(a) pressure ratio of the cycle *r* and specific heat ratio *r*.

(b) temperature ratio of the cycle $T_3/T_1$ and specific heat ratio *r*.

(c) pressure ratio, temperature ratio and the specific heat ratio.

(d) only on pressure ratio.

8.19 The work ratio of a closed cycle gas turbine plant with fixed top temperature of the cycle and fixed pressure ratio *r*

(a) increases with increase in the air inlet temperature to the compressor.

(b) increases with decrease in the air inlet temperature to the compressor.

(c) remains the same, irrespective of the inlet temperature to the compressor.

(d) none of the above.

8.20 The enthalpy–entropy characteristics at different pressures for a gas used in a gas turbine cycle are parallel. The power output from each cycle will be

(a) maximum.

(b) zero.

(c) dependent on the temperature range.

(d) dependent on the pressure range.

8.21 If the enthalpy–entropy lines at different pressures are more divergent for the working substance used in a gas turbine cycle, the power output for the same pressure ratio and same temperature ratio will be

(a) more. (b) less.

(c) unaffected. (d) unpredictable.

8.22 For a jet propulsion unit, ideally, the compressor work and turbine work are

(a) unequal. (b) equal.

(c) not related to each other. (d) unpredictable.

23. Greater the difference between jet velocity and aeroplane velocity

(a) greater the propulsive efficiency. (b) less the propulsive efficiency.

(c) the propulsive efficiency is unaffected. (d) none of the above.

# CHAPTER NINE

# Microscopic Approach



## 9.1 INTRODUCTION

We have so far discussed the macroscopic approach. As mentioned in Chapter 1, the other approach that has been rigorously followed by researchers is the microscopic approach earlier, wherein the matter is thought of being made from molecules. The theory developed on this basis by Bernoulli is known as the kinetic theory of gas. Later Clausius, Maxwell, and Boltzman made important contributions in its development. These scientists divided the study into two groups, namely, the kinetic theory and statistical thermodynamics. This chapter deals with these two approaches. Both of these theories were developed on the assumption that the system consists of a very large number of molecules at standard temperature and pressure, and that the molecules obey the laws of particle mechanics.

Proof of the existence of molecules and their movement is given by the phenomenon of diffusion, evaporation and expansion of gases. The kinetic theory led to the deduction of equation of state, and the prediction of many macroscopically observed quantities. It provides a theoretical basis to explain the transport phenomenon, such as molecular diffusion, that caused viscosity and laminar heat transfer. In the following sections, a theory is developed for gases, and is equally applicable for other phases. In gases, the distance is so large that these forces may be considered negligible. Figure 9.1 shows the variation of force with the intermolecular distance. We observe that if the distance between the molecules is more than $4d$, $d$ being the molecule diameter, then the force between them is negligible. Thus, the potential energy of the molecules in the absence of the gravitional field is zero. (On the other hand, if the distance is too less, then instead of attraction, there is repulsive force between the molecules.) At normal pressure and temperature the distance in gases is around 10 to 15 $d$. Hence the theory, based on the assumption of no force between molecules, holds true accurately for gases. Section 9.2 details the analysis procedure to determine pressure, temperature, distribution of velocity and speed, mean free path, and transport properties.

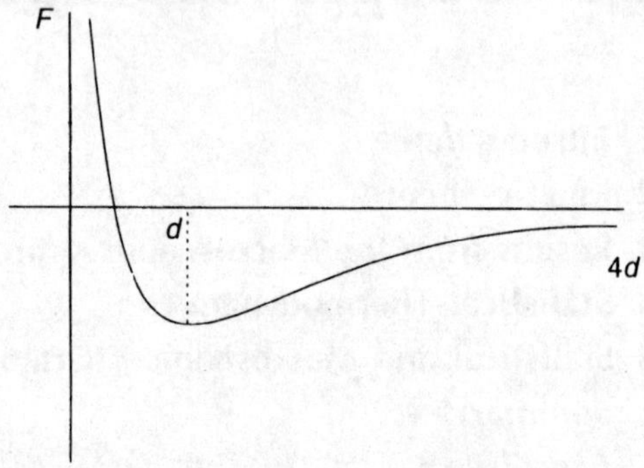


**Fig. 9.1** Variation of force with molecular distance

On the other hand, statistical thermodynamics developed on the line of quantum physics, determines the thermodynamic probability of the most probable macrostate, and relates all the macroscopic properties to this. This is discussed in Sections 9.4 and 9.5.

## 9.2 KINETIC THEORY

The kinetic theory is based on the following assumptions, which are known as the postulates of the kinetic theory of gases.

1. Gases consist of identical and rigid spherical spheres with negligible size compared to the mean free path (we shall study it later; mean free path is around 10 to 15 times the molecular diameter, whereas the diameter of a molecule is 2 to $3 \times 10^{-10}$ m). In the absence of external fields, these molecules are uniformly distributed in the space occupied by the gas. If $N$ is the number of molecules in a volume $V$, then the average number of molecules per unit volume $n$ is
$$n = N/V \text{ or } dn = n\, dV \tag{9.1}$$
$N$ is a very large number and its value in 1 kmole of gas at standard temperature and pressure (STP) is the Avogadro's number, $N_0$, which is equal to $6.023 \times 10^{26}$; the standard pressure is 760 mm Hg and temperature 0°C. 1 kmole of gas

under this condition occupies 22.4 $m^3$ of space. This gives the number of molecules in 1 $mm^3$ volume as $3 \times 10^{16}$.

We shall use the spherical coordinate system, shown in Fig. 9.2, in which $\theta$ is measured from the normal to the chosen surface, $\phi$ from some reference line drawn in the plane of the surface and radial distance measured from the origin of the reference system. These three chosen directions are perpendicular to each other, and molecule velocity $c$ is taken along the $r$ direction. We may also choose Cartesian coordinates with the same origin, and two of its axes coinciding with reference lines. Further notation $d^3N_{\theta\phi c}$ is used in our discussions to represent the number of molecules in $\theta\phi$ direction with velocity between $c$ and $c + dc$, $D^2N_{\theta\phi}$ for $\theta\phi$ molecules with any velocity $c$ from 0 to $\infty$ and $dN$ for molecules per unit spherical surface area having velocity $c$.

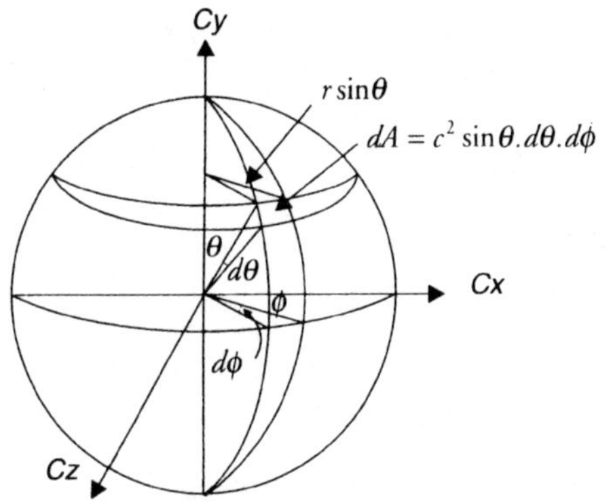


**Fig. 9.2** Spherical coordinates

2. Molecules move randomly in all directions along a straight line, with velocities ranging from 0 to $\infty$. Here, randomly means that all directions of molecular velocities are equally probable. If molecules having velocity between '$c$' and $c + dc$ are plotted around the origin (where $c = 0$), then each molecule will be represented by a unique point (negligible size assumption). These points will be distributed uniformly over the surface because of assumption 2. If $N_c$ is the total population of molecules with velocity $c$, then the number of molecules $dN$ per unit surface area of the sphere of radius $c$ is

$$dN = N_c/(4\pi c^2) \tag{9.2}$$

$d^2N_{\theta\phi}$, the number of molecules along $\theta$ and $\phi$ direction, will be that contained in an area element = $(c \sin\theta \cdot d\phi) \cdot (c \cdot d\theta)$ in the spherical surface of radius $c$. Therefore,

$$d^2N_{\theta\phi} = N_c/(4\pi\, c^2)\, c^2 \sin\theta \cdot d\theta \cdot d\phi$$

$$= (N_c/4\pi)\sin\theta \cdot d\theta \cdot d\phi$$

Dividing both sides by the gas volume $V$ and using the notation $n_c = N_c/V$,

$$d^2 n_c = n_c/(4\pi)\sin\theta \cdot d\theta \cdot d\phi \tag{9.3}$$

Similarly, we may write for $d^3N_{\theta\phi}$, which is given by

$$d^3N_{\theta\phi c} = (1/4\pi)\, dn_c \sin\theta \cdot d\theta \cdot d\phi\, dV \tag{9.4}$$

where, $dV$ is the volume of the $\theta\phi c$ cell, and $dn_c$ is the number of molecules per unit volume with speeds between $c$ and $c + dc$.

3. The molecules while moving, collide with each other, so that there is no loss of kinetic energy during collision. That is, the collision is perfect elastic. The particles change their velocity at each collision. Figure 9.3 shows collision between two elastic bodies of mass, say $m_1$ and $m_2$; $C$ is used for velocity, and subscripts $i$ and $f$ refer to before and after collision. Momentum and energy conservation principles give the final velocities of the two bodies

$$m_1 C_{1i} + m_2 C_{2i} = m_1 C_{1f} + m_2 C_{2f} \quad \text{(momentum)}$$

$$m_1 C_{1i}^2 + m_2 C_{2i}^2 = m_1 C_{1f}^2 + m_2 C_{2f}^2 \quad \text{(energy; 1/2 canals out)}$$

$$C_{1f} = \frac{2m_2 C_{2i} + (m_1 - m_2)C_{1i}}{m_1 + m_2} \quad \text{and} \quad C_{2f} = \frac{2m_1 C_{1i} - (m_1 - m_2)C_{2i}}{m_1 + m_2}$$

This gives for the case of equal masses

$$C_{1f} = C_{2i} \text{ and } C_{2f} = C_{1i} \tag{9.5}$$

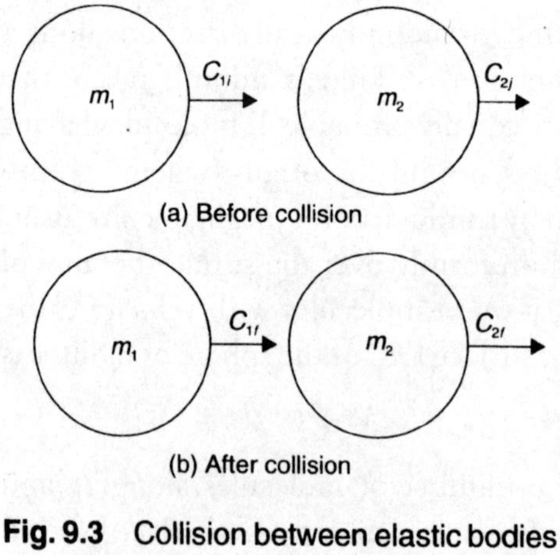


**Fig. 9.3** Collision between elastic bodies

This shows that the speed is interchanged during the collision between two bodies of equal mass. This means if the two molecules collide with each other when they are moving along a perpendicular direction, then one of them will have zero velocity and another the vector sum of their initial velocity. This means that the molecules, though initially may have the same velocity, will change their velocity due to collisions, and after some time, their velocities will be different. This supports postulate 2.

4. There is no force of attraction between the molecules, since the distance between them is too large.

### 9.2.1 Pressure Calculation

Molecules strike the surface of the container holding the gas, and transfer momentum to it, which exerts an impulse and hence pressure on it. We will calculate the magnitude of this pressure for which we first consider the molecules $d^3N_{\theta\phi c}$ striking the surface *dA* of the container wall along $\theta\phi$ direction, with velocities between *c* and *c* + *dc* in time *dt*. This will be the number contained in the cylinder of height *c dt* slanted along the $\theta\phi$ direction over the surface, as shown in Fig. 9.4(a). The volume of this cylinder is, $dV = (dA\cos\theta)\,cdt$. Therefore, the number of $\theta\phi c$ molecules in this cylinder by Eq. 9.4 is

$$d^3N_{\theta\phi c} = d^3N_{\theta\phi c}\,dV = \{(1/4\pi)\,dn_c \sin\theta \cdot d\theta \cdot d\phi\}\{dA\ \cos\theta)cdt\} \qquad (9.6)$$

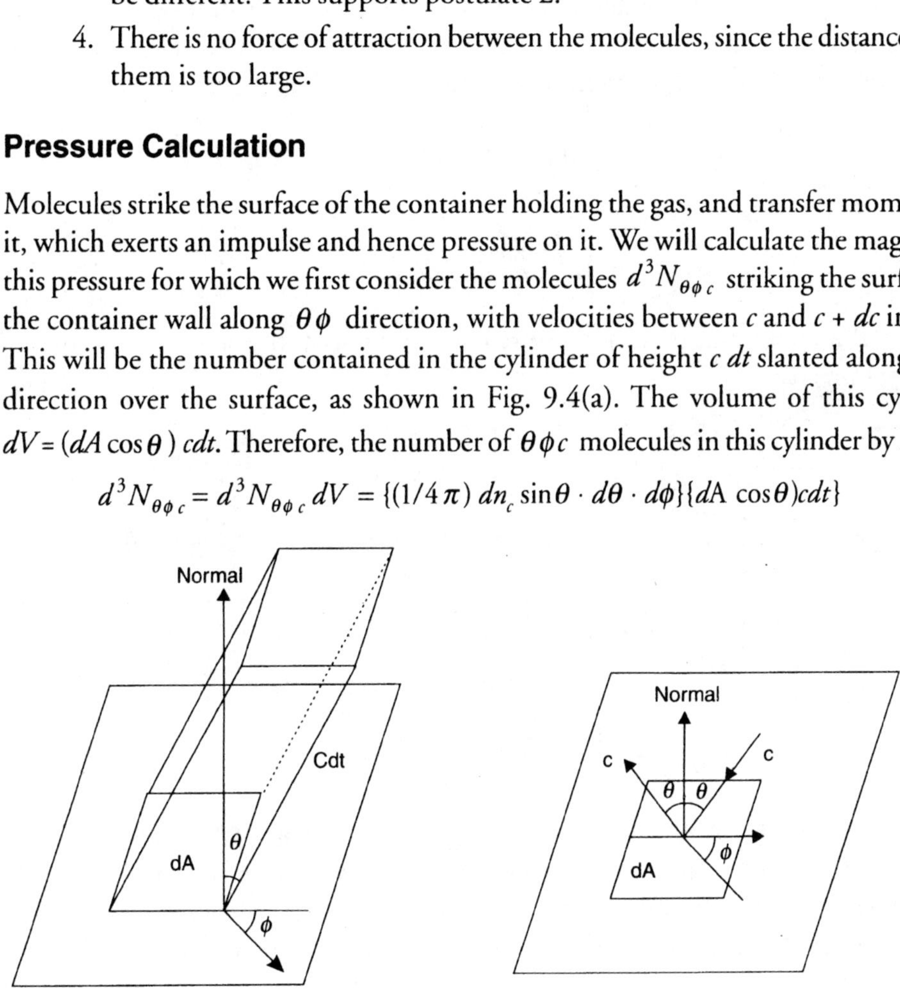


(a) Collisions with wall (b) Change in velocity in an elastic collision

**Fig. 9.4** Collision of $\theta\phi c$ collision of molecules with a wall

Since the molecules are hard and rigid particles, they follow the law of reflection when they strike any surface. Therefore, a molecule striking the surface at an angle $\theta$ with the normal to it, with speed *c* returns with speed *c* at an angle $\theta$ with the normal, as shown in Fig. 9.4(b). If $m$ is the mass of a molecule, then change in normal momentum is 2 mc cos $\theta$. (We are considering the normal component because this is responsible for the pressure exerted by the molecules on the container wall.) Consequently, the total change in normal momentum due to $\theta\phi c$ collisions in time $dt$ is

$$\{2mc\cos\theta\}\{(1/4\pi)\, c\; dn_c \sin\theta \cdot \cos\theta \cdot d\theta \cdot d\phi\, dA\; dt\} \tag{9.7}$$

But the molecules strike the surface $dA$ with any possible velocity from any direction, and we have to consider all such collisions made in time $dt$, during which a force $dF$ acts on the surface that develops impulse $dF.dt$ on it. This is obtained by integrating Eq. 9.7 for $\theta$ from 0 to $\pi/2$, $\phi$ from 0 to $2\pi$ and $c$ from 0 to $\infty$. This gives

$$dF \cdot dt = (m/2\pi)\, dA \cdot dt \int_0^{\infty} c^2\, dn_c \int_0^{\pi/2} \sin\theta \cos^2\theta\, d\theta \int_0^{2\pi} d\phi$$

$$= 1/3m\left[\int_0^{\infty} c^2\, dn_c\right] dA \cdot dt \tag{9.8}$$

But the average pressure $p$ exerted on the surface is the rate of change of the impulse per unit area.

$$\therefore\; p = dF/dA = 1/3m\left[\int_0^{\infty} c^2\, dn_c\right] = 1/3mn\, c_{rms}^2 \tag{9.9}$$

where, $c_{rms}^2$ is the root mean square velocity of the molecules defined as:

$$c_{rms}^2 = \frac{\int_0^{\infty} c^2\, dn_c}{\int_0^{\infty} dn_c} = \frac{\int_0^{\infty} c^2\, dn_c}{n} \tag{9.10}$$

Root mean square velocity thus is the average of the square of velocity of all molecules. Let the container contains 1 mole of the gas, then $mn = (M/No)\,(No/V) = M/V = \rho$. Here, $M$ is the molecular weight, $V$ is the volume occupied, $n$ is the number of molecules per unit volume, No is Avogadro's number and $\rho$ is the density of the gas. From Eq. 9.9

$$p = 1/3\,\rho\, c_{rms}^2 \text{ and } pV = (2/3)\left\{1/2m\,(N/V)\, V\, c_{rms}^2\right\} = 2/3U \tag{9.11}$$

where, $U$ is the total energy of gas = ½ $mNc_{rms}^2$, and $N$ is the total number of molecules in the gas. Note $n = N/V$ that gives the above result after substitution. Recall that as per the kinetic theory postulates, the molecules have no potential energy, and the only energy possessed by them is the kinetic energy. That is why the kinetic energy is taken as internal energy.

The perfect gas equation for $N$ moles of gas, where $N$ = mass of gas/$M$ = $mn/M$, is

$pV = NRT = (mn/M)\ RT$

Comparing it with Eq. 9.11, we get

$$T = 1/3\ (M/R)\ c_{rms}^2 = 1/3\ [m\ No/(k\ No)]\ c_{rms}^2 = 1/3\ (m/k)\ c_{rms}^2 \tag{9.12}$$

Since, $R = k$ No, where $k$ is the Boltzman constant = $1.38 \times 10^{-23}$ J/molecule K, $M = m\ No$, and $No$ is the Avogadro's number. We observe that temperature of the gas is related to the root mean square velocity of the molecules, and hence it is an intrinsic property of the gas. This gives

$$1/2\ m\ c_{rms}^2 = m \times (3/2)\ (k/m)\ T = 3/2\ kT \tag{9.13}$$

That is, the kinetic energy of a molecule is $3/2\ kT$. Or, temperature of the gas is related to the kinetic energy, or its internal energy $U$ and vice versa. Equation 9.13 envisages that the root mean square velocity is zero when $T = 0$. This defines the zero of the temperature scale, and coincides with the absolute zero of the thermodynamic temperature scale given by Kelvin, and accepted by the scientists as the thermodynamic temperature scale. Further, if there are two gases, represented by subscripts 1 and 2, at the same temperature $T$, then

$$T = 1/3\ (m_1/k)\ (c_{ms1}^2)_1 = 1/3\ (m_2/k)\ (c_{ms1}^2)_2$$

$$\rightarrow c_{rms1}/c_{rms2}$$

$$= \sqrt{(m_2/m_1)} = \sqrt{(M_2/M_1)}$$

That is, the root mean square velocity is inversely proportional to the mass of the molecule or the molecular weight of the gas. This fact has been used in the separation of heavier gases from lighter gases.

### 9.2.2 Velocity Distribution

We learnt before that molecules have a wide range of velocity, and would be interested to find out their distribution, which is the number of molecules at a time having velocity between $c_x$ and $c_x + dc_x$, $c_y$ and $c_y + dc_y$ and $c_z$ and $c_z + dc_z$, where $c_x$, $c_y$ and $c_z$ are the components of velocity $c$ along the $x$, $y$ and $z$ direction. Note that it is not the number of molecules having velocity between $c$ and $c + dc$, because the molecules are moving in different directions. If we plot the velocity of all the molecules in space, then all molecules having velocity between $c$ and $c + dc$ will lie on the strip formed by the radii $c$ and $c + dc$, but the number we are looking for will lie in a small region of this strip. We shall extend the notation introduced earlier to represent the number of molecules having velocity between $c_x$ and $c_x + dc_x$ by $dNc_x$, between $c_x$ and $c_x + dc_x$, $c_y$ and $c_y + dc_y$ and $c_z$ and $c_z + dc_z$ by $d^3N$. When $N$ is large, then $dNc_x$ and $d^3N$ will still be quite large. In such a situation, the function $dNc_x/N$ is a function of $c_x$ and $dc_x$. This holds true for other velocity components too. That is,

$$dNc_x/N = f(c_x)\,dc_x,\; dNc_y/N = f(c_y)\,dc_y,\; dNc_z/N = f(c_z)\,dc_z \tag{9.14}$$

In the theory of probability, $dNc_x/N$ is the probability of molecules having velocity between $c_x$ and $c_x + dc_x$. Therefore, the probability that the molecules have velocity between $c_x$ and $c_x + dc_x$, $c_y$ and $c_y + dc_y$, and $c_z$ and $c_z + dc_z$ is a product of the probability. Therefore,

$$d^3N = N f(c_x)\, f(c_y)\, f(c_z)\, dc_x\, dc_y\, dc_z \tag{9.15}$$

The density of the points in velocity space will be

$$\rho = d^3N/(dc_x\, dc_y\, dc_z) = N f(c_x)\, f(c_y)\, f(c_z) \tag{9.15a}$$

Since the velocity distribution is isotropic, the density is the same in any volume element, so that $d\rho = 0$. This gives after differentiation and dividing by $f(c_x)\, f(c_y)\, f(c_z)$

$$\frac{f'(c_x)}{f(c_x)}dc_x + \frac{f'(c_y)}{f(c_y)}dc_y + \frac{f'(c_z)}{f(c_z)}dc_z = 0 \tag{9.16}$$

Also, the velocity of the molecules falling in the volume element is the same. This yields the constraint equation

$$c_x^2 + c_y^2 + c_z^2 = 0 \quad \text{or} \quad c_x\, dc_x + c_y\, dc_y + c_z\, dc_z = 0 \tag{9.17}$$

Equation 9.17 shows that the three components of velocity are not independent. Therefore, we use Lagrange's multiplier method to find the solution of Eq. 9.16. Multiplying Eq. 9.17 by $\gamma$ and adding to Eq. 9.16, we get after rearrangement of terms

$$\left[\frac{f'(c_x)}{f(c_x)} + \lambda\, c_x\right]dc_x + \left[\frac{f'(c_y)}{f(c_y)} + \lambda\, c_y\right]dc_y + \left[\frac{f'(c_z)}{f(c_z)} + \lambda\, c_z\right]dc_z = 0 \tag{9.18}$$

Since $dc_x$, $dc_y$ and $dc_z$ cannot be zero, the quantity within the square brackets [ ] has to be zero. After setting to zero and integrating the resultant equations, this results in

$$f(c_x) = \alpha \exp(-\beta\, c_x^2),\; f(c_y) = \alpha \exp(-\beta\, c_y^2),\; f(c_z) = \alpha \exp(-\beta\, c_z^2) \tag{9.19}$$

where, $\alpha$ and $\beta = \lambda/2$ are the constants to be determined. Substituting the functions $f(c_x)$, etc. in Eq. 9.15, we get

$$d^3N = N\,\alpha^3 \exp(-\beta\, c^2)\, dc_x\, dc_y\, dc_z \tag{9.20}$$

and

$$d^3N/(dc_x\, dc_y\, dc_z) = N\,\alpha^3 \exp(-\beta\, c^2) \tag{9.20a}$$

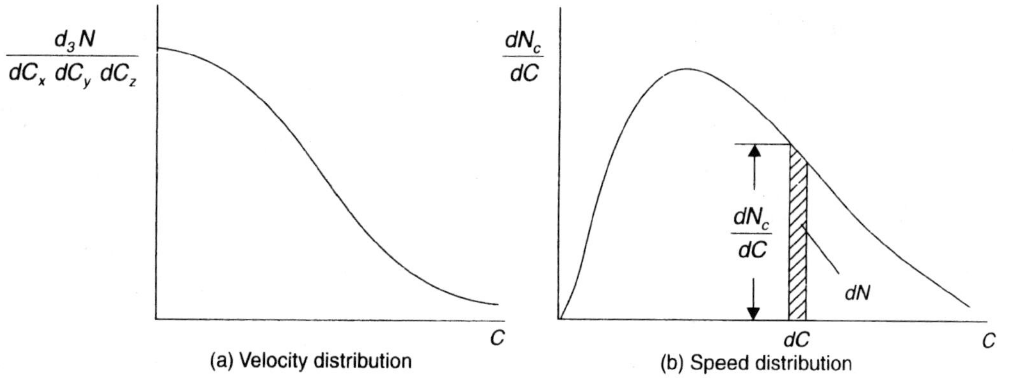


**Fig. 9.5** Maxwell velocity and speed distribution

The quantity $N\alpha^3 \exp(-\beta c^2)$ is known as the Maxwell velocity distribution function, and its variation with $c$ is shown in Fig. 9.5(a). This shows that the number of molecules per unit volume at zero velocity is maximum, and falls off rapidly as $c$ increases.

We may also find out the number of molecules having speed between $c$ and $c + dc$. This will be the number in the spherical shell formed by drawing two spheres of radii $c$ and $c + dc$, respectively. The volume of this shell is $4\pi c^2 dc$. The molecular density remains constant and may be obtained by substituting for functions $f(c_x)$, from Eq. 9.19 in Eq. 9.15(a). This gives the number of molecules $dn_c$, having speed between $c$ and $c + dc$

$$dN_c/dc = \{N\alpha^3 \exp(-\beta c^2)\}\, 4\pi c^2 dc = 4\pi\, N\alpha^3 c^2 \exp(-\beta c^2)\}dc \tag{9.21}$$

The ratio $dn_c/dc$ is called the speed ratio, and is plotted in Fig. 9.4(b). The number of molecules between speeds $c$ and $c + dc$ is obtained by measuring the area of the hatched region. We observe that the curve starts from 0, rises up to some maximum value, and then starts decreasing, reaching zero asymptotically. This means that the maximum number of molecules have speed, corresponding to the peak of the curve. In statistics, this speed is referred to as the mode value. This distribution can be explained from the superimposing effect of decrease in molecular density and increase in volume of the shell with increase in $c$. We may also infer that the molecular density at higher speed decreases very fast.

To determine constants $\alpha$ and $\beta$, we have two equations, namely, the total number of molecules is $N$ and that the root mean square velocity is $c_{rms}^2$. Determining these for the Maxwell speed distribution, we get

$$N = \int_0^\infty 4\pi\, N\alpha^3 c^2 \exp(-\beta c^2) dc \;\rightarrow\; \alpha = \sqrt{\beta/\pi} \tag{9.22}$$

$$c_{rms}^2 = \frac{3kT}{m} = \int_0^\infty c^2 dN_c/N = 4\pi\alpha^3 \int_0^\infty c^4 \exp(-\beta c^2) dc \;\rightarrow\; \beta = \frac{m}{2kT} \tag{9.23}$$

Substituting for $\alpha$ and $\beta$ in Eq. 9.21, we get

$$dN_c = \frac{4N}{\sqrt{\pi}}\left(\frac{m}{2kT}\right)^{3/2} c^2 \exp\left(-\frac{mc^2}{2kT}\right) dc \tag{9.24}$$

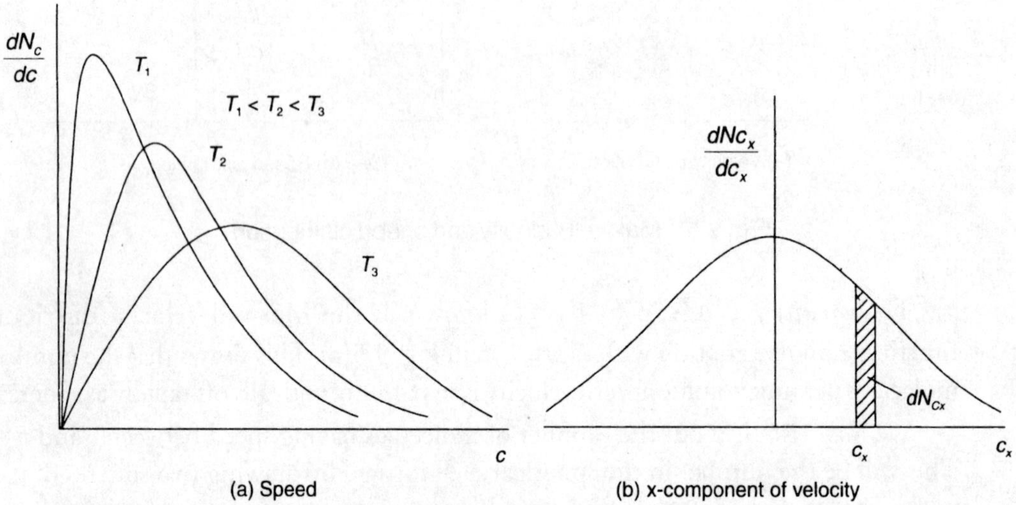


**Fig. 9.6** Effect of temperature on speed distribution

The distribution given by Eq. 9.24 is shown in Fig. 9.6(a) for different gas temperatures. It is observed that (i) the area under each curve for different isothermal is the same, since it represents the total number of molecules, (ii) as the temperature increases, the number of molecules having the most probable speed is decreased, but the most probable speed itself increases, (iii) the number of molecules within *dc* near the most probable speed decreases with increase in temperature, and (iv) there are a few molecules having velocity near zero or very high speed. Thus, by increasing the temperature, more number of molecules acquire high speed and results in increase in rates of chemical reaction or escaping of the gas molecules from planets, liquids in the evaporation process, or from solids in the melting process.

Similarly, we may find out the distribution function for *x*, *y* or *z* component velocity. We get the distribution function as

$$dNc_x = \frac{4N}{\sqrt{\pi}}\left(\frac{m}{2kT}\right)^{1/2} \exp\left(-\frac{mc^2}{2kT}\right) dc_x \tag{9.25}$$

The distribution of $dNc_x/dc_x$ is shown in Fig. 9.6(b). The curve is symmetrical about $c_x = 0$, and has maximum value of $N\,[m/(2\pi kT)]^{1/2}$.

### 9.2.3 Average Root Mean Square Velocity and Most Probable Speed

Since there are a large number of molecules having different velocity distributions, it will be desirable to use statistical tools for such distribution. The most commonly used statistical functions are the average $\bar{c}$, root mean square velocity $c^2_{rms}$, and the mode or most probable value $c_{mp}$ of the distribution. We may obtain useful expressions for these quantities using the Maxwell speed distribution function given by Eq. 9.24, and carrying out the integration in the limit $c = 0$ to $\infty$.

$$\bar{c} = \frac{\int_0^\infty c\,dN_c}{N} = \frac{4}{\sqrt{\pi}}\left(\frac{m}{2kT}\right)^{3/2}\int_0^\infty c^3 \exp\left(-\frac{mc^2}{2kT}\right)dc = \left[\frac{8kT}{\pi m}\right]^{1/2} \tag{9.26}$$

Root mean square velocity is given by

$$c^2_{rms} = \frac{\int_0^\infty c^2\,dN_c}{N} = \frac{4}{\sqrt{\pi}}\left(\frac{m}{2kT}\right)^{3/2}\int_0^\infty c^4 \exp\left(-\frac{mc^2}{2kT}\right)dc = \sqrt{\frac{3kT}{\pi m}} \tag{9.27}$$

The most probable speed corresponds to the peak of the speed distribution curve. Hence, at this point the first derivative w.r.t. $dc$ is zero. This gives after differentiating Eq. 9.21.

$$\frac{d}{dC}\left(\frac{dN_c}{dc}\right) = \frac{d}{dc}\left[4\pi N\,\beta^3 c^2 e^{-\beta c^2}\right] = 0 \rightarrow c_{mp} = \left[\frac{2kT}{m}\right]^{1/2} \tag{9.28}$$

The relative magnitude of the three speeds at a given temperature is

$$c_{mp} : \bar{c} : c_{rms} = 1 : 1.128 : 1.224 \tag{9.29}$$

**Example 9.1** Calculate the value of average, root mean square velocity and most probable speed of nitrogen at (a) 300 K, (b) 800 K. Also, compare the fraction of nitrogen molecules at 300 K, which have speed within ± 1% of $c_{mp}$. Given $M = 28$, $k = 1.38 \times 10^{-23}$ and $No = 6.025 \times 10^{26}$.

***Solution*** Given: $M = 28$ $No = 6.025 \times 10^{26}$

$\rightarrow m = M/No = 4.65 \times 10^{-26}$ kg

(a) At $T = 300$K, using Eqs. 9.26 to 9.28, we get $c_{rms} = (3kT/m)^{1/2}$

$= [(3 \times 1.38 \times 10^{-23} \times 300)/(4.65 \times 10^{-26})]^{1/2} = 517.5$ m/s

$c_{mp} = (2kT/m)^{1/2} = 422.0$ m/s and $\bar{c} = 476.0$ m/s

(b) At $T = 800$ K: $c_{mp} = (2kT/m)^{1/2} = 690$ m/s

$\therefore\ \bar{c} = 1.128 \times c_{mp} = 776$ m/s and $c_{rms} = 1.224 \times c_{mp} = 845$ m/s

Speed distribution is given by Eq. 9.21, which can be written in a finite difference form for a finite but small change as

$$dN c/N = \Delta\, Nc/N = f(c)\ \Delta c$$

where, $f(c) = 4\pi\ N\alpha^3 c^2 \exp(-\beta\ c^2)\}$. Writing for $2kT/m$ in terms of $c_{rmp}$ from Eq. 9.28 in 9.24, we get $f(c) = 4/\sqrt{\pi}.\ (1/c_{mp})^3.\ c^2.\ \exp(-c^2/c_{mp}^2)$.

We are given $c$ and $\Delta c = 0.02\ c$, whereas $c_{mp}$ and $c_{rms}$ have been determined above. Using these results and the above equation, and after writing for $\alpha$ and $\beta$ from Eqs. 9.22 and 9.23 and for $c_{mp}$ from Eq. 9.28. We get after simplification

(a) At $c = c_{mp} = 422$, $f(c) = (4/\sqrt{\pi}.) \times (1/422)^3 \times \exp(-1) = 3 \times 10^{-8}$

$$\therefore\ \ \Delta\, Nc/N = 0.02 \times 422 \times 3 \times 10^{-8} = 2.534 \times 10^{-7}$$

Many a time we also need to find out the number of molecules in a certain speed range. This can be obtained by integrating Eq. 9.21 in those speed limits. Integration can be done using the integration by parts technique. The result is

$$N_{0-c} = N\left[\operatorname{erf}(x) - \frac{2}{\sqrt{\pi}} x e^{-x^2}\right], \text{ where } x = \left[\frac{m}{2kT}\right]^{1/2} \text{ and}$$

$$\operatorname{erf}(x) = \frac{2}{\sqrt{\pi}} \int_0^x e^{-x^2}\, dx \qquad (9.30)$$

Error function tables are given in A1.12.

### 9.2.4 Energy Distribution Function

Like velocity and speed distribution functions, Maxwell and Boltzman also determined the energy distribution function. The translational kinetic energy $\varepsilon$ of a molecule of mass $m$ moving with speed $c$ is

$$\varepsilon = 1/2\ mc^2 \rightarrow d\varepsilon = mc\ dc = m\sqrt{(2\varepsilon/m)}\ dc$$

Changing $c$ in terms of $\varepsilon$ in Eq. 9.24 using the above relations, we get the number of molecules between $\varepsilon$ and $\varepsilon + d\varepsilon$ as

$$dN_\varepsilon = \frac{4N}{\sqrt{\pi}}\left(\frac{m}{2kT}\right)^{3/2} \frac{2\varepsilon}{m} \exp\left(-\frac{\varepsilon}{kT}\right) \frac{d\varepsilon}{(2m\varepsilon)^{1/2}}$$

$$= \frac{2N}{\sqrt{\pi}} \frac{\varepsilon^{1/2}}{(kT)^{3/2}} \exp\left(-\frac{\varepsilon}{kT}\right) d\varepsilon \tag{9.31}$$

Equation 9.31 gives the Maxwell-Boltzman energy distribution function $dN\varepsilon/d\varepsilon$. The most probable energy of the molecules can be obtained in the same way as we determined the most probable speed. This procedure gives on simplification:

$$\varepsilon_{mp} = 1/2\ kT \tag{9.32}$$

Similarly, we may determine the *rms* value of $c_x$, $c_y$ and $c_z$, as well as the average or mean translational kinetic energy of molecules with these velocity components. The result is

$$\bar{\varepsilon}_x = \bar{\varepsilon}_y = \bar{\varepsilon}_z = 1/2\ kT \text{ and } \bar{\varepsilon} = \bar{\varepsilon}_x + \bar{\varepsilon}_y + \bar{\varepsilon}_z = 3/2kT \tag{9.33}$$

where, $\bar{\varepsilon}$ is the mean total kinetic energy. This leads to an important observation that *the kinetic energy is equally divided in each direction*, called the degree of freedom. This is called as the principle of 'Equipartition of Energy', and may be extended to other degrees of freedom. The other degrees of freedom are introduced because of the rotation of the diatomic and higher molecules about the three axes. To understand, consider a molecule of diatomic gas. We may assume that the two atoms are connected with each other forming the shape of a dumbbell; this link between them may be rigid or may act like a spring, depending on the temperature of the gas. At low temperature, it will act as a rigid link. This model introduces three degrees of rotation about the three axes; one of these will be the line joining their centres. Since moment of inertia about such an axis will be very low, this rotational energy is neglected. So, for such a molecule, there are three translational and two rotational degrees of freedoms totalling to five. We further note that both translational and rotational energies are a function of the square of the velocity. At high temperature, however, it is not easy to determine the degree of freedom, as then the link between the atoms of a molecule have vibratonal energy also, which is difficult to determine.

## 9.3 RESULTS FROM THE MACROSCOPIC APPROACH

Now, let us compare some of the results obtained from the kinetic theory, and the macroscopic approach for gases and liquids. Let us first see the perfect gas equation, which is $pV = n\bar{R}T$. Combining Eqs. 9.11 and 9.13, we find out that according to the kinetic theory too, $pV = f(T)$. Therefore, it satisfies the Boyle's law. Now, consider a mixture of perfect gases having number of gases of molecular densities $\rho_1$, and $\rho_2$, and having *rms* velocity $c_{rms1}$, and $c_{rms2}$. The total pressure exerted by the gases by Eq. 9.9 is

$$P = 1/3\ \rho_1 (c^2_{rms})_1 + 1/3\ \rho_2 (c^2_{rms})_2 + 1/3\ \rho_3 (c^2_{rms})_3\ \ldots = p_1 + p_2 + p_3 + \ldots \tag{9.34}$$

which is Dalton's law of partial pressure. We may similarly verify that the kinetic theory satisfies the Avogadro's hypothesis, which states that equal volume of gases under similar conditions of temperature and pressure, contain the same number of molecules. Equality of pressure and volume of gases mean from Eq. 9.11 and 9.12

$$pV = 1/3\, m_1\, n_1\, (c^2_{rms})_1 = 1/3\, m_2\, n_2\, (c^2_{rms})_2 = \ldots \tag{9.35}$$

But from Eq. 9.12, $m c^2_{rms} = kT$ = constant for all the gases, since temperature is the same. Therefore, Eq. 9.35 is reduced to $n_1 = n_2 = \ldots$, which is Avogadro's hypothesis.

Thus, we observe that the kinetic theory of gases gives the same results as given by the macroscopic approach, at least for a perfect gas. This theory, of course, formed the basis for the van der Waal equation. Recall his arguments that the pressure exerted by the molecules on the container wall is reduced because of the force of attraction between the molecules at high pressure. This is justified because the intermolecular distance in such a case is less, and therefore, gravitational force exists between the molecules, as seen from Fig. 9.1.

We may also verify some of the other macroscopic results with the help of the kinetic theory of gases, and the process is discussed in the following sections.

### 9.3.1 Collision with a Moving Wall

Let us now explain the boundary work phenomenon or *pdV* work. In this case, the system boundary is moving. We may consider the gas contained in a cylinder piston arrangement, as shown in Fig. 9.7, to represent such a system as a simplified model, and that the piston is moving with velocity $u$. A $\theta\phi$ molecule is shown in the figure colliding with the piston surface, which is reflected in the direction $\theta'$. Let $c$ and $c'$ be the velocities before and after the collision. During the collision, the relative velocity along the normal to the piston will not change. The velocity of the molecule relative to the piston before and after the collision is: $c \cos\theta - u$ and $c' \cos\theta + u$. Equating the two, we get

$$c' \cos\theta' = c\cos\theta - 2u$$

The loss of *KE* in the collision is $= \frac{1}{2} m (c\cos\theta)^2 - \frac{1}{2} m (c' \cos\theta')^2 = 2muc\cos\theta$, neglecting the term $2\,mu^2$, since $u << c$. Equation 9.6 gives the number of such molecules. Therefore, the loss of *KE* in all the collisions per unit area and per unit time is

$$\int\limits_c \int\limits_\phi \int\limits_\theta (2mcu\cos\theta)\left[\frac{1}{4\pi} c\, dn_c \sin\theta \cos\theta\, d\theta\, d\phi\right]$$

$$= \frac{1}{2\pi} mu \int_0^\infty c^2\, dn_c \int_0^{\pi/2} \sin\theta \cos^2\theta\, d\theta \int_0^{2\pi} d\phi = \frac{1}{2\pi} mun\, c^2_{rms} \frac{1}{3} 2\pi = pu \tag{9.36}$$

If $A$ is the area of the piston, then the loss of $KE = pu.$ A $= Fu$ = Rate of work done on the piston, a result we get from macroscopic analysis. ($u.dA$) also represents the change in volume $dV$ of the system (gas confined in cylinder), and hence loss in $KE$ of molecules is $pdV$, which is the boundary work done.

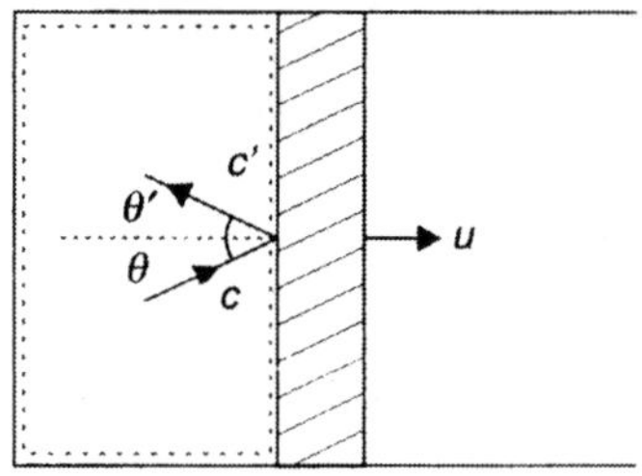


**Fig. 9.7** Collision with a moving wall

### 9.3.2 Specific Heats of a Gas

Continuing the discussions from Section 9.2.4, we can now find out the specific heats of a gas at constant volume and pressure. If $f$ represents the degree of freedom of a gas, then the mean total energy of $N$ molecules is nothing but the internal energy $U$. Thus:

$$U = N\varepsilon = ½ fkT \times N = ½ f(k \times No)T \times (N/No) = ½\ fn\bar{R}T$$

Here, $n$ is the number of moles of the gas. Or, molal internal energy, $u = ½\ f\bar{R}T$. Therefore, the molal specific heat at constant volume is

$$C_v = (\partial u/\partial T)_v = 1/2 f\bar{R},\ C_p = C_v + \bar{R} = (1 + f/2)\bar{R} \text{ and}$$
$$\gamma = C_p/C_v = (f + 2)/f \tag{9.37}$$

**Table 9.1** Specific heat results from experiments and the kinetic theory

| Atomicity | Kinetic theory | | Experimental results | | | |
|---|---|---|---|---|---|---|
| | $f$ | $\gamma$ | Gas | $C_p/R$ | $C_v/R$ | $\gamma$ |
| Monoatomic gas | 3 | 1.66 | He | 2.52 | 1.519 | 1.659 |
| | | | Ar | 2.52 | 1.509 | 1.670 |
| Diatomic gas | 5 | 1.4 | $H_2$ | 3.42 | 2.438 | 1.410 |
| | | | $O_2$ | 3.52 | 2.504 | 1.401 |
| | | | CO | 3.50 | 2.488 | 1.404 |
| | | | $N_2$ | 3.50 | 2.448 | 1.404 |
| Triatomic gas | 6 | 1.333 | $CO_2$ | 4.40 | 3.38 | 1.304 |

Table 9.1 compares the specific heat ratio obtained from the kinetic theory with the experimental values for a few gases of different atomicity at standard temperature and pressure. The theory gives results that are quite in agreement with the experimental results at STP. However, it is observed that specific heats, and hence their ratio, change with temperature. Figure 9.8 shows variation of specific heat at constant volume for hydrogen gas. This variation, as mentioned earlier too, is due to vibratonal energy of varying amounts added to the molecules at high temperature.

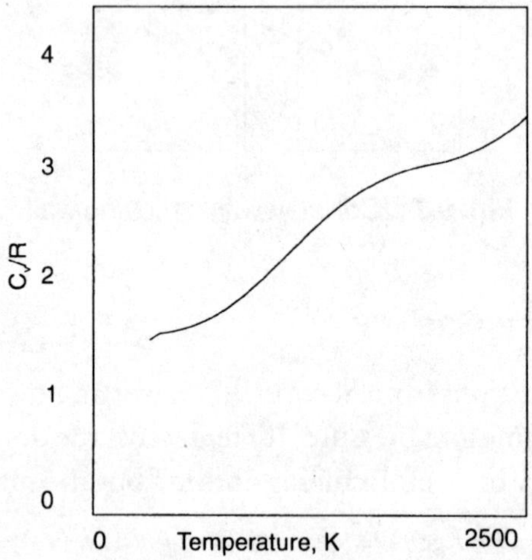


**Fig. 9.8** Specific heat as a function of $T$

### 9.3.3 Transport Processes in Gases

When gases flow over a surface, boundary layer formation occurs, resulting in a velocity gradient. There may similarly be temperature and concentration gradients. These gradients are developed because of the properties of gases like viscosity, thermal conductivity and diffusion. Existence of such properties can be explained by accounting the molecular property flux from all around to a given area in the flow region; the molecular properties respectively are momentum, energy and mass. Let us represent any molecular property by $\Gamma$, and its gradient in the normal direction $y$ as $d\Gamma/dy$. In order to find out the property flux to the area, we need to find out (i) the number of molecules, and (ii) the distance along the perpendicular direction from where the molecules arrive without collision. To determine these, we need to first learn the mean free path and the survival equation, discussed in the next paragraph.

Molecules, while moving, collide with each other and change their velocity, and in each collision, two molecules take part, and after each collision, two free paths are created. The distance travelled between two successive collisions, which will not be the same for all the collisions, is called the free path length. Free path length may theoretically vary from 0 to $\infty$. Therefore, mean free path length $\lambda$ is used, which is the statistical average of all the free path lengths. $\lambda$ was determined by taking a simplistic model of one molecule of $2d$ diameter moving with average velocity $\bar{c}$ through a point like stationary molecules. The idea is that when the molecules collide, the centre to centre distance is $2d$, where $d$ is the diameter of the molecule. The number of collisions made in time $dt$ will be equal to the number of points in a cylinder of cross-sectional area $\sigma = \pi/4 \times (2d)^2 = \pi d^2$ and height $\bar{c}\ dt$. The number of collisions made in a second is called collision frequency $z$. If $n$ is the molecular density, then

$$\lambda = \frac{\bar{c}t}{\pi d^2 \bar{c}tn} = \frac{1}{\pi d^2 n} = \frac{1}{\sigma n} \text{ and } z = n\sigma\bar{c} \tag{9.38}$$

Clausius obtained the mean free path length by assuming all the molecules to have the same velocity (which is not correct, since in the course of collision, as discussed before, molecules will change velocity). On the contrary, Maxwell used the velocity distribution (obtained by him) and got the mean free path. The results are

$$\lambda = (3/4)\{1/(\pi d^2 n)\} \text{ (Clausius) and } \lambda = (1/\sqrt{2})\{1/\sigma n)\} \text{ (Maxwell)} \tag{9.39}$$

Equation 9.38 can be extended to the case of electrons colliding with molecules of a gas. In this case, the distance from centre to centre would only be $d$, the diameter of the molecule, as the electron has no dimension. Thus, the collision cross-section area is $\sigma = \pi d^2/4$ and not $\pi d^2$, as in the previous case.

Having defined the mean free path length, we have to now find out the number of molecules that survive from the collision after travelling distance $y$ in a perpendicular direction. Let us say the population of molecules at the start is $No$, which decreases with $y$, and is $N$ after travelling distance $y$. If $P_c$ is the collision probability, then the number of collisions $dN$ made in distance $dy$ is

$$dN = -P_c\, N\, dy$$

The negative sign is used because the number of surviving molecules is decreasing. Integrating it and using the condition that $N = No$ at $y = 0$, we get

$$N = -No \exp(-P_c y) \tag{9.40}$$

Equation 9.40 is known as *survival equation*. The number of molecules having free path length between $y$ and $y + dy$ is obtained by differentiating Eq. 9.40 with respect to $y$. This gives

$$dN = No\, P_c \exp(-P_c y)\, dy \tag{9.41}$$

Now, the mean free path length can be obtained, as the number of molecules having free path length $y$ is known. Thus

$$\lambda = \frac{\int_0^\infty y\, dN}{No} = P_c \int_0^\infty y \exp(-P_c y)\, dy = \frac{1}{P_c} \rightarrow N = No \exp(-y/\lambda) \tag{9.42}$$

**Example 9.2** Calculate the mean free path for oxygen molecules at 273 K and 1.013 bar. What fraction of molecules have free paths greater than $2\lambda$? The effective diameter of $O_2$ molecule may be taken as $3.5 \times 10^{-10}$m. Assume Maxwell velocity distribution for calculation of $\lambda$.

***Solution*** The collision cross-section, $\sigma = \pi d^2 = 3.84 \times 10^{-19}\ \text{m}^2$

The number of molecules per unit volume is given by the equation,

$$p = \bar{R}\ T/V = k\ (No/V)\ T = nkT$$

$$n = (1.013 \times 10^5)/(1.38 \times 10^{-23} \times 273)$$

$$= 2.68 \times 10^{25}\ \text{molecules/m}^3$$

Using Eq. 9.39, we get

$$\lambda = (1/\sqrt{2}) \times (1/\sigma n) = 6.87 \times 10^{-8}\ \text{m}$$

Substituting $y/\lambda = 2$ in Eq. 9.42, we get

$$N/No = e^{-2} = 0.135$$

That is, 13.5% molecules have free path greater than twice the mean free path length.

Now, returning to the calculation of molecular flux, consider area $dA$ in a plane and a finite but small volume element $dV = \sin\theta\ d\theta\ d\phi\ dr$ at the point $(r, \phi, \theta)$, as shown in Fig. 9.9. This volume will have large number of molecules, as $n$ is quite large. The number of collisions made by the molecules in this volume element is ½ $zndV$; the factor ½ comes because two molecules collide at a time. However, each collision creates two free paths, and therefore, the number of free paths will be $zndV$. That is, $zndV$ number of molecules shoot out in all directions after the collision at point $(r, \phi, \theta)$. The number of molecules which moves towards the area $dA$ will be given by

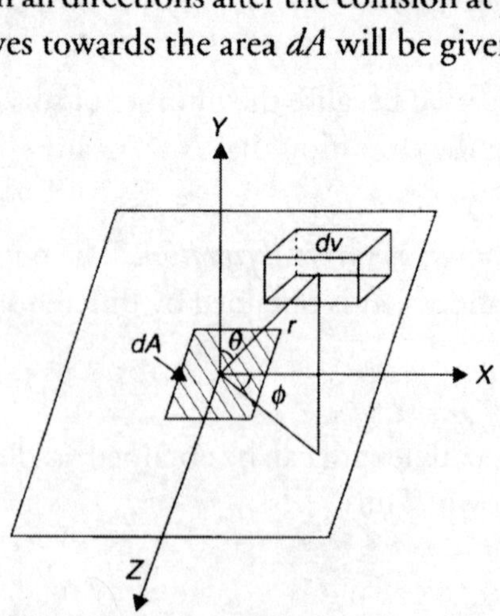


**Fig. 9.9** Calculation of molecular flux

$dNo$ = No. of molecules in unit solid angle × solid angle subtended by $dA$ at $(r, \phi, \theta)$.

that is, $dNo = (zndV/4\pi) \times (dA \cos\theta)/r^2$

Equation 9.42 gives the number of molecules $dN'$ reaching area $dA$ from $dV$ at distance $y = r$. This gives after substituting for $dV$ and simplifying

$$dN' = (1/4\pi)\, zn \sin\theta \cos\theta\, e^{-r/\lambda}\, dA\, d\theta\, d\phi\, dr \tag{9.43}$$

Molecules arrive at $dA$ from both the hemispheres, above and below the plane. Molecular flux, the number of molecules per unit area ($dA$ = 1), is then obtained by integrating Eq. 9.43 for $\phi$ = 0 to $2\pi$, $\theta$ = 0 to $\pi/2$ and $r$ = 0 to $\infty$. Integration of the equation gives

$$N = 1/4\, zn\lambda = (1/4\sigma n\bar{c})n(1/\sigma n) = 1/4\, n\bar{c} \tag{9.44}$$

after substituting for $z$ and $\lambda$ from Eq. 9.38.

Since the molecules are coming from all the planes above and below, the property associated with each of them will be different because of the property gradient. Therefore, we find out the average distance $\bar{y}$ from which these molecules may be assumed to come. This is obtained from

$$\bar{y} = \frac{\int y\, dN}{\int dN} = \frac{\dfrac{zn}{4\pi}\displaystyle\int_0^{2\pi} d\phi \int_0^{\pi/2} \sin\theta \cos^2\theta\, d\theta \int_0^{\infty} re^{-r/\lambda}\, dr}{1/4\, zn\lambda} = 2/3\lambda \tag{9.45}$$

Now, we know everything we need for calculation of transport of the property from above and below the plane. Let us represent these by $J_+$ and $J_-$. The flux of any molecular property is obtained by multiplying the number of molecules arriving at the plane given by Eq. 9.44, and the property value at $y = 2/3\lambda$. Thus we get

$$J_+ = -\frac{1}{4} n\bar{C} \left[\Gamma + d\Gamma/dy\right]_0 (2/3\lambda) \text{ and}$$

$$J_- = \frac{1}{4} n\bar{c} \left[\Gamma - d\Gamma/dy\right]_0 (2/3\lambda) \tag{9.46}$$

The negative sign appears with $J_+$ because the transport is along the negative $Y$ direction. The total transport then is:

$$J = J_+ + J_- = -1/3 n\bar{c}\lambda\, (d\Gamma/dy)_0 \tag{9.47}$$

The coefficient in Eq. 9.47 is called the *generalised transport coefficient*.

**Dynamic viscosity**. Viscosity of fluid results in variation of velocity in the direction perpendicular to the direction of flow, and the molecular property is the momentum $mu$, where $u$ is the velocity component in the direction of flow. That is, $\Gamma = mu$. Therefore, the coefficient, which is called dynamic viscosity $\mu$, is,

$$\mu = 1/3\, mn\, \bar{c}\, \lambda \tag{9.48}$$

Since average velocity increases with gas temperature, the dynamic viscosity of gases also increases with temperature. This equation also shows that viscosity does not depend on pressure.

**Thermal conductivity.** When the gas is flowing over a heated surface, then the temperature gradient also exists in the perpendicular direction. In this case, the molecular property is the energy, which by the principle of Equipartion of energy is $\Gamma = \frac{1}{2} fkT$, where $f$ is the degree of freedom for that condition. The transport coefficient is known as thermal conductivity $K$. It is

$$K = -1/6\, n\bar{c}\, \lambda\, f\, k \tag{9.49}$$

**Diffusion.** Diffusion is a process in which the molecules migrate, so that their concentration becomes uniform under equilibrium condition. Consider two different gases contained on either side of a partition made in a container, and let the partition be removed. The molecules of the two gases start moving to the other side, because there is different concentration of the molecules of both the gases, and this continues till the concentration is the same throughout the space. If $n_A$ is the number of molecules of $A$ in a volume having total molecules $n$ of both the gases, then $\Gamma = n_A/n$, which is the mole fraction. The transport coefficient is known as diffusion coefficient $D$, given by

$$D = 1/3\ n\bar{c}\lambda\ (n_A/n) = 1/3\ \bar{c}\lambda \tag{9.50}$$

These results for dynamic viscosity, thermal conductivity and diffusion coefficients agree with experimental results.

## 9.4 STATISTICAL THERMODYNAMICS

The kinetic theory of gases assumes uniform distribution of energy. However, in order to explain various phenomena like diffraction, reflection and refraction, related to light and electromagnetic waves, the dual behaviour of the electromagnetic waves or photons was recognised. Some of the phenomena could be explained as if these waves were made of particles, and others as if these were waves. Thus there were two thoughts on which research was centering around, and accordingly, classical and quantum physics were founded. While the kinetic theory is based on classical physics, statistical thermodynamics is developed on the concept of quantum physics, which assumes that the energy of the waves and photons is quantified. This means that the energy distribution among all the molecules is not continuous, but varies in step from some definite minimum value to any large value. These discrete values of energy are set by certain constraints on the system. For example, the translational energy of the particles are set by the volume and temperature of the system.

These energy levels may be very large in number, but there is some energy difference between the successive energy levels, however small or large it may be. Also, the difference between two successive energy levels may or may not remain the same throughout the energy levels domain. It is also not necessary that the molecules have to occupy all the energy levels. It is even possible that all the molecules of the gas may lie in only one energy level. We shall study later how to find the distribution of the molecules in these levels, and what is the effect of continuous collisions on this distribution, which principally, lies in the realm of statistical thermodynamics.

A number identifies each energy level. This number is called its quantum number. The lowest energy level is called the ground level, indicated by the number 0; the successive ones will be numbered as 1, 2, 3, and so on. Depending on the total energy of the molecules (internal energy of the system), some of the states will never be occupied. For example, if the total energy is, say, 20 $\varepsilon$, where $\varepsilon$ represents the difference of energy level between successive uniformly spaced levels, then all the energy levels having energy 21 $\varepsilon$ or more can never be occupied. Such levels are called *inaccessible levels.*

A molecule of gas has position and motion, which is represented by its momentum $p = mc$. Thus, each particle will have six coordinates to define its position in a 6-dimensional hyperspace formed by three components each, of position and momentum along three coordinate axes $X$, $Y$ and $Z$. The totality of such description for all the molecules is called phase space. Fortunately, we do not need such a description, as the total number of molecules having a particular momentum is needed to calculate the pressure or the number of molecules in a cube in space to give its density and so on. However, Heisenberg's uncertainty principle gives an idea of degeneracy of the energy level, which can be understood easily with the help of such hyperspace. Consider a small volume of size $dH = dx\, dy\, dz\, dp_x\, dp_y\, dp_z$, called cell, in the phase space. The uncertainty principle says that the momentum and the position of molecules in an energy level can be predicted with an accuracy given by

$$\Delta x\, \Delta p_x \approx h, \quad \Delta y\, \Delta p_y \approx h, \quad \Delta z\, \Delta p_z \approx h$$

The uncertainty in locating a particle in phase space is $\Delta x\, \Delta y\, \Delta z\, \Delta p_x\, \Delta p_y\, \Delta p_z \approx h^3$.

Here, $h$ is the Planck's constant = $6.624 \times 10^{-34}$ J-s. That is, the particle lies somewhere in a cube of size $h^3$. Hence, the given cell can be divided into $dH/h^3 = g$ compartments; $g$ is called the *degree of degeneracy* of that energy level. Degeneracy will, in general, be different for different energy levels. The importance of this is that the occupancy of different compartments gives different distributions, and hence it is to be counted separately. We shall discuss more about it later.

Schrodinger studied the wave phenomenon, and gave the wave equation and its solution. The results are summarised.

1. The total translational energy, $\varepsilon_{tr}$, per particle in any energy level is

$$\varepsilon_{tr} = (h^2/8m)[(K_x/a)^2 + (K_y/b)^2 + (K_z/c)^2] \tag{9.51}$$

where, $K_x$, $K_y$ and $K_z$ are the translational quantum numbers in the $X$, $Y$ and $Z$ direction, respectively, $m$ is the mass of the molecule, and $a$, $b$ and $c$, respectively are the dimensions of the container along the $X$, $Y$ and $Z$ directions. The quantum numbers have values 1, 2, 3, etc. This means that the energy of ground level will not be zero, as the minimum value of the quantum number is 1, and would instead be $3\,h^2/(8m\,L^2)$, where $L$ is the length of the cube; this is, however, very small—approximately of the order of $10^{-10}$ J/gm mole. Further, the energy difference between the levels is so small, that the energy distribution closely approximates continuous distribution. This justifies the assumption made in the kinetic theory of gases, since we mostly consider the translational energy mostly in the discussion.

2. The rotational energy in the $j^{th}$ energy level is given by

$$\varepsilon_{rot} = j(j+1)/(8\pi^2 I) \tag{9.52}$$

where, $I$ is the moment of inertia of the molecule, $j$ is rotational quantum number and the degeneracy of this level is $(2j + 1)$. The ground level has zero energy. Though the spacing of energy levels is much larger than that for translational energy levels, yet it is not very large, and the assumption of continuous energy distribution is reasonable.

3. The vibratonal energy of a linear harmonic oscillator is given by

$$\varepsilon_{vib} = (v + ½)h\nu \tag{9.53}$$

where, $\nu$ is the frequency of oscillations of the oscillator, and $v$ is the vibrational quantum number. The spacing of energy levels is large, usually 4 to 40 kJ/gm mole. This means that even at high temperatures, most particles will have vibrational energies associated with a few lower energy levels.

### 9.4.1 Distribution of Molecules in Energy Levels

For a given gas, the number of molecules and its energy is fixed. Under these constraints, the molecules will occupy different energy levels, and will jump from one level to another, or from one compartment to another compartment, or a different energy level, because of interactions among these. Thus, the distribution of molecules in the energy levels and within the compartments of an energy level will go on changing. Each distribution is called a microstate. However, there will be a number of such microstates in which the molecule would change, but not their numbers in any energy level or its compartments. That is, *all these microscopic states would refer to the same distribution, and is known as macrostate.* The number of microstates that form a macrostate is called *thermodynamic probability, W*. From the probability theory, the thermodynamic probability of any of the microstates is the same and is known as the postulate of 'equal a priori probability', in equilibrium state. That is, the system will pass through every conceivable microstate, and in the process, any microstate may repeat itself. We may visualise that the thermodynamic probability of all the macrostates would be different, and there will be one for which it shall be maximum. This state is the *most probable state,* and the basic problem that is addressed in statistical thermodynamics is to determine the thermodynamic probability of the most probable state, including the degeneracy of the energy levels.

To illustrate these concepts, let us consider a system consisting of four particles having total energy of four units, and that the energy levels are equispaced with an energy difference of 1, and ground level at value 0. The various ways in which these molecules can be distributed is given in Table 9.2. Let us consider the macrostate A. Let *a, b, c,* and *d* are the four particles. These can be arranged in the following four ways, taking any three in energy level 0 and 1 in energy level 4.

*abc*/*d* *bcd*/*a* *cda*/*b* and *dab*/*c*

**Table 9.2** Distribution of four particles with four energy units

| Energy level | Energy of that level | Macrostates | | | | | |
|---|---|---|---|---|---|---|---|
| | | A | B | C | D | E | $\Sigma W_i$ |
| 0 | 0 | 3 | 2 | 2 | 1 | 0 | |
| 1 | 1 | 0 | | 1 | 2 | 4 | |
| 2 | 2 | 0 | 2 | 1 | 1 | 0 | |
| 3 | 3 | 0 | 0 | 0 | 0 | 0 | |
| 4 | 4 | 1 | 0 | 0 | 0 | 0 | |
| $W_i$ | | 4 | 6 | 12 | 12 | 1 | 35 |

Thus, the thermodynamic probability of state *A* is 4, which is given in the column at the bottom. Similarly, we may determine the thermodynamic probability of other macrostates. Here, *abc*/*d*, *bcd*/*a*, are the microstates but all the four give the same distribution of molecules in energy levels 0 and 4. Here, we have assumed that the energy level has a degree of degeneracy, $g = 1$, that is, none of the energy levels is degenerating.

Coming back to the issue of determining thermodynamic probability, we have to also know the nature of particles being handled. These particles may be molecules in gases, photons in electromagnetic waves, or electrons in atoms or plasma, and so on. This subject was developed by many well-known authorities of the time, that culminated in three types of statistical distributions, which do not differ much. We shall now determine the thermodynamic probability for these distributions.

#### 9.4.1.1 Maxwell–Boltzman (MB) Statistics

Maxwell and Boltzman[1] determined the thermodynamic probability for the most probable state, on the assumption that the molecules of the system are distinguishable from each other, and that the degeneracy of an energy level $i$ is $g_i$. Let us consider $N$ distinguishable particles distributed in various energy levels, with $N_i$ number of particles in the $i^{th}$ energy level having degeneracy $g_i$. The number of ways $N_0$ particles out of a total of $N$ particles may be placed in level 0 is ${}^{N}C_{N_0}$. The number of particles now left is $N - N_0$, out of which $N_2$ are to be arranged in energy level 1. This can be done in ${}^{N-N_1}C_{N_2}$ ways, and so on. Therefore, the thermodynamic probability of the macrostate is

$$W = \prod_i W_i = {}^{N}C_{N_0} \; {}^{(N-N_0)}C_{N_1} \; {}^{(N-N_0-N_1)}C_{N_2} \dots = \frac{N!}{N_0!\, N_1!\, N_2! \dots}$$

$$= \frac{N!}{\prod_i N_i!} \tag{9.54}$$

where, $i$ varies from 0 to the number of the highest energy level; this has been taken as $\infty$ in a system having very large number of particles. It is justified because the number of energy levels involved is too large. Now, we consider distribution of $N_i$ particles in $g_i$ compartments of the energy level or cell. Since each particle will have $g_i$ choice, the number of ways in which $N_i$ distinguishable particles could be arranged among $g_i$ quantum states is $g_i^{N_i}$. Therefore, the thermodynamic probability of the macrostate would be a product of the probability determined on the basis of two considerations.

$$W = \frac{N!}{\prod_i N_i!} \prod_i g_i^{N_i} = N! \prod_i \frac{g_i^{N_i}}{N_i!} \tag{9.55}$$

Equation 9.55 gives the thermodynamic probability for MB distribution with distinguishable particles. However, this assumption may be relaxed. In that case, changing a particle by another particle in a compartment of energy level will not change the microstate, and hence all such states should not be counted. The $N$ particles can be

---

1. At the time Boltzman gave his results, the discrete theory of energy distribution was in the infancy stage, and he found distribution of molecules assuming continuous distribution of energy from 0 to $\infty$. This was true too, since he was dealing with translational energy of the molecules, for which, as mentioned earlier, the difference in energy of energy levels is so small, that it appears to be continuous.

arranged in $N!$ ways. Hence, thermodynamic probability for MB distribution with indistinguishable particles is

$$W = \frac{N!}{N!}\prod_i \frac{g_i^{N_i}}{N_i!} = \prod_i \frac{g_i^{N_i}}{N_i!} \tag{9.55a}$$

#### 9.4.1.2 Bose–Einstein (BE) Statistics

Bose and Einstein assumed the particles to be indistinguishable, but the energy levels as degenerated, like in the cases of photons of radiation, and gas molecules at high temperature. This therefore differs from the Maxwell-Boltzman statistics, because of the indistinguishability of the molecules from each other. We shall first find out the thermodynamic probability of distribution of $N_i$ particles in $g_i$ compartments of the cell $i$, which will be the number of ways of arranging $N_i + g_i$ variables. However, each such microscopic state will begin with the address of the compartment that can be permuted in $g_i$ ways. Therefore, the remaining $N_i + g_i - 1$ variables can be arranged in $(N_i + g_i - 1)!$ ways. But this number would include all those microstates that are produced by just changing the particle in a compartment, without changing the number of particles, since the particles are indistinguishable. This number would be $N_i!$, which is the number of ways $N_i$ particles may be arranged. For example, in case of distribution of say six particles in four compartments, the microstates a21 b435 c d6 and a34 b216 c d5 are identical, because there is change of the particles 2 and 1 in compartment a by particles 3 and 4, and like that for other compartments. Similarly, when the quantum states are also indistinguishable (no difference in compartments a or b), then arrangements like b435 a21 c d6 and c435 d21 b a6 (here, we changed the compartment keeping the same particles) are exactly the same. All such arrangements, which is $g_i!$, should be counted as one. Thus, the thermodynamic probability of energy level $g_i$ and the whole system is:

$$W_i = \frac{g_i(N_i + g_i - 1)!}{g_i! N_i} = \frac{(N_i + g_i - 1)!}{(g_i - 1)! N_i} \quad \text{and}$$

$$W = \prod_i W_i = \prod_i \frac{(N_i + g_i - 1)!}{(g_i - 1)! N_i} \tag{9.56}$$

#### 9.4.1.3 Fermi–Dirac (FD) Statistics

In BE statistics, there is no limit on the number of occupancy of a compartment. But when we consider distribution of electrons in metals, then we know that two electrons cannot occupy the same compartment, as they would repel each other. In such a situation, $g_i \geq N_i$, and only $N_i$ compartments will be occupied and $(g_i - N_i)$ compartments will lie vacant. To count the number of ways of filling $N_i$ particles in any $g_i$ compartment, we know the first particle will have $g_i$ choices, the second $(g_i - 1)$, the third $(g_i - 2)$ choices, and so on. Thus, the number of ways in which it is done is:

$$g_i(g_i - 1)(g_i - 2)(g_i - 3) \ldots [g_i - (N_i - 1)] = g_i!/(g_i - N_i)!$$

This distribution assumes that the particles are distinguishable. If it is not so, then it is immaterial which particle is the first or second and so on. All the microstates formed will be the same if the same compartments are filled by any particle. Therefore, $N_i!$ microstates will be the same macrostate. Thus, the thermodynamic probability of FD distribution for $i^{th}$ energy level and the whole system is:

$$W_i = \frac{g_i!}{N_i!(g_i - N_i)!} \text{ and } W = \prod_i W_i = \prod_i \frac{g_i!}{N_i!(g_i - N_i)!} \tag{9.57}$$

### 9.4.2 Thermodynamic Probability of the Most Probable State

In Section 9.4.1, we learnt how to find out the thermodynamic probability of a given distribution. Coming back to Table 9.2, we may verify its last row using Eq. 9.54. From the table, we note that states $C$ and $D$ are equally probable, and the two have 24 out of 35 possible microstates. We may explore the effect on thermodynamic probability, if both $N_i$ and the number of energy levels are increased to a large value. To start with let us consider the number of molecules, say six, having total energy of six units. This means there are seven energy levels. Table 9.3 lists all the possible macrostates and their thermodynamic probability. We observe the same trend as seen in the previous case, with more microstates associated with the macrostates like $F$, $G$, $H$ and I.

**Table 9.3** Distribution of six particles with six unit energy

| Energy level | Macrostates | | | | | | | | | | | $\Sigma W_i$ |
|---|---|---|---|---|---|---|---|---|---|---|---|---|
| | A | B | C | D | E | F | G | H | I | J | K | |
| 0 | 5 | 4 | 4 | 4 | 3 | 3 | 3 | 2 | 2 | 1 | | |
| 1 | | | 1 | | | | 1 | 2 | 3 | 2 | 4 | 6 |
| 2 | | | 1 | | 3 | 1 | | 2 | 1 | | | |
| 3 | | | | 2 | | 1 | | 1 | | | | |
| 4 | | | 1 | | | | 1 | | | | | |
| 5 | | 1 | | | | | | | | | | |
| 6 | 1 | | | | | | | | | | | |
| $W_i$ | 6 | 30 | 30 | 15 | 0 | 120 | 60 | 60 | 90 | 30 | 1 | 462 |

Likewise, the number of particles and the energy levels may be increased, and we may observe the changes made during this. The observations are given later.

We would also like to study the effect of change in number of particles on the thermodynamic probability of the most probable macrostate, since the system is passing spontaneously through different microstates with equal probability. Consider for example, 1550 particles are distributed in three energy levels: $N_0 = 1250$, $N_1 = 250$ and $N_2 = 50$ giving macrostate $A$. Using Eq. 9.54, we get

$$W_A = \frac{1550!}{1250!250!50!}$$

Now, consider another state $B$ close to state $A$, obtained by moving one molecule from cell 1 to cell 0, and another to cell 2, leaving distribution 1251, 248 and 51. For this distribution, the thermodynamic probability is

$$W_B = \frac{1550!}{1251!248!51!} = 0.924\,W_A$$

Consider a third distribution $C$ close to $A$, but having 1255, 240 and 55 as the distribution of particles in three cells; its thermodynamic probability is

$$W_C = \frac{1550!}{1255!240!55!} = 0.6\,W_A$$

We may continue to do so, and shall see that the thermodynamic probability of the states differing a little from the most probable state decreases very fast. This trend for the case of large number of particles and degeneracy is shown in Fig. 9.10. We may infer from the study that the distribution remains the same at or near the most probable state for most of the time.

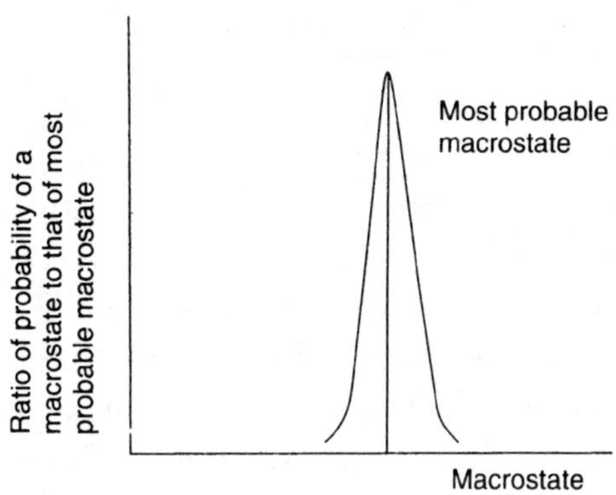


**Fig. 9.10** Probability of occurrence

In the event of collisions, one macrostate changes to another. Let us examine in how many ways such changes can take place. Table 9.4 reproduces Table 9.2 giving the distribution of four particles having a total energy of six units. Interaction of particles occurs in the following ways.

**Case 1.** $N_i$ particles in the same cell may interact in ${}^{N_i}C_2$ ways, since two particles collide with each other at a time. After interaction, any one or both the particles may go to the parent or new cell. The former can be done only in one way, but the latter in one or two ways, depending on whether they go to the same level or different levels.

**Case 2.** Particles from different levels may interact, say $N_i$ from the $i$th and $N_j$ from the $j$th level. These can interact in $N_i \times N_j$ ways. After each interaction,

(i) The particles may stay in the same levels, which they can do in only one way,

(ii) The particles may interchange their levels, which they can do in only one way, and

(iii) The particles may go to new levels, which they can do in one or two ways, depending upon whether they go to the same or different levels, respectively.

**Table 9.4** Transition from one to another macrostate

| Initial State | Energy levels | | | | | Final macrostate | | | | |
|---|---|---|---|---|---|---|---|---|---|---|
| | 0 | 1 | 2 | 3 | 4 | A | B | C | D | E |
| A | 3 | 0 | 0 | 0 | 1 | 9 | 3 | 6 | 0 | 0 |
| B | 2 | 0 | 2 | 0 | 0 | 2 | 10 | 2 | 4 | 0 |
| C | 2 | 1 | 0 | 1 | 0 | 2 | 1 | 11 | 4 | 0 |
| D | 1 | 2 | 1 | 0 | 0 | 0 | 2 | 4 | 11 | 1 |
| E | 0 | 4 | 0 | 0 | 0 | 0 | 0 | 0 | 12 | 6 |
| | | | | | Total | 13 | 16 | 23 | 31 | 7 |

From these considerations, we can find out the number of ways a macrostate can change to another. For example, consider transformation of state $A$ to $C$. This is possible only if one particle from 0 and another from level 4 interact, and these move to levels 1 and 3, respectively after the interaction. Thus, the possible number of ways, by case 2, is $2 \times 3 \times 1 = 6$ (the factor 2 comes because after interaction, the particles go to two different energy levels, and this is done in two ways, as discussed). Similarly, consider the transition from $C$ to $C$, then the total number of ways it is possible is the sum of the following.

(i) The particles in 0 level interact and stay there. This can happen in $1 \times 2c_2 = 1$ way.

(ii) Particles from 0 level interact with particles from level 1, and after interaction, either stay in the same cell or interchange their position. This can be done in $2 \times 1 \times 1 + 2 \times 1 \times 1 = 4$ ways.

(iii) Similarly, for interaction between particles from levels 0 and 3, as discussed above, can be done in four ways.

(iv) Lastly, consider the interaction between particles in levels 1 and 3. After interaction, these particles may either stay in the same level or interchange, both of which can be done in one way. Thus the possible number of ways for this interaction is $1 \times 1 \times 1 + 1 \times 1 \times 1 = 2$ ways.

So, the total number of ways of transition from $C$ to $C$ is $1 + 4 + 4 + 2 = 11$. This way, the number of possible ways for transition from one state to another state can be determined, which are indicated in Table 9.4 in the column under final macrostate. Here, the number 9 at the top left corner indicates the number of ways state A can change to state A. We observe from this table that,

(i) The thermodynamic probability of transition from a state to the same state is more (diagonal elements in the table).

(ii) The number of ways for transition from any state to $C$ or $D$ is more. Recall our earlier discussion that thermodynamic probability of two states $C$ and $D$ is maximum and are more probable.

The above analysis brings out the following important results.

1. The number of macrostates and microstates increase sharply with increase in $N$ and $U$.
2. A group of a few macrostates contain the majority of microstates. As $N$ and $U$ increase the number of macrostates in this group decreases. Figure 9.10 shows the ratio of probabilities of occurrence of a given macrostate with respect to the probability of the most probable macrostate for a system comprising of a very large number of molecules. This figure shows that the system remains almost close to the most probable macrostate.
3. When the number of particles and energy levels is large, then the interactions among particles go on changing the macrostate. However, these interactions tend to keep the distribution of the particles close to the most probable one.
4. The number of molecules may be large, but the thermodynamic probability of all the macrostates will be a finite number for each equilibrium state of the system. The total sum of the thermodynamic probability of all macrostates has a certain fixed value for each equilibrium state and, therefore, satisfies the requirement of a property.

5. Though the thermodynamic probability of the most probable macrostate $W_{mp}$ as well as $W_{total}$ increases with increase in N, the ratio $W_{mp}/W_{total}$, however, decreases. This can be seen from Table 9.5 developed for a system having different number of particles, and for which the allowed energy levels are 0, 1, and 2.

**Table 9.5** Comparison of system behaviour at microscopic level

| System | $N$ | $U$ | Distribution $n_0$ | $n_1$ | $n_2$ | $W_{mp}$ | $W_{total}$ | $W_{mp}/W_{total}$ | log $W_{mp}$ | log $W_{total}$ | $\frac{\log W_{mp}}{\log W_{total}}$ |
|---|---|---|---|---|---|---|---|---|---|---|---|
| A | 62 | 14 | 50 | 10 | 2 | $1.4 \times 10^{14}$ | $4.25 \times 10^{16}$ | $3.3 \times 10^{-3}$ | 14.15 | 16.62 | 0.851 |
| B | 310 | 70 | 250 | 50 | 10 | $6.95 \times 10^{75}$ | $1.13 \times 10^{80}$ | $5.25 \times 10^{-5}$ | 75.78 | 80.05 | 0.948 |
| C | 1550 | 350 | 1250 | 250 | 50 | $4.4 \times 10^{386}$ | $5.7 \times 10^{395}$ | $7.7 \times 10^{-10}$ | 386.64 | 395.7 | 0.978 |

The most probable state is the equilibrium state and for a system having large number of particles, we may write

$$\text{Log } W_{mp} = \text{Log } W_{total}$$

## 9.4.3 Entropy

We know from classical thermodynamics that entropy of an isolated system is maximum under the equilibrium condition, and from statistical thermodynamics, we have just seen that the system exhibits the most probable state when it is in equilibrium condition. This suggests that there is some functional relation between the two and may be written as

$$S = f(W)$$

Let us consider two systems of probabilities $W_1$ and $W_2$, and entropies $S_1$ and $S_2$, so that

$$S_1 = f(W_1) \text{ and } S_2 = f(W_2)$$

Then, for the combined system,

$$S = S_1 + S_2 \text{ and } W = W_1 . W_2$$

$$\rightarrow S = f(W) = f(W_1 . W_2) = f(W_1) + f(W_2)$$

This suggests that the function should be of the form

$$S = k \ln W \tag{9.58}$$

where, $k$ is constant and is found to be the Boltzman constant. Thermodynamic probability is a measure of the order or disorder of the system, and it is seen from Eq. 9.58, that the entropy and disorder are related to each other. This also means that any process or cause that increases the system entropy also increases the degree of disorder. Thus, heat addition increases the degree of disorder, as it increases the entropy of the system. The largest order will be when $W = 1$ or entropy = 0. This happens when all the particles occupy the same cell in the phase space. Thus, more is the spread of the points in the phase space, then more is the disorder. For an isothermal expansion process, the increase in volume increases the disorder and hence the entropy, while in constant volume heating, increase in temperature increases disorder. However, in an isentropic expansion process, the decrease in temperature counters the increase in disorder due to expansion, and ultimately, there is no change in disorder.

From the above discussions, it means that the system has maximum entropy in the most probable macrostate. Does this mean that the system will never transit to other macrostate having less thermodynamic probability and hence less entropy? The second law says yes, but we know it is not so. This suggests that the second law is the law of probability.

### 9.4.4 Distribution Law

We have studied so far how to calculate thermodynamic probability for MB, BE or FD distributions, and that the system has a tendency to remain close to the most probable state, which is the equilibrium state. But we still do not know what the distribution pattern will be in the most probable state, which is the purpose of studying statistical thermodynamics. Let us now address this question. We shall demonstrate the process for MB distribution, which may be followed for others too.

We assume that the system has a large number of particles, so that the $N_i$ is also large enough to use Stirling approximation, and that the system is isolated, so that the energy and the number of particles are fixed, and that puts a restriction on our solution. These restrictions are expressed mathematically as

$$N = \Sigma N_i \;\rightarrow\; \Sigma \delta N_i = 0, \text{ and} \tag{9.59}$$

$$U = \Sigma N_i \varepsilon_i \;\rightarrow\; \Sigma \delta N_i \varepsilon_i = 0 \tag{9.60}$$

where, $N_i$ and $\varepsilon_i$ are the number of particles and the energy of the energy level $i$, respectively. Taking log of both sides of Eq. 9.55 for MB distribution, using Stirling's approximate formula,

$$\ln N! = N \ln N - N$$

true for large value of $N$, we get

$$\begin{aligned} \ln W &= \ln N! - \Sigma \ln N_i! + \Sigma N_i \ln g_i \\ &= (N \ln N - N) - \Sigma(N_i \ln N_i - N_i) + \Sigma N_i \ln g_i \\ &= N \ln N - \Sigma N_i \ln (N_i / g_i) \end{aligned} \tag{9.60a}$$

Since the system remains close to the most probable state at which the thermodynamic probability is maximum, then for the most probable state, the first differential of $W$ has to be zero. This gives

$$\partial \ln W = \partial/\partial N_i (ln\, W)\delta N_i = 0$$
$$\rightarrow \Sigma[\ln (N_i/g_i) + 1]\, \delta N_i = 0$$

or
$$\Sigma \ln (N_i/g_i)\, \delta N_i = 0 \qquad (9.61)$$

To find the solution to the Eq. 9.61 with the constraint equations 9.59 and 9.60, the method of Lagrange undetermined multiplier is used. Multiplying Eq. 9.59 by $\alpha$ and 9.60 by $\beta$ and adding to Eq. 9.61 results in

$$\Sigma[\ln (N_i/g_i) + \alpha + \beta\varepsilon_i]\, \delta N_i = 0$$

Since all $\delta N_i$ cannot be zero, the quantity within the summation sign has to be zero. Thus,

$$\ln (N_i/g_i) + \alpha + \beta\varepsilon_i = 0$$
$$\rightarrow N_i/g_i = \exp - (\alpha + \beta\varepsilon_i) = 1/\{A \exp (\beta\varepsilon_i)\} \qquad (9.62)$$

where, $A$ is a constant = $e^{\alpha}$. To find unknown $\alpha$, we use Eq. 9.59. Thus for MB distribution

$$N = \Sigma\, g_i/\{A \exp (\beta\varepsilon_i)\} = (1/A)\, \Sigma\, g_i \exp - (\beta\varepsilon_i) = Z/A, \text{ where}$$
$$Z = \Sigma\, g_i \exp - (\beta\varepsilon_i) \qquad (9.63)$$

$Z$ is called the *partition function*. The MB distribution can be written as

$$N_i = [N\, g_i \exp - (\beta\varepsilon_i)]/[\Sigma\, g_i \exp - (\beta\varepsilon_i)]$$
$$= N\, g_i \exp - (\beta\varepsilon_i)/Z \qquad (9.64)$$

and, $\ln N_i = \ln N - \ln Z + \ln g_i - (\beta\varepsilon_i)$ (9.65)

We use Eq. 9.58 to calculate $\beta$. Substituting for $\ln W$ from Eq. 9.60(a), and $\ln N_i$ in 9.58, and using Eq. 9.65 for $\ln N_i$ we get

$$S = k \ln W = k[N \ln N - \Sigma N_i \ln (N_i/g_i)]$$
$$= k[N \ln N + \Sigma(N_i \ln g_i - N_i \ln N_i)]$$
$$= K\,[N \ln N + \Sigma N_i (\ln g_i - \ln N + \ln Z - \ln g_i + \beta\varepsilon_i)]$$
$$= k\,(N \ln Z + \beta\, U) \qquad (9.66)$$

simplified by using $\Sigma N_i \ln N = N \ln N$ and $\Sigma \beta N_i \varepsilon_i = \beta U$. From the $T\,dS$ equation, we know,

$$1/T = (\partial S/\partial U)_v = k\beta$$

Since Eq. 9.66 gives on differentiation $(\partial S/\partial U)_v = k\beta$. Therefore,

$$\beta = 1/(kT) \text{ and } N_i = (Ng_i/Z) \exp - (\varepsilon_i/kT) \qquad (9.67)$$

Eq. 9.67 gives the distribution of particles in different energy levels corresponding to the equilibrium state for MB distribution. The results for other distributions are:

BE statistics: $N_i = (Ng_i/Z)/[\exp(\varepsilon_i/kT) - 1]$ (9.68)

FD statistics: $N_i = (Ng_i/Z)/[\exp(\varepsilon_i/kT) + 1]$ (9.69)

We note that BE statistics reduce to MB statistics when $N_i >> g_i$.

### 9.4.5 Partition Function

The partition function was defined in Eq. 9.63 as

$$Z = \Sigma g_i \exp - (\beta\varepsilon_i) \tag{9.70}$$

We shall show later that the macroscopic quantities are expressible in terms of the partition function, and hence their determination is required. We studied earlier the energy associated with *one particle* in different energy levels for three modes, namely, translational, rotational and vibrational. For these cases, we shall now determine the partition function, that will involve a triple integral.

**Translational mode.** For such a mode $g_i$ is = 1. Eq. 9.51 gives the value of $\varepsilon_i$. Substituting it in Eq. 9.70 and replacing the summation sign by an integral, as the spacing between the energy levels is small, we get

$$Z_{tr} = \Sigma\Sigma\Sigma \exp\left[-\frac{\beta h^2}{8m}\left(\frac{K_x^2}{a^2} + \frac{K_y^2}{b^2} + \frac{K_z^2}{c^2}\right)\right]$$

$$= \int_0^\infty \exp\left(-\frac{\beta h^2 K_x^2}{8ma^2}\right) dk_x \int_0^\infty \exp\left(-\frac{\beta h^2 K_y^2}{8mb^2}\right) dk_y \int_0^\infty \exp\left(-\frac{\beta h^2 K_z^2}{8mc^2}\right) dk_z$$

$$Z_x = \int_0^\infty \exp\left(-\frac{\beta h^2 K_x^2}{8ma^2}\right) dk_x = \frac{a}{h}\left(\frac{2\pi m}{\beta}\right)^{1/2}$$

Similarly $Z_y = \dfrac{b}{h}\left(\dfrac{2\pi m}{\beta}\right)^{1/2}$

and $Z_z = \dfrac{c}{h}\left(\dfrac{2\pi m}{\beta}\right)^{1/2}$

$$\therefore\ Z_{tr} = \frac{abc}{h^3}\left(\frac{2\pi m}{\beta}\right)^{3/2} = \frac{V}{h^3}\left(\frac{2\pi m}{\beta}\right)^{3/2} = \frac{V}{h^3}(2\pi mkT)^{3/2}$$

where, $V$ is the volume of the container having sides $a$, $b$ and $c$, and $h$ is the Planck's constant. The translational partition function depends on the temperature of the gas.

**Rotational mode.** For this mode, $g_i = (2i + 1)$. Substituting for $\varepsilon_i$ from Eq. 9.52 in Eq. 9.70, we get

$$Z_{rot} = \int_0^{\infty} (2i+1)\exp\left[\frac{-i(i+1)h^2\beta}{8\pi^2 I}\right] di = \frac{8\pi^2 I}{h^2\beta} = \frac{8\pi^2 IkT}{h^2}$$

Integration is done easily by assuming $i^2 + i = x$, so that $(2i + 1)\,di = dx$

**Vibrational mode.** Similarly, the partition function for this mode is obtained by noting that for this mode, $g_i = 1$.

$$Z_{vib} = \sum_{n=0}^{\infty} \exp\left[-\frac{(n+1/2)h\nu}{kT}\right]$$

$$= e^{-h\nu/2kT}\left[1 + e^{-h\nu/kT} + e^{-2h\nu/kT} + e^{-3h\nu/kT} + \ldots\right]$$

$$= e^{-h\nu/2kT}\left[1 - \exp\left(\frac{h\nu}{kT}\right)\right]^{-1}$$

Similarly, the partition function for any other mode of energy can be determined. Degeneracy of an energy level and the partition function are multiplicative, and therefore, the total partition function per particle can then be obtained by multiplying the respective partition function of the energy modes.

$$g = g_{trans} \cdot g_{rot} \cdot g_{vib} \cdots \tag{9.71}$$

$$Z = Z_{trans} \,.\, Z_{rot} \,.\, Z_{vib} \cdots \tag{9.72}$$

The total partition function for the totality of the $N$ distinguishable and indistinguishable molecules is then given by:

$Z_{total} = Z^N$ for distinguishable, and $Z_{total}$
$= Z^N/N!$ for indistinguishable molecules (9.73)

### 9.4.6 Thermodynamic Properties

The macroscopic properties used in thermodynamics can now be determined in terms of the partition function. For example, the internal energy of the system is given by

$$U = \Sigma N_i \varepsilon_i = (N/Z)\cdot \Sigma g_i \exp(-\beta\varepsilon_i)\cdot \varepsilon_i$$

Differentiating $Z$ given by Eq. 9.70 w.r.t. $\beta$, while treating volume constant (so that $\varepsilon_i$ is constant), we get

$$\ln Z = \ln \Sigma g_i \exp(-\beta\varepsilon_i) \rightarrow \left[\frac{\partial(\ln Z)}{\partial\beta}\right]_V = \frac{-\Sigma g_i \varepsilon_i \exp(-\beta\varepsilon_i)}{\Sigma g_i \exp(-\beta\varepsilon_i)}$$

$$= \frac{-\Sigma g_i \varepsilon_i \exp(-\beta \varepsilon_i)}{Z}$$

Comparing it with the expression for $U$ and substituting $\beta = 1/(kT)$, we get

$$U = -N\left(\frac{\partial \ln Z}{\partial \beta}\right)_V = NkT^2\left(\frac{\partial \ln Z}{\partial T}\right)_V$$

$$= N\bar{R}T^2\left(\frac{\partial \ln Z}{\partial T}\right) \rightarrow C_v = \frac{1}{n}\left(\frac{\partial U}{\partial T}\right) = \frac{\bar{R}}{T^2}\left[\frac{\partial^2 \ln Z}{\partial (1/T)^2}\right]_v \tag{9.74}$$

Equation 9.58 gives entropy in terms of the thermodynamic probability of the most probable macrostate. We can now express it in terms of the partition function as well. Substituting for $W$ from Eq. 9.55. (a) and using Stirling approximation,

$$\ln W = \Sigma[N_i \ln g_i - (N_i \ln N_i - N_i)]$$

$$= \Sigma N_i [\ln (g_i/N_i + 1] \tag{9.75}$$

Writing Eq. 9.67 in the form $g_i/N_i$ and taking its log, we get

$$\ln (g_i/N_i) = \ln (Z/N) + \beta\varepsilon_i \tag{9.76}$$

$$S = k \ln W = k\Sigma N_i [\ln (Z/N) + \beta\varepsilon_i + 1]$$

$$= kN \ln (Z/N) + k\beta U + kN$$

$$= kN[\ln (Z/N) + 1] + U/T$$

$$= n\bar{R}\ [\ln (Z/N) + T\ (\partial \ln Z/\partial T)_v + 1] \tag{9.77}$$

For the whole system, from Eq. 9.73,

$$Z_{\text{total}} = Z^N/N! \rightarrow \ln Z_{\text{total}} = N \ln Z - N \ln N + N$$

$$= N\,[\ln (Z/N + 1]$$

$$S = k\,[ln\ Z_{\text{Total}} + T\ (\partial \ln Z_{\text{total}}/\partial T)_v] \tag{9.77a}$$

The Helmholtz function $F$ is $F = U - TS$. Substituting for U and S from Eq. 9.74 and 9.77, we get

$$F = n\bar{R}T^2\left(\frac{\partial \ln Z}{\partial T}\right)_V - NkT\left[ln\left(\frac{Z}{N}\right) + T\left(\frac{\partial \ln Z}{\partial T}\right)_V + 1\right]$$

$$= -n\bar{R}T\left[\ln\left(\frac{Z}{N}\right) + 1\right] = -kT \ln Z_{\text{total}} \tag{9.78}$$

Pressure is defined in terms of Helmholtz function as $p = -(\partial F/\partial V)_T$. From Eq. 9.78

$$p = n\bar{R}T[\partial \ln Z/\partial V]_T \tag{9.79}$$

Now, the other properties enthalpy and Gibbs function can similarly be obtained from their definition. We get after simplification and using relation

$$\partial/\partial T = (\partial/\partial \ln T)/T \text{ and } \partial/\partial V = (\partial/\partial \ln V)/V$$

$$H = U + pV = n\bar{R}T[(\partial \ln Z/\partial \ln T)_V + (\partial \ln Z/\partial \ln V)_T] \quad (9.80)$$

$$G = U + pV - TS = n\bar{R}T[\ln(Z/N) + 1 - V(\partial \ln Z/\partial V)_T] \quad (9.81)$$

## 9.5 STATISTICAL AND MACROSCOPIC THERMODYNAMICS

We mentioned in Chapter 1 that microscopic and macroscopic analyses are different ways of analysis and should give the same results. We compared some results from the kinetic theory with those from macroscopic analysis to prove the statement, and shall continue it for the results from statistical thermodynamic analysis. Let us consider gas molecules contained in a cube of dimensions *a*, *b* and *c* along directions *x*, *y* and *z* respectively, of the coordinate axes. Let $F_i$ is the force exerted by the $i^{th}$ molecule, and *da* is the displacement along the *x* direction. If $F_{av}$ is the average force, and *Fi* that of the $i^{th}$ molecule in the *x* direction, then:

Work done by the $i^{th}$ particle = $F_i\, da$ = decrease in energy $d\varepsilon_i$

that is, $d\varepsilon_i = -F_i\, da \rightarrow F_i = -\partial\varepsilon_i/\partial a$

$$F_{av} = \Sigma N_i F_i/\Sigma N_i = \Sigma F_i \exp(-\varepsilon_i/kT)/\Sigma \exp(-\varepsilon_i/kT)$$

$$= -\Sigma(\partial\varepsilon_i/\partial a)\exp(-\varepsilon_i/kT)/Z$$

or $\quad Z = \Sigma \exp(-\varepsilon_i/kT) \rightarrow (\partial Z/\partial a)_T = -(1/kT)\,\Sigma \exp(-\varepsilon_i/kT)\,(\partial\varepsilon_i/\partial a)_T$

$$F_{av} = (ZkT/Z)\,(\partial Z/\partial a)_T = kT\,(\partial Z/\partial a)_T$$

$$dW = NF_{av}\, da = NkT\,(\partial Z/\partial a)_T\, da = n\bar{R}T\,(\partial Z/\partial a)_T = pdV$$

using Eq. 9.79. Here, we multiplied $F_{av}\, da$ by *N*, where *N* is the number of molecules in the container, because $F_{av}$ is the average force exerted by a single molecule. The result obtained by such analysis is the same as what we get from macroscopic analysis for a reversible process. We may similarly proceed to find the relation for heat transfer for a reversible process. We know from macroscopic analysis that

$$dQ = dU + dW = n\bar{R}d\left[T^2\left(\frac{\partial \ln Z}{\partial T}\right)_V\right] + n\bar{R}T\left[T^2\left(\frac{\partial \ln Z}{\partial V}\right)_T\right]dV$$

Since $d\ln Z = \left(\dfrac{\partial \ln Z}{\partial T}\right)_V dT + \left(\dfrac{\partial \ln Z}{\partial V}\right)_T dV$

$$dQ = n\bar{R}T\left[T^2\left(\frac{\partial \ln Z}{\partial V}\right)_V\right] - n\bar{R}T\left[\left(\frac{\partial \ln Z}{\partial V}\right)_T\right]dT + n\bar{R}T\, d\ln Z$$

$$= n\bar{R}T\, d\left[\ln Z + T\left(\frac{\partial \ln Z}{\partial V}\right)_v\right] = Tds$$

In these equations, $Z$ refers to the whole system, and hence, Eq. 9.77 (a) has been used for $S$. This is the result obtained from the second law of thermodynamics for a reversible process.

We studied the concepts of heat and work transfer in Chapter 1. We may add another perspective to these. Consider a system, the total energy of system, $U = \Sigma N_i \varepsilon_i$. This gives:

$$dU = d\,\Sigma N_i \varepsilon_i = \Sigma N_i\, d\varepsilon_i + \varepsilon_i\, dN_i \qquad 9.82$$

By keeping external constraints fixed, $\varepsilon_i$ does not change, i.e. $d\varepsilon_i = 0$. Also, there is no work transfer in such a case, and the change in internal energy is due to heat transfer alone. That means $\varepsilon_i\ dNi$ is a measure of heat transfer. In other words, heat transfer causes change in the distribution of molecules in the most probable macrostate. Comparing the Eq. 9.82 with the non-flow energy equation, we infer that the term $N_i\ d\varepsilon_i$ should then represent he work transfer component. That is, work transfer changes the energy of the energy levels, and not the distribution of the molecules. Since heat transfer is changing, the distribution and hence the entropy is increased or the disorder is increased, whereas the work transfer does not increase or decrease the disorder.

We may also verify some other macroscopic results.

## SUMMARY

Microscopic analysis is based on the statistical analysis of the behaviour of all the particles in a system, subjected usually to some constraints like constant number of particles and the system energy. These particles may be molecules of a gas, photons, or electrons and so on. We did not discuss electrons and photons in this chapter, and concentrated mainly on the molecules. The analysis is divided into two kinds of studies, which were developed simultaneously. These were the kinetic theory and statistical thermodynamics, the latter gained importance as the subject of quantum physics gained momentum. The basic difference in the two approaches is that while the kinetic theory assumes the molecules to be distributed in energy levels, which are treated as continuous, statistical thermodynamics followed the concept of quantum physics, that the energy levels are discrete and not continuous. Both the theories got the macroscopic results in their own way, and were found to agree with the macroscopic results, as was expected. These theories gave an explanation for many physically observed phenomena such as viscous force, molecular heat and mass transfer, that is, the transport phenomena, and found expression for the constants involved in the transportation of molecular mass, momentum and energy. Statistical thermodynamics gave a new perspective to work and heat transfer, and also that the second law is the law of probability.

In the kinetic theory, we consider the collision between two particles at a time, and the velocity distribution in the particles from which all the results were predicted. The chances of collision of three particles at a time is feeble because of the distances between the molecules. Maxwell got the expression for speed distribution, that gives the number of molecules having velocity in the range $c$ to $c + dc$. He and the others also determined the number of molecules which would travel a distance without colliding with another molecule. This is known as the number of survived molecules. This number would go on decreasing as the distance travelled increased. The concept of mean free path was also developed as the statistical average of the free paths, which is the distance travelled by a molecule between successive collisions.

On the other hand, in statistical thermodynamics, the probability of the occurrence of a macrostate was determined, and on that basis, it was found that the system in the equilibrium state remains almost in the most probable macrostates or close to it. Not only this, the total thermodynamic probability of all the possible macrostates tend to a fixed number in equilibrium state. But as the number of particles increases, the ratio of the logarithm of the thermodynamic probability for the most probable macrostate, and the logarithm of the thermodynamic probability of all the macrostates approaches 1. This means that the thermodynamic probability of the most probable macrostate represents some property of the system. It is related to entropy, from where entropy is linked to the concept of degree of disorder of the system. Depending on the nature of the particles, three statistical distributions, namely MB, BE and FD statistics, were proposed. These statistics give distribution of the molecules in different energy levels for the most probable macrostate, and hence the entropy of the system. During the analysis, an important function, the partition function of a particle and the whole system, emerged. All the macroscopic properties are linked to this function or its derivative. Therefore, the methods to determine it for different modes of energy that constitute any energy level, have been presented. This basically involves the use of integration or series summation techniques.

## EXERCISES

### Review Problems

9.1 Find out the velocity of two molecules moving in a perpendicular direction after their collision.

9.2 Verify Eqs. 9.32 and 9.33.

9.3 What is the importance of the kinetic theory?

9.4 What do you understand by elastic collision? Derive an expression for the velocity of *two* molecules after collision.

9.5 State the basic assumptions of the kinetic theory of gases.

9.6 Give some idea about the size of the molecules, the number of molecules in one mol, length of free path and the number of collisions per second of a molecule. What is Avogadro's Number?

9.7 Sketch the curve for the force of attraction between molecules against the intermolecular distance.

9.8 Explain the macroscopic concept of pressure by the kinetic theory of gases, and derive an expression for the pressure on the stationary wall.

9.9 Derive the following laws from the kinetic theory of gases and statistical thermodynamics: (a) Boyles law, (b) Dalton's law, and (c) Avogadro's law.

9.10 Explain the macroscopic concept of temperature by the kinetic theory of gases, and establish a relation between the temperature and motion of molecules.

9.11 Derive the relation $c_{rms1}/c_{rms2} = \sqrt{m_2/m_1}$, where subscripts 1 and 2 refer to two gases. What is the practical application of this?

9.12 Derive the relation $pV^{\gamma}$ = constant for ideal gases from kinetic theory of gas.

9.13 What do you understand by distribution of molecular velocity? Explain the Maxwellian distribution of molecular velocities with the help of sketches. What is the effect of temperature on it?

9.14 Define the average, root mean square and most probable velocities, and show that $c_{mp} : \overline{C} : c_{rms}$ :: 1: 1.128: 1.224.

9.15 What is degree of freedom? How many degrees of freedom may a diatomic molecule have at (a) room temperature, (b) very low temperature, and (c) very high temperature?

9.16 Explain the principle of equipartition of energy.

9.17 Derive the specific heats of monoatomic and diatomic gases from the kinetic theory of gases as well as statistical thermodynamics. Why are the molal specific heats of two diatomic gases quite similar? How do these results compare with the experimental values?

9.18 Define mean free path and derive an expression for the same. What is the approximate ratio of mean free path in the earth's atmosphere at sea level to the diameter of the molecules?

9.19 Derive an expression for the number of molecules travelling a distance without collision.

9.20 What is meant by transport phenomena in the fluids? What are the various transport phenomena and what properties do they give rise to?

9.21 Define molecular flux and derive an expression for the same.

9.22 What is generalised transport coefficient and derive an expression for it.

9.23 Prove that dynamic viscosity is proportional to mean molecular speed, mean free path, the nature of fluid and its distribution in space.

9.24 Define thermal conductivity. Derive the relation: $K = 1/6\ \bar{c}\ n\lambda f\ k$, where the notations have the usual meaning.

9.25 Explain the phenomenon of diffusion with the help of the kinetic theory. Derive the expression for diffusion coefficient.

9.26 Does molecular density change in the diffusion process? If no, what is the property that is changing?

9.27 Explain the phenomenon of evaporation with the help of the theory of distribution of molecular velocities.

9.28 Why do the molecules of smoke from a chimney not disperse out immediately into the atmosphere in spite of their high velocities?

9.29 We might reasonably assume that the heat of vaporisation for a fluid would be the energy necessary to increase the velocity of the molecules from an essentially zero value in the liquid state to the *rms* value, representative of the mean energy, in the vapour state. Check the validity of this assumption for vaporisation of water at 1 atm pressure.

9.30 What fractions of molecules have speeds (a) greater than $c_{mp}$, (b) less than $c_{mp}$, (c) greater than some arbitrary velocity $c$?

9.31 Consider a reversible adiabatic expansion of an ideal gas in a cylinder. The piston is assumed to move with some velocity $u$, such that $u \ll c$. For a molecule impinging on the piston at an incident angle $\phi$ show that the decrease in kinetic energy of the molecule is $2\ muv\cos\phi$, where $v$ is the velocity at impact. By a suitable integration, obtain an expression for the kinetic energy lost by all the molecules, and show that this energy is equal to the work done on the piston by the expanding gas.

9.32 A container has two identical chambers separated by an appropriate membrane. One chamber is filled with a gas and the other chamber is evacuated. A small pinhole of area $A$ is punched in the membrane. Derive an expression for the time required for the pressure to drop to three-fourths of its original value.

9.33 Derive an expression for the rms value of free path in terms of the mean free path. What is the most probable free path?

9.34 What is the advantage of statistical methods?

9.35 What is the difference between classical and quantum physics? Explain how Planck's law has helped in the development of the latter.

9.36 What do you mean by energy level and their degeneracy? Give their physical significance.

9.37 How is quantum number related with energy levels? Comment on the statement that energy of ground level is not always zero.

9.38 How are energy levels represented on the diagram? What is the spacing between them?

9.39 What is meant by allowed energy levels? Can a system of particles have infinite number of energy levels?

9.40 Comment on the statement that distribution of energy in energy levels can be assumed as continuum for the translational and the rotational mode, but not for the vibrational mode.

9.41 Define and explain the terms: (a) probability (b) thermodynamic probability, (c) quantum state, (d) microstate, and (e) macrostate.

9.42 What is the postulate of 'equal probability'? Does the macrostate of the system change continuously? What is most probable macrostate?

9.43 Comment on the statement that a measuring instrument will read the property associated with the most probable macrostate.

9.44 It is observed that as the number of particles and energy of the system increases, the percentage of microstates in the most probable macrostate goes on decreasing. Irrespective of this fact, we use $W_{mp}$ instead of $W_{total}$ while calculating entropy. Justify it.

9.45 How does the transition from macrostate to macrostate occur? Does each interaction of particles give rise to a new macrostate?

9.46 Why do we assume binary interaction (i.e. only two particles interacting at a time)? Are we justified in doing so? Why?

9.47 How will you count the number of ways of transition, which may take place from one macrostate to another?

9.48 Show clearly the difference between three statistics. Find out the thermodynamic probability of a macrostate for each of these statistics. What will happen if $g_i = 1$?

9.49 What is the limitation of MB distribution? Find out the distribution law for FD statistics.

9.50 BE and FD statistics in the limit give the Maxwellian distribution. What are these conditions?

9.51 What is the method of Lagrange's undetermined multiplier and when is it used? Determine $\beta$ for three statistics studied, where $\beta$ is the Lagrange constant in the distribution law.

9.52 How is the partition function defined? Determine it for translational, vibrational and rotational modes. What is the partition function for the whole system of $N$ particles?

9.53 How is entropy related to thermodynamic probability? Define the relation.

9.54 Express pressure, $p$, in terms of the derivative of the partition function with respect to volume.

## Problems for Practice

9.1 The speed of 10 particles in m/s are 0, 1.0, 2.0, 3.0, 3.0, 3.0, 4.0, 4.0, 5.0 and 6.0. Calculate the average, *rms* and the most probable speed. [*Ans.* 3.1 m/s, 3.5 m/s, 3 m/s]

9.2 Calculate the value of $c_{rms}$ and $c_m$ for oxygen at 300 K. Compare the fraction of oxygen molecules which have speeds within $\pm 1$ per cent of (a) the most probable speeds, and (b) the *rms* speed. [*Ans.* 445 m/s, 483 m/s, 395 m/s, 0.0167,0.0184].

9.3 Find the collision frequency and the mean free path of nitrogen molecules having $c_{mp} = 455$ m/s. The radius of the nitrogen molecule is $1.8 \times 10^{-10}$ m; the number of molecules per m$^3$ is $3 \times 10^{25}$. [*Ans.* $5.56 \times 10^9$ collisions, $8 \times 10^{-8}$ m]

9.4 Calculate the number of collisions per second and the mean free path for molecules of diameter $2.5 \times 10^{-8}$ cm, having *rms* speed of $10^5$ cm/s. Given that the number of molecules per cc is $3 \times 10^{19}$. [*Ans.* $8 \times 10^8$ collisions/s, $1.2 \times 10^{-5}$ cm]

9.5 Find the specific heats at constant volume and constant pressure (a) for hydrogen, (b) for oxygen. Assume perfect gas behaviour and $\gamma = 1.4$ for both gases. Given Boltzman constant $k = 1.3803 \times 10^{-23}$ J/molecule K.

9.6 Calculate the thermal conductivity of oxygen at 1 atm and 300 K.

9.7 Calculate the dynamic viscosity of oxygen at 300 K and 1 atm pressure, given: diameter of oxygen molecule = $3.5 \times 10^{-10}$ m. [*Ans.* $2.05 \times 10^{-5}$ $N$ s/m$^2$].

9.8 The position of an electron moving in the $x$-direction is known to be within 0.01 mm. Find the order of magnitude of uncertainty in the velocity of the electron.

9.9 For a cubical box of side 1 cm, what is the lowest energy level and the corresponding velocity for an electron and for a proton?

9.10 What is the translational energy level of an oxygen molecule contained in a cube of 1 cm side, if the quantum numbers are $n_x = n_y = n_z = 10^8$?

9.11 Consider a tri-atomic water molecule (atomic weight 18.02). It is contained in a cube of 10 cm side. Find the kinetic energy of the molecule if its translational quantum numbers are $n_x = 10^9$ $n_y = 10^{10}$ and $n_z = 10^{11}$. What are the wavelengths associated with each of the quantum numbers?

9.12 Calculate the rotational energy levels for a diatomic hydrogen molecule for the first five values *of j* (i.e. $j = 0, 1, 2, 3$, and 4). The mass of a hydrogen atom can be considered to be equal to that of a proton $(r = 0.742$ A).

# APPENDIX ONE

# Tables

**Table A1.1** Properties of gases at 25 °C, 100 kPa[a] (SI units)

| Gas | Chemical formula | Molecular mass | $R$ (kJ/kg-K) | $\rho$ (kg/m³) | $C_{p0}$ (kJ/kg-K) | $C_{v0}$ (kJ/kg-K) | $\gamma = \frac{C_p}{C_v}$ |
|---|---|---|---|---|---|---|---|
| Steam | $H_2O$ | 18.015 | 0.4615 | 0.0231 | 1.872 | 1.410 | 1.327 |
| Acetylene | $C_2H_2$ | 26.038 | 0.3193 | 1.05 | 1.699 | 1.380 | 1.231 |
| Air | — | 28.97 | 0.287 | 1.169 | 1.004 | 0.717 | 1.400 |
| Ammonia | $NH_3$ | 17.031 | 0.4882 | 0.694 | 2.130 | 1.642 | 1.297 |
| Argon | Ar | 39.948 | 0.2081 | 1.613 | 0.520 | 0.312 | 1.667 |
| Butane | $C_4H_{10}$ | 58.124 | 0.1430 | 2.407 | 1.716 | 1.573 | 1.091 |
| Carbon dioxide | $CO_2$ | 44.01 | 0.1889 | 1.775 | 0.842 | 0.653 | 1.289 |
| Carbon monoxide | CO | 28.01 | 0.2968 | 1.13 | 1.041 | 0.744 | 1.399 |
| Ethane | $C_2H_6$ | 30.07 | 0.2765 | 1.222 | 1.766 | 1.490 | 1.186 |
| Ethanol | $C_2H_5OH$ | 46.069 | 0.1805 | 1.883 | 1.427 | 1.246 | 1.145 |
| Ethytlene | $C_2H_4$ | 29.054 | 0.2964 | 1.138 | 1.548 | 1.252 | 1.237 |
| Helium | He | 4.003 | 2.0771 | 0.1615 | 5.193 | 3.116 | 1.667 |
| Hydrogen | $H_2$ | 2.016 | 4.1243 | 0.0813 | 14.209 | 10.085 | 1.409 |
| Methane | $CH_4$ | 16.043 | 0.5183 | 0.648 | 2.254 | 1.736 | 1.299 |
| Methanol | $CH_3OH$ | 32.042 | 0.2595 | 1.31 | 1.405 | 1.146 | 1.227 |
| Neon | Ne | 20.183 | 0.4120 | 0.814 | 1.03 | 0.618 | 1.667 |
| Nitric oxide | NO | 30.006 | 0.2771 | 1.21 | 0.993 | 0.716 | 1.387 |
| Nitrogen | $N_2$ | 28.013 | 0.2968 | 1.13 | 1.042 | 0.745 | 1.400 |
| Nitrous oxide | $N_2O$ | 44.013 | 0.1889 | 1.775 | 0.879 | 0.690 | 1.274 |
| *n*-Octane | $C_8H_{18}$ | 114.23 | 0.07279 | 0.092 | 1.711 | 1.638 | 1.044 |

*(Contd.)*

| Gas | Chemical formula | Molecular mass | $R$ (kJ/kg-K) | $\rho$ (kg/m³) | $C_{p0}$ (kJ/kg-K) | $C_{v0}$ (kJ/kg-K) | $\gamma = \frac{C_p}{C_v}$ |
|---|---|---|---|---|---|---|---|
| Oxygen | $O_2$ | 31.999 | 0.2598 | 1.292 | 0.922 | 0.662 | 1.393 |
| Propane | $C_3H_8$ | 44.094 | 0.1886 | 1.808 | 1.679 | 1.490 | 1.126 |
| R-12 | $CCl_2F_2$ | 120.914 | 0.06876 | 4.98 | 0.616 | 0.547 | 1.126 |
| R-22 | $CHClF_2$ | 86.469 | 0.09616 | 3.54 | 0.658 | 0.562 | 1.171 |
| R-32 | $CF_2H_2$ | 52.024 | 0.1598 | 2.128 | 0.822 | 0.662 | 1.242 |
| R-125 | $CHF_2CF_3$ | 120.022 | 0.06927 | 4.918 | 0.791 | 0.721 | 1.097 |
| R-134a | $CF_3CH_2F$ | 102.03 | 0.08149 | 4.20 | 0.852 | 0.771 | 1.106 |
| Sulphur dioxide | $SO_2$ | 64.059 | 0.1298 | 2.618 | 0.624 | 0.494 | 1.263 |
| Sulphur trioxide | $SO_3$ | 80.053 | 0.10386 | 3.272 | 0.635 | 0.531 | 1.196 |

[a] Or saturation pressure if it is less than 100 kPa.

**Table A1.2** Equations of state constants (SI units)

| Gas | van der Waals constants[a] | | |
|---|---|---|---|
| | R, KJ/(kg · K) | a, kN · m⁴/kg² | b × 10³, m³/kg |
| Air | 0.287 | 0.162 | 1.25 |
| Ammonia | 0.488 | 1.457 | 2.18 |
| Butane ($n$) | 0.143 | 0.410 | 2.00 |
| Carbon dioxide | 0.189 | 0.189 | 1.00 |
| Carbon monoxide | 0.297 | 0.187 | 1.44 |
| Helium | 2.077 | 0.211 | 5.81 |
| Hydrogen | 4.124 | 6.125 | 13.24 |
| Methane | 0.518 | 0.889 | 2.68 |
| Nitrogen | 0.297 | 0.174 | 1.37 |
| Oxygen | 0.260 | 0.135 | 1.00 |
| Propane | 0.189 | 0.482 | 2.06 |
| Regrigerant-12 | 0.069 | 0.073 | 0.81 |
| Water vapor | 0.462 | 1.704 | 1.69 |

[a] These values of $R$, $a$, and $b$ can be obtained from

$R = \Re/M$, $a = (27/64)(RT_c)^2/p_c$ and $b = RT_c/(8p_c)$ where $M$ is the molecular mass and $T_c$ and $p_c$ are the temperature and pressure at the critical state.

**Table A1.3** Critical point data

| Substance | Formula | Molec. Mass | Temp. (K) | Press. (MPa) | Vol. ($m^3/kg$) |
|---|---|---|---|---|---|
| Ammonia | $NH_3$ | 17.031 | 405.5 | 11.35 | 0.00426 |
| Argon | Ar | 39.948 | 150.8 | 4.87 | 0.00188 |
| Bromine | $Br_2$ | 159.808 | 588 | 10.30 | 0.000796 |
| Carbon dioxide | $CO_2$ | 44.01 | 304.1 | 7.38 | 0.00212 |
| Carbon monoxide | CO | 28.01 | 132.9 | 3.50 | 0.00333 |
| Chlorine | $Cl_2$ | 70.906 | 416.9 | 7.98 | 0.00175 |
| Fluorine | $F_2$ | 37.997 | 144.3 | 5.22 | 0.00174 |
| Helium | He | 4.003 | 5.19 | 0.227 | 0.0143 |
| Hydrogen (normal) | $H_2$ | 2.016 | 33.2 | 1.30 | 0.0323 |
| Krypton | Kr | 83.80 | 209.4 | 5.50 | 0.00109 |
| Neon | Ne | 20.183 | 44.4 | 2.76 | 0.00206 |
| Nitric oxide | NO | 30.006 | 180 | 6.48 | 0.00192 |
| Nitrogen | $N_2$ | 28.013 | 126.2 | 3.39 | 0.0032 |
| Nitrogen dioxide | $NO_2$ | 46.006 | 431 | 10.1 | 0.00365 |
| Nitrous oxide | $N_2O$ | 44.013 | 309.6 | 7.24 | 0.00221 |
| Oxygen | $O_2$ | 31.999 | 154.6 | 5.04 | 0.00229 |
| Sulphur dioxide | $SO_2$ | 64.063 | 430.8 | 7.88 | 0.00191 |
| Water | $H_2O$ | 18.015 | 647.3 | 22.12 | 0.00317 |
| Xenon | Xe | 131.30 | 289.7 | 5.81 | 0.000982 |
| Acetylene | $C_2H_2$ | 26.038 | 308.3 | 6.14 | 0.00433 |
| Benzene | $C_6H_6$ | 78.114 | 562.2 | 4.89 | 0.00332 |
| *n*-Butane | $C_4H_{10}$ | 58.124 | 425.2 | 3.80 | 0.00439 |
| Chlorodifluoroethane (142b) | $CH_3CClF_2$ | 100.495 | 410.3 | 4.25 | 0.00230 |
| Chlorodifluoromethane (22) | $CHClF_2$ | 86.469 | 369.3 | 4.97 | 0.00191 |
| Dichlorofluoroethane (141) | $CH_3CCl_2F$ | 116.95 | 481.5 | 4.54 | 0.00215 |
| Dichlorotrifluoroethane (123) | $CHCl_2CF_3$ | 152.93 | 456.9 | 3.66 | 0.00182 |
| Difluoroethane (152a) | $CHF_2CH_3$ | 66.05 | 386.4 | 4.52 | 0.00272 |
| Difluoromethane (32) | $CF_2H_2$ | 52.024 | 351.3 | 5.78 | 0.00230 |
| Ethane | $C_2H_6$ | 30.070 | 305.4 | 4.88 | 0.00493 |
| Ethyl alcohol | $C_2H_5OH$ | 46.069 | 513.9 | 6.14 | 0.00363 |
| Ethylene | $C_2H_4$ | 28.054 | 282.4 | 5.04 | 0.00465 |
| *n*-Heptane | $C_7H_{16}$ | 100.205 | 540.3 | 2.74 | 0.00431 |

*(Contd.)*

| Substance | Formula | Molec. Mass | Temp. (K) | Press. (MPa) | Vol. ($m^3/kg$) |
|---|---|---|---|---|---|
| *n*-Hexane | $C_6H_{14}$ | 86.178 | 507.5 | 3.01 | 0.00429 |
| Methane | $CH_4$ | 16.043 | 190.4 | 4.60 | 0.00615 |
| Methyl alcohol | $CH_3OH$ | 32.042 | 512.6 | 8.09 | 0.00368 |
| *n*-Octane | $C_8H_{18}$ | 114.232 | 568.8 | 2.49 | 0.00431 |
| Pentafluoroethane (125) | $CHF_2CF_3$ | 120.022 | 339.2 | 3.62 | 0.00176 |
| *n*-Pentane | $C_5H_{12}$ | 72.151 | 469.7 | 3.37 | 0.00421 |
| Propane | $C_3H_8$ | 44.094 | 369.8 | 4.25 | 0.00454 |
| Propene | $C_3H_6$ | 42.081 | 364.9 | 4.60 | 0.00430 |
| Tetrafluoroethane (134a) | $CF_3CH_2F$ | 102.03 | 374.2 | 4.06 | 0.00197 |

**Table A1.4** Constant pressure specific heat ideal gas temperature relations (metric units)

$$\bar{c}_p = \frac{\text{KJ}}{\text{kgmole K}} \quad \theta = \frac{\text{T(Kelvin)}}{100}$$

| Gas | | Range (R) | Max error % |
|---|---|---|---|
| $N_2$ | $\bar{c}_p = 39.060 - 512.79\theta^{-1.5} + 1072.7\theta^{-2} - 820.40\theta^{-3}$ | 300–3500 | 0.43 |
| $O_2$ | $\bar{c}_p = 37.432 + 0.020102\theta^{1.5} - 178.57\theta^{-1.5} + 236.88\theta^{-2}$ | 300–3500 | 0.30 |
| $H_2$ | $\bar{c}_p = 56.505 - 702.74\theta^{-0.75} + 1165.0\theta^{-1} - 560.70\theta^{-1.5}$ | 300–3500 | 0.60 |
| CO | $\bar{c}_p = 69.145 - 0.70463\theta^{0.75} - 200.77\theta^{-0.5} + 176.76\theta^{-0.75}$ | 300–3500 | 0.42 |
| OH | $\bar{c}_p = 81.546 - 59.350\theta^{0.25} + 17.329\theta^{0.75} - 4.2660\theta$ | 300–3500 | 0.43 |
| NO | $\bar{c}_p = 59.283 - 1.7096\theta^{0.5} + 70.613\theta^{-0.5} + 74.889\theta^{-1.5}$ | 300–3500 | 0.34 |
| $H_2O$ | $\bar{c}_p = 143.05 - 183.54\theta^{0.25} + 82.751\theta^{0.5} - 3.6989\theta$ | 300–3500 | 0.43 |
| $CO_2$ | $\bar{c}_p = -3.7357 + 30.529\theta^{0.5} - 4.1034\theta - 0.024198\theta^2$ | 300–3500 | 0.19 |
| $NO_2$ | $\bar{c}_p = 46.045 + 216.10\theta^{-0.5} - 363.66\theta^{-0.75} + 232.550\theta^{-2}$ | 300–3500 | 0.26 |
| $CH_4$ | $\bar{c}_p = -672.87 + 439.74\theta^{0.25} - 24.875\theta^{0.75} + 323.88\theta^{-0.5}$ | 300–2000 | 0.15 |
| $C_2H_4$ | $\bar{c}_p = -95.395 + 123.15\theta^{0.5} - 35.641\theta^{0.5} + 182.77\theta^{-3}$ | 300–2000 | 0.07 |
| $C_2H_6$ | $\bar{c}_p = 6.895 + 17.26\theta - 0.6402\theta^2 - 0.00728\theta^3$ | 300–1500 | 0.83 |
| $C_3H_8$ | $\bar{c}_p = -4.042 + 30.46\theta - 1.571\theta^2 - 0.03171\theta^3$ | 300–1500 | 0.40 |
| $C_4H_{10}$ | $\bar{c}_p = 3.954 + 37.12\theta - 1.833\theta^2 + 0.03498\theta^3$ | 300–1500 | 0.54 |

**Table A1.5** Thermochemical properties of substances at 25 °C, 100 kPa

| Name | Molar mass M kg/kmol | Phase | $\bar{h}_{fg}$ kJ/kmol | Boiling point (1 atm), °C | Enthalpy of combustion at 25 °C, 100 kPa, $H_2O$ (*l*) $\Delta h_c$ kJ/kg | $H_2O$ (*l*) $\Delta \bar{h}_c$ kJ/kmol | $H_2O$ (*g*) $\Delta h_c$ kJ/kg | $H_2O$ (*g*) $\Delta \bar{h}_c$ kJ/kmol | Enthalpy of formation $\Delta \bar{h}^\circ_{f,ref}$ kJ/kmol | Gibbs energy formation $\Delta \bar{g}^\circ_{f,ref}$ kJ/kmol | Absolute entropy $\Delta \bar{s}^\circ_{f,ref}$ kJ/kmol-K |
|---|---|---|---|---|---|---|---|---|---|---|---|
| Hydrogen, $H_2$ | 2.016 | gas | — | 252.77 | −141779 | ** | −119959 | −241826 | 0 | 0 | 130.680 |
| Water, $H_2O$ | 180.015 | gas | 43986 | 100.00 | — | — | — | — | −241826 | −228582 | 188.834 |
| ** | ** | ** | — | — | — | — | — | — | −285830 | −237141 | 69.950 |
| Carbon, C | 12.011 | solid | — | — | −32766 | −393552 | −32766 | −393552 | 0 | 0 | ** |
| Carbon monoxide CO | 28.010 | gas | — | 191.45 | −10104 | −283022 | −10104 | −283022 | −110530 | −137163 | 197.650 |
| Carbon dioxide $CO_2$ | 44.010 | gas | — | −78.48 | — | — | — | — | −393522 | −394389 | 213.795 |
| Nitrogen, $N_2$ | 28.013 | gas | — | −195.81 | — | — | — | — | 0 | 0 | 191.609 |
| Oxygen, $O_2$ | 31.999 | gas | — | −182.96 | — | — | — | — | 0 | 0 | 205.147 |
| Methane, $CH_4$ | 16.043 | gas | — | −161.52 | −55496 | −890303 | −50012 | −802231 | −74873 | −50768 | 186.251 |
| Acetylene, $C_2H_2$ | 26.038 | gas | — | −84.00 | −49914 | −1299650 | −48224 | −1255660 | 226731 | 248163 | 200.958 |
| Ethylene, $C_2H_4$ | 28.054 | gas | — | −103.71 | −50303 | −1411200 | −47167 | −1323220 | 52467 | 68421 | 219.330 |
| Ethane, $C_2H_6$ | 30.069 | gas | 5021 | −88.61 | −51903 | −1560693 | −47514 | −1428730 | −83851 | −31951 | 229.447 |
| Propene, $C_3H_6$ | 42.080 | gas | — | −17.70 | −48902 | −2057400 | −45766 | −1925840 | 19710 | 62719 | 265.007 |
| Propene, $C_3H_8$ | 44.023 | gas | 14820 | −42.08 | −50331 | −2219230 | −46341 | −2043290 | −104673 | −24398 | 270.696 |

*(Contd.)*

| Name | Molar mass M kg/kmol | Phase | $\bar{h}_{fg}$ kJ/kmol | Boiling point (1 atm), °C | Enthalpy of combustion at 25 °C, 100 kPa $H_2O$ (*l*) $\Delta h_c$ | $H_2O$ (*l*) $\Delta \bar{h}_c$ | $H_2O$ (*g*) $\Delta h_c$ | $H_2O$ (*g*) $\Delta \bar{h}_c$ | Enthalpy of formation $\Delta \bar{h}^\circ_{f,ref}$ | Gibbs energy formation $\Delta \bar{g}^\circ_{f,ref}$ | Absolute entropy $\Delta \bar{s}^\circ_{f,ref}$ |
|---|---|---|---|---|---|---|---|---|---|---|---|
| | | | | kJ/kg | kJ/kmol | kJ/kg | kJ/kmol | kJ/kmol | kJ/kmol | kJ/kmol-K | |
| *n*-Batane, $C_4H_{10}$ | 58.123 | gas | 21066 | –0.51 | –49509 | –2877620 | –45725 | –2657690 | –125652 | –16560 | 310.464 |
| *n*-Pentane, $C_5H_{12}$ | 72.150 | gas | 26426 | 36.07 | –61198 | –3556980 | –56657 | –3293060 | –146712 | –8770 | 350.120 |
| *n*-Hexane, $C_6H_{14}$ | 86.177 | gas | 31552 | 68.73 | –48680 | –4195050 | –45107 | –3887150 | –166942 | –150 | 389.777 |
| *n*-Heptane, $C_7H_{16}$ | 100.203 | gas | 36547 | 98.42 | –48439 | –4853710 | –44927 | –4501820 | –187654 | 8151 | 428.887 |
| *n*-Octane, $C_8H_{18}$ | 114.230 | gas | 41484 | 125.67 | –48253 | –5511900 | –44787 | –5116030 | –208824 | 15921 | 468.243 |
| *n*-Octane, $C_8H_{18}$ | 114.230 | liquid | 41484 | 125.67 | –47889 | –5470420 | –44424 | –5074540 | –250308 | 5939 | 362.583 |
| Benzene, $C_6H_6$ | 78.113 | gas | 33849 | 80.10 | –42268 | –3301680 | –40579 | –3169730 | 82935 | 129661 | 269.760 |
| Benzene, $C_8H_6$ | 78.113 | liquid | 33849 | 80.10 | –41835 | –3267830 | –40145 | –3535880 | 49086 | 124532 | 173.433 |
| Methanol, $CH_3OH$ | 32.042 | gas | 39790 | 64.79 | –23851 | –264242 | –21104 | –676234 | –200940 | –163240 | 239.880 |
| Ethanol, | 32.042 | liquid | 39790 | 64.70 | –22666 | –726272 | –19920 | –838264 | –238910 | –166640 | 127.270 |
| $C_2H_5OH$ | 46.069 | gas | 42560 | 78.29 | –30597 | –1409534 | –27732 | –1277572 | –234950 | –167730 | 280.640 |
| Ethanol, $C_2H_5OH$ | 46.069 | liquid | 42560 | 78.29 | –29673 | –1367024 | –26808 | –1235012 | –277510 | –174350 | 160.100 |
| Hydrazine, $N_2H_4$ | 32.045 | liquid | 44760 | 114.15 | –19418 | –622250 | –16673 | –534278 | 50626 | 149440 | 121.544 |

*(Contd.)*

| Name | Molar mass M kg/kmol | Phase | $\bar{h}_{fg}$ kJ/kmol | Boiling point (1 atm), °C | Enthalpy of combustion at 25 °C, 100 kPa $H_2O$ (*l*) | | $H_2O$ (*g*) | | Enthalpy of formation | Gibbs energy formation | Absolute entropy |
|---|---|---|---|---|---|---|---|---|---|---|---|
| | | | | | $\Delta h_c$ kJ/kg | $\Delta \bar{h}_c$ kJ/kmol | $\Delta h_c$ kJ/kg | $\Delta \bar{h}_c$ kJ/kmol | $\Delta \bar{h}^\circ_{f,ref}$ kJ/kmol | $\Delta \bar{g}^\circ_{f,ref}$ kJ/kmol | $\Delta \bar{s}^\circ_{f,ref}$ kJ/kmol-K |
| Hydrogen, H | 1.008 | gas | — | — | –358076 | –360905 | –336256 | –338912 | 217999 | 203278 | 114.716 |
| Nitrogen, N | 14.007 | gas | — | — | — | — | — | — | 472680 | 455540 | 153.300 |
| Oxygen, O | 15.999 | gas | — | — | — | — | — | — | 249170 | 231736 | 161.058 |
| Hydroxyl, OH | 17.007 | gas | — | — | — | — | — | — | 38987 | 34277 | 183.708 |

**Table A1.6** Air table

| $T$ K | $T$ °C | $h$ kJ/kg | $p_r$ | $v_r$ | $\phi$ kJ/kg.K |
|---|---|---|---|---|---|
| 200 | –73.15 | 201.87 | 0.3414 | 585.82 | 6.3000 |
| 210 | –63.15 | 211.94 | 0.4051 | 518.39 | 6.3491 |
| 220 | –53.15 | 221.99 | 0.4768 | 461.41 | 6.3959 |
| 230 | –43.15 | 232.04 | 0.5571 | 412.85 | 6.4406 |
| 240 | –33.15 | 242.08 | 0.6466 | 371.37 | 6.4833 |
| 250 | –23.15 | 252.12 | 0.7458 | 335.21 | 6.5243 |
| 260 | –13.15 | 262.15 | 0.8555 | 303.92 | 6.5636 |
| 270 | –3.15 | 272.19 | 0.9761 | 276.61 | 6.6015 |
| 280 | 6.85 | 282.22 | 1.1084 | 252.62 | 6.6380 |
| 290 | 16.85 | 292.25 | 1.2531 | 231.43 | 6.6732 |
| 300 | 26.85 | 302.29 | 1.4108 | 212.63 | 6.7072 |
| 310 | 36.85 | 312.33 | 1.5823 | 195.92 | 6.7401 |
| 320 | 46.85 | 322.37 | 1.7682 | 180.98 | 6.7720 |
| 330 | 56.85 | 332.42 | 1.9693 | 167.57 | 6.8029 |
| 340 | 66.85 | 342.47 | 2.1865 | 155.50 | 6.8330 |
| 350 | 76.85 | 352.54 | 2.4204 | 144.60 | 6.8621 |
| 360 | 86.85 | 362.61 | 2.6720 | 134.73 | 6.8905 |
| 370 | 96.85 | 372.69 | 2.9419 | 125.77 | 6.9181 |
| 380 | 106.85 | 382.79 | 3.2312 | 117.60 | 6.9450 |
| 390 | 116.85 | 392.89 | 3.5407 | 110.15 | 6.9713 |
| 400 | 126.85 | 403.01 | 3.8712 | 103.31 | 6.9969 |
| 410 | 136.85 | 413.14 | 4.2238 | 97.619 | 7.0219 |
| 420 | 146.85 | 423.29 | 4.5993 | 97.318 | 7.0464 |
| 430 | 156.85 | 433.45 | 4.9989 | 86.619 | 7.0703 |
| 440 | 166.85 | 443.62 | 5.4234 | 81.130 | 7.0937 |
| 450 | 176.85 | 453.81 | 5.8739 | 76.610 | 7.1166 |
| 460 | 186.85 | 464.02 | 6.3516 | 72.423 | 7.1390 |
| 470 | 196.85 | 474.25 | 6.8575 | 68.538 | 7.1610 |
| 480 | 206.85 | 484.49 | 7.3927 | 64.929 | 7.1826 |
| 490 | 216.85 | 494.76 | 7.9584 | 61.370 | 7.2037 |
| 500 | 226.85 | 505.04 | 8.5558 | 59.340 | 7.2245 |
| 510 | 236.85 | 515.34 | 9.1861 | 55.619 | 7.2449 |
| 520 | 246.85 | 525.66 | 9.8506 | 52.489 | 7.2650 |

*(Contd.)*

| $T$ K | $T$ °C | $h$ kJ/kg | $p_r$ | $v_r$ | $\phi$ kJ/kg.K |
|---|---|---|---|---|---|
| 530 | 256.85 | 536.01 | 10.551 | 50.232 | 7.2345 |
| 540 | 266.85 | 546.37 | 11.287 | 47.843 | 7.3040 |
| 550 | 276.85 | 556.76 | 12.062 | 45.598 | 7.3231 |
| 560 | 286.85 | 567.16 | 12.877 | 43.488 | 7.3418 |
| 570 | 296.85 | 577.59 | 13.732 | 41.509 | 7.3603 |
| 580 | 306.85 | 588.04 | 14.630 | 39.645 | 7.3785 |
| 590 | 316.85 | 598.52 | 15.572 | 37.889 | 7.3964 |
| 600 | 326.85 | 609.02 | 16.559 | 36.234 | 7.4140 |
| 610 | 336.85 | 619.54 | 17.593 | 34.673 | 7.4314 |
| 620 | 346.85 | 630.08 | 18.676 | 33.198 | 7.4486 |
| 630 | 356.85 | 640.65 | 19.810 | 31.802 | 7.4655 |
| 640 | 366.85 | 651.24 | 20.995 | 30.483 | 7.4821 |
| 650 | 376.85 | 661.85 | 22.234 | 29.235 | 7.4986 |
| 660 | 386.85 | 672.49 | 23.528 | 28.052 | 7.5148 |
| 670 | 396.85 | 683.15 | 24.880 | 26.929 | 7.5309 |
| 680 | 406.85 | 693.84 | 26.291 | 25.864 | 7.5467 |
| 690 | 416.85 | 704.55 | 27.763 | 24.853 | 7.5623 |
| 700 | 426.85 | 715.28 | 29.298 | 23.892 | 7.5778 |
| 710 | 436.85 | 726.04 | 30.898 | 22.979 | 7.5931 |
| 720 | 446.85 | 736.82 | 32.565 | 22.110 | 7.6081 |
| 730 | 456.85 | 747.63 | 34.301 | 21.282 | 7.6230 |
| 740 | 466.85 | 758.46 | 36.309 | 20.494 | 7.6378 |
| 750 | 476.85 | 769.32 | 37.989 | 19.743 | 7.6523 |
| 760 | 486.85 | 780.19 | 39.945 | 19.026 | 7.6667 |
| 770 | 496.85 | 791.10 | 41.978 | 18.343 | 7.6810 |
| 780 | 506.85 | 802.02 | 44.092 | 17.690 | 7.6951 |
| 790 | 516.85 | 812.97 | 46.288 | 17.067 | 7.7090 |
| 800 | 526.85 | 823.94 | 48.568 | 16.472 | 7.7228 |
| 810 | 536.85 | 834.94 | 50.935 | 15.903 | 7.7365 |
| 820 | 546.85 | 845.96 | 53.392 | 15.358 | 7.7500 |
| 830 | 556.85 | 857.00 | 55.941 | 14.837 | 7.7634 |
| 840 | 566.85 | 868.06 | 58.584 | 14.338 | 7.7767 |
| 850 | 576.85 | 879.15 | 61.325 | 13.861 | 7.7898 |
| 860 | 586.85 | 890.26 | 64.165 | 13.403 | 7.8028 |
| 870 | 596.85 | 901.39 | 67.107 | 12.964 | 7.8156 |

*(Contd.)*

| $T$ K | $T$ °C | $h$ kJ/kg | $p_r$ | $v_r$ | $\phi$ kJ/kg.K |
|---|---|---|---|---|---|
| 880 | 606.85 | 912.54 | 70.155 | 12.544 | 7.8284 |
| 890 | 616.85 | 923.72 | 73.310 | 12.140 | 7.8410 |
| 900 | 626.85 | 934.91 | 76.576 | 11.753 | 7.8535 |
| 910 | 636.85 | 946.13 | 79.956 | 11.381 | 7.8659 |
| 920 | 646.85 | 957.37 | 83.452 | 11.024 | 7.8782 |
| 930 | 656.85 | 968.63 | 87.067 | 10.681 | 7.8904 |
| 940 | 666.85 | 979.90 | 90.805 | 10.352 | 7.9024 |
| 950 | 676.85 | 991.20 | 94.667 | 10.035 | 7.9144 |
| 960 | 686.85 | 1002.52 | 98.659 | 9.7305 | 7.9262 |
| 970 | 696.85 | 1013.86 | 102.78 | 9.4376 | 7.9380 |
| 980 | 706.85 | 1025.22 | 107.04 | 9.1555 | 7.9496 |
| 990 | 716.85 | 1036.60 | 111.43 | 8.8845 | 7.9612 |
| 1000 | 726.85 | 1047.99 | 115.97 | 8.6229 | 7.9727 |
| 1010 | 736.85 | 1059.41 | 120.65 | 8.3713 | 7.9840 |
| 1020 | 746.85 | 1070.84 | 125.49 | 8.1281 | 7.9953 |
| 1030 | 756.85 | 1082.30 | 130.47 | 7.8945 | 8.0065 |
| 1040 | 766.85 | 1093.77 | 135.60 | 7.6696 | 8.0175 |
| 1050 | 776.85 | 1105.26 | 140.90 | 7.4521 | 8.0285 |
| 1060 | 786.85 | 1116.76 | 146.36 | 7.2424 | 8.0394 |
| 1070 | 796.85 | 1128.28 | 151.98 | 7.0404 | 8.0503 |
| 1080 | 806.85 | 1139.82 | 157.77 | 6.8454 | 8.0610 |
| 1090 | 816.85 | 1151.38 | 163.74 | 6.6569 | 8.0716 |
| 1100 | 826.85 | 1162.95 | 169.88 | 6.4752 | 8.0822 |
| 1110 | 836.85 | 1174.54 | 176.20 | 6.2997 | 8.0927 |
| 1120 | 846.85 | 1186.15 | 182.71 | 6.1299 | 8.1031 |
| 1130 | 856.85 | 1197.77 | 189.40 | 5.9662 | 8.1134 |
| 1140 | 866.85 | 1209.40 | 196.29 | 5.8077 | 8.1237 |
| 1150 | 876.85 | 1221.06 | 203.38 | 5.6544 | 8.1339 |
| 1160 | 886.85 | 1232.72 | 210.66 | 5.5065 | 8.1440 |
| 1170 | 896.85 | 1244.41 | 218.15 | 5.3633 | 8.1540 |
| 1180 | 906.85 | 1256.10 | 225.85 | 5.2247 | 8.1639 |
| 1190 | 916.85 | 1267.82 | 233.77 | 5.0905 | 8.1738 |
| 1200 | 926.85 | 1279.54 | 241.90 | 4.9607 | 8.1836 |
| 1210 | 936.85 | 1291.28 | 250.25 | 4.8352 | 8.1934 |
| 1220 | 946.85 | 1303.04 | 258.83 | 4.7135 | 8.2031 |

*(Contd.)*

| *T* K | *T* °C | *h* kJ/kg | $p_r$ | $v_r$ | $\phi$ kJ/kg.K |
|---|---|---|---|---|---|
| 1230 | 956.85 | 1314.81 | 267.64 | 4.5957 | 8.2127 |
| 1240 | 966.85 | 1326.59 | 276.69 | 4.4815 | 8.2222 |
| 1250 | 976.85 | 1338.39 | 285.98 | 4.3709 | 8.2317 |
| 1260 | 986.85 | 1350.20 | 295.51 | 4.2638 | 8.2411 |
| 1270 | 996.85 | 1362.03 | 305.30 | 4.1598 | 8.2504 |
| 1280 | 1006.85 | 1373.86 | 315.33 | 4.0592 | 8.2597 |
| 1290 | 1016.85 | 1385.71 | 325.63 | 3.9616 | 8.2690 |
| 1300 | 1026.85 | 1397.58 | 336.19 | 3.8669 | 8.2781 |
| 1310 | 1036.85 | 1409.45 | 347.02 | 3.7750 | 8.2872 |
| 1320 | 1046.85 | 1421.34 | 358.13 | 3.6858 | 8.2963 |
| 1330 | 1056.85 | 1433.24 | 369.51 | 3.5994 | 8.3052 |
| 1340 | 1066.85 | 1445.16 | 381.19 | 3.5153 | 8.3142 |
| 1350 | 1076.85 | 1457.08 | 393.15 | 3.4338 | 8.3230 |
| 1360 | 1086.85 | 1469.02 | 405.40 | 3.3547 | 8.3318 |
| 1370 | 1096.85 | 1480.97 | 417.96 | 3.2778 | 8.3406 |
| 1380 | 1106.85 | 1492.93 | 430.82 | 3.2032 | 8.3493 |
| 1390 | 1116.85 | 1504.91 | 444.00 | 3.1306 | 8.3579 |
| 1400 | 1126.85 | 1516.89 | 457.89 | 3.0602 | 8.3663 |
| 1410 | 1136.85 | 1528.89 | 471.30 | 2.9917 | 8.3751 |
| 1420 | 1146.85 | 1540.89 | 485.44 | 2.9252 | 8.3836 |
| 1430 | 1156.85 | 1552.91 | 499.92 | 2.8605 | 8.3920 |
| 1440 | 1166.85 | 1564.94 | 514.74 | 2.7975 | 8.4004 |
| 1450 | 1176.85 | 1576.98 | 529.90 | 2.7364 | 8.4087 |
| 1460 | 1186.85 | 1589.03 | 545.42 | 2.6768 | 8.4170 |
| 1470 | 1196.85 | 1601.09 | 561.29 | 2.6190 | 8.4252 |
| 1480 | 1206.85 | 1613.17 | 577.53 | 2.5626 | 8.4334 |
| 1490 | 1216.85 | 1625.25 | 594.13 | 2.5079 | 8.4415 |
| 1500 | 1226.85 | 1637.34 | 611.12 | 2.4545 | 8.4496 |
| 1510 | 1236.85 | 1649.44 | 628.48 | 2.4026 | 8.4577 |
| 1520 | 1246.85 | 1661.56 | 646.24 | 2.3521 | 8.4657 |
| 1530 | 1256.85 | 1673.68 | 664.38 | 2.3029 | 8.4736 |
| 1540 | 1266.85 | 1685.81 | 682.94 | 2.2350 | 8.4815 |
| 1550 | 1276.85 | 1697.95 | 701.89 | 2.2083 | 8.4894 |
| 1560 | 1286.85 | 1710.10 | 721.27 | 2.1629 | 8.4972 |
| 1570 | 1296.85 | 1722.26 | 741.06 | 2.1186 | 8.5050 |

*(Contd.)*

| $T$ K | $T$ °C | $h$ kJ/kg | $p_r$ | $v_r$ | $\phi$ kJ/kg.K |
|---|---|---|---|---|---|
| 1580 | 1306.85 | 1734.43 | 761.29 | 2.0754 | 8.5127 |
| 1590 | 1316.85 | 1746.61 | 781.94 | 2.0334 | 8.5204 |
| 1600 | 1326.85 | 1758.80 | 803.04 | 1.9924 | 8.5280 |
| 1610 | 1336.85 | 1771.00 | 824.39 | 1.9525 | 8.5356 |
| 1620 | 1346.85 | 1783.21 | 846.60 | 1.9135 | 8.5432 |
| 1630 | 1356.85 | 1795.42 | 869.06 | 1.8756 | 8.5507 |
| 1640 | 1366.85 | 1807.64 | 892.00 | 1.8386 | 8.5582 |
| 1650 | 1376.85 | 1819.88 | 915.41 | 1.8025 | 8.5656 |
| 1660 | 1386.85 | 1832.12 | 939.31 | 1.7673 | 8.5730 |
| 1670 | 1396.85 | 1844.37 | 963.70 | 1.7329 | 8.5804 |
| 1680 | 1406.85 | 1856.63 | 988.59 | 1.6994 | 8.5877 |
| 1690 | 1416.85 | 1868.89 | 1013.99 | 1.6667 | 8.5950 |
| 1700 | 1426.85 | 1881.17 | 1039.9 | 1.6348 | 8.6022 |
| 1725 | 1451.85 | 1911.89 | 1107.0 | 1.5583 | 8.6201 |
| 1750 | 1476.85 | 1942.66 | 1177.4 | 1.4863 | 8.6378 |
| 1775 | 1501.85 | 1973.48 | 1251.4 | 1.4184 | 8.6553 |
| 1800 | 1526.85 | 2004.34 | 1329.0 | 1.3544 | 8.6726 |
| 1825 | 1551.85 | 2035.25 | 1410.4 | 1.2940 | 8.6897 |
| 1850 | 1576.85 | 2066.21 | 1495.6 | 1.2370 | 8.7065 |
| 1875 | 1601.85 | 2097.20 | 1584.9 | 1.1830 | 8.7231 |
| 1900 | 1626.85 | 2128.25 | 1678.4 | 1.1320 | 8.7396 |
| 1925 | 1651.85 | 2159.33 | 1776.2 | 1.0838 | 8.7558 |
| 1950 | 1676.85 | 2190.45 | 1878.4 | 1.0381 | 8.7719 |
| 1975 | 1701.85 | 2221.61 | 1985.3 | 0.99481 | 8.7878 |
| 2000 | 1726.85 | 2252.82 | 2096.9 | 0.95379 | 8.8035 |
| 2025 | 1751.85 | 2284.05 | 2213.5 | 0.91484 | 8.8190 |
| 2050 | 1776.85 | 2315.33 | 2335.1 | 0.87791 | 8.8344 |
| 2075 | 1801.85 | 2346.64 | 2461.9 | 0.84284 | 8.8495 |
| 2100 | 1826.85 | 2377.99 | 2594.2 | 0.80950 | 8.8646 |
| 2125 | 1851.85 | 2409.38 | 2732.0 | 0.77782 | 8.8794 |
| 2150 | 1876.85 | 2440.80 | 2875.6 | 0.74767 | 8.8941 |
| 2175 | 1901.85 | 2472.25 | 3025.0 | 0.71901 | 8.9087 |
| 2200 | 1926.85 | 2503.73 | 3180.6 | 0.69169 | 8.9230 |
| 2225 | 1951.85 | 2535.25 | 3342.5 | 0.66567 | 8.9373 |
| 2250 | 1976.85 | 2566.80 | 3510.8 | 0.64088 | 8.9514 |

*(Contd.)*

| *T* K | *T* °C | *h* kJ/kg | $p_r$ | $v_r$ | $\phi$ kJ/kg.K |
|---|---|---|---|---|---|
| 2275 | 2001.85 | 2598.38 | 3685.8 | 0.61723 | 8.9654 |
| 2300 | 2026.85 | 2630.00 | 3867.6 | 0.59468 | 8.9752 |
| 2325 | 2051.85 | 2661.64 | 4056.5 | 0.57315 | 8.9929 |
| 2350 | 2076.85 | 2693.32 | 4252.6 | 0.55260 | 9.0064 |
| 2375 | 2101.85 | 2725.02 | 4456.2 | 0.53297 | 9.0198 |
| 2400 | 2126.85 | 2756.75 | 4667.4 | 0.51420 | 9.0331 |
| 2425 | 2151.85 | 2788.51 | 4886.5 | 0.49627 | 9.0463 |
| 2450 | 2176.85 | 2820.31 | 5113.7 | 0.47911 | 9.0593 |
| 2475 | 2201.85 | 2852.12 | 5349.2 | 0.46269 | 9.0722 |
| 2500 | 2226.85 | 2883.97 | 5593.2 | 0.44697 | 9.0851 |
| 2525 | 2251.85 | 2915.84 | 5846.0 | 0.43192 | 9.0977 |
| 2550 | 2276.85 | 2947.74 | 6107.7 | 0.41751 | 9.1103 |
| 2575 | 2301.85 | 2979.67 | 6378.7 | 0.40369 | 9.1228 |
| 2600 | 2326.85 | 3011.63 | 6659.2 | 0.39044 | 9.1351 |
| 2625 | 2351.85 | 3043.61 | 6949.4 | 0.37773 | 9.1474 |
| 2650 | 2376.85 | 3075.61 | 7249.5 | 0.36554 | 9.1595 |
| 2675 | 2401.85 | 3107.64 | 7559.8 | 0.35385 | 9.1715 |
| 2700 | 2426.85 | 3139.70 | 7880.6 | 0.34261 | 9.1895 |
| 2725 | 2451.85 | 3171.78 | 8212.1 | 0.33183 | 9.1953 |
| 2750 | 2476.85 | 3203.88 | 8554.7 | 0.32146 | 9.2070 |
| 2775 | 2501.85 | 3236.01 | 8908.4 | 0.31150 | 9.2186 |
| 2800 | 2526.85 | 3268.16 | 9273.7 | 0.30193 | 9.2302 |
| 2825 | 2551.85 | 3300.33 | 9650.9 | 0.29272 | 9.2416 |
| 2850 | 2576.85 | 3332.53 | 10040.0 | 0.28386 | 9.2530 |
| 2875 | 2601.85 | 3364.75 | 10441.6 | 0.27534 | 9.3642 |
| 2900 | 2626.85 | 3396.99 | 10855.8 | 0.26714 | 9.2754 |
| 2925 | 2651.85 | 3429.25 | 11283.0 | 0.25924 | 9.2865 |

**Table A1.7** Properties of ideal gases. $\bar{h}$ in kJ/k mol, $\bar{\phi}$ in kJ/k mol-K

| | CO | | $CO_2$ | | $H_2$ | | $H_2O$ | | $N_2$ | |
|---|---|---|---|---|---|---|---|---|---|---|
| $T$, K | $\bar{h}$ | $\bar{\phi}$ | $\bar{h}$ | $\bar{\phi}$ | $\bar{h}$ | $\bar{\phi}$ | $\bar{h}$ | $\bar{\phi}$ | $\bar{h}$ | $\bar{\phi}$ |
| 200 | 5804.9 | 185.991 | 5951.8 | 199.980 | 5667.8 | 119.303 | 6626.8 | 175.506 | 5803.1 | 179.944 |
| 298.15 | 8671.0 | 197.653 | 9364.0 | 213.795 | 8467.0 | 130.680 | 9904.0 | 188.834 | 8670.0 | 191.609 |
| 300 | 8724.9 | 197.833 | 9432.8 | 214.025 | 8520.4 | 130.858 | 9966.1 | 189.042 | 8723.9 | 191.789 |
| 400 | 11646.2 | 206.236 | 13366.7 | 225.314 | 11424.9 | 139.212 | 13357.0 | 198.792 | 11640.4 | 200.179 |
| 500 | 14601.4 | 212.828 | 17668.9 | 234.901 | 14348.6 | 145.736 | 16830.2 | 206.538 | 14580.2 | 206.737 |
| 600 | 17612.7 | 218.317 | 22271.3 | 243.284 | 17278.6 | 151.078 | 20405.9 | 213.054 | 17564.2 | 212.176 |
| 700 | 20692.6 | 223.063 | 27120.0 | 250.754 | 20215.1 | 155.604 | 24096.2 | 218.741 | 20606.6 | 216.865 |
| 800 | 23845.9 | 227.273 | 32172.6 | 257.498 | 23166.4 | 159.545 | 27907.2 | 223.828 | 23715.2 | 221.015 |
| 900 | 27070.6 | 231.070 | 37395.9 | 263.648 | 26141.9 | 163.049 | 31842.5 | 228.461 | 26891.8 | 224.756 |
| 1000 | 30359.8 | 234.535 | 42763.1 | 269.302 | 29147.3 | 166.215 | 35904.6 | 232.740 | 30132.2 | 228.169 |
| 1100 | 33705.1 | 237.723 | 48248.2 | 274.529 | 32187.4 | 169.112 | 40094.1 | 236.732 | 33428.8 | 231.311 |
| 1200 | 37099.6 | 240.676 | 53836.7 | 279.391 | 35266.4 | 171.791 | 44412.4 | 240.489 | 36778.0 | 234.225 |
| 1300 | 40537.1 | 245.428 | 59512.8 | 283.934 | 38386.7 | 174.289 | 48851.4 | 244.041 | 40173.0 | 236.942 |
| 1400 | 44012.0 | 246.003 | 65263.1 | 288.195 | 41549.8 | 176.633 | 53403.6 | 247.414 | 43607.8 | 239.487 |
| 1500 | 47519.3 | 248.422 | 71076.3 | 292.206 | 44756.1 | 178.845 | 58061.6 | 250.628 | 47077.1 | 241.881 |
| 1600 | 51054.8 | 250.704 | 76943.0 | 295.992 | 48005.7 | 180.942 | 62818.1 | 253.697 | 50576.6 | 244.139 |
| 1700 | 54614.9 | 252.862 | 82855.4 | 299.576 | 51297.9 | 182.937 | 67666.1 | 256.636 | 54102.5 | 246.276 |
| 1800 | 58196.5 | 254.909 | 88807.3 | 302.978 | 54631.5 | 184.843 | 72599.1 | 259.455 | 57651.4 | 248.305 |
| 1900 | 61796.9 | 256.856 | 94793.9 | 306.215 | 58004.8 | 186.666 | 77610.7 | 262.165 | 61220.9 | 250.235 |
| 2000 | 65413.9 | 258.711 | 100811.2 | 309.301 | 61416.1 | 188.416 | 82694.7 | 264.772 | 64808.5 | 252.075 |
| 2100 | 69045.8 | 260.483 | 106856.4 | 312.250 | 64863.3 | 190.098 | 87845.6 | 267.285 | 68412.5 | 253.833 |
| 2200 | 72691.0 | 262.179 | 112927.3 | 315.075 | 68344.2 | 191.717 | 93057.9 | 269.710 | 72031.5 | 255.517 |
| 2300 | 76348.3 | 263.805 | 119022.2 | 317.784 | 71856.8 | 193.279 | 98326.7 | 272.052 | 75664.0 | 257.132 |

**Table A1.8** Log $K_p$ for seven ideal gas reactions, $p° = 100$ kPa

| $T$, K | $H_2 \rightarrow 2H$ | $N_2 \rightarrow 2N$ | $O_2 \rightarrow 2O$ | $CO_2 \rightarrow CO + \frac{1}{2}O_2$ | $H_2O \rightarrow OH + \frac{1}{2}H_2$ | $\frac{1}{2}N_2 + \frac{1}{2}O_2 \rightarrow NO$ | $CO_2 + H_2 \rightarrow CO + H_2O$ |
|---|---|---|---|---|---|---|---|
| 298.15 | −71.30 | −159.7 | −81.28 | −45.05 | −46.10 | −15.15 | −4.950 |
| 300 | −70.83 | −158.7 | −80.73 | −44.74 | −45.80 | −15.07 | −4.905 |
| 400 | −51.73 | −117.4 | −58.91 | −32.43 | −33.48 | −11.14 | −3.215 |
| 500 | −40.26 | −92.63 | −45.82 | −25.03 | −26.09 | −8.784 | −2.193 |
| 600 | −32.62 | −76.12 | −37.10 | −20.10 | −21.16 | −7.210 | −1.506 |
| 700 | −27.16 | −64.33 | −30.86 | −16.57 | −17.64 | −6.086 | −1.014 |
| 800 | −23.06 | −55.48 | −26.19 | −13.92 | −15.00 | −5.243 | −0.642 |
| 900 | −19.88 | −48.60 | −22.55 | −11.86 | −12.95 | −4.587 | −0.352 |
| 1000 | −17.33 | −43.10 | −19.64 | −10.21 | −11.31 | −4.063 | −0.120 |
| 1100 | −15.18 | −38.54 | −17.21 | −8.843 | −9.922 | −3.633 | 0.040 |
| 1200 | −13.40 | −34.75 | −15.20 | −7.739 | −8.784 | −3.275 | 0.152 |
| 1300 | −11.89 | −31.54 | −13.49 | −6.802 | −7.821 | −2.972 | 0.251 |
| 1400 | −10.60 | −28.79 | −12.93 | −6.004 | −6.996 | −2.712 | 0.330 |
| 1500 | −9.474 | −26.41 | −10.76 | −5.315 | −6.280 | −2.487 | 0.397 |
| 1600 | −8.492 | −24.32 | −9.657 | −4.711 | −5.654 | −2.290 | 0.456 |
| 1700 | −7.626 | −22.48 | −8.680 | −4.175 | −5.102 | −2.116 | 0.510 |
| 1800 | −6.856 | −20.85 | −7.811 | −3.697 | −4.611 | −1.962 | 0.560 |
| 1900 | −6.168 | −19.39 | −7.033 | −3.268 | −4.172 | −1.824 | 0.608 |
| 2000 | −5.548 | −18.07 | −6.334 | −2.879 | −3.777 | −1.699 | 0.652 |
| 2100 | −4.987 | −16.88 | −5.701 | −2.527 | −3.419 | −1.587 | 0.692 |
| 2200 | −4.477 | −15.79 | −5.125 | −2.207 | −3.094 | −1.484 | 0.729 |
| 2300 | −4.012 | −14.81 | −4.600 | −1.917 | −2.797 | −1.391 | 0.761 |
| 2400 | −3.585 | −13.90 | −4.118 | −1.652 | −2.525 | −1.306 | 0.788 |
| 2500 | −3.192 | −13.06 | −3.675 | −1.412 | −2.274 | −1.227 | 0.810 |
| 2600 | −2.830 | −12.29 | −3.266 | −1.194 | −2.043 | −1.154 | 0.828 |

*(Contd.)*

| $T$, K | $H_2 \to 2H$ | $N_2 \to 2N$ | $O_2 \to 2O$ | $CO_2 \to CO + \frac{1}{2}O_2$ | $H_2O \to OH + \frac{1}{2}H_2$ | $\frac{1}{2}N_2 + \frac{1}{2}O_2 \to NO$ | $CO_2 + H_2 \to CO + H_2O$ |
|---|---|---|---|---|---|---|---|
| 2700 | −2.495 | −11.58 | −2.887 | −0.995 | −1.829 | −1.087 | 0.840 |
| 2800 | −2.183 | −10.92 | −2.536 | −0.813 | −1.631 | −1.025 | 0.849 |
| 2900 | −1.893 | −10.30 | −2.208 | −0.646 | −1.446 | −0.967 | 0.855 |
| 3000 | −1.622 | −9.729 | −1.903 | −0.491 | −1.273 | −0.913 | 0.859 |
| 3100 | −1.369 | −9.191 | −1.617 | −0.347 | −1.111 | −0.863 | 0.863 |
| 3200 | −1.131 | −8.686 | −1.349 | −0.208 | −0.960 | −0.815 | 0.869 |
| 3300 | −0.908 | −8.213 | −1.097 | −0.073 | −0.818 | −0.771 | 0.881 |
| 3400 | −0.698 | −7.767 | −0.860 | −0.062 | −0.684 | −0.729 | 0.900 |
| 3500 | −0.501 | −7.346 | −0.637 | −0.202 | −0.558 | −0.690 | 0.929 |

**Table A1.9** Log $K_f$ for substances, $p^\circ$ = 100 kPa

| T, K | $CO_2$ | CO | H | OH | $H_2O$ | N | O | NO |
|---|---|---|---|---|---|---|---|---|
| 298.15 | 69.09 | 24.04 | −35.65 | −6.001 | 40.10 | −79.85 | −40.64 | −15.15 |
| 300 | 68.66 | 23.92 | −35.42 | −5.959 | 39.84 | −79.34 | −40.37 | −15.07 |
| 400 | 51.54 | 19.11 | −25.86 | −4.266 | 29.21 | −58.70 | −29.46 | −11.14 |
| 500 | 41.26 | 16.23 | −20.13 | −3.250 | 22.84 | −46.32 | −22.91 | −8.784 |
| 600 | 34.41 | 14.31 | −16.31 | −2.573 | 18.59 | −38.06 | −18.55 | −7.210 |
| 700 | 29.51 | 12.94 | −13.58 | −2.089 | 15.55 | −32.16 | −15.43 | −6.086 |
| 800 | 25.83 | 11.91 | −11.53 | −1.726 | 13.28 | −27.74 | −13.09 | −5.243 |
| 900 | 22.97 | 11.11 | −9.940 | −1.444 | 11.51 | −24.30 | −11.28 | −4.587 |
| 1000 | 20.68 | 10.47 | −8.666 | −1.218 | 10.09 | −21.55 | −9.822 | −4.063 |
| 1100 | 18.80 | 9.961 | −7.592 | −1.039 | 8.883 | −19.27 | −8.605 | −3.633 |
| 1200 | 17.25 | 9.511 | −6.700 | −0.8920 | 7.892 | −17.37 | −7.598 | −3.275 |
| 1300 | 15.92 | 9.119 | −5.945 | −0.7678 | 7.053 | −15.77 | −6.746 | −2.502 |
| 1400 | 14.78 | 8.771 | −5.298 | −0.6612 | 6.334 | −14.40 | −6.015 | −2.712 |

*(Contd.)*

| T, K | $CO_2$ | CO | H | OH | $H_2O$ | N | O | NO |
|---|---|---|---|---|---|---|---|---|
| 1500 | 13.79 | 8.480 | –4.737 | –0.5689 | 5.711 | –13.20 | –5.382 | –2.487 |
| 1600 | 12.93 | 8.220 | –4.246 | –0.4881 | 5.166 | –12.16 | –4.829 | –2.290 |
| 1700 | 12.17 | 7.993 | –3.813 | –0.4168 | 4.685 | –11.24 | –4.340 | –2.116 |
| 1800 | 11.49 | 7.793 | –3.428 | –0.3535 | 4.258 | –10.42 | –3.905 | –1.962 |
| 1900 | 10.88 | 7.617 | –3.084 | –0.2968 | 3.875 | –9.693 | –3.517 | –1.824 |
| 2000 | 10.34 | 7.460 | –2.774 | –0.2458 | 3.531 | –9.034 | –3.167 | –1.699 |
| 2100 | 9.846 | 7.319 | –2.494 | –0.1996 | 3.219 | –8.439 | –2.850 | –1.587 |
| 2200 | 9.398 | 7.190 | –2.239 | –0.1576 | 2.936 | –7.897 | –2.563 | –1.484 |
| 2300 | 8.988 | 7.072 | –2.006 | –0.1193 | 2.677 | –7.403 | –2.300 | –1.391 |
| 2400 | 8.613 | 6.961 | –1.793 | –0.08421 | 2.440 | –6.949 | –2.059 | –1.306 |
| 2500 | 8.268 | 6.856 | –1.596 | –0.05189 | 2.222 | –6.532 | –1.838 | –1.227 |
| 2600 | 7.949 | 6.756 | –1.415 | –0.02206 | 2.021 | –6.147 | –1.633 | –1.154 |
| 2700 | 7.654 | 6.660 | –1.247 | 0.005554 | 1.835 | –5.791 | –1.444 | –1.087 |
| 2800 | 7.380 | 6.567 | –1.091 | 0.03120 | 1.662 | –5.460 | –1.268 | –1.025 |
| 2900 | 7.125 | 6.479 | –0.9464 | 0.05508 | 1.501 | –5.152 | –1.104 | –0.967 |
| 3000 | 6.887 | 6.396 | –0.8111 | 0.07736 | 1.350 | –4.864 | –0.9514 | –0.913 |
| 3100 | 6.665 | 6.318 | –0.6844 | 0.09821 | 1.210 | –4.595 | –0.8084 | –0.863 |
| 3200 | 6.456 | 6.248 | –0.5657 | 0.1178 | 1.078 | –4.343 | –0.6744 | –0.815 |
| 3300 | 6.260 | 6.187 | –0.4542 | 0.1361 | 0.9538 | –4.106 | –0.5485 | –0.771 |
| 3400 | 6.075 | 6.138 | –0.3492 | 0.1534 | 0.8372 | –3.883 | –0.4301 | –0.729 |
| 3500 | 5.901 | 6.104 | –0.2503 | 0.1697 | 0.7272 | –3.673 | –0.3184 | –0.690 |

**Table A1.10** Saturated mercury (metric units)

| Sat. press., MPa | Sat. temp., °C | Specific volume, m³/kg | Enthalpy, kJ/kg | | | Entropy, kJ/(kg.K) | | |
|---|---|---|---|---|---|---|---|---|
| | | $v_g$ | $h_f$ | $h_{fg}$ | $h_g$ | $s_f$ | $s_{fg}$ | $s_g$ |
| 0.00006 | 109.2 | 259.6 | 15.13 | 297.20 | 312.33 | 0.0466 | 0.7774 | 0.8240 |
| 0.00007 | 112.3 | 224.3 | 15.55 | 297.14 | 312.69 | 0.0477 | 0.7709 | 0.8186 |
| 0.00008 | 115.0 | 197.7 | 15.93 | 297.09 | 313.02 | 0.0487 | 0.7654 | 0.8141 |
| 0.00009 | 117.5 | 176.8 | 16.27 | 297.04 | 313.31 | 0.0496 | 0.7604 | 0.8100 |
| 0.00010 | 119.7 | 160.1 | 16.58 | 297.00 | 313.58 | 0.0503 | 0.7560 | 0.8063 |
| 0.0002 | 131.9 | 83.18 | 18.67 | 296.71 | 315.38 | 0.0556 | 0.7271 | 0.7827 |
| 0.0004 | 151.5 | 43.29 | 20.93 | 296.40 | 317.33 | 0.0610 | 0.6981 | 0.7591 |
| 0.0006 | 161.8 | 29.57 | 22.33 | 296.21 | 318.54 | 0.0643 | 0.6811 | 0.7454 |
| 0.0008 | 169.4 | 22.57 | 23.37 | 296.06 | 319.43 | 0.0666 | 0.6690 | 0.7356 |
| 0.0010 | 175.5 | 18.31 | 24.21 | 295.95 | 320.16 | 0.0685 | 0.6596 | 0.7281 |
| 0.002 | 195.6 | 9.570 | 26.94 | 295.57 | 322.51 | 0.0744 | 0.6305 | 0.7049 |
| 0.004 | 217.7 | 5.013 | 29.92 | 295.15 | 325.07 | 0.0806 | 0.6013 | 0.6819 |
| 0.006 | 231.6 | 3.438 | 31.81 | 294.89 | 326.70 | 0.0843 | 0.5842 | 0.6685 |
| 0.008 | 242.0 | 2.632 | 33.21 | 294.70 | 327.91 | 0.0870 | 0.5721 | 0.6591 |
| 0.010 | 250.3 | 2.140 | 34.33 | 294.54 | 328.87 | 0.0892 | 0.5627 | 0.6519 |
| 0.02 | 278.1 | 1.128 | 38.05 | 294.02 | 332.07 | 0.0961 | 0.5334 | 0.6295 |
| 0.04 | 309.1 | 0.5942 | 42.21 | 293.43 | 335.64 | 0.1034 | 0.5039 | 0.6073 |
| 0.06 | 329.0 | 0.4113 | 44.85 | 293.06 | 337.91 | 0.1078 | 0.4869 | 0.5947 |
| 0.08 | 343.9 | 0.3163 | 46.84 | 292.78 | 339.62 | 0.1110 | 0.4745 | 0.5855 |
| 0.1 | 356.0 | 0.2581 | 48.45 | 292.55 | 341.00 | 0.1136 | 0.4649 | 0.5785 |
| 0.2 | 397.1 | 0.1377 | 53.87 | 291.77 | 345.64 | 0.1218 | 0.4353 | 0.5571 |
| 0.3 | 423.8 | 0.095 51 | 57.38 | 291.27 | 348.65 | 0.1268 | 0.4179 | 0.5447 |
| 0.4 | 444.1 | 0.073 78 | 60.03 | 290.89 | 350.92 | 0.1305 | 0.4056 | 0.5361 |
| 0.5 | 460.7 | 0.060 44 | 62.20 | 290.58 | 352.78 | 0.1334 | 0.3960 | 0.5294 |
| 0.6 | 474.9 | 0.051 37 | 64.06 | 290.31 | 354.37 | 0.1359 | 0.3881 | 0.5240 |

*(Contd.)*

| Sat. press., MPa | Sat. temp., °C | Specific volume, m³/kg | Enthalpy, kJ/kg | | | Entropy, kJ/(kg.K) | | |
|---|---|---|---|---|---|---|---|---|
| | | $v_g$ | $h_f$ | $h_{fg}$ | $h_g$ | $s_f$ | $s_{fg}$ | $s_g$ |
| 0.7 | 487.3 | 0.044 79 | 65.66 | 290.08 | 355.74 | 0.1380 | 0.3815 | 0.5195 |
| 0.8 | 498.4 | 0.039 78 | 67.11 | 289.87 | 36.98 | 0.1398 | 0.3757 | 0.5155 |
| 0.9 | 508.5 | 0.035 84 | 68.42 | 289.68 | 358.10 | 0.1415 | 0.3706 | 0.5121 |
| 1.0 | 517.8 | 0.032 66 | 69.61 | 289.50 | 359.11 | 0.1429 | 0.3660 | 0.5089 |
| 1.2 | 534.4 | 0.027 81 | 71.75 | 289.19 | 360.94 | 0.1455 | 0.3581 | 0.5036 |
| 1.4 | 549.0 | 0.024 29 | 73.63 | 288.92 | 362.55 | 0.1478 | 0.3514 | 0.4992 |
| 1.6 | 562.0 | 0.021 61 | 75.37 | 288.67 | 364.04 | 0.1498 | 0.3456 | 0.4954 |
| 1.8 | 574.0 | 0.019 49 | 76.83 | 288.45 | 365.28 | 0.1515 | 0.3405 | 0.4920 |
| 2.0 | 584.9 | 0.017 78 | 78.23 | 288.24 | 366.47 | 0.1531 | 0.3359 | 0.4890 |
| 2.2 | 595.1 | 0.016 37 | 79.54 | 288.05 | 367.59 | 0.1546 | 0.3318 | 0.4864 |
| 2.4 | 604.6 | 0.015 18 | 80.75 | 287.87 | 368.62 | 0.1559 | 0.3280 | 0.4839 |
| 2.6 | 613.5 | 0.014 16 | 81.89 | 287.70 | 369.59 | 0.1571 | 0.3245 | 0.4816 |
| 2.8 | 622.0 | 0.013 29 | 82.96 | 287.54 | 370.50 | 0.1583 | 0.3212 | 0.4795 |
| 3.0 | 630.0 | 0.012 52 | 83.97 | 287.39 | 371.36 | 0.1594 | 0.3182 | 0.4776 |
| 3.5 | 648.5 | 0.010 96 | 86.33 | 287.04 | 373.37 | 0.1619 | 0.3115 | 0.4734 |
| 4.0 | 665.1 | 0.009 78 | 88.43 | 286.73 | 375.16 | 0.1641 | 0.3056 | 0.4697 |
| 4.5 | 680.3 | 0.008 85 | 90.35 | 286.44 | 376.79 | 0.1660 | 0.3004 | 0.4664 |
| 5.0 | 694.4 | 0.008 09 | 92.11 | 286.18 | 378.29 | 0.1678 | 0.2958 | 0.4636 |
| 5.5 | 707.4 | 0.007 46 | 93.76 | 285.93 | 379.69 | 0.1694 | 0.2916 | 0.4610 |
| 6.0 | 719.7 | 0.006 93 | 95.30 | 285.70 | 381.00 | 0.1709 | 0.2878 | 0.4587 |
| 6.5 | 731.3 | 0.006 48 | 96.75 | 285.48 | 382.23 | 0.1723 | 0.2842 | 0.4565 |
| 7.0 | 742.3 | 0.006 09 | 98.12 | 285.28 | 383.40 | 0.1736 | 0.2809 | 0.4545 |
| 7.5 | 752.7 | 0.005 75 | 99.42 | 285.08 | 384.50 | 0.1748 | 0.2779 | 0.4527 |

**Table A1.11** Compressed water (metric units)

| Temp. °C | $v$ m³/kg | $u$ KJ/kg | $h$ KJ/kg | $s$ KJ/(kg · K) | $v$ m³/kg | $u$ KJ/kg | $h$ KJ/kg | $s$ KJ/(kg · K) | $v$ m³/kg | $u$ KJ/kg | $h$ KJ/kg | $s$ KJ/(kg · K) |
|---|---|---|---|---|---|---|---|---|---|---|---|---|
| | $p$ = 5 MPS (263.99 °C) | | | | $p$ = 10 Mpa (311.06 °C) | | | | $p$ = 15 Mpa (34.24 °C) | | | |
| Sat. | .001285 9 | 1147.8 | 1154.2 | 2.9202 | .001452 4 | 1393.0 | 1407.6 | 3.3596 | .001658 1 | 1585.6 | 1610.5 | 3.6848 |
| 0 | .000997 7 | .04 | 5.04 | .0001 | .000995 2 | .09 | 10.04 | .0002 | .000992 8 | .15 | 15.05 | .0004 |
| 20 | .000999 5 | 83.65 | 88.65 | .2956 | .000997 2 | 83.86 | 93.33 | .2945 | .000995 0 | 83.06 | 97.99 | .0004 |
| 40 | .001005 6 | 166.95 | 171.97 | .5705 | .001003 4 | 166.35 | 176.38 | .5686 | .001001 3 | 165.76 | 180.78 | .2934 |
| 60 | .001014 9 | 250.23 | 255.30 | .8285 | .001012 7 | 249.36 | 259.49 | .8258 | .001010 5 | 248.51 | 263.67 | .5666 |
| 80 | .001026 8 | 333.72 | 338.85 | 1.0720 | .001024 5 | 332.59 | 342.83 | 1.0688 | .001022 2 | 331.48 | 346.81 | .8232 |
| 100 | .001041 0 | 417.52 | 422.72 | 1.3030 | .001038 5 | 416.12 | 426.50 | 1.2992 | .001036 1 | 414.74 | 430.28 | 1.0656 |
| 120 | .001057 6 | 501.80 | 507.09 | 1.5233 | .0010549 | 500.08 | 510.64 | 1.5189 | .001052 2 | 498.40 | 514.19 | 1.2955 |
| 140 | .001076 8 | 586.76 | 592.15 | 1.7343 | .001073 7 | 584.68 | 595.42 | 1.7292 | .001070 7 | 582.66 | 598.72 | 1.5145 |
| 160 | .001098 8 | 672.62 | 678.12 | 1.9375 | .001095 3 | 670.13 | 681.08 | 1.9317 | .001091 8 | 667.71 | 684.09 | 1.7242 |
| 180 | .001124 0 | 759.63 | 765.63 | 2.1341 | .001119 9 | 756.65 | 767.84 | 2.1275 | .001115 9 | 753.76 | 770.50 | 1.9260 |
| 200 | .001153 0 | 848.1 | 853.9 | 2.3255 | .001148 0 | 844.5 | 856.0 | 2.3178 | .001143 3 | 841.0 | 858.2 | 2.1210 |
| 220 | .001186 6 | 938.4 | 944.4 | 2.5128 | .001180 5 | 934.1 | 945.9 | 2.5039 | .001174 8 | 929.9 | 947.5 | 2.3104 |
| 240 | .001226 4 | 1031.4 | 1037.5 | 2.6979 | .001218 7 | 1026.0 | 1038.1 | 2.6872 | .001211 4 | 1020.8 | 1039.0 | 2.4953 |
| 260 | .001274 9 | 1127.9 | 1134.3 | 2.8830 | .001264 5 | 1121.1 | 1133.7 | 2.8699 | .001255 0 | 1114.6 | 1133.4 | 2.6771 |
| 280 | | | | | .001321 6 | 1220.9 | 1234.1 | 3.0548 | .001308 4 | 1212.5 | 1232.1 | 3.8576 |
| 300 | | | | | .001397 2 | 1328.4 | 1328.4 | 3.2469 | .001377 0 | 1316.1 | 1337.3 | 3.2260 |
| 320 | | | | | | | | | .001472 4 | 1431.1 | 1453.2 | 3.4247 |
| 340 | | | | | | | | | .001631 1 | 1567.5 | 1591.9 | 3.6546 |
| Sat | .002036 | 1785.6 | 1826.3 | 4.0139 | | | | | | | | |
| 0 | .000990 4 | .19 | 20.01 | .0004 | .000985 6 | .25 | 29.82 | .0001 | .000976 6 | .20 | 49.03 | –.0014 |
| 20 | .000992 8 | 82.77 | 102.62 | .2923 | .000988 6 | 82.17 | 111.84 | .2899 | .000980 4 | 81.00 | 130.02 | .2848 |
| 40 | .000999 2 | 165.17 | 185.16 | .5646 | .001995 1 | 164.04 | 193.89 | .5607 | .000987 2 | 161.86 | 211.21 | .5527 |
| 60 | .001008 4 | 247.68 | 267.85 | .8206 | .001004 2 | 246.06 | 276.19 | .8154 | .000996 2 | 242.98 | 292.79 | .8052 |
| 80 | .001019 9 | 330.40 | 350.80 | 1.0624 | .001015 6 | 328.30 | 358.77 | 1.0561 | .001007 3 | 324.34 | 374.70 | 1.0440 |

*(Contd.)*

| Temp. °C | $v$ m³/kg | $u$ KJ/kg | $h$ KJ/kg | $s$ KJ/(kg · K) | $v$ m³/kg | $u$ KJ/kg | $h$ KJ/kg | $s$ KJ/(kg · K) | $v$ m³/kg | $u$ KJ/kg | $h$ KJ/kg | $s$ KJ/(kg · K) |
|---|---|---|---|---|---|---|---|---|---|---|---|---|
| | $p$ = 5 MPS (263.99 °C) | | | | $p$ = 10 Mpa (311.06 °C) | | | | $p$ = 15 Mpa (34.24 °C) | | | |
| 100 | .001037 7 | 413.39 | 434.06 | 1.2917 | .001029 0 | 410.78 | 441.66 | 1.2844 | .001020 1 | 405.88 | 456.89 | 1.2703 |
| 120 | .001049 6 | 496.76 | 517.76 | 1.5102 | .001044 5 | 493.59 | 524.93 | 1.5018 | .001034 8 | 487.65 | 539.39 | 1.4857 |
| 140 | .001067 8 | 580.69 | 602.04 | 1.7193 | .001062 1 | 576.88 | 608.75 | 1.7098 | .001051 5 | 569.77 | 622.35 | 1.6915 |
| 160 | .001088 5 | 665.35 | 687.12 | 1.9204 | .001082 1 | 660.82 | 693.28 | 1.9096 | .001070 3 | 652.41 | 705.92 | 1.8891 |
| 180 | .001112 0 | 750.95 | 773.20 | 2.1147 | .001104 7 | 745.59 | 778.73 | 2.1024 | .001091 2 | 735.68 | 790.25 | 2.0794 |
| 200 | .001133 8 | 837.7 | 860.5 | 2.3031 | .001130 2 | 831.4 | 865.3 | 2.2893 | .001114 6 | 819.7 | 875.5 | 2.2634 |
| 220 | .001169 3 | 925.9 | 949.3 | 2.4870 | .001159 0 | 918.3 | 953.1 | 2.4711 | .001140 8 | 904.7 | 961.7 | 2.4419 |
| 240 | .001204 6 | 1016.0 | 1040.0 | 2.6674 | .001192 0 | 1006.9 | 1042.6 | 2.6490 | .001170 2 | 990.7 | 1049.2 | 2.6158 |
| 260 | .001246 2 | 1108.6 | 1133.5 | 2.8459 | .001230 3 | 1097.4 | 1134.3 | 2.8243 | .001203 4 | 1078.1 | 1138.2 | 2.7860 |
| 280 | .001296 5 | 1204.7 | 1230.6 | 3.0248 | .001275 5 | 1190.7 | 1229.0 | 2.0086 | .001241 5 | 1167.2 | 1229.3 | 2.9537 |
| 300 | .001359 6 | 1306.1 | 1333.3 | 3.2071 | .001330 4 | 1287.9 | 1327.8 | 3.1741 | .001286 0 | 1258.7 | 1323.3 | 3.1200 |
| 320 | .001443 7 | 1415.7 | 1444.6 | 3.3979 | .001399 7 | 1390.7 | 1432.7 | 3.5339 | .001338 8 | 1353.3 | 1420.2 | 3.2868 |
| 340 | .001568 4 | 1539.7 | 1571.0 | 3.6075 | .001492 0 | 1501.7 | 1546.5 | 3.5426 | .001403 2 | 1452.0 | 1522.1 | 3.4557 |
| 360 | .001822 6 | 1702.8 | 1739.3 | 3.8772 | .001626 5 | 1626.6 | 1675.4 | 3.7494 | .001483 8 | 1556.0 | 1630.2 | 3.6291 |
| 380 | | | | | .001869 1 | 1781.4 | 1837.5 | 4.0012 | .001588 4 | 1667.2 | 1746.6 | 3.8101 |

**Table A1.12** The error function

| $\eta$ | erf $\eta$ | $\eta$ | erf $\eta$ | $\eta$ | erf $\eta$ |
|---|---|---|---|---|---|
| 0.00 | 0.00000 | 0.76 | 0.71754 | 1.52 | 0.96841 |
| 0.02 | 0.02256 | 0.78 | 0.73001 | 1.54 | 0.97059 |
| 0.04 | 0.04511 | 0.80 | 0.74210 | 1.56 | 0.97263 |
| 0.06 | 0.06762 | 0.82 | 0.75381 | 1.58 | 0.97455 |
| 0.08 | 0.09008 | 0.84 | 0.76514 | 1.60 | 0.97635 |
| 0.10 | 0.11246 | 0.86 | 0.77610 | 1.62 | 0.97804 |
| 0.12 | 0.13746 | 0.88 | 0.78669 | 1.64 | 0.97962 |
| 0.14 | 0.15695 | 0.90 | 0.79691 | 1.66 | 0.98110 |
| 0.16 | 0.17901 | 0.92 | 0.80677 | 1.68 | 0.98249 |
| 0.18 | 0.20094 | 0.94 | 0.81627 | 1.70 | 0.98379 |
| 0.20 | 0.22270 | 0.96 | 0.82542 | 1.72 | 0.98500 |
| 0.22 | 0.24430 | 0.98 | 0.83423 | 1.74 | 0.98613 |
| 0.24 | 0.26570 | 1.00 | 0.84270 | 1.76 | 0.98719 |
| 0.26 | 0.28690 | 1.02 | 0.85084 | 1.78 | 0.98817 |
| 0.28 | 0.30788 | 1.04 | 0.85865 | 1.80 | 0.98909 |
| 0.30 | 0.32863 | 1.06 | 0.86614 | 1.82 | 0.98994 |
| 0.32 | 0.34913 | 1.08 | 0.87333 | 1.84 | 0.99074 |
| 0.34 | 0.36936 | 1.10 | 0.88020 | 1.86 | 0.99147 |
| 0.36 | 0.38933 | 1.12 | 0.88079 | 1.88 | 0.99216 |
| 0.38 | 0.40901 | 1.14 | 0.89308 | 1.90 | 0.99279 |
| 0.40 | 0.42809 | 1.16 | 0.89910 | 1.92 | 0.99338 |
| 0.42 | 0.44749 | 1.18 | 0.90484 | 1.94 | 0.99392 |
| 0.44 | 0.46622 | 1.20 | 0.91031 | 1.96 | 0.99443 |
| 0.46 | 0.48466 | 1.22 | 0.91553 | 1.98 | 0.99489 |
| 0.48 | 0.50272 | 1.24 | 0.92050 | 2.00 | 0.995322 |
| 0.50 | 0.52050 | 1.26 | 0.92524 | 2.10 | 0.997020 |
| 0.52 | 0.53790 | 1.28 | 0.92973 | 2.20 | 0.98137 |
| 0.54 | 0.55494 | 1.30 | 0.93401 | 2.30 | 0.998857 |
| 0.56 | 0.57162 | 1.32 | 0.93806 | 2.40 | 0.999311 |
| 0.58 | 0.58792 | 1.34 | 0.94191 | 2.50 | 0.999593 |
| 0.60 | 0.60386 | 1.36 | 0.94656 | 2.60 | 0.999764 |
| 0.62 | 0.61941 | 1.38 | 0.94902 | 2.70 | 0.999866 |
| 0.64 | 0.63459 | 1.40 | 0.95228 | 2.80 | 0.999925 |

*(Contd.)*

| $\eta$ | erf $\eta$ | $\eta$ | erf $\eta$ | $\eta$ | erf $\eta$ |
|---|---|---|---|---|---|
| 0.66 | 0.64938 | 1.42 | 0.95638 | 2.90 | 0.999959 |
| 0.68 | 0.66278 | 1.44 | 0.95830 | 3.00 | 0.999978 |
| 0.70 | 0.67780 | 1.46 | 0.96105 | 3.20 | 0.999994 |
| 0.72 | 0.69143 | 1.48 | 0.96365 | 3.40 | 0.999998 |
| 0.74 | 0.70468 | 1.50 | 0.96610 | 3.60 | 1.000000 |

Pressure
0.020
0.05
0.10
0.15
0.20
0.3
0.4
0.5
0.75
500°C
450°C
400°C
350°C
300°C
250°C
200°C
150°C
Constant Specific Volume
Constant Tem
10 bar
9
8
7
6
5
4
3
2
1 bar
0.5
0.4
0.3
0.2
0.1 bar
Saturation Line
Constant Dryness Fraction
Constant Pressure

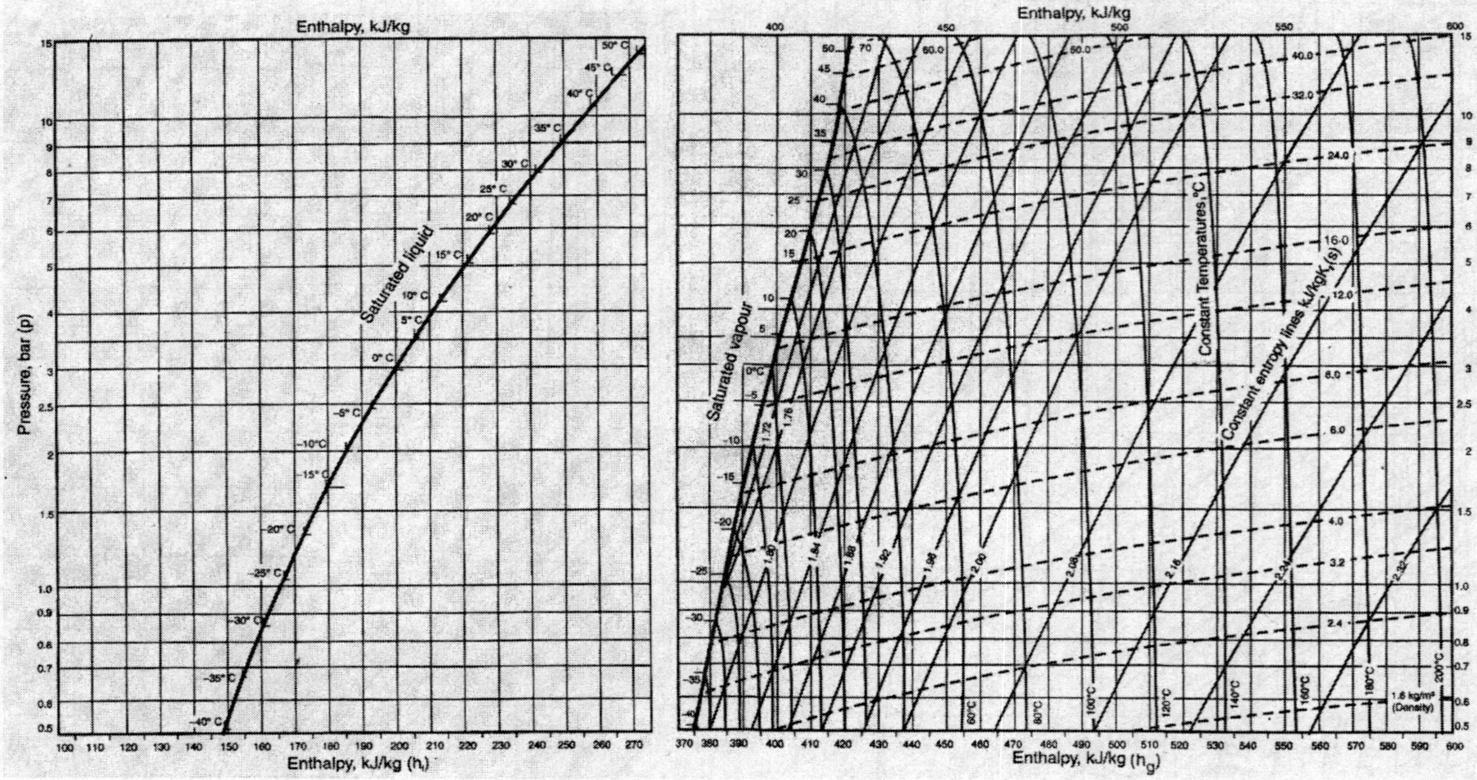


Pressure-Enthalpy Diagram, Refrigerant, 134a

*Note*: The reference point used for the chart is different than that used in the R-134a tables. Therefore, problems should be solved using all property data either from the tables or from the chart, but not from both.

**Fig. A1.14** *P-h* diagram for refrigerant–134a

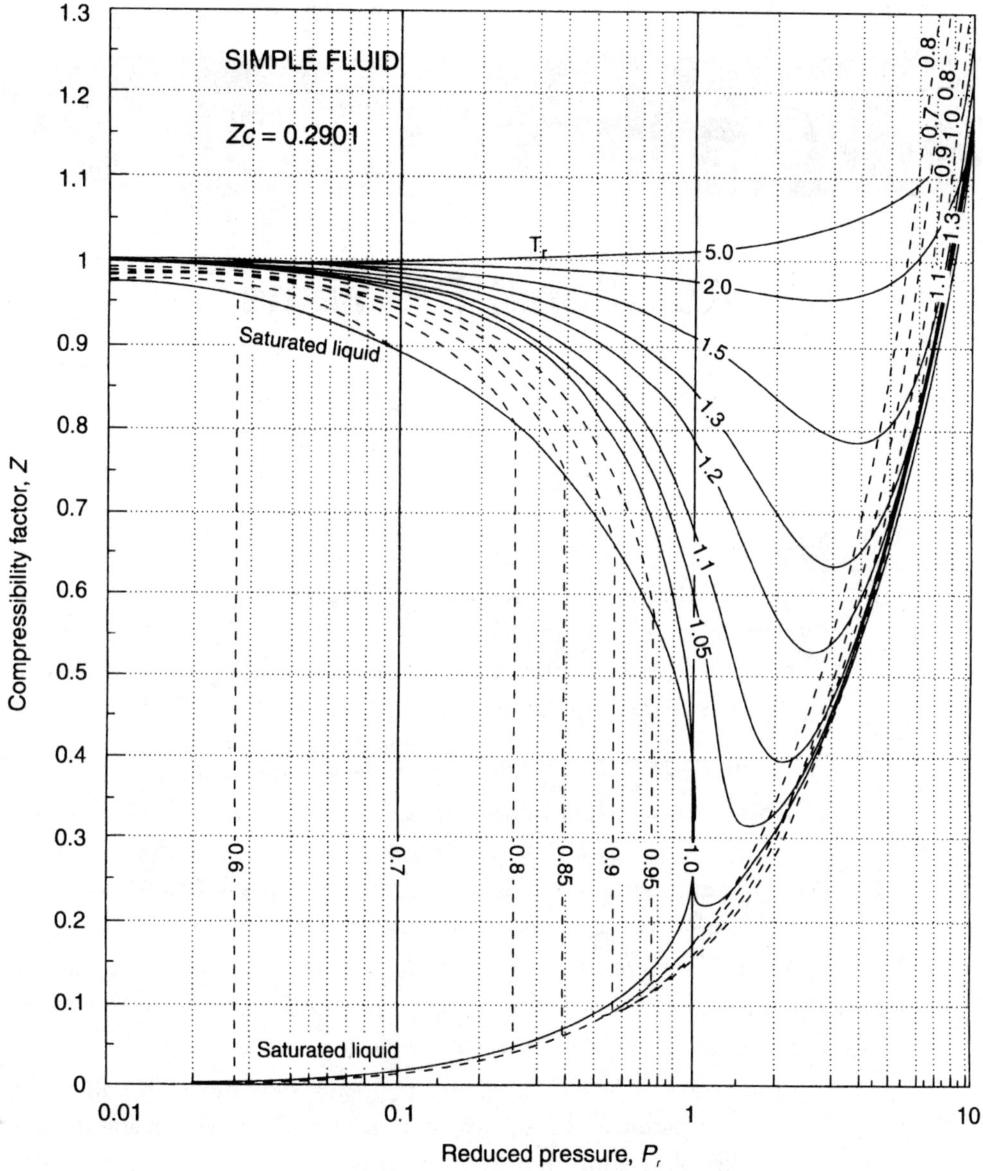


**Fig. A1.15** Nelson–Obert generalised compressibility chart—intermediate pressures.

# APPENDIX TWO

# Additional Problems

## Chapter 1

### Example AP 1.1

Specify whether the following systems are open or closed when in operation. (a) Water wheel (b) Scooter engine (c) Boiler in steam plant (d) Liquid cooling system of an automobile (e) Electric fan (f) Motor car battery (g) Air compressor

*Solution*

(a) *Water wheel.* It is a device that converts kinetic energy of water into mechanical work. Water enters at one end and leaves at another; therefore, the system is an open system. Flow is steady, as conditions do not change at any of its point with time.

(b) *Scooter engine.* In scooter engine, air is sucked in and the burnt gases are left out. Thus, there is flow across its boundaries and hence it is an open system. However, the flow is not steady across it.

(c) *Boiler in steam plant.* It is also an open system in which water enters at some section and steam leaves from some other section and the process takes place continuously without change in any of the properties. So, the flow is steady.

(d) *Liquid cooling system.* Liquid is used to cool the automobile for which it is pumped from its storage location around the engine and other parts to take away heat and finally goes to the radiator where it is cooled. Thus, the system is a steady flow open system.

(e) *Electric fan.* It is a device, which is run by electricity. It has blades that rotate, as a result of which air is pushed down. So, there is mass flow across its boundaries and hence is an open system.

(f) *Motor car battery*. Battery is used to start the engine and provide electricity for lights. The interaction between system (battery) and the surrounding is through wires, which carry electric current. There is no mass flow across the system boundary. Hence, battery is a closed system. However, during the process, chemical reaction occurs that produces some gases, which escape to the atmosphere. This however, maybe ignored.

(g) *Air compressor*. It is a typical system. It is used to raise air pressure for which air is sucked into its cylinder during suction stroke. During this period the system is open. After air has been sucked it is compressed during which, there is no mass flow across the system boundary and hence is closed. Similarly, when air starts flowing out from compressor, the system is again open.

**Example AP 1.2**

In each of the following examples, determine whether the system shown in italics is open or closed, homogeneous or heterogeneous and adiabatic or diathermal (a) small metal *sphere* rotating within a rigid non-conducting hemispherical bowl eventually comes to rest, (b) A *mixture of oxygen and hydrogen* in combustible proportions in a rigid non-conducting envelope is ignited by an infinitesimal electric work.

(a) Sphere is a system and the bowl is its surrounding. No mass flow occurs across the system boundary and hence it is closed. When system comes to rest, its kinetic energy is converted into internal energy raising its temperature. Because of this, energy transfer will take place across the sphere to the inside of the bowl due to temperature difference. Hence, the system is diathermal. (It is common for anyone to say the system is adiabatic because the bowl is non-conducting.) This difference should be understood clearly. The surrounding is adiabatic and not the system. Lastly, the system is homogeneous, since it is made of metal.

(b) In combustion systems, where air and fuel is mixed prior to combustion chamber, the mixture is homogeneous prior to combustion. However, after the mixture is ignited, the combustion proceeds and the composition no longer remains same throughout. Therefore, the mixture becomes heterogeneous. That is, composition of mixture is the same everywhere. Since a non-conducting envelope envelops it, the system is adiabatic. For the same reason there is no mass flow across the system boundary and hence it is a closed system.

**Example AP 1.3**

A layer of gas between two metal plates, which are at different temperatures, is suddenly isolated from its surroundings with reference to mass and energy transfer. Discuss whether the system is in thermodynamic equilibrium or not.

***Solution*** The gas trapped between the two plates is the system and the two plates are its surroundings. Since the air is in contact with the plates initially, there will be a temperature gradient across its thickness with the layers in contact with plates having their respective temperatures. When the system is isolated, there is no more energy transfer across the system but still redistribution of energy would occur within the system till the temperature is same everywhere. Therefore, the moment system is isolated, it is not in thermal and hence, thermodynamic equilibrium. However, once the temperature is equalised then it is in thermodynamic equilibrium.

**Example AP 1.4**

Check whether the differential equation $2(x^3 + y^3)\,dx + 3y^2(x^2 + y^2)\,dy$ can be integrated? Show that its line integral from (0,0) to (1,1) is independent of path.

***Solution*** If the integral of a given differential exists then the integrand is called the point function for which the condition to be satisfied is

$$\left[\frac{\partial}{\partial x}\left\{3y^2\left(x^2+y^2\right)\right\}\right]_y = \left[\frac{\partial}{\partial y}\left\{2\left(x^3+y^3\right)\right\}\right]_x$$

Thus, the condition is satisfied and hence the integrand is a point function. The integrand is

$$= \int_{0,0}^{1,1}\left\{2x\left(x^2+y^2\right)\right\}\partial x + 3y^2\left(x^2+y^2\right)\partial y = 2 \qquad \text{(A)}$$

This integration is done along the line joining (0,0) and (1,1).

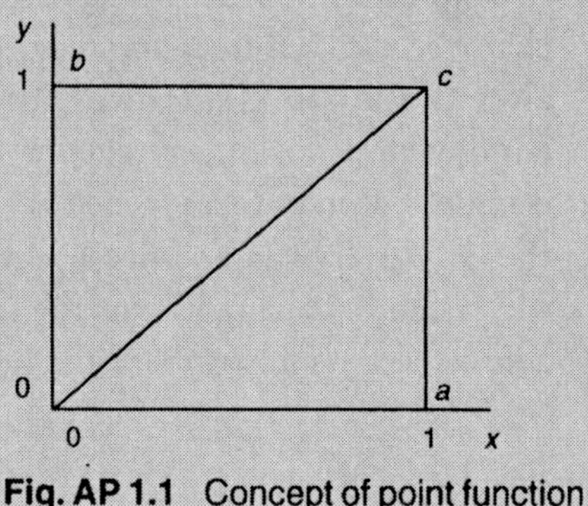


**Fig. AP 1.1** Concept of point function

We also know that $\int_0^1 y\,dx = \int_0^a y\,dx + \int_a^1 y\,dx$. Using this fact we can determine the integral in Eq. A along the path *oac* and then *obc*. You may verify that in both the cases, the integral is 2, which was expected, since change in point function depends only on the end states and not on how is this change made.

**Example AP 1.5**

Check whether the following quantities are thermodynamic properties? (i) $vdp/T + pdv/T$, (ii) $dT/T + pdv/T$, (iii) $pdv + v\,dp$, and (iv) $pdv$.

***Solution*** The necessary condition for the quantity $Mdx + Ndy$ to represent some system property is

$$(\partial M/\partial y)_x = (\partial N/\partial x)_y. \tag{A}$$

(i) Here, $x = p$, $M = v/T = R/p$, $y = v$ and $N = p/T = R/v$. We observe that condition A is satisfied and hence it is system property.

(ii) Here, $x = T$, $M = 1/T$, $y = v$ and $N = p/T = R/v$. We observe that condition A is satisfied and hence it is system property.

(iii) In the same way, we observe that $pdv + vdp$ is system property.

(iv) There is only one independent variable and hence we have to see whether its integral depends on path. For example, we may plot path on *p*-*v* diagram. Then, $pdv$ is nothing but area of the curve under it with $v$ axis. We have already seen in the text that this area depends on the type of the path and hence is a path function and not a property.

**Example AP 1.6**

The following table lists some of the parameters belonging to a system. Put ✓ in the appropriate place.

| Property | Intensive | Extensive | Intrinsic | Extrinsic |
|---|---|---|---|---|
| Pressure | ✓ | | ✓ | |
| Volume | | ✓ | ✓ | |
| Entropy | | ✓ | ✓ | |
| Specific entropy | ✓ | | ✓ | |
| Specific Volume | ✓ | | ✓ | |

*(Contd.)*

| Property | Intensive | Extensive | Intrinsic | Extrinsic |
|---|---|---|---|---|
| Temperature | ✓ | | ✓ | |
| Internal energy | | ✓ | ✓ | |
| Mass | | ✓ | ✓ | |
| Specific enthalpy | ✓ | | ✓ | |
| Weight | | ✓ | | ✓ |
| Area | | ✓ | ✓ | |
| Velocity | ✓ | | | ✓ |
| Elevation | ✓ | | | ✓ |
| PE and KE | | ✓ | | ✓ |
| Density | ✓ | | ✓ | |
| Charge | ✓ | | ✓ | |

**Example AP 1.7**

An insulated copper vessel contains a quantity of oil at a temperature of 15°C. A steel block of mass 0.1 kg at a temperature of 800°C is dropped into oil. When the temperature has equalised the temperature of oil is 16.5°C. Determine the magnitude and sign of heat transfer for each of the following systems (i) the vessel, (ii) the steel block, and (iii) the oil.

***Solution*** Let c kJ/kg K be the specific heat of the steel block. From physics course, we know that the amount of heat transferred to oil is

$$Q = 0.1 \times c \times (800 - 16.5) = 78.35c \text{ kJ}$$

(i) *Vessel.* Since vessel is insulated and the energy transfer is taking place within it, there is no energy interaction with its surrounding. Therefore, heat transfer is = 0.

(ii) *Steel block.* With this as system, the water in the vessel becomes surrounding and hence heat is flowing from the system to its surrounding and is negative by the sign convention. Heat transfer is –78.35c kJ.

(iii) *Oil.* For this system, the steel block and the vessel boundary are the surroundings. There is no energy interaction across the vessel boundary. However, there is heat transfer from the steel block. By sign convention it is positive. The net heat transfer to the system is

$$0 + 78.35c \text{ kJ} = 78.35c \text{ kJ}$$

**Example AP 1.8**

A constant gas thermometer containing helium gives reading of gas pressure $p$ as 1000 and 1366 mm of mercury at the ice and steam point of water, respectively. (a) Express the gas thermometer Fahrenheit temperature in terms of gas pressure $p$. (b) The temperature thermometer when left in the atmosphere registers 1070 mmHg. Determine the atmospheric temperature.

***Solution*** We may use linear polynomial to express temperature $t$ with pressure $p$ read by the thermometer, since only two fixed points are given,

$$t = a + b\,p \tag{A}$$

(a) The given fixed points give values of $a$ and $b$; the relation is

$$t = -460 + 0.492\,p$$

(b) Substituting for $p = 1070$, we get,

$$t_a = \text{atmospheric temperature} = 67°\text{F}$$

**Example AP 1.9**

(a) Given that the ice point is 0°C and 32°F and that the steam point is 100°C and 212 °F on two temperature scales—the Celsius and the Fahrenheit scales. (a) Find a relationship between these scales. (b) Find the temperature at which the Celsius and Fahrenheit scales have the same numerical value. (c) Is there any temperature at which the Kelvin and absolute Fahrenheit scales have the same numerical value? (d) On a hot summer day the temperature may be 110°F and on a cold day it might be –5°F. The average normal human body temperature is 98.6°F. Convert these temperature to °C, K and °R.

***Solution***

(a) Let the relation between temperature on Celsius and Fahrenheit scales and some property $x$ of the thermometer be $t = a\,x + b$, where, constants $a$ and $b$ on the two scales are determined from the given fixed points at which points value of $x$ is $x_i$ and $x_s$, respectively. Note that $x_i$ and $x_s$ will be independent of the temperature scale as these represent change in some property of the measuring instrument. We thus get

$$t_c = 100\,(x - x_i)/(x_s - x_i) \text{ and}$$

$$t_F = 180\,(x - x_i)/(x_s - x_i) + 32$$

$$\rightarrow t_c/100 = (t_F - 32)/180 = (x - x_i)/(x_s - x_i)$$

or $t_c/5 = (t_F - 32)/9$

(b) Let this temperature be $t$. Then,

$$t/5 = (t - 3)/9 \rightarrow t = -40°$$

(c) Let the temperature be $T$. On Celsius scale, $t_c = T - 273$ and on absolute Fahrenheit scale it is $T - 460$. (Like Kelvin scale, there is another absolute Fahrenheit scale on which temperature is expressed in Rankine and written as $R$ such that $T = 460 + t_F$. $R$) Substitution gives, $T = -0.75°$

(d) 110°F = 110 + 460 = 570 R and

$$t_c = 5 \times (110 - 32)/9 = 43.3°C$$
$$= 43.3 + 273 = 316.3 \text{ K}$$

−5°F = −5 + 460 = 455 R and

$$t_c = 5 \times (-5 - 32)/9 = -20.6°C$$
$$= -20.6 + 273 = 252.4 \text{ K}$$

98.6°F = 98.6 + 460 = 558.6 R and

$$t_c = 5 \times (98.6 - 32)/9 = 37°C$$
$$= 37 + 273 = 310 \text{ K}$$

**Example AP 1.10**

Newton's second law is $F \propto m\,a$, where $m$ is mass and $a$ is acceleration. This can be written as, $F = m\,a/g_c$, where $1/g_c$ is a constant of proportionality whose value and unit will depend on the system of units used. Find out its value and unit in some of the system of units used all over the world and tabulate the results.

***Solution*** The following table gives the results in different units. The constant of proportionality is used to make Newton's second law of motion consistent w.r.t. dimensions when four fundamental dimensions are used. The following table lists the units used for these fundamental dimensions in the different system of units as well as the value of $g_c$.

| System of unit | Mass | Force | Length | Time | $g_c$ |
|---|---|---|---|---|---|
| British engineering | ibm | ibf | ft | s | 32.174 ibm-ft/(ibf-s$^2$) |
| | Ibm | poundal | ft | s | |
| | Gm | 1 | cm | s | 1 ibm-ft/(poundal-s$^2$) |
| | slug | dyne | ft | s | |

*(Contd.)*

| System of unit | Mass | Force | Length | Time | $g_c$ |
|---|---|---|---|---|---|
| | kg | ibf | m | s | 1 gm-cm/(dyne-s$^2$) |
| Metric or MKS | kg | N | m | s | 1 slug-ft/(ibf-s$^2$) |
| | | kgf | | | 1 kg-m/(N-s$^2$) |
| | | | | | 9.80665 kg-m/(kgf-s$^2$) |

***Note***: lbm is used to represent 1 lb mass; lbm is used to differentiate it from lbf used for force in British engineering system, defined as the force acting on a mass of 1 lb when placed on equator.

***Example***: We shall demonstrate the procedure by taking one example from, say the last row in the table. In this system, 1 kgf force is defined as the force acting on 1 kg mass when placed in earth's field at equator, where $g$ = 9.80665 m/s$^2$. Using the Newton's law, we write

$$1 \text{ kgf} = (1/g_c)\ 1 \text{ kg} \times g \text{ m/s}^2 \rightarrow g_c = 9.80665 \text{ kg-m/(kgf-s}^2)$$

Thus this value of $g_c$ makes the Newton's law dimensionally consistent. From the above table, we may also get the following results, using the given equation.

(i) Newton is a force that acts on 1 kg mass to accelerate it by 1 m/s$^2$.

(ii) Slug is mass of a chunk such that when acted upon by 1 lbm force would be accelerated by 1 ft/s$^2$.

(iii) 1 poundal is a force that would accelerate 1 lb mass by $g$ ft/s$^2$ at the equator.

**Example AP 1.11**

Set equivalence between (a) 1 kgm and chunk, where chunk is defined as unit of mass, m$^3$/s$^2$ in LT system and (b) between 1 N and push, where push is defined as unit of force, m$^4$/s$^4$, in LT system.

***Solution*** In text we mentioned that it is not necessary to have four or three fundamental dimensions. We may even take two fundamental dimensions. However, the number of equations required to get the derived dimensions increase, since we need $M$, $L$ and $T$ to get any other dimension. In the system of four fundamental dimensions we do not need any relation, whereas in the system of three fundamental dimensions, we needed one, the Newton's law of motion. This means, we shall need one more such relation if we want to adopt two fundamental dimensions $L$ and $T$. The second relation used for the purpose is the Newton's law of gravitation in which also, like Newton's second law of motion, we take constant of proportionality as unity. Then the two relations are written as

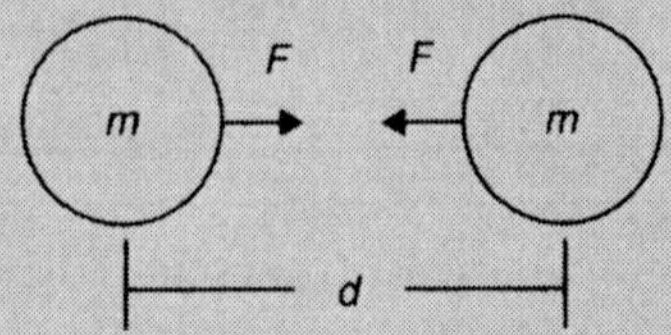


**Fig. AP 1.2** LT system of units

$F = m \times a$ and $F = m \times m/d^2$

where, $d$ is distance between two masses of mass $m$ each and $a$ is acceleration produced in each mass due to force of attraction (gravitational force) $F$ between the masses. From these equations, we get

$$m = a.\ d^2 \rightarrow [M] = [a.\ d^2] = [LT^{-2}.\ L^2] = [L^3\ T^{-2}] \qquad \text{(A)}$$

*MKS Sytem.* In MKS engineering system of units (where mass, length, and time are measured in kg, m and s), gravitational force between two masses is given by

$$F = G \times m \times m/d^2, \text{ where, } G = 6.67 \times 10^{-11} \text{m}^3/\text{kg-s}^2$$

Let $m = 1$ kg then $F = 1 \times a = G/d^2 \rightarrow a = 6.67 \times 10^{-11}/d^2)$ (B)

*LT System.* In proposed LT system, $a$ would be $= x^2/d^2$, where 1 kg $\equiv x$ chunk (C)

Equating value of acceleration from Eq. B and C, we get

$x = 0.8167 \times 10^{-5}$

or 1 kg = $0.8167 \times 10^{-5}$ chunk

Similarly, 1 $N \equiv$ 1 kg $\times$ 1 m/s$^2$

$= (0.8167 \times 10^{-5}) \times 1\ (\text{m/s}^2)$

$= 0.8167 \times 10^{-5}\ \text{m}^4/\text{s}^4$ (push)

**Example AP 1.12**

1 kg of mass is accelerated with a force of 10 lbf. Determine the acceleration of mass in cm/s$^2$ and ft/s$^2$.

***Solution*** By definition of lbf, the unit of force, 1 lbf

$$= 1 \text{ lb} \times g \text{ ft/s}^2$$
$$= (453.6 \times 10^{-3} \text{ kg}) \times 981 \text{ cm/s}^2$$
$$= 445 \text{ kg-cm/s}^2$$

$\therefore$ acceleration of 1 kg mass = 10 lbf/1 kg

$$= 10 \times 445 \text{ (kg-cm/s}^2\text{)/1 (kg)}$$
$$= 4450 \text{ cm/s}^2$$

In ft/s$^2$, 1 lbf = 1 lb × $g$ ft/s$^2$ and 1 lb = 0.4536 kg

$\therefore$ acceleration of 1 kg mass = 10 lbf/1 kg

$$= 10 \times 32.2 \text{ (lb ft/s}^2\text{)/(1/0.4536) (lb)}$$
$$= 138 \text{ ft/s}^2$$

**Example AP 1.13**

If a mass is weighed on a beam at a location where $g$ = 900 cm/s$^2$, what is the reading on the balance? The spring scale originally is calibrated in a region where $g$ = 981 cm/s$^2$.

*Solution* A beam balance reads mass whereas the spring balance reads the weight. Since mass is the property of the chunk, it is same irrespective of where it is measured. On the other hand, the force acting on this mass would change from place to place, and hence the spring balance reading, since acceleration due to gravity depends on the place.

Spring balance reading where

$$g = 900 \text{ cm/s}^2 = 1 \text{ (kg)} \times 900 \text{ (cm/s}^2\text{)} = 900 \text{ kg-cm/s}^2$$

But the reading on spring balance scale will be marked 1 for the place where $g$ = 981 cm/s$^2$

$\therefore$ spring balance reading = 900/981 = 0.917 kg

A spring balance gives reading in kg, which is nothing but kgf. That is, if a mass of 1 kg is attached to its hook then it records 1 kg though it measures force.

**Example AP 1.14**

Viscosity of air in poise is given by the relation,

$$\mu = 1.65 \times 10^{-4} + 2.5 \times 10^{-7}\, T$$

where $T$ is temperature in °F. Poise is used in CGS system of units where force, distance and time are measured in dyne, cm and s. Convert the above expression into its appropriate form when the units of force, distance, and time are (a) lbf, ft, s and (b) kgf, m, s.

***Solution*** Conversion factors are

$g$ ft/s$^2$ = 981 cm/s$^2$ and 453.6 $g$ = 1 lb

1 ft = 2.54 × 12 cm = 30.48 cm or

1 cm = 1/30.48 ft

By definition, 1 dyne = 1 gm × 1 cm/s$^2$ and

1 lbf = 1 lb × g ft/s$^2$

∴ 1 dyne/1 lbf = (gm/lb) × {(cm/s$^2$)/($g$ ft/s$^2$)}

→ 1 dyne = {1/(453.6 × 981)} lbf

and 1 poise = 1 dyne × s/cm$^2$

= {1/(453.6 × 981)}/(1/30.48$^2$)

= 2.09 × 10$^{-3}$ lbf-s/ft$^2$

∴ $\mu$ in new units = $\mu$ (given) × 2.09 × 10$^{-3}$

= 3.45 × 10$^{-7}$ + 5.22 × 10$^{-10}$ T lbf s/ft$^2$

(b) In this system, 1 kgf = 1 kg × $g$ m/s$^2$

= 1000 (gm) × 981 (cm/s$^2$)

= 981000 dyne

→ 1 dyne = 1/981000 kgf

1 poise = 1 dyne × s/cm$^2$

= (1/981000)/{(1/100)}2

= 1.02 × 10$^{-2}$ kgf-s/m$^2$

∴ $\mu$ = (1.65 × 10$^{-4}$ + 2.5 x 10$^{-7}$ T) × 1.02 × 10$^{-2}$

= 1.68 × 10$^{-6}$ + 2.55 × 10$^{-9}$ $T$ kgf-s/m$^2$

***Note***: In these expressions, temperature is measured in °F. If temperature is given in °C then we may use the conversion relation in the two temperature scales

$C/5$ = (F – 32)/9 → $T$°F = 1.8 $C$ + 32°C and can be used in these relations.

**Example AP 1.15**

The hourly rate of heat loss from unit area of a horizontal cylinder of diameter d ft and having a surface temperature °F in air at 50°F may be taken as

$$q = \frac{0.226(t-59)^{5/9}}{d^{1/4}} \text{ btu/(ft}^{7/4} - \text{hr} - \text{F}^{5/4})$$

If now $d$ is measured in $m$, $t$ in °C and heat transfer in kcal, find out the new equation for $q$. How will the result change if $q$ is measured in kJ?

***Solution*** Dimensions of $[q] = [HL^{-2}T^{-1}]$, where $H$ is dimension of heat, considered as fundamental dimension in heat transfer problems. (You may study in the Heat Transfer course that temperature has been considered as fundamental dimension and then heat transfer becomes secondary or derived dimension. But it is common to see heat transfer also as a fundamental dimension.)

Since heat transfer has been chosen as fundamental dimension, we have to define unit too. Btu is that amount of heat which raises temperature of 1 lb of water by 1°F whereas 1 kcal increases temperature of 1 kg water by 1°C.

The point to recognise is that the physical quantity does not depend on the system of units used; only the number representing it changes. For example, if a plate has length 1m, we may express it as 100 cm, 1000 mm, or 0.001 km. That is, depending upon the unit used for measurement of length, the number representing this length is changed. This can be expressed in the following way

$$n_1 u_1 = n_2 u_2 = \ldots, \text{etc.}$$

Where *ns* represent the numeral part and *us* the unit. In the given problem the constant is 0.226 in btu/(ft$^{7/4}$ – hr – F$^{5/4}$) and is to be obtained in kcal/(m$^{7/4}$ – hr – C$^{5/4}$). Using conversion factor for different dimensions, we get

$$n_2 = n_1 u_1/u_2 = n_1 \text{ (btu/kcal)} \times (\text{m/ft})^{7/4} \times (°\text{C}/°\text{F})^{5/4}$$

Conversion factors are: 1 btu = 0.251 kcal, 1m = 3.2808 ft, and $F = (1.8C + 32)$. Substituting these, we get

$$n_2 = 0.226 \times 0.251 \times (3.2808)^{7/4}/(1.8C + 32)^{5/4}$$
$$= 0.056726 \times 7.998/(1.8C + 32)^{5/4}$$
$$= 0.4537/(1.8C + 32)^{5/4} \quad \text{(A)}$$

If kJ is used for heat transfer, then we know 1 kJ = 4.187 kcal and $C = K - 273$. From Eq. A

$$n_2 = 0.10836/(1.8K - 523.4)^{5/4}$$

## Chapter 2

### Example AP 2.1

A piston cylinder contains 25 g of saturated water vapour that is maintained at a constant pressure of 300 kPa. A resistance heater within the cylinder is turned on and passes a current of 0.2 A for 5 min from a 120 V source. At the same time, a heat loss of 3.7 kJ occurs. Determine the final temperature.

***Solution*** Given $m$ = 25, $g$ = 0.025 kg of saturated $H_2O$ vapour, $p$ = 300 kPa, I = 0.2 A, $V$ = 120 V, $t$ = 5 min and $Q_L$ = 3.7 kJ. From steam table at 300 kPa for saturated steam, $h_1$ = 2725.3 kJ/kg.

Since the system is closed, nfee can be used for analysis. Heat generated within cylinder = 0.2 × 120 × 5 × 60/1000 = 7.2 kJ. In fact, this is the electrical work transfer to the system; heat generation is only a physical way in which it is passed on to steam. In a thermodynamic sense, it is work transfer to the system. The other way of work transfer, since pressure is constant as well in the process, is the boundary work, which is $p\,(V_2 - V_1)$. Using nfee, we get

$$-Q_L = (U_2 - U_1) + p\,(V_2 - V_1) + 7.2$$
$$= m \times (h_2 - h_1) + 7.2$$
$$\rightarrow h_2 = 2865.3 \text{ kJ/kg}$$

Steam is, therefore, superheated after 5 minutes; the final temperature from steam table is 200°C.

### Example AP 2.2

A rigid tank is divided into two partitions. Initially, one side of the tank contains 5 kg of $H_2O$ at 200 kPa and 25°C and the other side is evacuated. The partition is now removed and the water expands into the entire tank. The water is allowed to exchange heat with the surroundings until the temperature in the tank returns to the initial value of 25°C. Determine (i) the volume of tank, (ii) the final pressure and (iii) the heat transfer to the process.

***Solution*** Given: $m$ = 5 kg, $p_1$ = 2 bar, $T_1$ = 25°C

(i) **Volume of tank.** Let us consider the entire tank space within six walls of the container as system, then the system is a closed system.[1] Its

---

1. If we choose water contained in the partition as system then its volume cannot be considered constant as we have done in solving this problem. Try to solve this problem taking this as a system.

volume does not change after the partition is removed. Since the condition of steam in one partition is given, we can find out the specific volume and hence volume of steam as well as that of the whole container. We observe that water is compressed liquid[2] since saturation pressure at 25°C is only 0.03169 bar. For such liquid enthalpy, volume, entropy etc., are almost same as that of the saturated liquid at 25°C. From steam table,

$$v_1 \approx 0.001 \text{ m}^3/\text{kg}$$

$$\rightarrow V = \text{container volume}$$

$$= 2 \times m \times v_1 = 0.01 \text{ m}^3$$

Here, factor of two is used because the container is partitioned with equal volumes and steam is initially contained in only half the portion.

(ii) **Final pressure.** Pressure would decrease when partition is removed because of expansion in the vacuum space. Let us assume that initial temperature of the container contents is restored, otherwise there is not enough data to determine the final state. Since mass does not change, the specific volume in the final state is, $v_2 = 0.01/5 = 0.002$ m³/kg, which is a little more than the volume 0.001003 of saturated liquid at 25°C. Therefore, $H_2O$ in final state us wet steam at 25°C. From steam table, saturation pressure is = 3.169 kPa.

(iii) **Heat transfer to the process.** Let us find out the quality of steam in the final state using values at 25°C from steam table.

$$v_2 = v_{f2} + x_2 (v_{g2} - v_{f2})$$

$$\rightarrow x_2 = 2.3 \times 10^{-5} \text{ and}$$

$$u_2 = u_{f2} + x_2 u_{fg2} = 104.93 \text{ kJ/kg}$$

During the process, work transfer is zero, since expansion is within the system and the only interaction with the surrounding is heat transfer (given).

Applying nfee, we get

$$Q = m \times (u_2 - u_1) = 0.25 \text{ kJ}$$

The positive sign indicates that heat is transferred to the system. This means after unrestrained expansion, temperature of water has decreased. This type of expansion process is known as *unrestrained expansion* process.

2. See Example AP4.5 for further comments on compressed liquid.

**Example AP 2.3**

R-134a is to be cooled in a condenser by cold water. The refrigerant enters the condenser with a mass flow rate of 6 kg/min at 1 MPa and 70°C and leaves at 35°C. The cold water enters at 300 kPa and 15°C and leaves at 25°C. Neglecting any pressure drops, determine the mass flow rate of the cooling water, and the heat transfer from refrigerant to the water.

***Solution*** This problem can be solved by taking either the heat exchanger or a coil through which refrigerant or water is flowing as a control volume; our results must be the same. Let us take the coil through which refrigerant is flowing as control volume and use subscripts $w$ and $r$ for water and refrigerant and 1 and 2 for inlet and outlet conditions, respectively. Under assumption of steady state, energy balance gives,

Heat given by refrigerant = heat taken by water

or $$m_r (h_1 - h_2)_r = m_w (h_2 - h_1)_w \qquad \textbf{(A)}$$

From R–134a table, we read: $h_1$ = 30.34 and $h_2$ = 98.78 kJ/kg. But for water $h_1 = c_p T_1$, where $T_1$ is in °C, etc. This gives $h_1$ = 62.99 and $h_2$ = 104.89 kJ/kg. Substituting in A, we get $m_w$ = 29.15 kg/min.

*Heat transfer from refrigerant to the water is*

$$m_r (h_1 - h_2)_r = 1221 \text{ kJ/min}$$

**Example AP 2.4**

3 kg of air in a rigid container changes its state from 5 bar and 60°C to 15 bar while being stirred. The heat absorbed is 200 kJ. Determine the final temperature, change in internal energy and work done.

***Solution*** Given: $m$ = 3 kg, $p_1$ = 5 bar, $T_1$ = 333 K, $p_2$ = 15 bar, $Q$ = 200 kJ.

System (air in container) is a closed system, which is subjected to a constant volume process. Hence, boundary work is 0. Using the fact that air is, perfect gas, we can find out final temperature

$$T_2 = (p/p_2), T_1 = 999\text{K}$$

Change in internal energy = $\Delta U$

$$= m (u_2 - u_1)$$

$$= m\, c_v (T_2 - T_1), \text{ where,}$$

$c_v$ for air = 0.71 kJ/kg K = 1432.1 kJ

Non-flow energy equation $\rightarrow$ 200 – $W$ = $\Delta U$ or

$W$ = stirring work = –1232 kJ

Negative sign shows that work is done on the system.

**Example AP 2.5**

The energy (in joule) of a closed system can be expressed as

$$E = 100 + 50\ T + 0.04\ T^2$$

The heat absorbed (in joule) is given by $Q = 5000 + 20T$; the temperature in these relations is in K. Calculate the work done during a process when the temperature rises from 500 to 1000 K.

***Solution*** Given: $E = 100 + 50\ T + 0.04\ T^2$, $Q = 5000 + 20T$, $T_1 = 500$ k and $T_2 = 1000$ K.

System is closed. Therefore, $Q - W = \Delta E$. Substituting it for $\Delta E$ in nfee, and then integrating from $T_1$ to $T_2$ w.r.t $T$, we get

$$W = \int(Q - \Delta E)$$
$$= \int(4900 - 30T - 0.04T^2)\ dT$$
$$= -20467 \text{ kJ}$$

Negative sign means work is done on the system.

**Example AP 2.6**

An ideal gas requires 1150 kJ/kg of heat to raise its temperature from 20 °C to 100°C when heated at constant pressure. When heated at constant volume, the heat requirement is 825 kJ/kg for the same temperature range. Determine the specific heats at constant pressure and constant volume, adiabatic exponent, characteristic gas constant and the molecular weight of the gas.

***Solution*** Given: $q_p = 1150$ kJ/kg, $q_v = 825$ kJ/kg, $T_1 = 293$ K, $T_2 = 373$ K.

We know that

$$q_p = c_p\ dT = c_p \times (T_2 - T_1) = 1150$$
$$\rightarrow c_p = 14.375 \text{ kJ/kg K}$$
$$q_v = c_v\ dT = c_v \times (T_2 - T_1) = 825$$
$$\rightarrow c_v = 10.312 \text{ kJ/kg K}$$

and $$\gamma = c_p / c_v = 1.394$$

Characteristic gas constant is given by

$$R = c_p - c_v = 4.063 \text{ kJ/kg K}$$

$$\text{Molecular weight} = \overline{R}/R = 8.3143/R = 2.046$$

*Ideal Gas?* Check if $p_1 V1/T_1 = p_2 V_2/T_2$. If yes, the gas is ideal otherwise it is not.

**Example AP 2.7**

A 25 cm diameter cylinder is fitted with a frictionless piston, which is loaded with 1200 N weight. Initially, the piston is 30 cm above the bottom position and the space beneath the piston contains $CO_2$ at 290 K. The gas is supplied heat from an electric heater of 1 W power. Presuming that there is no heat loss to the surroundings, evaluate the distance through which the piston moves outward during the process for an hour. Take barometric pressure to be 1 bar. For $CO_2$, $c_p$ = 820 J/kg K.

***Solution*** Given:

$V_1$ = initial volume of $CO_2$ = $\pi/4 \times (0.25)^2 \times 0.3 = 0.0147$ m$^3$,

$T_1$ = 290 K

$p_1 = p_2 = p_a$ + weight/piston area = 124458 N/m$^2$

$Q = 1 \times 3600 = 3600$ J, $c_p$ = 820 J/kg K, and

$R = \overline{R}/M = 83143/44 = 189$ J/kg K (molecular weight of $CO_2$ = 44).

From Victor Mayer relation, $c_p - c_v = R$, we get

$c_v$ = 631 J/kg K

Perfect gas equation gives

$m = p_1 V_1/(R T_1) = 1224458 \times 0.0147/(189 \times 290) = 0.0334$ kg.

Also, $Q = m \times c_p \times (T_2 - T_1) \rightarrow T_2 = 417.62$ K.

Non-flow energy equation $\rightarrow Q - W = m c_v (T_2 - T_1)$

or $W = 3600 - 0.0334 \times 631 \times (417.62 - 290) = 910.36$ J

But $W = p_1 (V_2 - V_1) = p_1 \times A \times l$, where $l$ is the distance moved by the piston.

or $l = 910.36/(124458 \times 0.04909) = 0.149$ m

**Discussion.** Work done in moving against 1200 N load by distance $l$ is $= 1200 \times l = 178.81$ J whereas the system is doing work (910.36 J). This difference is because the system also works against atmospheric pressure and is not utilised in a useful manner, as explained in Section 1.2.9.2.

**Example AP 2.8**

Air initially at 60 kPa, 800 K and occupying a volume of 0.1 $m^3$ is compressed isothermally until the volume is halved and subsequently it goes further compression at constant pressure till the volume is halved again. Sketch the process on p-v plot and make calculations for work done and total heat interaction for the two processes. Assume ideal gas behaviour for air and take $c_p$ = 1.005 kJ/kg K.

***Solution*** Refer Fig. AP 2.1 for the state points.

Given: We would arrange data in the following way for convenience. A dash shows the value is not known. Also, take $R$ = 0.287 kJ/kg K for air.

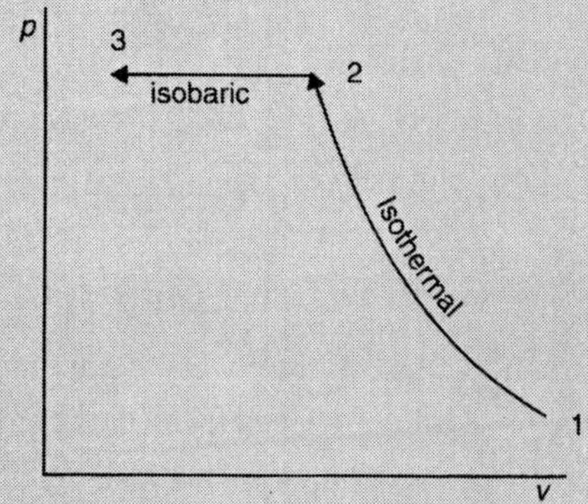


**Fig. AP 2.1** $T = c$ and $p = c$ processes

| State | p, bar | V, $m^3$ | T, K |
|---|---|---|---|
| 1 | 0.6 | 0.1 | 800 |
| 2 | – | 0.05 | 800 |
| 3 | $p_2$ | 0.002<br>5 | – |

From state 1, using perfect gas equation, we can find mass of air

$$m = p_1 V_1/(R T_1) = (0.6 \times 10^5) \times 0.1/\{(0.287 \times 10^3) \times 800\}$$
$$= [0.6 \times 0.1/(0.287 \times 800)] \times 100 = 0.0261 \text{ kg}$$

*Note*: Factor 100 comes, as discussed in Chapter 1, when we use in equation of state, $p$ in bar, $v$ in $m^3$ and R in kJ/kg K. We shall not carry out such calculations henceforth.

**Process 12.** We see from table $T_1 = T_2 \Rightarrow p_1 V_1 = p_2 V_2 \rightarrow p_2 = 1.2$ bar, and

$$W_{12} = \text{work done} = m\,R\,T_1\,ln\,(v_2/v_1) = -4.159 \text{ kJ}.$$

In this process, change in internal energy is zero. Hence heat transfer

$$= W_{12} = -\,4.159 \text{ kJ}$$

The negative sign in both expressions shows that work is done and heat is rejected by the system during isothermal compression.

**Process 23.** $p_3 = p_2 \Rightarrow T_3 = T_2\,(V_3/V_2) = 400$ K

$$W_{23} = p_2\,(V_3 - V_2) = -3 \text{ kJ}$$

$$Q_{23} = W_{23} + m\,c_v\,(T_3 - T_2)$$

$$= m\,c_p\,(T_3 - T_2) = -13.49 \text{ kJ}$$

During both the processes

$$W_{13} = W_{12} + W_{23} = -7.159 \text{ and}$$

$$Q_{13} = Q_{12} + Q_{23} = -17.649 \text{ kJ}$$

**Example AP 2.9**

2 kg of an ideal gas is compressed adiabatically from pressure 100 kPa and 220 K to a final pressure of 400 kPa. Make calculations for (i) initial volume, (ii) final volume and temperature, (iii) work performed, (iv) heat exchange, and (v) change in internal energy. Take $c_p = 1$ and

$$c_v = 0.707 \text{ kJ/kg K}$$

***Solution*** Given: $m = 2$ kg, $p_1 = 1$ bar, $T_1 = 220$ K, $p_2 = 4$ bar, $c_p = 1$ and $c_v = 0.707$ kJ/kg K

From given $c_p$ and $c_v$, we get, $R = 0.293$ kJ/kg K and $\gamma = c_p/c_v = 1.414$.

(i) **Initial volume.** $V_1 = mR\,T_1/p_1 = 1.29$ m$^3$

(ii) **Final volume and temperature.** Using process equation $pV^{\gamma} = c$, we get, $V_2 = 0.484$ m$^3$

We may also write process equation in the form $p \propto T^{\gamma/(\gamma-1)}$

$$\rightarrow T_2 = 330.4 \text{ K}$$

Final temperature can also be obtained using equation of state to give, $T_2 = p_2\,V_2/R = 330.4$ K

(iii) Work performed $= mR(T_1 - T_2)/(\gamma - 1) = -\,156.04$ kJ

(iv) Heat exchange = 0, since the process is adiabatic (no heat transfer).

(v) Change in internal energy $= Q - W = -W = 156.04$ kJ

**Example AP 2.10**

3 kg of air kept at an absolute pressure of 100 kPa and temperature of 300 K is compressed polytropically until the pressure and temperature become 1500 kPa and 500 K, respectively. Evaluate the polytropic exponent, the final volume, the work transfer and the heat interaction. Take $R = 0.287$ kJ/kg K.

***Solution*** Given: $m = 3$ kg, $p_1 = 1$ bar, $T_1 = 300$ K, $p_2 = 15$ bar, $T_2 = 500$K, $R = 0.287$ kJ/kg K

*Polytropic exponent.* Using process equation $p \propto T^{n/(n-1)}$ we may write,

$$p_2/p_1 = (T_2/T_1)^{n/(n-1)}$$

$$\rightarrow 15/1 = (500/300)^{n/(n-1)} \text{ or}$$

$$n/(n-1)\, ln\,(5/3) = ln\, 15 \rightarrow n = 1.23$$

*Final volume.* $V_1 = mR\,T_1/p_1 = 2.583$ m³. Process equation gives

$$V_2 = V_1\,(p_2/p_1)^{1/n} = 0.287 \text{ m}^3$$

(Alternately, $V_2 = m\,R\,T_2/p_2 = 0.287$ m³)

$$\text{Work transfer} = W = m\,R(T_1 - T_2)/(n-1)$$

$$= -748.7 \text{ kJ}$$

$$\text{Heat interaction} = W + m\,c_v\,(T_2 - T_1)$$

$$= -318.2 \text{ kJ}$$

Hence, heat is rejected and work is done in the process.

**Example AP 2.11**

1 kg of air at a pressure of 7 bar and temperature of 360 K undergoes a reversible process which may be represented by $pv^{1.1}$ = constant. If the final pressure is 1.4 bar, determine (i) final specific volume and temperature, (ii) work done and heat transfer. How the work and heat transfer would be affected if the process were irreversible and 15 kJ of work is lost due to friction? Take $R = 0.287$ kJ/kg K and $\gamma = 1.4$.

***Solution*** Given: $m = 1$ kg, $p_1 = 7$ bar, $T_1 = 360$ K, $p_2 = 1.4$ bar, $n = 1.1$, $R = 0.287$ kJ/kg K, and $\gamma = 1.4$

From given data, we can determine

$$V_1 = m\,R\,T_1/p_1 = 0.1476 \text{ m}^3\text{/kg and}$$

$$c_v = R/(\gamma - 1) = 0.7175 \text{ kJ/kg K}$$

(i) Final specific volume and temperature.

$$V_2 = V_1 (p_2/p_1)^n = 0.637 \text{ m}^3/\text{kg, and}$$

$$T_2 = m R T_2/p_2 = 310.73 \text{ K}$$

(ii) Work done and heat transfer.

$$W = m R (T_1 - T_2) (n - 1) = 141.405 \text{ kJ/kg.}$$

$$\Delta U = m c_v (T_2 - T_1)$$

$$= -35.35 \text{ kJ/kg}$$

$$\therefore Q = W + \Delta U \text{ (by nfee)}$$

$$= 106.055 \text{ kJ/kg}$$

**Irreversible process.** Internal energy being the property of the system, the change between given states remains same. However, due to frictional losses, work delivered to the surroundings is reduced by the loss given in the problem.

$$\therefore W' = \text{Work output in the irreversible process}$$

$$= W - 15 = 116.405 \text{ kJ/kg}$$

and by nfee, $Q' = W' + \Delta U = 91.055$ kJ/kg

**Example AP 2.12**

A cylinder closed at both ends is divided into two compartments by a frictionless and freely moving piston; both the cylinder and piston are perfectly insulated. One compartment contains $N_2$ and the other contains air. Initially, both air and $N_2$ are at 1 bar, 298 K and 0.02 m$^3$. An electric heater on the side of air compartment heats the air till the volume of $N_2$ is reduced to 0.01 m$^3$. Find out the final temperature of air and heat supplied to it. For air $c_v = 0.715$ kJ/kg K and $R = 0.287$ kJ/kg K and for $N_2$ $\gamma = 1.4$.

***Solution*** A schematic diagram is given in Fig. AP 2.2. Air on heating pushes piston to its right and compresses $N_2$ thereby increasing its temperature and pressure. In the new condition, pressure on both sides of the piston would be same as there is no friction between piston and cylinder. There is no heat transfer to $N_2$ because the piston is perfectly adiabatic. Let subscripts 1 refer to initial and 2 and 3 to final conditions of nitrogen and air, respectively.

**For nitrogen.** $V_1 = 0.02$ and $V_2 = 0.01$ m$^3$, and $p_1 = 1$ bar. The process is

$pv^\gamma$ = constant.

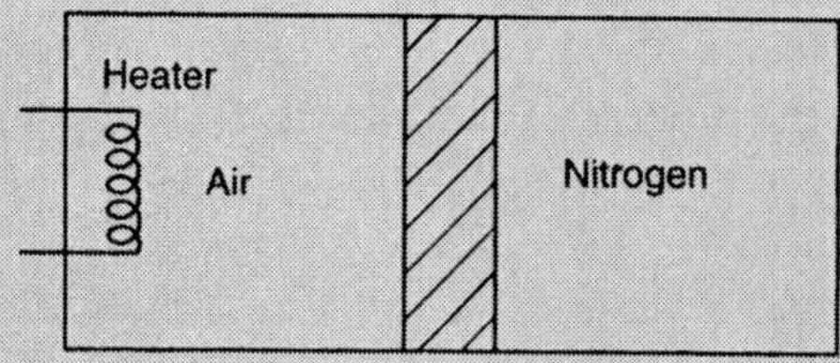


**Fig. AP 2.2**

$\therefore\ p_2 = p_1\ (V_1/V_2)^{\gamma} = 2.639$ bar $= p_3$, the pressure on air side,

and $W_{12}$ = work done on

$N_2 = (p_1 V_1 - p_2 V_2)/(\gamma - 1)$

$= -1597.5$ J

**For air.** $m = p_1\ V_1/R\ T_1$

$= \{1 \times 0.02/(0.287 \times 298)\} \times 100$

$= 0.02378$ kg

$V_3 = 0.02 + 0.01 = 0.03$ m$^3$, and

$T_3 = p_3\ V_3/(m\ R) = 1159.84$ K

$\therefore\ \Delta U = m\ c_v\ (T_3 - T_1) = 14.739$ kJ

Considering air as the system, we note work done by it on its surrounding (nitrogen) = $W = 1.5975$ kJ. Using nfee

$Q = W + \Delta U = 1.5975 + 14.739 = 16.3365$ kJ

**Example AP 2.13**

The piston cylinder arrangement of an engine contains 0.04 m$^3$ of a gas initially at 24 bar and 2000 K. The gas is allowed to expand to a volume of 0.16 m$^3$ and a pressure of 4 bar. The variation in specific heats with temperature is of the form: $c_p = a + 0.0001$ T and $c_v = b + 0.0001$ T and at 2000 K, the values of $c_p$ and $c_v$ are 1.095 and 0.795 kJ/kg K, respectively. Determine (i) index of expansion and final temperature, (ii) work transfer, (iii) change in internal energy of the system, (iv) heat flow, and (v) polytropic specific heat.

***Solution*** Given: $p_1 = 24$ bar, $T_1 = 2000$ K, $V_1 = 0.04$ m$^3$, $V_2 = 0.16$ m$^3$, $p_2 = 4$ bar, $c_p = a + 0.0001$ T, $c_v = b + 0.0001$ and $R = c_p - c_v = a - b$

From given $c_p$ and $c_v$ at 2000 K, we get $a = 0.895$, $b = 0.595$ and

$R = a - b = 0.3$ kJ/kg K

$m$ = mass of gas $= p_1\ V_1/(R\ T_1) = 0.16$ kg

(i) **Index of expansion and final temperature.** $n$ can be obtained from the process equation $pv^n = C$.

$$\rightarrow p_2/p_1 = (V_1/V_2)^n \text{ or}$$
$$n = 1.293$$

Also, $T \propto p^{(n-1)/n}$ or

$$T_2 = T_1 (p/p_1)^{(n-1)/n}$$
$$= 1332.6 \text{ K}$$

(ii) **Work transfer.**

$$mR(T_1 - T_2)/(n-1) = 109.215 \text{ kJ}$$

(iii) **Change in internal energy of the system.**

$$= m \int_{T_1}^{T_2} (b + 0.0001T)\, dT = -81.33 \text{ KJ}$$

Alternately, we may determine $c_v$ at mean temperature and then determine change in internal energy in the usual way.

(iv) Heat flow = $W + \Delta U$ (by nfee) = 27.855 kJ

(v) **Polytropic specific heat.** Mean temperature = (2000 + 1332.6)/2

$$= 1666.3 \text{ K}$$

At this temperature, mean $c_p$ = 1.0616, $c_v$ = 0.7617 kJ/kg K and

$$\gamma = c_p/c_v = 1.394$$

Polytropic specific heat = $c_n = c_v(\gamma - n)/(1 - n)$ = 0.2625 kJ/kg K

**Example AP 2.14**

A turbine power plant operating under steady state receives 1 kg/s superheated steam at 3.0 MPa and 300°C. The steam enters the turbine with a velocity of 10 m/s at an elevation of 5 m above the ground. The turbine discharges wet steam of quality 0.85 at 50 kPa pressure with velocity 50 m/s at an elevation of 10 m above the ground. The energy losses from the turbine casing are estimated to be 10 kJ/s. Find out the power output of the turbine. If the changes in the kinetic and potential energies are ignored, how much error will be introduced?

***Solution*** Given: $p_1$ = 30 bar, $T_1$ = 300°C, $m$ = 1 kg/s, $C_1$ = 10 m/s, $z_1$ = 5m, $x_2$ = 0.85, $p_2$ = 0.5 bar, $C_2$ = 50 m/s, $z_2$ = 10 m,

$Q_L$ = heat loss = –10 kJ/s. Assume steady flow.

From steam table, $h_1$ = 2993.5 kJ/kg and

$$h_2 = h_f + x_2 h_{fg} = 2300.08 \text{ kJ/kg}$$

Steady flow energy equation is

$$q - w = (h_2 - h_1) + (C_2^2 - C_1^2)/2000 + g(z_2 - z_1)/1000$$

Substituting the values while taking $q = Q_L$, we get, $w = 682.17$ kW

Here, $\Delta(KE + PE) = (C_2^2 - C_1^2)/2000 + g(z_2 - z_1)/1000 = 1.249$ kW

$\therefore$ error made if these energies are neglected

$$= 1.249/682.17 = 0.183\%$$

This shows that not much error creeps in by neglecting change in kinetic and potential energies in actual cases.

**Example AP 2.15**

A throttling calorimeter is used to measure the quality of steam in a pipeline where the pressure is 0.5 MPa. The conditions in the calorimeter are $p = 0.1$ MPa and $t = 110°C$. Determine the quality o f the steam in pipeline.

***Solution*** Given: $p_1 = 5$ bar, wet steam at $x_1$, $p_2 = 1$ bar, $T_2 = 110°C$.

Expansion in throttling calorimeter is isenthalpic, since there is no work and heat transfer in the process and flow through calorimeter is continuous and steady. The sfee gives $h_2 = h_1$

From steam table at $p = 1$ bar and $t = 110°C$, $h_2 = 2696.24$ bar (by interpolation) and at 5 bar, $h_f = 640.23$, $h_{fg} = 2108.5$ kJ/kg

$$\therefore\ 640.23 + 2108.5\, x_1 = 2696.24, \text{ or } x_1 = 0.975$$

**Example AP 2.16**

A 2 m$^3$ insulated tank contains helium ($\gamma = 1.67$) at 0.1 MPa and 300 K. If this tank is connected to a high pressure line carrying helium at 3 MPa and 500 K, helium rushes into the tank till the pressure in the tank rises to 3 MPa. Assuming that helium behaves as an ideal gas determine the quantity of helium that flows into the cylinder. Also, estimate the temperature of the gas.

***Solution*** Let us use subscripts $i$ and $f$ to represent the condition inside the tank before beginning and after end of the filling process and 1 for the section at inlet to tank. Filling of tank is the transient process in which conditions inside the tank go on changing with time and there is no mass outflow or any work and heat transfer. Hence, Eq. 2.5 can be used neglecting changes in kinetic and potential energies. The equation reduces to

$$0 - 0 = 0 - m_1 h_1 + dE/dt, \text{ or}$$

$$h_1 \int m_1\, dt = \int dE$$
$$= E_f - E_i = m_f u_f - m_i ui \qquad \text{(A)}$$

where, $dE/dt$ represents time rate of change of energy of the control volume and $m_1$ is the mass inflow rate. We note that $\int m_1\, dt$ is the net mass flow to the tank during the process and hence, by mass balance, is equal to $m_f - m_i$. Writing $h_1 = c_p\, T_1$ and $u = c_v\, T$ in Eq. A

$$c_p T_1\, (m_f - m_i) = c_v\, (m_f T_f - m_i\, T_i)$$

or
$$c_p T_1 \left( \frac{p_f V}{R\, T_f} - \frac{p_i V}{R\, T_i} \right) = \left( \frac{p_f V}{R} - \frac{p_i V}{R} \right)$$

$$\rightarrow \gamma T_1 \left( \frac{p_f}{T_f} - \frac{p_i}{T_i} \right) = p_f - p_i \qquad \text{(B)}$$

Given: $V = 2\ \text{m}^3$, $p_i = 1$ bar, $T_i = 300$ K, $p_f = 30$ bar, $\gamma = 1.67$, $T_1 = 500$ K. Substituting in Eq. B, we get

$T_f = 788.15$ K

$R$ for helium = 8.3143/2 = 4.15715 kJ/kg K

$\therefore\ m_f - m_i = p_f\, V/(R\, T_f) - p_i\, V/(R\, T_i) = 1671$ kg

**Example AP 2.17**

Suppose, an ideal gas at $p_i$ and $T_i$ contained in a rigid and insulated tank is allowed to escape by opening a valve till the pressure is reduced to $p_f$. The state of the gas leaving the tank continuously changes and it cannot be assumed that the gas leaves with constant enthalpy. Show that the expansion in the tank is isentropic.

***Solution*** Evacuation of tank is also a transient process and Eq. 2.5 can be applied with no heat and work transfer and mass inflow. Let $T$ be temperature at any time when mass inside tank is $m$ and is reducing at the rate $dm/dt$, then Eq. 2.5 is reduced to

$$C_p\, T\, dm/dt = dE = d\,(m\, c_v\, T)/dt \text{ or}$$
$$\gamma T\ \ dm/dt = m\, dT/dt + T\, dm/dt$$
$$\rightarrow (\gamma - 1)\ T\, dm/dt = m\, dT/dt$$

This gives on integration w.r.t. time $t$

$$(\gamma - 1)\, ln\, (m_f\, /m_i) = ln\, (T_f / T_i), \text{ or}$$
$$(\gamma - 1)\, ln\, [\{p_f\, V/(RT_f)\}/\{p_i\, V/(RT_i)\}] = ln\, (T_f / T_i)$$

$$\rightarrow T_f/T_i = (p_f/p_i)^{(\gamma-1)/\gamma}$$

This shows that the expansion in tank is isentropic.

## Chapter 3

### Example AP 3.1

A system absorbs 100 kJ heat from a source at 500°C and does 50 kJ of work. If it is required to restore the system to its original state, is it possible to do so by an adiabatic process?

***Solution*** In the first process internal energy of the system would change that can be obtained from nfee.

$$\Delta U = Q - W = 100 - 50 = 50 \text{ kJ}$$

In second process, $Q = 0$ (adiabatic process). To restore the system state, $\Delta U = -50$ kJ, nfee gives

$$0 - W = -50 \text{ or } W = 50 \text{ kJ}$$

If the second process completes the cycle then the net work output in the cycle is 50 + 50 = 100 kJ. This violates Kelvin–Planck statement. Therefore, the suggested process is not possible.

### Example AP 3.2

A reversible engine operating between a reservoir at 600 K and the ambient temperature 300 K drives a refrigerator operating between 240 K and ambient temperature. Determine the ratio of energy rejected by both the devices to the atmosphere to the energy absorbed by the engine from the reservoir at 600 K.

***Solution*** **For engine.** $T_1 = 600$ K, $T_2 = 300$ K,

$$\eta = 1 - T_2/T_1 = W/Q_1 \rightarrow W = 0.5\, Q_1 \text{ and} \tag{A}$$

heat rejected by the engine to atmosphere $= Q_2 = Q_1 - W = 0.5\, Q_1$

where, $Q_1$ is the amount of heat received by the engine from the reservoir.

**For refrigerator.** $T_c = 300$ K and $T_e = 240$ K

C.O.P. of heat pump $= T_c/(T_c - T_e) = Q_c/W = 2\, Q_c/Q_1$; using $W$ from Eq. A, since engine output is used to run refrigerator.

$$\rightarrow Q_c = \text{heat rejected by refrigerator}$$

$$= 0.5 \times Q_1 \times 300/(300 - 240) = 2.5\, Q_1$$

$\therefore$ heat rejected by engine + heat rejected by refrigerator

$$= 3\,Q_1$$

and $(Q_2 + Q_r)/Q_1 = 3$

**Example AP 3.3**

A reversible heat engine receives heat from a high temperature reservoir $T$ and rejects heat to a low temperature sink at 560°C. A second reversible engine receives the heat rejected by first engine and rejects heat to a cold reservoir at 4°C. Find the temperature $T$ (i) for equal thermal efficiency of the two engines, (ii) for the same work output of the two engines.

***Solution*** Given: $T_2 = 833$ K, $T_3 = 277$ K

(i) For equal thermal efficiency of the two engines. This requires

$$1 - T_2/T = 1 - T_3/T_2$$

$$\rightarrow T = 2505 \text{ K}$$

(ii) For the same work output of the two engines. Let $Q$ be the amount of heat extracted from source at $T$. Then,

$$W_1 = (1 - T_2/T)\,Q$$

$Q_r$ = heat rejected by first engine is equal to heat received by the second engine, which is $Q - W_1$, and

$$W_2 = (1 - T_2/T_3)\,(Q - W_1)$$

The required condition demands

$$W_1 = W_2 \rightarrow T = 1308.61 \text{ K}$$

**Example AP 3.4**

A reversible heat engine operates with three reservoirs at 300 K, 400 K, and 1200 K. It absorbs 1200 kJ energy as heat from the reservoir at 1200 K and delivers 400 kJ work. Determine the heat interactions with the other two reservoirs.

***Solution*** Given: $T_1 = 1200$ K, $T_2 = 400$ K and

$$T_3 = 300 \text{ K},\ Q_1 = 1200 \text{ kJ},\ W = 400 \text{ kJ}$$

Let $Q_2$ and $Q_3$ be the heat interactions with other reservoirs. Considering the engine as the system executing integral number of cycles in the period it gets 1200 kJ heat, Joule's law gives

$$Q_3 + Q_2 + 1200 = 400$$

$$\rightarrow Q_3 + Q_2 = -800 \qquad \text{(A)}$$

We further assume that the whole process is reversible so that $\int \delta Q/T$ over the cycle is 0. Thus

$$1200/T_1 + Q_2/T_2 + Q_3/T_3 = 0$$

$$\rightarrow 4\,Q_3 + 3\,Q_2 = -1200 \qquad \text{(B)}$$

Solving Eqs A and B, we get, $Q_2 = -2000$ and $Q_3 = 1200$ kJ. That is, the engine absorbs 1200 kJ of heat from 300 K and rejects 2000 kJ to reservoir at 400 K.

**Example AP 3.5**

The following processes consritute a cycle (i) addition of heat at constant volume so that volume rises from 6 bar to 15 bar, (ii) polytropic expansion ($n$ = 1.3) till the system reaches original temperature of 110°C, and (iii) isothermal compression till the system reaches the original state. Calculate (i) pressure at the end of polytropic process, (ii) change in entropy during each of the process per kg of air. Show the cycle on p-v and T-s diagrams.

***Solution*** Given: $p_1 = 6$ bar, $v_1 = v_2$, $p_2 = 15$ bar, $T_3 = T_1 = 383.15$ K, $n = 1.3$, $m = 1$ kg. Cycle is shown in Fig. AP 3.5.

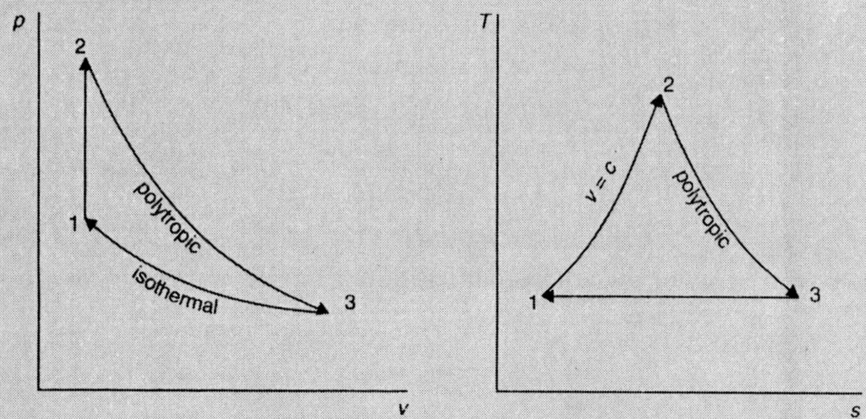


**Fig. AP 3.1** System subjected to three-process cycle

(i) **Pressure at the end of polytropic process.** From equation of state,

$$v_1 = R\,T_1/p_1 = (0.287 \times 1000) \times 383.15/(6 \times 10^5) = 0.18 \text{ m}^3/\text{kg}$$

$$T_2 = T_1\,(p_2/p_1) = 957.875 \text{ K}$$

$$p_3 = p_2\,(T_3/T_2)^{(n-1)/n} = 0.283 \text{ bar}$$

(ii) **Change in entropy during each of the process.** From Eq. 4.25

$$s_2 - s_1 = c_v \, ln \, (p_2/p_1) = 0.657 \text{ kJ/kg K}$$

$$s_3 - s_2 = c_p \, ln \, (T_3/T_2) - R \, ln \, (p_3/p_2) = 0.219 \text{ kJ/kg K}$$

and $s_1 - s_3 = -R \, ln \, (p_1/p_3) = -0.876$ kJ/kg K

**Example AP 3.6**

A 5 kg steel block at 1000°C is quenched in 5 kg water at 30°C in an open container. Calculate the entropy change of steel, $H_2O$ and the universe. The specific heat of steel is 0.5 kJ/kg K.

***Solution*** Given for water: $T_1 = 303$ K, $m_w = 5$ kg and
for steel, $m_s = 5$ kg, $T_3 = 1273$ K

$T_2$ = final temperature of steel and water will depend on how much heat is given by the steel block. The temperature of water at the most will be $T_{sat}$ at 1 atm = 373 K, since steam will be formed if more heat is transferred from the steel block. Let $M$ kg of steam be formed in this process. Energy balance gives

heat given by steel block = heat received by $H_2O$

or $m_s \times 0.5 \times (1273 - 373) = 5 \times 4.187 \times (100 - 30) + M \, h_{fg}$

where $h_{fg}$ is read from steam table at 1 bar and is = 2257 kJ/kg

or $M$ = mass of steam formed = 0.3473 kg

$$\Delta S \text{ (steel)} = m_s \times c_p \times ln \, (373/1273) = -3.0682 \text{ kJ/kg K}$$

$$\Delta S \, (H_2O) = 5 \times 4.187 \, ln \, (373/303) + M \times h_{fg}/373$$

$$= 6.4505 \text{ kJ/kg K}$$

Change in entropy of $H_2O$ is sum of change during sensible heat transfer and latent heat transfer (during which temperature remains constant at 373 K).

$$\Delta S \text{ (universe)} = \Delta S \text{ (steel)} + \Delta S \, (H_2O) = 3.823 \text{ kJ/kgK}$$

A positive sign indicates that heat transfer from steel block to water is an irreversible process.

**Example AP 3.7**

Air is flowing through a horizontal and insulated duct. The pressure and temperature are measured at two different sections A and B. These are $p_A = 2$ bar, $T_A = 423$ K, $p_B = 1.8$ bar, $T_B = 403$ K, respectively. Determine the flow direction, assuming air to be an ideal gas.

***Solution*** Given: $p_A = 2$ bar, $T_A = 423$ K, $p_B = 1.8$ bar, $T_B = 403$ K

Let flow be from A to B. By second law, $S_B$ must be > $S_A$

But $S_B - S_A = C_p\ [ln\ (T_B/T_A) - R\ ln\ (p_B/p_A) = -0.5329$ kJ/mole K

Negative sign shows that the direction of flow assumed is wrong and hence, flow is from station B to A.

**Example AP 3.8**

An inventor claims to have designed a flow device which gives equal amount of hot air at 1 bar, 400 K and cold air at 1 bar 200 K when the device is fed with air at 5 bar and 300 K. It is also claimed that the device does not require energy either in the form of heat or work. Comment on his claim. Take $C_p$ for air = $3.5\bar{R}$.

***Solution*** Given: $p_1 = 5$ bar, $T_1 = 300$ K, $p_2 = 1$ bar, $T_2 = 400$ K, $p_3 = 1$ bar, $T_3 = 200$ K, $C_p = 3.5\bar{R}$. Here, subscripts 1,2, and 3 refer to inlet, hot and cold air streams.

Let us assume 1 mole of air enters the device to produce 0.5 mole of hot and 0.5 mole of cold streams. The claim will be justified if the entropy of the universe is increasing. Therefore, we have to find out the entropy change. For determination of change in entropy, we may consider 0.5 mole of air changing from state 1 to 2 and another 0.5 mole from state 1 to 3.

$$\begin{aligned}\Delta S_{12} &= \text{entropy change from state 1 to 2}\\ &= 0.5 \times [C_p ln\ (400/300) - \bar{R} ln(1/5)]\\ &= 10.8761\ \bar{R}\ \text{kJ/K}\end{aligned}$$

and

$$\begin{aligned}\Delta S_{13} &= \text{entropy change from state 1 to 3}\\ &= 0.5 \times [C_p ln(200/300) - \bar{R} ln(1/5)]\\ &= 0.7911\ \bar{R}\ \text{kJ/K}\end{aligned}$$

$$\therefore\ \Delta S = \Delta S_{12} + \Delta S_{13} = 11.6672\ \text{kJ/K}$$

Since entropy is increasing in this process, the claim of the inventor is justified. This device is called vortex tube and is used for cooling electronic components, especially in those applications where compressed air is available. We may experiment over increasing moles of the cold air so long $\Delta S > 0$ to get its maximum mass.

**Example AP 3.9**

An insulated tank of 1 m$^3$ volume contains air at 1 bar and 300 K. This tank is connected to a high pressure line in which air at 10 bar and 600 K flows. The tank is quickly filled with air by opening the valve between the tank and the high-pressure line. If the final pressure of air in the tank is 10 bar, determine the mass of air that entered the tank and the change in the entropy associated with the filling process.

***Solution*** Given: $p_i$ = 1 bar, $T_i$ = 300 K, $p_f$ = 10 bar, $T_1$ = 600 K,

$p_1$ = 10 bar, $\gamma$ = 1.4, $C_p\ (1 - 1/\gamma) = \overline{R} \rightarrow C_p = 3.5\ \overline{R}$

Here, subscripts *i*, *f* and 1 refer to the initial and final conditions inside the tank and at the inlet to tank, respectively. The mass and energy balance gives the net mass inflow and the final state, as has been done earlier (Example AP2.16). Following the procedure detailed there, we get, mass of air entering the tank = 128.869 mole and final temperature of air in the tank = $T_f$ = 711.87 K. Also, the initial mass of air in the tank is 40.093 mole.

For the purpose of finding out change in entropy, (a) 40.093 mole air is changing from $p_i$, $T_i$ to $p_f$ and $T_f$ while (b) 128.869 mole of air is changing from $p_1$, $T_1$ to $p_f$ and $T_f$.

$$\therefore\ \Delta S = \Delta S_a + \Delta S_b$$

$$= 40.093\ [C_p ln(T_f/T_i) - \overline{R} ln(p_f/p_i)]$$

$$+ 128.869\ [C_p ln(T_f/T_1) - R\ ln\ (p_f/p_1)]$$

$$= 881.717\ \text{J/K}$$

## Chapter 4

### Example AP 4.1

A mass of 2 kg of saturated water is completely vaporised at a constant pressure of 100 kPa. Determine (i) the volume change and (ii) the amount of energy added to the water.

***Solution*** Given: $m$ = 2 kg, $p$ = 100 kPa

From steam table at 100 kPa = 1 bar, we read

$v_f$ = 0.001043 and $v_g$ = 1.694 m³/kg, $h_f$ = 417.4 and $h_g$ = 2675 kJ/kg

(i) Change in volume = $m \times (v_g - v_f)$

$$= m \times v_g\ (v_f <<< v_g) = 3.388\ \text{m}^3$$

(ii) **Amount of energy added to the water.** Let us find out how constant pressure may be maintained during the vaporisation process. One of the methods would be to fill water in an open vessel and heat it, like preparing tea. Steam will be formed. The other method maybe that water enters a heat exchanger at one end and comes out as saturated vapour at the other end because of heat transferred to it. In both the cases, we observe the system (container or heat exchanger) is an open system. We assume that this process is going on continuously so that steady state can be assumed (for simplification). Application of sfee then gives $Q = m \times (h_g - h_f) = 4515.2$ kJ

We may heat water at constant pressure by assuming to be contained in a vessel fitted with a piston. As heat is supplied, volume of $H_2O$ increases from $v_f$ to $v_g$ and the piston moves against surrounding pressure $p$, which for a reversible process has to be the system pressure. $H_2O$ contained in the vessel is a closed system and the non-flow energy equation results in

$$q = (ug - u_f) + p\,(v_g - v_f) = h_g - h_f \text{ and}$$
$$Q = m \times (h_g - h_f) = 4515.2 \text{ kJ}$$

We may also think of a process in which water is filled in a vessel of volume 3.388 m$^3$ and is heated till it is completely converted into saturated vapour. The system is a closed system and nfee gives the desired result. However, we have to understand that while water converts into vapour, its volume increases against surrounding pressure, which in this case is constant. Therefore $W$ = work done is $p\,(v_g - v_f)$ kJ/kg. Applying nfee, we get

$$Q = m \times [p\,(v_g - v_f) + (u_g - u_f)]$$
$$= m\,(h_g - h_f) = 4515.2 \text{ kJ}$$

**Discussion.** In last method, our answer to amount of work transfer normally is zero because the vessel is of constant volume. Remember, the system is water and not the container and as discussed work is done by the surrounding on the system (expanding water surface) and is supplied by the surroundings in the form of heat. That is why, the amount of heat required to change water into steam is not changed in internal energy but its enthalpy. Some authors have referred internal work to the work done on expanding water surface.

**Example AP 4.2**

A rigid tank contains 20 kg of $H_2O$ at 90°C. If 16 kg of liquid is present and the rest is vapour, find out the pressure, quality of steam and volume of the container.

***Solution*** Given: $m = 20$, $m_w = 16$, and $m_s = 4$ kg (by mass balance)

From steam table at 90°C, $p = 0.7041$ bar, $v_f = 0.001036$ and

$$v_g = 2.361 \text{ m}^3/\text{kg}$$

Pressure of steam is 0.7041 bar

$x$ = dryness fraction of steam = $m_w/(m_w + m_s) = 0.2$

Volume of container = $m\,(x\,v_g) = 20 \times 0.2 \times 2.361 = 9.444 \text{ m}^3$

**Example AP 4.3**

An 80 L vessel contains 4 kg of refrigerant 134a at a pressure of 160 kPa. Determine (i) the temperature of the refrigerant, (ii) the quality, (iii) the specific enthalpy, and (iv) the volume occupied by the vapour phase.

***Solution*** Given: $V = 80 \text{ L} = 80 \times 10^{-3} = 0.08 \text{ m}^3$, $m = 4$ kg,

$$p = 160 \text{ kPa} = 160/100 = 1.6 \text{ bar.}$$

From R-134a table at 1.6 bar: $v_f = 0.0007435$ and $v_g = 0.1229 \text{ m}^3/\text{kg}$, $h_f = 29.78$ and $h_{fg} = 208.18$ kJ/kg

(i) Saturation temperature at 1.6 bar = –15.62°C

(ii) Specific volume of vapour in vessel

$$v = V/m = 0.02 \text{ m}^3/\text{kg}$$
$$= v_f + x\,(v_g - v_f) \rightarrow x = 0.158$$

(iii) Specific enthalpy, $h = h_f + x\,h_{fg} = 62.7$ kJ/kg

(iv) Volume occupied by the vapour phase

$$= x\,m\,v_g = 0.158 \times 4 \times 0.1229$$
$$= 0.0777 \text{ m}^3 = 77.7 \text{ L}$$

**Example AP 4.4**

Determine the temperature of $H_2O$ at a state of $p$ = 0.5 Mpa and $h$ = 2890 kJ/kg.

***Solution*** Given: $p$ = 0.5 Mpa and $h$ = 2890 kJ/kg.

First step is to find out whether $H_2O$ is liquid, wet, or superheated vapour. For this, we read enthalpy (since this is given in the question) at $p$ = 5 bar for saturated states. We read: $h_f = 640.2$ and $h_g = 2749$ kJ/kg. We note $h = 2890 > h_g$ and hence, $H_2O$ exists as superheated steam. From superheated steam table, we read following values.

| | $t$°C | $h$, kJ/kg |
|---|---|---|
| | 200 | 2855.4 |
| | 250 | 2960.7 |
| Difference | 50 | 105.3 |

For a change in $h = 105.3$, change in temperature is 50°C. Therefore, for a change in $h$ of $(h - h_g)$ the change in temperature would be $\{(50/105.3) \times (h - h_g)\} = 16.4$. This procedure is called linear interpolation.

$\therefore$ steam temperature = 200 + 16.4 = 216.4°C

**Example AP 4.5**

Determine the internal energy of $H_2O$ at 80°C and 5 MPa, using (a) wet steam table and (b) compressed water table. What is the error involved in first process?

***Solution*** Given: $p = 5$ MPa and $t = 80$°C.

From wet steam table, we note that the saturation temperature at 5 MPa is 264°C, which is more than the given temperature. This means $H_2O$ is in liquid state. Such state is called compressed liquid state for which tables have been made to give properties. However, these are not commonly available commercially and we have to work out using wet steam tables assuming $H_2O$ to be saturated liquid at given temperature, since internal energy, volume and enthalpy of liquid do not change significantly with pressure. From wet steam table at 80°C, we read $u = 334.86$ kJ/kg. From the tables for compressed liquid, $u = 333.72$ kJ/kg.

Error made in reading values from wet steam table

$$= (334.86 - 333.72)/333.72 = 0.34\%$$

Note saturation pressure at 80°C is 0.04739 MPa, which is much lesser than the given pressure. Even then, error is only 0.34% that justifies use of wet steam table.

**Discussion.** This state of water is also referred to as undercooled water for the reason that at 5 MPa the saturation temperature is 264°C. The compressed or undercooled are therefore, relative terms and their use depends on whether temperature or pressure is under consideration.

**Example AP 4.6**

0.05 kg of steam at 15 bar is contained in a rigid vessel of volume 0.0076 m$^3$. What is the temperature of the steam? If the vessel is cooled, at what temperature will the steam be just dry saturated. Cooling continues until the pressure in the vessel is 11.0 bar.

***Solution*** Given: $p_1 = 15$ bar, $V = 0.0076$ m$^3$, $m = 0.05$ kg, specific volume of steam = $V/m = 0.152 > 0.132$, $v_g$ at 15 bar.

Hence, steam is superheated. From superheated steam table, we note that steam has this volume at temperature 250°C.

At saturation condition, $v_{g2} = 0.152$ m$^3$/kg. We look for $v_2 = 0.152$ in the column of $v_g$ in wet steam table; $v_g$ is 0.157 and 0.0.151 at pressure of 12.5 bar and 13 bar, respectively. Therefore, pressure is close to 13 bar at which condition, $T_2 = 191.6$°C.

**Process 23 in which further cooling takes place.** This cooling is different from the above process because steam is condensing out and therefore pressure would decrease. At conclusion of the process, the liquid and vapour will be in equilibrium. However, specific volume of the wet vapour would remain same, since the system is closed and the vessel is rigid. So, in final state, $p_3 = 11$ bar and $v_3 = 0.152$. we can find out d.f., since $v_g$ at 11 bar is 0.178 m$^3$/kg.

$$\therefore\ x = v_3/v_{g3} = 0.8539$$

**Example AP 4.7**

Considering $H_2O$ at 5 MPa and 80°C as undercooled water, determine its enthalpy and entropy using wet steam table.

***Solution*** Given: $p = 5$ MPa and $t = 80$°C $= 353$ K, $c_p = 4.187$ kJ/kgK.

From steam table at 5 MPa, $h_f = 1154$ kJ/kg and $s_f = 2.92$ kJ/kgK, $T_{sat} = 264$°C $= 537$ K. We may heat water at constant pressure 5 Mpa to increase its temperature from 80°C to saturation temperature 264°C for which heat required per kg of water is $c_p \times (264 - 80)$ kJ/kg. But this is increase in ethalpy.

$$\begin{aligned}\therefore\ q &= \text{heat added}\\ &= c_p \times (264 - 80)\\ &= h_f - h \rightarrow h = 383.6 \text{ kJ/kg}\end{aligned}$$

From second law of thermodynamics, we have $ds = c_p\, dT/T$, which on integration from 353 to 537 K gives the change in entropy during heat addition process. Since $c_p$ can be assumed to remain constant during this interval of integration, we get

$$s_f - s = c_p\, ln\,(T_{sat}/T) \rightarrow s = 1.1634 \text{ kJ/kg K}$$

**Discussion.** This procedure is also used for superheated vapours in the region close to the saturation line though the error will be more. But in absence of proper tables, there is no other way, too. The vapour tables, available commercially, give properties for wet vapour and in addition either $c_p$ at different pressure and temperature or properties of superheated vapour; the former is more common. Take for example, the refrigerants, where $c_p$ is given.

**Example AP 4.8**

An insulated container has 1 kg supercooled liquid water at –10°C. Suppose, a negligible amount of ice is added as a seed to aid solidification, calculate the mass of water, which solidifies, and the change in the entropy associated with the process. $C_p$ for water = 4.2 kJ/kg K and the latent heat of fusion of water at 0°C is 333.43 kJ/kg.

***Solution*** Supercooled state is a meta-stable state and can be brought to the normal state by creating some kind of infinitesimal disturbance, which is done by adding ice as seed. This brings water temperature to 0°C. Let $X$ be the mass of ice formed. Energy balance gives

$$4.2\,(273.15 - 263.15) = 333.43\,X \text{ or}$$

$$X = 0.126 \text{ kg}$$

The supercooled water first changes to water at 0°C and then to ice at 0°C. Let the change in entropy during the two processes $\Delta S_1$ and $\Delta S_2$, then total change in entropy is

$$\Delta S = \Delta S_1 + \Delta S_2$$

$$= 4.2\, ln\,(273.15/263.15) + 0.126 \text{ x } 333.43/273.15$$

$$= 2.8406 \text{ J/K}$$

**Example AP 4.9**

Refrigerant R-11 at 0.2571 bar and 10°C of superheat is compressed isentropically to 3.111 bar. Determine the final state.

***Solution*** Given: $p_1$ = 0.2571 bar, $T_1$ = 10°C superheat, $p_2$ = 3.111.

From the table for R-11, we observe that $T_{sat,1}$ = –10°C, $h_g$ = 383.77 kJ/kg, $s_g$ = 1.69922 kJ/kg and that $c_p$ is given at the saturation temperature for different degree of superheat. We get, $c_p$ = 0.554 kJ/kg K from the table.

$$\therefore\; h_1 = h_g + c_p \times 10 = 389.31 \text{ kJ/kg and}$$

$$s_1 = s_g + c_p\, ln\,(273/263)$$

$$= 1.71989 \text{ kJ/kg K}$$

Process is isentropic $s_1 = 1.71989 = s_{g,2} + c_p \ln\,(T_2/T_{sat})$

From table, we read at 3.111 bar,

$$T_{sat} = 60°\text{C} = 333 \text{ K},\; h_{g,2} = 418.73 \text{ and } s_{g,2} = 1.67109$$

Substitution results in

$$c_p \ln (T_2/333) = 0.0488 \quad \text{(A)}$$

You may now appreciate the problem. $C_p$ can be read only if $T_2 - T_{sat}$ is known. Therefore, such an equation is to be solved by trial and error method. It is still difficult because $c_p$ is tabulated for a few degrees of superheat. In table it is given as 0.646 for 30°C superheat. Taking this value, we get $T_2$ = 359.13 K. (Hint: if you have the $p$ –$h$ chart for this refrigerant then on the chart point 1 is located and a vertical line is drawn to intersect $p$ = 3.111 bar line. Read the temperature and then proceed to solve Eq. A.)

**Example AP 4.10**

Two boilers A and B are delivering steam in equal proportion by mass to a common main. Both the boilers operate at a pressure of 14 bar, which is also the pressure of steam in the main. The boiler A supplies steam at 300°C. If the temperature of the resulting mixture is 235°C, estimate the quality of steam supplied by the boiler B. Also, find out the change in entropy during this adiabatic mixing process. Assume $c_p$ for superheated steam as 2.25 kJ/kgK.

***Solution*** Given: $p$ = 14 bar, $T_1$ = 300°C, $T_3$ = 235°C, where subscripts 1, 2 and 3 refer to boilers A, B and the main line, respectively. In this process the two streams are mixing adiabatically and hence the mass and enthalpy will be conserved; we are conserving enthalpies because the steam is flowing into and out from the steam main. Let 1 kg of mass be coming from each boiler then mass leaving mains is 2 kg by mass conservation.

**From steam table.** $h_1$ = 3024.05, $h_3$ = 2877.5 kJ/kg. Writing enthalpy balance equation

$$h_1 + h_2 = 2\,h_3 \rightarrow h_2 = 2730.95 \text{ kJ/kg}$$

$$< h_g = 2787.8 \text{ at } 14 \text{ bar}$$

∴ steam is wet in boiler 2. let $x$ be its dryness fraction, $h_f$ = 830.1, $h_{fg}$ = 1957.7 (from steam table)

$$h_2 = h_f + h_{fg}\,x \rightarrow x = 0.9713$$

**Entropy calculation.** From steam table we can find entropy of steam at conditions in boilers A and B. We shall find it out for superheated steam using $c_p$ and values of saturated steam from wet steam table.

At 14 bar $\quad s_g = 6.4651,\ s_f = .2837,\ s_{fg} = 4.1814$ kJ/kg K

$$T_{sat} = 195°\text{C} = 468 \text{ K}$$

$$s_1 = s_g + c_p \, ln\,(573/468) = 6.9205 \text{ kJ/kg K}$$
$$s_2 = s_f + x\,s_{fg} = 6.345 \text{ kJ/kg K}$$
and
$$s_3 = s_g + c_p \, ln\,(508/468) = 6.65 \text{ kJ/kg K}$$

$\therefore$ change in entropy $= 2 \times s_3 - s_1 - s_2 = 0.0345$ kJ/kg K

Increase in entropy shows that adiabatic mixing of two streams is an irreversible process.

**Example AP 4.11**

A cylinder contains steam at 1 bar and temperature 150°C. The steam is compressed reversibly and isothermally to a state where the specific volume is 0.25 m³/kg. Find the change of internal energy, change of entropy and work done per kg of steam.

***Solution*** Given: $p_1 = 1$ bar, $T_1 = 150°\text{C} = T_2$, $v_2 = 0.25$ m³/kg.

From steam table at $p_1 = 1$ bar, $T_{sat} = 99.62°\text{C} \Rightarrow$ steam is superheated at state 1, since $T_1 > T_{sat}$. We read $u_1 = 2582.8$ kJ/kg and

$s_1 = 7.613$ kJ/kg K

**At state.** $v_g$ at 150°C = 0.3928 m³/kg $\Rightarrow$ steam is wet since $v_2 < v_g$.

$\therefore\ x_2 = v_2/v_g = 0.718$, $u_2 = h_{f2} + x_2\, h_{fg2} = 2016.4$ kJ/kg and similarly,

$s_2 = 5.428$ kJ/kg K

**Change of internal energy** $= u_2 - u_1 = -566.4$ kJ/kg

Negative sign indicates decrease in internal energy

**Change of entropy** $= s_2 - s_1 = -2.185$ kJ/kg K

**Work done per kg of steam.** System is closed. Therefore, nfee gives: $q - w = du$. In this equation only $du$ is known. However, if we note that the process is reversible then heat transfer can be related to change in entropy by the relation

$$\int \delta q = \int T ds = T \int ds$$

since the process is isothermal, too. This gives

$$q = T\,(s_2 - s_1) = -924.3 \text{ kJ/kg and}$$
$$w = q - du = -358 \text{ kJ/kg}$$

**Discussion.** Work input does not affect change in entropy but heat rejected during the process is responsible for its decrease. On the other hand, the change in internal energy depends on both heat and work transfer.

**Example AP 4.12**

A rigid vessel of 2 m$^3$ volume is filled with superheated steam at 20 bar and 300°C. The vessel is cooled until steam is just dry saturated. Calculate the mass of steam in the vessel, the final pressure of steam and the amount of heat transfer. Represent the process on T-s diagram.

***Solution*** Given: $p_1$ = 20 bar, $T_1$ = 300°C, V = 2 m$^3$.

At 20 bar, saturation temperature $T_{sat}$ = 212.4°C, from steam table. Steam is superheated at 300°C. From steam table, we read that

$$v_1 = 0.1255 \text{ m}^3/\text{kg and } u_1 = 2773 \text{ kJ/kg}$$

$$m = \text{mass of steam in vessel} = V/v_1 = 16.0 \text{ kg}$$

**Final pressure.** Process is constant volume, $\rightarrow v_2 = v_1$ and dry saturated. Look in the $v_g$ column of the wet steam table to locate 0.1255 value or a value just less and just above this so that actual condition can be determined by linear interpolation. The new pressure is 16 bar.

Also, $u_2$ = 2596.

**Heat transfer.** System is closed.

Non-flow energy equation gives

$$q = u_2 - u_1 = -177 \text{ kJ/kg}$$

$$= -177 \times m = -2832 \text{ kJ}$$

Process is shown in Fig. AP 4.2, where 1 is the initial point and 2 is final point.

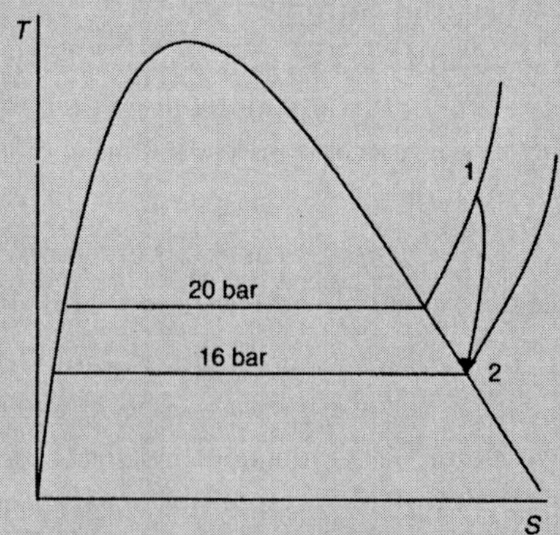


**Fig. AP 4.2** Constant volume cooling

**Example AP 4.13**

Determine difference in specific heat at constant pressure and constant volume for $H_2O$ at 0°C. We are given $\beta = 0.207 \times 10^{-6}\ K^{-1}$ and $\kappa = 45.9 \times 10^{-11}\ m^2/N$, where $\beta$ and $\kappa$ are coefficient of volume expansion and isothermal compressibility, respectively for water.

***Solution*** $\beta$ and $\kappa$ are defined in terms of $p$, $v$ and $T$ or their derivative by Eq. 4.15 and difference $c_p - c_v$ is given by Eq. 4.21. Specific volume of water at 0°C from steam table is 0.001002 m³/kg and $T = 273$ K. Substituting values in Eq. 4.21, we get

$$c_p - c_v = 2.74 \times 10^{-8} \text{ kJ/kg K}$$

We observe that difference is negligible and hence we do not speak of specific heat at constant volume or pressure in case of liquids and solids. But in gases, the difference is reasonable.

**Example AP 4.14**

Air at 300 K and 200 kPa is heated at constant pressure to 600 K. Determine the change in internal energy if air per unit mass, using (i) gas tables, (ii) functional forms of specific heat, and (iii) the average specific heat value.

***Solution*** $p_1 = 200$ kPa, $T_1 = 300$ K, $T_2 = 600$ K, and $p_2 = p_1$, air.

(i) **Gas tables.** From gas tables for air, we observe that all the properties are tabulated with temperature as an independent variable. We read $u_1 =$ 214.7 at 300 K and $u_2 = 434.78$ kJ/kg at 600 K. Thus,

$$\Delta u = u_2 - u_1 = 220.71 \text{ kJ/kg}$$

(ii) **Functional forms of specific heat.** We know for a gas, $du = c_v\, dT$ whose integration between two temperatures gives the change in internal energy. This needs a relation between $c_v$ and $T$ and is given in the appendix. A polynomial is usually used to express such relation. The third-degree polynomial is

$$c_v = (a - \overline{R}) + bT + cT^2 + dT^3 \tag{A}$$

where, $a = 28.11$, $b = 0.1967 \times 10^{-2}$, $c = 0.4802 \times 10^{-5}$ and

$d = -1.966 \times 10^{-9}$ in units consistent with that of $c_v$ (kJ/kg-mol K),

and $\overline{R} = 8.3143$ kJ/kg-mol K. Substituting in Eq A and integration from 300 K to 600 K gives, noting that molecular weight of air is 28.97

$$\Delta u = \int_{300}^{600} \left\{ \left(a - \overline{R}\right) + bT + cT^2 + dT^3 \right\} dT$$

$= 6446$ kJ/kgmole

$= 6446/28.97 = 222.5$ KJ/kg

Error in calculated value

$= (222.5 - 220.71)/220.71$

$= 0.8\%$, negligible

(iii) **Average specific heat value.** The other method of expressing dependence of $c_p$ and $c_v$ in gases on temperature is to tabulate the values at different temperatures and such tables are available. From such a table, we may read $c_v$ at the mean temperature $(T_1 + T_2)/2 = 450$ K; $c_v$ at this temperature is = 0.733 kJ/kg K.

$\therefore$ $\Delta u = c_v (T_2 - T_1) = 220.0$ kJ/kg

Error in calculated value = (220 – 220.71)/220.71 = 0.4%, close to the exact result. This closeness in the result is because of linear variation of $c_v$ with temperature. It is, therefore, normal practice to take values at mean temperature. However, this is true only for a small temperature range. You may also try taking $c_v$ at 300k or 600 K and find out how much error is introduced.

**Example AP 4.15**

Tank A of volume 1 m$^3$ containing 5 kmole of a gas at 600 K is brought into contact through a diathermal wall with a tank B of volume 1 m$^3$ containing 3 kmole of the same gas at 400 K. Determine the equilibrium temperature of the gas and the internal energy in tank A and the tank B if the gas obeys the equation of state $1/T = 3N\overline{R}/(2U)$.

***Solution*** Given for A: $V = 1$ m$^3$, $N_A = 5$, $T_A = 600$ K, and for B: $N_A = 5$, $V = 1$ m$^3$, $T_B = 400$ K

$\therefore$ $U_{A1} = 3\ N_A\ R\ T_A/2 = 3 \times 5 \times 8.314 \times 600/2 = 37.413$ kJ and similarly, $U_{B1} = 14.965$ kJ

When tanks A and B come in contact through diathermal wall, heat transfer would occur till both the gases acquire same temperature, T. Let us assume there is no energy transfer to the surrounding atmosphere so that considering both tanks A and B as system, the total change in internal energy is zero. The heat transfer only readjusts the energy distribution. This gives following two equations.

$T_A = T_B \rightarrow [2U/(3NR)]_A = [2U/(3NR)]_B$, or $5\,U_B = 3\,U_A$ (A)

and $U_A + U_B = U_{A1} + U_{B1} = 52.278$ (B)

Solving Eq. A and B, we get, $U_A = 32.736$ and $U_B = 19.64$ kJ; $T = 525$ K

**Example AP 4.16**

The vapour pressure of benzene is 53.329 kPa at 60.6°C whereas the normal boiling point is 80.1°C. Assume that benzene behaves like an ideal gas and estimate the latent heat of vaporisation of benzene.

***Solution*** Given: $p_1 = 53.329$ kPa, $T_1 = 333.75$ K, $p_2 = 100$ kPa, $T_2 = 353.25$ K, $R = 8.314$ kJ/kmoleK

Clausius–Clapeyron equation gives

$$dp/dT = h_{fg}\,p/(R\,T^2), \text{ or}$$

$$\int dp/p = (h_{fg}/R) \int dT/T^2$$

$$\rightarrow ln\,(p_2/p_1) = (h_{fg}/R)\,(1/T_1 - 1/T_2) \text{ or}$$

$$h_{fg} = 32.266 \text{ kJ/mole}$$

**Example AP 4.17**

On a hill the boiling point of water is 91°C. The normal boiling point is 100°C with latent heat of vaporisation of 2257 kJ/kg. Assuming that the atmosphere is isothermal at 25°C with pressure at sea level 1.01325 bar, estimate the height of the hill.

***Solution*** Pressure and temperature along the saturation curve are related by the relation, derived in Example AP 4.16,

$$\ln\,(p_2/p_1) = (h_{fg}/R)\,(1/T_1 - 1/T_2) = -0.3236 \quad \text{(A)}$$

Atmosphere is isothermal and therefore $pv = p_0\,v_0$, where subscript 0 refer to the sea level. Consider a volume element of air of height $dh$, as shown in Fig. AP 4.14. Forces acting on it are shown and are self-explanatory. Equilibrium demands

$(p + dp)\,A + mg = p\,A$, where $A$ is the cross-sectional area of the element.

or $dp = -\rho\,g\,dh = -g\,dh/v = -gp\,dh/(p_0\,v_0)$ (isothermal atmosphere given)

After integration from $p_1$ to $p_2$ while $h$ varies from o to H, the height of the hill, we get

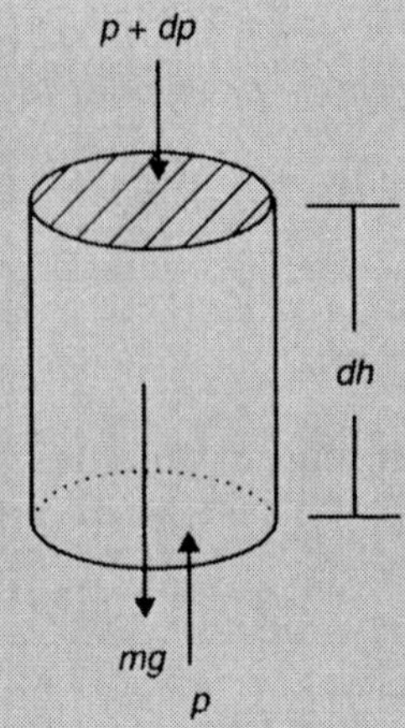


**Fig. AP 4.3** Pressure variation with height

$\ln (p_2/p_1) = -gH\{M/(Ru\ T_0)\}$, where $M = 28.97$ is molecular weight of air or $-0.03236 = -9.81 \times H X 28.97/(8.314 \times 298.15)$

$\rightarrow H = 2822.5$ m

**Example AP 4.18**

An air conditioner delivers 2 m³ of air at 5°C, which is completely saturated with water vapour. If this air is mixed with 5 m³ air at 35°C and of 40% RH in a room, determine the temperature, the specific humidity and the relative humidity of the air after mixing.

***Solution*** $V_1 = 2$ m³, $t_{db1} = 5$°C, $\phi_1 = 1$, $V_2 = 5$ m³, $t_{db2} = 35$°C, $\phi_2 = 0.4$.

We can locate points 1 and 2 on psychrometric chart and read

$$\omega_1 = 0.0054,\ v_1 = 0.7941 \text{ m}^3/\text{kg},\ \omega_2 = 0.0138,\ v_2 = 0.89216 \text{ m}^3/\text{kg}$$

This gives $m_1 = V_1/v_1 = 2.5186$ and $m_2 = V_2/v_2 = 5.6044$ kg. Join points 1 and 2 and divide the line such that $13/23 = m_2/m_1$. This locates point 3, the condition of air after adiabatic mixing. We read from chart

$$t_{db3} = 26\text{°C},\ \phi_3 = 0.55 \text{ and } \omega_3 = 0.0113 \text{ kgw/kga}$$

**Example AP 4.19**

A cooling tower is to be installed to supply 1000 kg/min of cold water at 27°C. Water enters the cooling tower at the rate of 1000 kg/min at 45 °C. The air enters the cooling tower at 30°C with 30% $\phi$ and leaves at 35°C with 80% $\phi$. Determine the make up water required and the airflow rate.

***Solution*** Let subscripts 1 and 2 refer to the entrance and exit of air while 3 and 4 for water, respectively.

Given: $T_4 = 27°C$, $m_3 = 1000$ kg/min, $T_3 = 45°C$, $t_{db1} = 30°C$, $\phi_1 = 0.3$, $t_{db2} = 35°C$, $\phi_2 = 0.8$

We can locate air condition at entrance and exit of the tower on the psychrometric chart and read: $\omega_1 = 0.0078$ kgw/kga, $h_1 = 51$ kJ/kga, $\omega_2 = 0.029$ kgw/kga, $h_2 = 113.25$ kJ/kga

From steam table: $h_f$ at $T_3$ and $T_4$ is: $h_{f3} = 188.45$, $h_f = 113.25$ kJ/kg.

Mass and energy balance on minute basis gives

$$1000 - m_4 = m_a(\omega_2 - \omega_1)$$

where $m_a$ = mass flow rate of air in kga/min (A)

$$m_3 h_3 - m_4 h_4 = m_a(h_2 - h_1) \quad \text{(B)}$$

There are two equations in two unknowns, $m_4$ and $m_a$ to give

$$m_a = 1328.6 \text{ kga/min and } m_4 = 971.83 \text{ kg/min}$$

$\therefore$ make up water required $= 1000 - m_4 = 28.17$ kg/min

## Chapter 5

### Example AP 5.1

A mixture of gases having volumetric composition as given in table is used to cool highly inflammable oil from 250°C to 50°C. $C_p$ of oil is 1.2 kJ/kg K. Determine the flow rate of the gas mixture to cool 3000 kg/h of the oil if the gas mixture enters the heat exchanger at 30°C and leaves at 50°C. The molar heat capacities are given in table.

***Solution***

| Gas | Vol % | $x_i$ | $C_{pi}$ | $X_i c_{pi}$ |
|---|---|---|---|---|
| $N_2$ | 30 | 0.3 | 29.1783 | 8.7535 |
| He | 50 | 0.5 | 20.7860 | 10.393 |
| $CO_2$ | 20 | 0.2 | 37.1444 | 7.42889 |
| | | | | $C_p = 26.5754$ |

Calculations for finding out $C_p$ are given in table, which are self-evident. If $N$ moles of mixture are used per hour then energy balance gives

$$N \times 26.5754 \times (50 - 30) = 3000 \times 1.2 \times (250 - 50)$$

$$\rightarrow N = 1354.644 \text{ kmol/h}$$

**Example AP 5.2**

A mixture of 50% propane and 50% $O_2$ by volume is compressed reversibly and adiabatically from 1 bar and 300 K to 10 bar. Determine the final temperature of the mixture and the work required to compress 100 mole of the gas mixture. The molar specific heats are $C_p$ = 74.0565 and $C_v$ = 65.7442 kJ/kmole K for propane and 29.4912 and 21.1776 kJ/kmole K for $O_2$.

***Solution*** Given: $T_1 = 300$ K, $p_1 = 1$ bar, $p_2 = 10$ bar, $N = 100$ mole

$$C_{p,\,\text{mix}} = \sum x_i\, C_{pi} = 0.5 \times 74.0565 + 0.5 \times 29.4912 = 51.7739$$

$$C_{v,\text{mix}} = 43.4608 \text{ kJ/kmole K} \rightarrow \gamma = 1.19$$

Using the relation $T \propto p^{(\gamma - 1)/\gamma}$, we get

$$T_2 = (p_2/p_1)^{(\gamma - 1)/\gamma}\, T_1 = 433.3 \text{ K}$$

System is closed and the process is isentropic. The nfee gives,

$$W = -\Delta U = -N\,C_{v,\text{mix}}\,(T_2 - T_1) = -579.33 \text{ kJ}$$

**Example AP 5.3**

A 5 m$^3$ tank contains 30% $h_2$, 30% Argon, and 40% methane by volume at 10 bar and 300 K. It is necessary to change the composition of the gas mixture such that it contains 60% by volume methane. Determine the amount of methane to be added and the final pressure of the mixture in the tank.

***Solution*** Given: $p_1 = 10$ bar, $T_1 = 300$ K, $V = 5$ m$^3$

| Gas | xi | $P_i$, Kpa | $N_i = p_i$ V/(RT) |
|---|---|---|---|
| $H_2$ | 0.3 | 300 | 0.6014 |
| A | 0.3 | 300 | 0.6014 |
| $CH_4$ | 0.4 | 400 | 0.8019 |

Let the new number of moles of methane in the mixture is $N$, then

$$X\,CH_4 = 0.6 \text{ (given)} = N/(0.6014 + 0.6014 + N)$$

$$\rightarrow N = 1.804 \text{ kmole}$$

$\therefore$ amount of $CH_4$ added = 1.804 – 0.8019 = 1.0023.

Total number of moles = 0.6014 + 0.6014 + 1.804 = 3.0068.

Final pressure = $N\bar{R}T/V = 3.0068 \times (8.314 \times 10^3) \times 300/5 \times 10^{-5}$

$= 15$ bar

**Example AP 5.4**

A rigid insulated tank of volume 1 $m^3$ contains He at 1 bar and 300 K. The tank is connected through a valve to a line carrying $CO_2$ at 4 bar and 600 K. The valve is opened and the $CO_2$ enters the tank till the pressure inside the tank is 4 bar. Then, the valve is closed and the tank is isolated. Determine (i) the final temperature of the mixture in the tank, (ii) the composition of the gas mixture in the tank. Given: $C_v$ for He = 12.4717 kJ/kmole K and for $CO_2$ 28.8354 kJ/kmole K, $C_p$ for $CO_2$ = 37.1444 kJ/kmole K.

***Solution*** Given: $V = 1$ $m^3$, $p_i = 10^5$ N/$m^2$, $T_i = 300$ K, $p_f = 4$ bar, $T_1 = 600$ K, $R = 8.314$ kJ/kmole K, $C_v$ for He = 12.4717 kJ/kmole K and for $CO_2$ 28.8354 kJ/kmole K, $C_p$ for $CO_2$ = 37.1444 kJ/kmole K.

(i) **Final temperature of the mixture in the tank.** The process is transient which results in increase in internal energy. Application of energy equation with subscripts $i$ and $f$ for conditions in tank before and after filling and 1 at the entrance of tank, we get

$$(N\,h)_1 = U_f - U_i \text{ or } (N\,C_p)_{CO2}\,T_1 = U_f - U_i \qquad \text{(A)}$$

But $U_i = (N\,C_v\,T_i)_{He} = (p_i\,V\,C_v/R)_{He} = 150$ kJ,

$$U_f = \{(N\,C_v)_{CO_2} + (N\,C_v)_{He}\}\,T_f\,,\ N_{He}$$

$$= p_i\,V/(R\,T_i) = 0.0401 \text{ kmole,}$$

$$T_f = p_f\,V/[R \times N_{CO_2} + N_{He}]$$

$$= 48.112/(N_{CO_2} + N_{He})$$

Substituting these in Eq. A and solving, we get

$$N_{CO_2} = 0.0372 \text{ kmole and}$$

$$T_f = 622.41 \text{ K}$$

(ii) **Composition of the gas mixture in the tank.**

N = Total moles in tank = $N_{CO_2} + N_{He}$

∴ Composition is $CO_2$ = 32.83 and $H_e$ = 67.17%

**Example AP 5.5**

It is desired to separate 1 kmole air at 0.1 MPa and 300 K into its pure components $N_2$ and $O_2$ at the same temperature and pressure. Calculate the minimum work required.

***Solution*** Since temperature is same, change in internal energy is zero. Assuming the separation process to be reversible requiring $\delta Q$ and $\delta w$

amount of heat and work, we can write from first and second law of thermodynamics

$$\delta Q - \delta W = TdS - pdV = 0$$

or $\int pdV$ = work done = $\int TdS = T\Delta S$, where, $\Delta S$ is the change in entropy during the process.

We know that change in entropy is given by Eq. 5.5 (a).

$\therefore$ $\Delta S$ (mixing) $= -\bar{R} \sum x_i \ln x_i$

$= -8.3143$ [0.21 ln (0.21) + 0.79 ln (0.79)]

= 4.273 kJ/moleK

here, composition of air is taken as 21% $O_2$ and 79% $N_2$ by mole.

$\therefore$ $\Delta S$ (separation) = – 4.273 kJ/moleK, and

minimum work done = $T\,\Delta S$

= 300 × 4.273 = 1281.9 kJ

## Chapter 6

### Example AP 6.1

Acetylene is burned with air and the resulting combustion products have the following volumetric composition on dry basis. $CO_2$ = 10.89%, CO = 3.63%, $O_2$ = 3.63%, and $N_2$ = 81.85%. Determine the actual air fuel ratio and the per cent excess air used for the combustion of acetylene.

***Solution*** We first write the combustion equation with air to give 100 mole of dry products.

$X\,C_2H_2 + a\,O_2 + 3.76\,a\,N_2$

$\rightarrow 10.89\,CO_2 + 3.63\,CO + 3.63\,O_2 + 81.85\,N_2 + b\,H_2O$

Let us balance the mass.

Nitrogen balance gives

3.76 a = 81.85 or a = 21.77

Carbon balance gives

$2X = 10.89 + 3.63 \rightarrow X = 7.26$, and

Hydrogen balance gives: $2X = 2b \rightarrow b = X = 7.26$

$\therefore$ actual air fuel ratio = mole air/mole fuel = $4.76a/X = 14.27$

In text, we expressed air fuel ratio as ratio of mass. However, we can express it in terms of ratio of mole also. For calculating theoretical air fuel ratio, we will write the stoichiometric equation, which is

$$C_2H_2 + a\,O_2 + 3.76\,a\,N_2 \rightarrow 2\,CO_2 + H_2O + 3.76a\,N_2$$

Balance of mass of oxygen $\rightarrow a = 2 + ½ = 2.5$

$\therefore$ theoretical air fuel ratio = 4.76a/1 = 11.9

Percentage of excess air = (14.27 – 11.9)/11.9 = 19.92%

## Chapter 7

### Example AP 7.1

In a 15 MW steam power plant operating on ideal reheat cycle, steam enters the HP turbine at 150 bar and 600°C. The condenser is maintained at a pressure of 0.1 bar. If the moisture content at the LP exit is 10.4%, determine: (i) reheat pressure, (ii) thermal efficiency, (iii) specific steam consumption, and (iv) rate of pump work in kW. Assume steam is reheated to the initial temperature.

***Solution*** Refer Fig. AP 7.1 for state points.

Given: $p_1$ = 150 bar, $T_1$ = 600°C, $p_4$ = 0.1 bar, $x_4$ = 1 – 0.104 = 0.896, $P$ = 15 MW.

Expansion in turbine is shown on T-s diagram, since points 1 and 4 can be located from given data. Draw vertical line 43 to meet 600°C line at point 3. This determines reheat pressure $p_3$. Point 2 at the HP turbine exit is point of intersection of $s = c$ line through point 1 with $p_3$ curve. We may also show the processes 12 and 34 on h-s diagram (not shown in figure) to read enthalpy at these points or may use steam table. We get

$h_1 = 3580$, $h_2 = 3140$, $h_3 = 3675$, $h_4 = 2335$ kJ/kg. Also,

$h_5 = 191.8$ kJ/kg from steam table

(i) **Reheat pressure** = $p_3$ = 40 bar

(ii) **Thermal efficiency.** $w_t = h_1 - h_2 + h_3 - h_4 = 1780$ kJ/kg and

$q = h_1 - h_4 + h_3 - h_2 = 3923.2$ kJ/kg

$\therefore \eta_i = w_t/q = 45.37\%$

(iii) **Specific steam consumption**

$m_s = P/w_t = 8.47$ kg/s

$\therefore$ ssc $= m_s \times 3600/P = 3600/w_t = 2.0225$ kg/kWh

(iv) Rate of pump work $= m_s \times v \times (p_1 - p_4)$

$= 8.47 \times 0.001 \times (150 - 0.1) \times 100 = 1.26$ kW

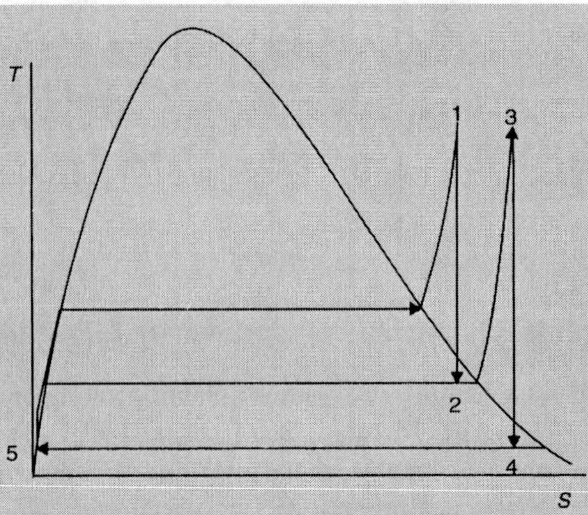


**Fig. AP 7.1** Reheat pressure

**Example AP 7.2**

A steam power plant developing 120 MW of electrical output is equipped with reheating and regenerative feed heating arrangement consisting of two feed heaters—one surface type on HP side and another direct contact type on LP side. The steam conditions before the steam stop valve are 100 bar and 530°C. A pressure drop of 5 bar takes place due to throttling in valves. Steam exhausts from HP turbine at 25 bar. A small quantity of steam is bled off at 25 bar for HP heater for feed heating and the remaining is reheated in a reheater to 550°C and the steam enters the LP turbine at 22 bar for further expansion up to 0.05 bar. The drain from the HP heater is led to the LP heater and the combined feed from the LP heater is pumped to the HP feed heater and finally to the boiler with the help of a boiler feed pump. The component efficiencies are: turbine efficiency = 0.85, pump efficiency = 0.90, generator efficiency = 96%, boiler efficiency = 0.9, and mechanical efficiency = 0.95. It may be assumed that the feed-water is heated up to the saturation temperature at the prevailing pressure in feed heater. Work out the following: (i) sketch the feed heating system and show the process on T-s diagram, (ii) amounts of steam bled off, (iii) overall thermal efficiency of turbo-generator, and (iv) specific steam consumption. [AMIE Summer, 1998]

***Solution***

(i) From given data for steam, the steam state at different points are shown in Fig. AP 7.2 both on h-s and T-s diagrams. Figure also shows the schematic with the numbers marked at different state points. Processes 1 to 6 can be drawn using the efficiencies and other data given in the problem. We read following values from it.

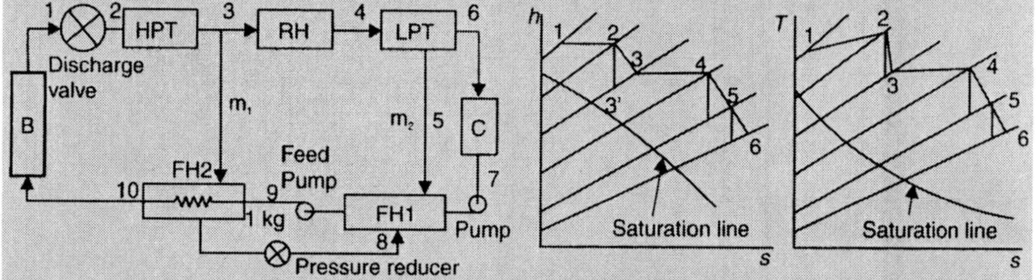


**Fig. AP 7.2** Reheating and regenerative cycle

$h_1 = h_2 = 3460$, $h_3 = 3111.5$, $h_4 = 3585$, $h_5 = 3207$

$h_6 = 2466$ kJ/kg and from steam table

$h_7 = 137.8$, $h_8 = h_{10} = 962$, $h_9 = 670.4$ kJ/kg

To get state 3, we consider isentropic expansion in HPT from 2 to 3' and measure $h_2 - h_3'$. Then, $h_2 - h_3 = 0.85 \times (h_2 - h_3')$ and thus we locate point 3 on the h-s chart. Here, 0.85 is the isentropic efficiency of HP turbine. Like this all other points can be located.

(ii) **Amounts of steam bled off.** Let $m_1$ kg steam be bled off after HP turbine. Making energy balance for the HP feed heater, we get:

$$m_1 h_3 + h_9 = m_1 h_8 + h_{10} \rightarrow m_1 = 0.13566 \text{ kg}$$

Energy balance for contact type heater (FH1) gives

$m_2 h_5 + (1 - m_1 - m_2) h_7 + m_1 h_8 = h_9$, where $m_2$ kg of steam is bled off from LP turbine. This gives, $m_2 = 0.1371$ kg.

$w_p$ = Pump work

$= [(1 - m_1 - m_2)(p_5 - p_6) + 1 \times (p_1 - p_5)] \times 0.001 \times 100/\eta_m$

$= 10.925$ kJ/kg

$w_t = (h_2 - h_3) + (1 - m_1)(h_4 - h_5) + (1 - m_1 - m_2)(h_5 - h_6)$

$= 1214.10$ kJ/kg

$w_e$ = Electrical output

$= (w_t - w_p) \times \eta_m \times \eta_g$

$= 1040$ kJ/kg

$\therefore$ $m_s$ = mass flow rate of steam at HP inlet = $P/w_e = 4.155 \times 10^5$ kg/h

Amount of steam bled off for HP heater = $m_1 \times m_s = 5.6367 \times 10^4$ kg/h, and for LP heater = $m_2 \times m_s = 5.697 \times 10^4$ kg/h

(iii) **Overall thermal efficiency of turbo-generator.** Heat input to boiler per kg of steam

$$q = [1 \times (h_1 - h_{10}) + (1 - m_1)(h_4 - h_3)]/\eta_b$$
$$= 3301.7 \text{ kJ/kg}$$
$$\therefore \ \eta_0 = w_t/q = 31.48\%.$$

(iv) **Specific steam consumption** $= 4.155 \times 10^5/(120 \times 10^3)$
$= 3.4625$ kg/kWh

**Example AP 7.3**

In a single heater regenerative cycle the steam enters the turbine at 30 bar, 400°C and the exhaust pressure is 0.10 bar. The feed water heater is a direct contact type, which operates at 5 bar. Find (i) the efficiency and the steam rate of the cycle and (ii) the increase in mean temperature of heat addition, efficiency and steam rate as compared to Rankine cycle without regeneration.

***Solution*** Refer Fig. AP 7.3 for the cycle and schematic. You can draw the cycle on h-s diagram and read the following values from it.

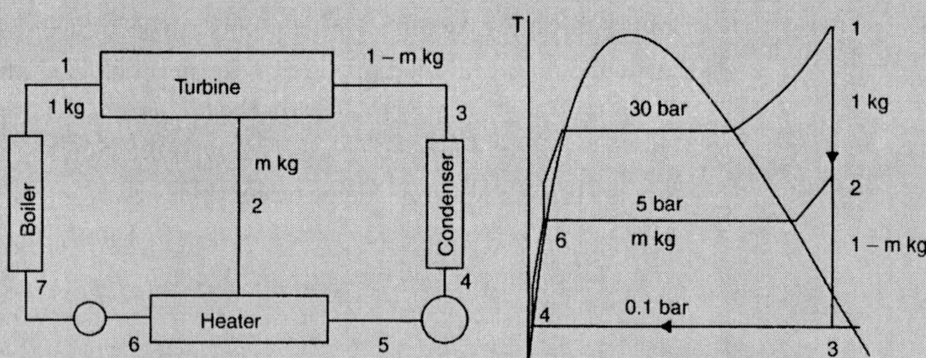


**Fig. AP 7.3**

$h_1 = 3230.9$, $h_2 = 2796$, and $h_3 = 2192.2$ kJ/kg, $s_1 = 6.921$ kJ/kg K. Also, from steam table $h_4 = h_{f,4} = 191.8$, $h_6 = h_{f,6} = 640.1$ kJ/kg, $s_4 = 0.649$, and $s_6 = 1.8604$ kJ/kg K, and $h_5 = h_4$ (neglecting pump work).

(i) **Efficiency and the steam rate of the cycle.**

Let $m$ kg of steam be bled off for heater. Energy balance gives

$$mh_2 + (1 - m)\,h_5 = h_6 \rightarrow m = 0.172 \text{ kg}$$

$$\therefore\ w_t = 1 \times (h_1 - h_2) + (1 - m)\,(h_2 - h_3)$$
$$= 934.8 \text{ kJ/kg},\ q = h_1 - h_6 = 2590.8$$
$$\eta_t = w_t/q = 36.08\% \text{ and steam rate}$$
$$= \text{ssc} = 3600/w_t = 3.85 \text{ kg/kWh}$$

(ii) **Increase in mean temperature of heat addition, efficiency and steam rate.** Mean temperature of heat addition = $q/(s_1 - s_6)$ = 512 K = 239°C

Mean temperature of heat addition without regeneration,

$$(h_1 - h_4)/(s_1 - s_4) = 484.5 \text{ K} = 211.5°\text{C}$$

$\therefore$ Increase in mean temperature = 239 – 211.5 = 27.5°C

Work done without regeneration = $w_t' = h_1 - h_3$ = 1038.7 kJ/kg

$\therefore$ $\eta_t'$ = new overall efficiency = $w_t'/(h_1 - h_4)$ = 34.18%

→ Increase in $\eta_t$ = 36.08 = 34.18 = 1.9%

Steam rate without regeneration = $3600/w_t'$ = 3.46

→ Increase in steam rate = 3.85 – 3.46 = 0.39

**Example AP 7.4**

A cyclic steam power plant is to be designed for a steam temperature at turbine inlet of 360°C and an exhaust pressure of 0.08 bar. After isentropic expansion of steam in the turbine, the moisture content at the turbine exhaust is not to exceed 15%. Determine the greatest allowable steam pressure at turbine inlet, and calculate the Rankine cycle efficiency for these steam conditions. Estimate also the mean temperature of heat addition.

***Solution*** Refer Fig. AP 7.4 where effect of change in boiler pressure is shown by taking two pressures $p$ and $p'$ bar. It is observed that for same initial steam temperature, dryness fraction of steam becomes less after expansion as boiler pressure increases. Therefore, there is a maximum boiler pressure; say it is p bar, to ensure that wetness of steam at point 2 is limited to 15%. You may now locate point 2 on the T-s chart, since x and p are given at this point. Through point 2 draw vertical line 21 to intersect $t$ = 360°C line at point 1. Read pressure from the chart. This gives the maximum pressure = 16.832 bar.

From steam table: $h_1$ = 3165.54, $h_2$ = 2216.52, and

$h_3$ = 173.88 kJ/kg, $s_1$ = 7.0833 and $s_3$ = 0.5926 kJ/kg K, and we neglect pump work.

$\therefore\ w_t = h_1 - h_2 = 949,\ q = h_1 - h_3 = 2990$ kJ/kg and

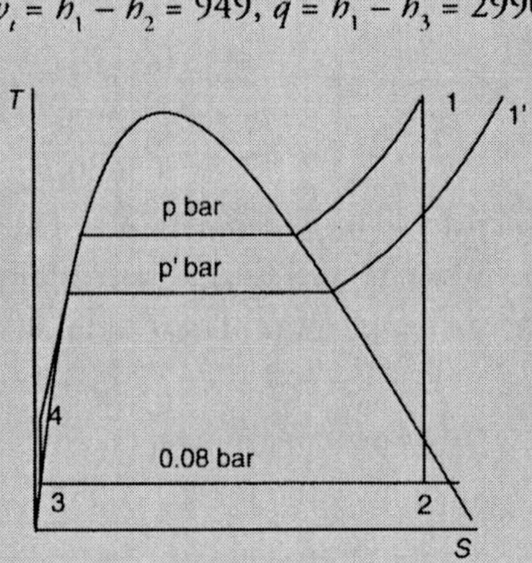


**Fig. AP 7.4** Effect of $p_b$ on d.f.

$$\eta_{th} = w_t/q = 31.68\%$$

Mean temperature of heat addition $= q/(s_1 - s_3) = 460.66$ K

$= 187.51$°C

**Discussion.** Though the maximum temperature achieved in the cycle is 360°C but the whole heat is not added at this temperature. Rather, it is added at a mean temperature of 187.51°C, which is much less than the maximum temperature attained by steam.

**Example AP 7.5**

In a cogeneration plant, the power load is 5.6 MW and the heating load is 1.163 MW. Steam is generated at 40 bar and 500°C and is expanded isentropically through a turbine to a condenser pressure of 0.06 bar. The heating load is supplied by extracting steam from the turbine at 2 bar, which is condensed in the process heater to saturated liquid at 2 bar and then pumped back to the boiler. Compute (i) the steam generation capacity of the boiler in *t/h*, (ii) the heat input to the boiler in kW, (iii) the fuel burning rate of the boiler in *t/h* if coal of C.V. 25 MJ/kg is burned and the boiler efficiency is 0.88, (iv) the heat rejected to the condenser, (v) the rate of flow of cooling water in the condenser if the temperature rise is 6°C. Neglect pump work.

***Solution*** Refer Fig. AP 7.5.

Given: $p_1 = 40$ bar, $T_1 = 500$°C, $p_2 = 2$ bar, $p_3 = p_4 = 0.06$ bar, $P = 5.6$ MW, $Q$ = heating load = 1.163 MW, C.V. 25 MJ/kg, $\eta_b = 0.88$.

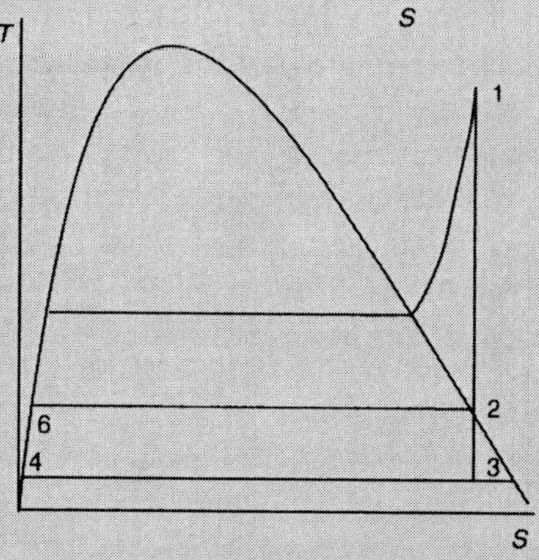


**Fig. AP 7.5** Cogeneration

From steam table and using process equation $s_1 = s_2 = s_3$, we determine

$h_1 = 3445.3$, $h_2 = 2706.7$, $h_3 = 180.6$, $h_4 = h_{f,4} = 151.5$,

$h_6 = h_{f,6} = 504.7$ kJ/kg

(i) **Steam generation capacity of the boiler.** Let $m_1$ kg of steam be extracted for process heating and m kg is supplied to turbine. Heat supplied by steam to process is

$$Q = m_1 (h_2 - h_6) \rightarrow m_1 = 0.528 \text{ kg/s} = 1901.4 \text{ kg/h}$$

$$P = m (h_1 - h_2) + (m - m_1) (h_2 - h_3) \rightarrow m = 16.73 \text{ t/h}$$

(ii) **Heat input to the boiler.** Heat is required to heat $m_1$ kg of water from process heater and $m - m_1$ kg from the condenser.

$$\therefore\ Q_{added} = m_1 (h_1 - h_6) + (m - m_1) (h_1 - h_4) = 15.111 \text{ MW}$$

(iii) **Fuel burning rate of the boiler** = $Q_{added}/(\text{C.V. X } \eta_b) = 2.473$ t/h.

(iv) **Heat rejected to the condenser** = $Q_{rej} = (m - m_1) (h_3 - h_4) = 8.367$ MW.

(v) **Rate of flow of cooling water in the condenser** = $Q_{rej}/(4.187 \times 6)$

= 333.05 kg/s.

$\therefore$ Volume flow rate of cooling water = $333.05/\rho$

= $333.05/1000 = 0.333$ m$^3$/s

## Example AP 7.6

A certain factory has an average electrical load of 1500 kW and requires 3.5 MJ/s for heating purpose. It is proposed to instal a single extraction pass out turbine to operate under the following conditions.

Initial pressure = 15 bar, initial temperature = 300°C, condenser pressure = 0.1 bar. Steam is extracted between the two turbines at 3 bar, 0.96 dry and is isobarically cooled without subcooling in process heaters to provide heating load. The internal efficiency of the turbine is 0.8 and the boiler efficiency is 0.85 when using oil of calorific value 44 MJ/kg. If 10% steam is used for auxiliaries, calculate the steam requirement per day. Assume that the condensate from the heaters and that from the condenser mix in a separate vessel (hot well) before being pumped to the boiler.

***Solution*** Refer Fig. AP 7.6. It shows the arrangement used for pass out turbine and the expansion process on T-s diagram. Pass out turbine is used in process industries when the electrical load is more and the amount of steam expanding in back-pressure turbine is not sufficient to meet the electrical load. In such case, whole steam is first expanded in HP turbine from which the amount of steam required for the process heat is extracted and the remaining steam is then expanded further in the LP turbine. The boiler feed back arrangement is also shown.

We know how to draw the line 123 on Mollier and T-s charts from the known isentropic efficiencies. The values read from the Mollier chart are: $h_1$ = 3037.3, $h_2$ = 2638.7, and $h_3$ = 2245.14 kJ/kg. From steam table: $h_4$ = 191.83 and $h_5$ = 561.47 kJ/kg. Let $m$ be the amount of steam required to give process heat and $m_s$ is he amount of steam entering the HP turbine then

$$m\,(h_2 - h_5) = 3.5 \times 1000 \rightarrow m = 1.685 \text{ kg/s, and}$$

$$m_s\,(h_1 - h_2) + (m_s - m)\,(h_2 - h_3) = 1500 \rightarrow m_s = 2.73 \text{ kg/s}$$

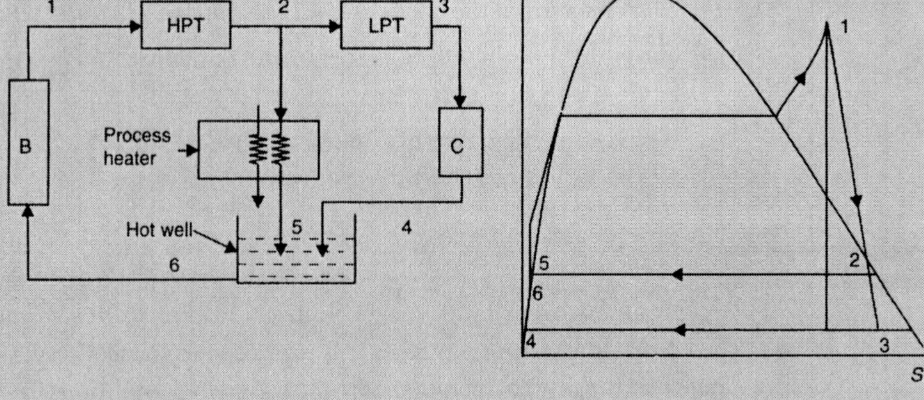


**Fig. AP 7.6** Pass out turbine

We may now find out the condition of water coming out from the hot well by making its energy balance.

$$(m_s - m)\, h_4 + m\, h_5 = m_s\, h_6 \rightarrow h_6 = 420 \text{ kJ/kg}$$

Since 10% steam is consumed in the auxiliaries, the amount of steam produced = 1.1 × $m_s$ kg/s. If $m_f$ kg of fuel is used per second then

$$\eta_b = 1.1 \times m_s \times (h_1 - h_6)/(m_f \times \text{C.V.})$$

$$\rightarrow m_f = 756.56 \text{ kg/h} = 756.56 \times 24/1000 = 18.16 \text{ t/day}$$

## Chapter 8

**Example AP 8.1**

A simple open cycle gas turbine power plant works with a pressure ratio of 12:1, the inlet pressure and temperature to the compressor are 1.013525 bar and 15°C. The temperature of hot gas at entry to the turbine is 757°C. Find for 1 kg (a) the isentropic compression work, (b) the heat supplied by fuel in the combustor, (c) the isentropic work developed by the turbine, (d) the net shaft work, (e) work ratio and (f) the overall thermal efficiency. For air, $c_p$ = 1.005 kJ/kg K and $\gamma$ = 1.4.

***Solution*** Given: $R$ = 12, $p_1 = p_4$ = 1.013525 bar, $T_1$ = 288 K, $T_3$ = 1030 K, $\gamma$ = 1.4.

Refer Fig. AP 8.8 (b) for the notations. Compressor is an open system and the flow through it is steady and therefore, sfee can be used to find out the amount of work transfer. It is true for combustor and the turbine, too. We assume that changes in kinetic energies are negligible as these are not given or the temperature is the stagnation temperature and not the static temperature so that we do not need to bother changes in kinetic energies. Further, processes 12 and 34 are isentropic one for which the process equation is $T \propto p^{(\gamma-1)/\gamma}$. Using this relation, we get

$$T_2 = T_1\, (p_2/p_1)^{(\gamma-1)/\gamma} = 585.8 \text{ K and}$$

$$T_4 = T_3\, (p_4/p_3)^{(\gamma-1)/\gamma} = 506.4 \text{ K}$$

(a) Isentropic compression work = $w_c = c_p\,(T_2 - T_1)$ = 299.3 kJ/kg.

(b) Heat supplied by fuel in the combustor = $q = c_p\,(T_3 - T_2)$

= 446.4 kJ/kg

(c) Isentropic work developed by the turbine = $w_t = c_p\,(T_3 - T_4)$

= 526.2 kJ/kg

(d) Net shaft work = $w_{net} = w_t - w_c = 226.9$ kJ/kg

(e) Work ratio = $w_{net}/w_t = 0.4312$

(f) Overall thermal efficiency = $w_{net}/q = 50.83\%$

**Example AP 8.2**

A theoretical gas turbine receives air at 1 bar and 27°C. The maximum allowable temperature is 727°C and the pressure ratio is 5. Determine per kg of air the work of compression, the work of expansion and the net cycle work. Also, determine the thermal efficiency of the cycle. If the mass flow rate is 5 kg/s, what is the power output? How is the work output and efficiency augmenting if the maximum allowable temperature is raised somehow to 817°C? Make comments.

***Solution*** Given: $R = 5$, $p_1 = p_4 = 1.0$ bar, $T_1 = 300$ K, $T_3 = 1000$ K, $\gamma = 1.4$, $T_3$, new = 1090 K.

This is similar to Example AP 8.1 and hence you should be able to do it. However, the difference is that there are two temperatures given at the entrance of the turbine. Therefore, this problem is to be solved again taking this new temperature and then we can find out how much is the percentage increase in work output and the efficiency of the cycle. You should be able to justify this increase.

**Example AP 8.3**

A gas turbine plant with an exhaust heat exchanger has the following data: turbine and compressor pressure ratio = 10, minimum and maximum cycle temperature = 300 K and 1500 K, flow rate through turbine and compressor 10 kg/s, thermal ratio[3] of the heat exchanger = 0.8, turbine efficiency = 0.85, compressor efficiency = 0.82. Mass flow of fuel may be neglected. Take for air, $c_p = 1.005$ kJ/kg and $\gamma = 1.4$. Determine (a) power developed, (b) thermal efficiency of the plant, and (c) efficiencies of the ideal joule cycle with heat exchanger and Carnot's cycle. Recalculate (a) and (b) when thermal ratio of the heat exchanger is unity.

***Solution*** Given: $R = 10$, $T_1 = 300$ K, $T_3 = 1500$ K, $\gamma = 1.4$,

$$m = 10 \text{ kg/s}, \ \varepsilon = 0.8, \ \eta_t = 0.85, \ \eta_c = 0.82.$$

3. Different names, such as efficiency ratio, thermal ratio or heat exchanger effectiveness are in use.

Refer Fig. AP 8.1, which shows Brayton cycle with heat exchanger for 100% efficient compressor, turbine and heat exchanger. Temperature at different points can be obtained easily using the procedure that has been used in number of questions

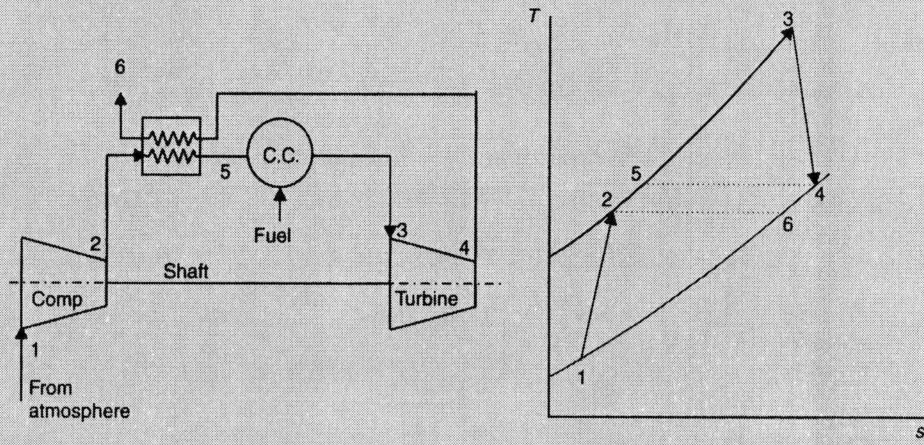


**Fig. AP 8.1** Brayton cycle with heat exchanger

$$T_2' = T_1\,(R)^{(\gamma-1)/\gamma} \text{ and } \eta_c = (T_2' - T_1)/(T_2 - T_1)$$
$$\rightarrow T_2 = 640.6 \text{ K and similarly, } T_4 = 885 \text{ K}$$

$T_5$ is determined by considering effectiveness of the heat exchanger defined as

$$\varepsilon = (T_5 - T_2)/(T_4 - T_2)$$
$$\rightarrow T_5 = 836.12 \text{ K}$$

Energy balance of the heat exchanger results in:

$$T_5 - T_2 = T_4 - T_6 \rightarrow T_6 = 689.48 \text{ K.}$$

(a) **Power developed** = $m\,c_p\,(T_3 - T_4) = 2757.7$ kW

(b) **Thermal efficiency of the plant.** We may determine thermal efficiency either as ratio of work output to the heat input or in the following way, too.

$Q_s$ = Heat supplied per kg

$= c_p\,(T_3 - T_5) = 667.2$ kJ/kg

and $Q_r$ = heat rejected

$= c_p\,(T_6 - T_1) = 391.43$ kJ/kg

$\therefore\ \eta_{th} = (Qs - Qr)/Qs = 41.33\%$

(c) **Efficiencies of the ideal Joule cycle with heat exchanger.** You should show that following relation gives efficiency of the heat exchanger cycle with perfect heat exchange

$$\eta_{th} = 1 - c/t, \text{ where } c = (R)^{(\gamma-1)/\gamma} \text{ and}$$
$$t = T_3/T_1, \rightarrow \eta_{th} = 61.36\%$$

Efficiency of Carnot cycle working between $T_3$ and $T_1$ is

$$= 1 - T_1/T_3 = 80\%$$

When thermal ratio of the heat exchanger is unity.

In this case, $T_5 = T_4 = 885$ K and $T_6 = T_2 = 640.6$ K

$P$ = Power developed

$$= mc_p (T_3 - T_4) = 2757 \text{ kW}$$
$$Q_s = m\, c_p (T_3 - T_5) = 6180 \text{ kW and}$$
$$\eta_{th} = P/Q_s = 44.6\%$$

**Example AP 8.4**

Pressure and temperature at inlet to compressor and turbine are 1.0, 15°C and 5.85 bar, 670°C whereas these at their outlet are 6.0 bar, 250°C and 1.03 bar, 370°C, respectively. Calculate the isentropic efficiency of compressor and turbine. Also, determine the power developed by the plant with an air flow rate of 24.5 kg/s and the work ratio. Take

$$c_p = 1.005 \text{ kJ/kg K and } \gamma = 1.4$$

***Solution*** For compressor: $R = 6/1 = 6$, $T_1 = 288$ K

$$\rightarrow T_2' = T_1 (R)^{(\gamma-1)/\gamma} = 480.5 \text{ K}, T_2 = 523 \text{ K}$$
$$\therefore \eta_c = (T_2' - T_1)/(T_2 - T_1) = 81.9\%$$

Similarly, for turbine: $R = 5.85/1.03 = 5.68$, $T_3 = 943$ K

$$\rightarrow T_4' = T_3 (1/R)^{(\gamma-1)/\gamma} = 574.1 \text{ K}, T_4 = 643 \text{ K}$$
$$\therefore \eta_c = (T_3 - T_4)/(T_2 - T_4') = 81.3\%$$
$$w_c = c_p (T_2 - T_1) = 236.2, w_t = c_p (T_3 - T_4) = 301.5$$

and $$w_{net} = w_t - w_c = 65.3 \text{ kJ/kg}$$

$$\therefore P = \text{power developed} = m \times w_{net} = 1600 \text{ kW}$$

and work ratio $= w_{net}/w_t = 0.217$

We note that work ratio of this cycle is very poor. This means most of the turbine work is consumed internally in the cycle by the compressor. This is the main problem of gas turbine cycles. However, when used with some other cycle to form a combined or a binary cycle then the overall efficiency is increased because of the coupling of the two cycles, as discussed in the binary vapour cycle in Chapter 7.

**Example AP 8.5**

Single shaft gas turbine uses two stage compression with perfect intercooling, the work of compression being equal. The inlet air pressure and temperature are 1.0 bar and 21°C. The maximum temperature of the cycle is 788°C and the overall pressure ratio is 6. The isentropic efficiency of compression of both the compressor is 0.85. A single stage turbine with an isentropic efficiency of 0.83 performs the expansion process to 1.0 bar. The heat transfer efficiency of the regenerator is 0.75. Assuming air as the working fluid, determine (a) the overall thermal efficiency and (b) work ratio. If the mass flow rate of air is 20.3 kg/s, what is the power output of the cycle? Determine the quantities when helium is used in place of air. Take $c_p$ = 5.193 kJ/kg K and $\gamma$ = 1.67 for helium.

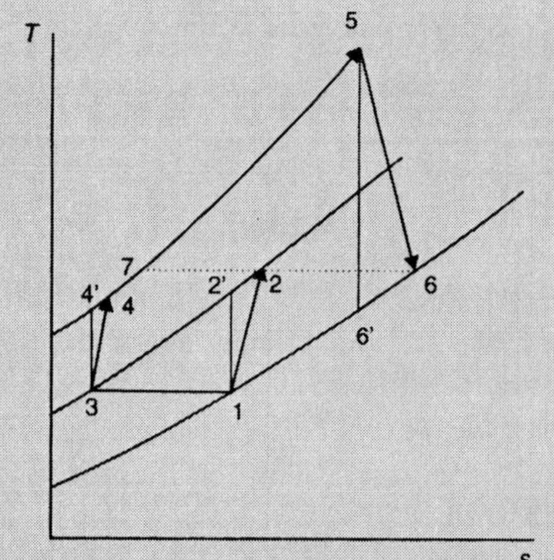


**Fig. AP 8.2** Multi-compression

***Solution*** Given: $R = 6$, $T_1 = T_3 = 294$ K, $p_1 = 1.0$ bar, $T_5 = 1061$K, $\eta_c = 0.85$, $\eta_t = 0.83$, $p_6 = 1$ bar, $\gamma = 1.4$, $\varepsilon = 0.75$, $m = 20.3$ kg/s.

Refer Fig. AP 8.2. We are given perfect intercooling and work done in both the compressors is equal. This gives the condition

$$p_2/p_1 = p_4/p_3 = p_4/p_2 \rightarrow p_2/p_1 = p_4/p_3 = \sqrt{R} = 2.45 \text{ bar}$$

Using isentropic relation $T \propto p^{(\gamma-1)/\gamma}$, we get

$$T_2' = T_1\,(p_2/p_1)^{(\gamma-1)/\gamma} = 379.8 \text{ and}$$

$$\eta_c = (T_2' - T_1)/(T_2 - T_1) \rightarrow T_2 = 379.8 \text{ K}$$

Similarly we get, $T_4 = 394.9$ K, $T_6 = 708.2$ K

Heat Exchanger. $\varepsilon = 0.75 = (T_7 - T_4)(T_6 - T_4)$

$$\rightarrow T_7 = 629.9 \text{ K}$$

Recall the maximum temperature that can be achieved in heat exchanger is $T_6$ and that defines the heat exchanger effectiveness or efficiency.

$$w_c = 2\, c_p\, (T_2 - T_1) = 202.8 \text{ kJ/kg}$$

$$w_t = c_p\, (T_5 - T_6) = 354.6 \text{ kJ/kg and}$$

$$w_{\text{net}} = w_t - w_c = 151.8 \text{ kJ/kg}$$

$$q = c_p\, (T_5 - T_7) = 433.3 \text{ kJ/kg}$$

$$\therefore\ \eta_{\text{th}} = w_{\text{net}}/q = 35\% \text{ and}$$

work ratio = $w_{\text{net}}/w_t = 0.428$

$$P = \text{power produced} = m \times w_{\text{net}} = 3081.5 \text{ kW}$$

**When working substance is He.** The procedure is same. The difference comes because of different values of $c_p$ and $\gamma$. Using these values, we get (verify)

$$T_2 = 443.6 \text{ K},\ T_4 = 443.6 \text{ K},\ T_6 = 609.5 \text{ K},\ T_7 = 568 \text{ K}$$

$$w_c = 1553.7,\ w_t = 2344.6,\ w_{\text{net}} = 790.9 \text{ and}$$

$$q = 2560.1 \text{ kJ/kg}$$

$$\therefore\ \eta_{\text{th}} = 30.9\%, \text{ work ratio} = 0.337 \text{ and } P = 16055 \text{ kW}$$

**Discussion.** While comparing the results for the two working fluids, we observe that (i) both efficiency and the work ratio decrease whereas (ii) the net power output of the cycle increases 5 times when helium is used as working substance. However, the cycle has to be a closed cycle, fuel has to be burned externally and the heat then would be transferred to the working fluid in the heat exchanger. That is, in place of a simple combustion chamber we shall also have the heat exchanger, built within the combustion chamber, and arrangement for supplying air to it for combustion of fuel.

**Example AP 8.6**

A gas turbine plant has an output of 100 MW at the generator terminals. Its data is: air compressor inlet pressure and temperature $p_1 = 1.013$ bar, $T_1 = 310$ K, compressor pressure ratio = 8, compressor efficiency = 0.85, turbine inlet pressure = 0.98 × compressor exit pressure, turbine exit pressure = 1.02 bar, C.V. of fuel = 40 MJ/kg, combustion efficiency = 0.97, mechanical efficiency = 0.97, generator efficiency = 0.98. Take $\gamma$ = 1.33 and $R$ = .287 kJ/kg for gases. Determine (a) gas flow rate, (b) fuel air ratio, (c) air flow rate, (d) thermal efficiency of the power plant, (e) overall efficiency, and (f) ideal joule cycle efficiency.

***Solution*** Given: $P$ = 100 MW, $R$ = 8, $T_1$ = 310 K, $p_1$ = 1.013 bar, $T_3$ = 1350 K, $\eta_c$ = 0.85, $p_3$ = 0.98 $X p_2$ = 7.942 bar, $p_4$ = 1.02 bar, $\eta_t$ = 0.90, $\gamma$ = 1.33 for gases, C.V. = 40 MJ/kg, $\eta_b$ = 0.98, $\eta_m$ = 0.97, $\eta_g$ = 0.98, $c_{pa}$ = 1.005 and $c_{pg}$ = 1.157 kJ/kg K.

Refer Fig. 8.11.

For air, $(\gamma - 1)/\gamma = (1.4 - 1)/1.4 = 0.2857$ and for gas,

$(\gamma - 1)/\gamma = (1.33)-1)/1.33 = 0.248$

**Compressor.** Using isentropic equation for compression process 12 with $\gamma$ = 1.4 for air and definition of isentropic efficiency, we get

$$T_2 = 605.776 \text{ K}$$

**Turbine.** For turbine $R = p_3/p_4 = 7.942/1.02 = 7.786$ and $\gamma$ = 1.33. Like compressor, we obtain $T_4$ = 865.6 K

$$\therefore\ w_c = c_{pa}(T_2 - T_1) = 297.26 \text{ kJ/kg}$$

$$w_t = c_{pg}(T_3 - T_4) = 560.44 \text{ kJ/kg}$$

$$w_{\text{net}} = w_t - w_c = 263.18 \text{ kJ/kg}$$

and

$$w_g = \eta_m \times \eta_g \times w_{\text{net}} = 250.19 \text{ kJ/kg}$$

(a) **Gas flow rate.** $m_g \times w_g = 100 \times 1000 \rightarrow m_g = 399.7$ kg/s

(b) **Fuel air ratio.** Energy and mass balance of the combustion chamber gives

$$(m_f \times \text{C.V.}) \times \eta_b + m_a c_{pa} T_2 = m_g c_{pg} T_3 \quad \text{(A)}$$

and

$$m_g = m_a + m_f \quad \text{(B)}$$

Solving these two equations, we get $f = m_f / m_a = 0.0253$

(c) **Air flow rate.** Eq. B gives $m_a = m_g/(1 + f) = 389.84$ kg/s

(d) **Thermal efficiency of the power plant** = $(m_g \times w_{\text{net}})/(m_f \times \text{C.V.})$ = 26.66%

(e) **Overall efficiency** = $P/(m_f \times \text{C.V.})$ = 25.28%

(f) **Ideal joule cycle efficiency** = $1 - (1/R)^{(\gamma - 1)/\gamma}$

$$= 44.78\% \text{ (taking } \gamma = 1.4)$$

**Example AP 8.7**

An ideal steam and gas turbine combined power cycle operates according to the following specifications. For the gas turbine plant, maximum cycle temperature = 1077°C, minimum cycle temperature = 16°C, compressor pressure ratio = 10, temperature of air leaving the steam generator = 150°C. For air $c_p$ = 1.045 kJ/kg and $\gamma$ = 1.39. For the steam turbine cycle, pressure of steam at turbine inlet = 40 bar, temperature of steam at turbine inlet = 400°C, pressure of open feed heater = 0.075 bar. Neglect pump work. Determine: (i) the thermal efficiency of gas turbine plant, (ii) the thermal efficiency of the steam plant, (iii) the thermal efficiency of the combined cycle, (iv) the flow rates of fluids in each cycle, (v) the power output of the steam plant and (vi) the available energy utilisation compared with a conventional gas turbine plant.

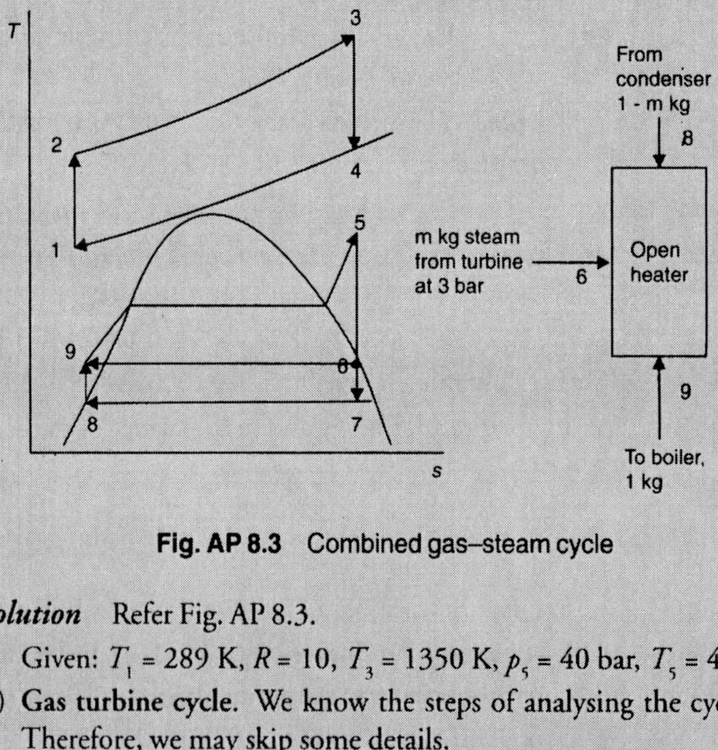


**Fig. AP 8.3** Combined gas–steam cycle

***Solution*** Refer Fig. AP 8.3.

Given: $T_1$ = 289 K, $R$ = 10, $T_3$ = 1350 K, $p_5$ = 40 bar, $T_5$ = 400°C

(i) **Gas turbine cycle.** We know the steps of analysing the cycle now. Therefore, we may skip some details.

$$T_2 = T_1\,(R)^{(\gamma-1)/\gamma} = 551.4 \text{ K and } T_4 = T_3\,(1/R)^{(\gamma-1)/\gamma} = 707.6 \text{ K}$$

$$\therefore\ w_{\text{net}} = c_p\,[(T_3 - T_4) - (T_2 - T_1)] = 397.1 \text{ kJ/kg},\ q = c_p\,(T_3 - T_2)$$

$$= 834.5 \text{ kJ/kg and } \eta_{\text{th}} = w_{\text{net}}/q = 47.59\%$$

(ii) **Steam turbine cycle.** $p_5$ = 40 bar, $T_5$ = 400°C $\rightarrow$ $h_5$ = 3215.7

$h_8 = 168.8$ and $h_9 = h_{f,\,3\,\text{bar}} = 561.5$ kJ/kg

From isentropic expansion of steam to pressure $p_6 = 3$ bar and then to $p_7 = 0.075$ bar, we can find out enthalpies at these conditions. These are $h_6 = 2636.2$ and $h_7 = 2111.5$ kJ/kg.

Steam and condensate flow to the open heater is shown in the figure. Water comes at state 8 (saturated liquid at $p_7$) from the condenser and steam at state 6 from turbine. The two mix and then flow to boiler. In T-s diagram we have ignored the details of pumping. We will also neglect pump work. Let $m$ be the mass of steam bled then the mass of condensate by mass conservation would be $1 - m$ kg. Making energy balance of the heater, we get

$$(1 - m)\, h_8 + m\, h_6 = 1 \times h_9 \rightarrow m = 0.1591 \text{ kg}$$

$$w_{ts} = 1 \times (h_5 - h_6) + (1 - m) \times (h_6 - h_7)$$

$$= 1012.7 \text{ kJ/kg},\ q = h_5 - h_9 = 2654.2 \text{ and}$$

$$\eta_{\text{th}} = 38.46\%$$

(iii) **Thermal efficiency of the combined cycle.**

$m$ = mass of air required per kg of steam = $(h_5 - h_9)/(h_4 - h_1) = 8.79$ kg.

$$Q = m \times q = 7334.1$$

$$W = m \times w_{\text{net}} + w_{ts} = 4502.65 \text{ and}$$

$$\eta_{\text{th}} = W/Q = 61.4\%$$

(v) **Power output of the steam plant.** $1 \times w_{ts} = 1012.7$ kW

(vi) **Available energy utilisation compared with a conventional gas turbine plant**

Available energy can be determined using Eq. 3.6. this gives

$b_{23}$ = available energy of heat supplied in the combustion chamber/kg of air

$$= q - T_0 (s_3 - s_2) = 834.5 - c_p\, T_0 \ln(T_3/T_2)$$

$$= 564.085 = 564.085 \text{ kJ/kg}$$

$\therefore$ percentage available energy used in gas turbine = $w_t/b_{23} = 72.5\%$

percentage available energy used in combined turbine

$$= (w_{\text{net}} + w_{ts}/m)/b_{23} = 94.36\%$$

**Example AP 8.8**

A 100 MW gas turbine plant having thermal efficiency 26.66% supplies 400 kg/s of exhaust gas at 59.6°C to the heat recovery boiler of a bottoming steam power plant. Steam is generated at a pressure of 90 bar with a pinch point of 20°C. Condenser pressure = 0.1 bar. If the gas temperature in the stack is 176°C, determine (i) capacity of the boiler in kg of steam per hour, (ii) temperature of steam at the entry to turbine and approved point temperature, (iii) steam turbine plant output, turbine efficiency and thermal efficiency, and (iv) the thermal efficiency of the combined cycle plant. Take $c_{pg}$ = 1.157 kJ/kg for the exhaust gases and dryness fraction of steam at turbine exit as $x = 0.9$.

***Solution*** Given: $P$ = 100 MW, $m_g$ = 400 kg/s, $T_4$ = 865.6 K

$\eta_{th,g}$ = 26.66%, $p_b$ = 90 bar, $p_c$ = 0.1 bar

$x = 0.9$, $t_{pp}$ = pinch point = 20°C

Refer Fig. AP 8.4 for variation of temperature in the waste heat recovery boiler.

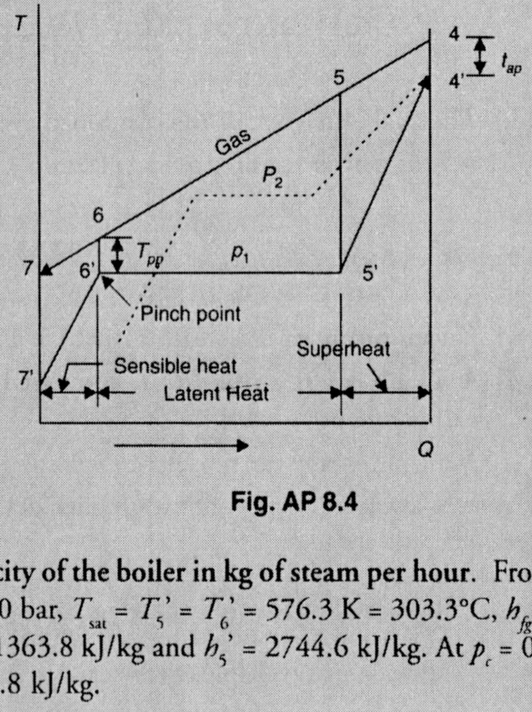


**Fig. AP 8.4**

(i) **Capacity of the boiler in kg of steam per hour.** From steam table at $p_b$ = 90 bar, $T_{sat} = T_5' = T_6'$ = 576.3 K = 303.3°C, $h_{fg}$ = 1380.8 kJ/kg, $h_6'$ = 1363.8 kJ/kg and $h_5'$ = 2744.6 kJ/kg. At $p_c$ = 0.1 bar, $h_7'$ (= $h_f$) = 191.8 kJ/kg.

Also $T_6 = T_6' + t_{pp}$ (by definition of pinch point) = 323.3°C

Considering economiser portion, energy given by gases = sensible heat taken by water.

$\rightarrow m_g c_{pg} (T_6 - T_7) = m_s (h_6' - h_7')$, where $m_s$ is mass of steam formed.

or $m_s/m_g = 0.1454 \rightarrow m_s = 58.16$ kg/s

and boiler capacity[4] = 58.16 × 3600/1000

= 209.376 tonnes/hr

(ii) **Temperature of steam at the entry to turbine and approach point.** We may similarly make energy balance for evaporator and superheater to give

$$m_g c_{pg} (T_5 - T_6) = m_s (h_5' - h_6') \rightarrow T_5 = 496.55°\text{C}$$

$$m_g c_{pg} (T_4 - T_5) = m_s (h_4' - h_5') \rightarrow h_4' = 3506.7 \text{ kJ/kg}$$

From steam table, $t_4' = 540°\text{C}$

Approach point temperature

$$t_{ap} = t_4 - t_{4'} = 52.6°\text{C}$$

(iii) **Steam turbine plant output, turbine efficiency and thermal efficiency.** We have learnt how to analyse a vapour power cycle in Chapter 7. We know condition of steam at the inlet and outlet of the turbine and therefore the enthalpies at these points. Let us represent the point after expansion in turbine by symbol '1'. From steam table at $p_c = 0.1$ bar and $x_1 = 0.9$, $h_1 = 2350$ kJ/kg.

∴ actual enthalpy drop = $w_{st} = h_4' - h_1 = 1156.7$ kJ/kg of steam.

power output = $m_s \times w_{st} = 67.27$ MW

Also, by equating $s_4'$ to $s_1$, we may get $h_1' = 2150$ kJ/kg for isentropic expansion in the turbine. this gives isentropic heat drop

$$= h_4' - h_1' = 1356.7 \text{ kJ/kg}$$

∴ $\eta_t$ = isentropic efficiency of turbine

$$= (h_4' - h_1)/(h_4' - h_1') = 0.8526$$

Heat supplied to steam = $q = h_4' - h_7' = 3314.9$ kJ/kg

$$\therefore \eta_{th} = w_{st}/q = 34.89\%$$

(iv) **Thermal efficiency of the combined cycle plant.**

For gas turbine plant,

$$P = m_g \times w_{gt} \rightarrow w_{gt} = 250 \text{ kJ/kg}$$

∴ $q_{cc}$ = Heat supplied to combined plant

$$= w_{gt}/h_{th,\,g} = 937.73 \text{ kJ/kg of air, and}$$

$\eta_t$ = thermal efficiency of combined cycle

$$= (w_{gt} + w_{st}/m_s)/q_{cc} = 44.6\%$$

4. Tonne is unit of mass and is equal to 1000 kg and t is the symbol used for it.

**Example AP 8.9**

In a combined gas–steam turbine power plant, the exhaust gas from the open gas turbine cycle is supplied to steam generator of the steam cycle at which additional fuel is burned in the gas. The pressure ratio for the gas turbine is 7.5, the air inlet temperature is 15°C and the maximum cycle temperature is 750°C. Combustion of additional fuel raised the gas temperature to 750°C, and the gas leaves the steam generator at 100°C. The steam is supplied to the turbine at 50 bar, 600°C and the condenser pressure is 0.1 bar. The total power output of the plant is 200 MW. The calorific value of the fuel is 43.3 MJ/kg. Determine (i) the flow rate of air and steam required, (ii) the power output of the gas and steam turbines, (iii) the thermal efficiency of the combined plant, and (iv) the air fuel ratio. Take $c_p$ = 1.11 kJ/kg and $\gamma$ = 1.33 for combustion gases and 1.005 kJ/kg, 1.4 for air, respectively.

***Solution*** Given for gas cycle: $R = 7.5$, $T_1 = 288$ K, $T_3 = 1023$ K,

$$T_5 = 1023 \text{ K}, T_9 = 373 \text{ K}$$

For steam cycle: $p_6 = 50$ bar, $T_6 = 600$°C, $p_7 = 0.1$ bar.

Other data, $P_t$ = Total power = 100 MW, C.V. = 43.3 MW, $c_{pg} = 1.11$ and $c_{pa} = 1.005$ kJ/kg

Refer Fig. AP 8.5 for gas and steam cycles. Figure also shows the schematic of the plant.

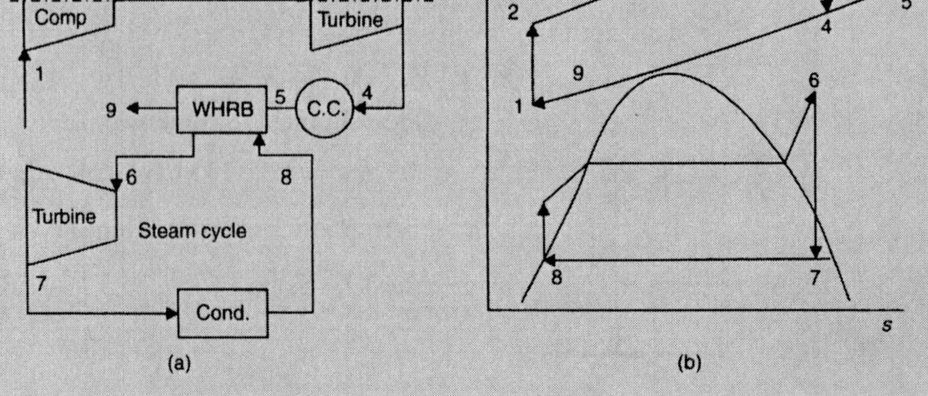


**Fig. AP 8.5** Combined cycle: (a) Block (b) T-s diagram

**Gas turbine cycle.** $T_2 = T_1\,(R)^{(\gamma-1)/\gamma} = 512$ K

$$T_4 = T_3/(R)^{(\gamma-1)/\gamma} = 620.52 \text{ K, and}$$

$$\therefore\ (w_{net})_{gt} = c_{pg}\,(T_3 - T_4) - c_{pa}\,(T_2 - T_1) = 221.47 \text{ kJ/kg of air}$$

**Steam cycle.** From steam table: $h_6 = 3670$, $h_7 = 2305$ and

$$h_8 = 192 \text{ kJ/kg}$$

$$\therefore\ (w_{net})_{st} = h_6 - h_7 = 1365 \text{ kJ/kg of steam}$$

(i) **Flow rate of air and steam required.** These can be obtained from the heat balance of the heat recovery boiler and the total power output of the plant. Gases from the second combustion chamber enter the boiler at condition 5 and leaves at 9 whereas water enters at condition 8 and leaves as steam at 6. Let $m_a$ and $m_s$ be the mass flow rate of gas and water through their respective lines. Energy balance of the recovery boiler gives

$$m_a\,c_{pg}\,(T_5 - T_9) = m_s\,(h_6 - h_8) \qquad \text{(A)}$$

There are two unknowns in Eq. A, namely $m_a$ and $m_s$. We obtain another equation from the total power output of the plant. This gives

$$P_t = 200 \times 1000 = m_a \times (w_{net})_{gt} + m_s \times (w_{net})_{st} \qquad \text{(B)}$$

These two equations give, $m_a = 398.34$ and $m_s = 81.9$ kg/s

(ii) **Power output of the gas and steam turbines.** Power output from

steam plant = $m_s \times (w_{net})_{st} = 111.78$ MW and

gas plant = 20 – 111.78 = 88.22 MW

(iii) **Thermal efficiency of the combined plant.**

$Q$ = Heat added to the cycle

$$= m_a\,[(c_{pg}\,T_3 - c_{pa}\,T_2) + c_{pg}\,(T_5 - T_4)] = 400.2 \text{ MW}$$

$$\therefore\ \eta_{th} = Pt/Q = 50\%$$

(iv) **Air fuel ratio.** This is obtained by making energy and mass balance of the first combustion chamber. This results in A/F = 42.7

**Example AP 8.10**

The following data refer to a combined cycle plant. Inlet condition of air to the compressor = 1 bar, 25°C, compressor pressure ratio = 8, maximum temperature at inlet to the gas turbine = 900°C, pressure drop in combustion chamber = 3% of compressor delivery pressure, compressor efficiency = 0.88, turbine efficiency = 0.88,
calorific value of fuel = 44.43 MJ/kg, fuel used = liquid octane ($C_8H_{18}$), specific heat of air and gas = 1.006 and 1.148 kJ/kg K, specific heat ratio

for air and gas = 1.33 and 1.4, condition of steam at inlet to steam turbine = 40 bar, 425°C, condenser pressure = 0.04 bar, feed water temperature to WHRB 170.4°C, steam turbine efficiency = 0.82, pressure drop in WHRB = 0.05 bar, steam flow rate = 29.235 kg/s. Determine (i) percentage of excess air, (ii) total power output and overall efficiency, (iii) lost coefficient, and (iv) exergetic efficiency. Also, draw exergy balance sheet. Assume $\psi = \Delta G_0/\Delta H_0 = 1.0401 + 0.1728\ (h/c)$, where $h/c$ is mass ratio of hydrogen to carbon in fuel and pitch point temperature difference is 30°C.

***Solution*** Given for gas cycle: $p_1$ = 1 bar, $R = 8 \rightarrow p_2$ = 8 bar, $T_1$ = 298 K, $T_3$ = 1173 K, $p_3 = 0.97 \times p_2$ = 7.76 bar, $p_4$ = 1 + 0.05 = 1.05 bar,

$$\eta_c = \eta_t = 0.88$$

For steam cycle: $p_6$ = 40 bar, $T_6$ = 425°C, $p_7$ = 0.04 bar,

$$\eta_{st} = 0.82,\ m_s = 29.235\text{ kg/s}$$

Other data: C.V. = 43.3 MW, $c_{pg}$ = 1.148 and $c_{pa}$ = 1.006 kJ/kg, fuel $C_8H_{18}$.

Refer Fig. AP 8.6 for gas and steam cycles.

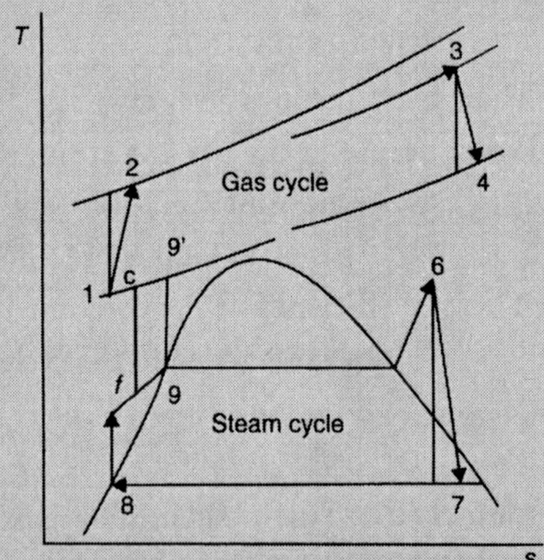


**Fig. AP 8.6** Combined cycle

**Gas turbine cycle.** We have been determining temperature at the end of compression and expansion processes by using isentropic process equation and isentropic efficiency relation. Determining value of the index $n$ of the process can also do this. For an adiabatic process, $n$ is related to $\gamma$ by the relations,

$(n-1)/n = (\gamma - 1)/(\gamma\,\eta_c)$ for compressor and $(\gamma - 1)\,\eta_t/\gamma$ for turbine and also $T_2/T_1 = (p_2/p_1)^{(n-1)/n}$, etc.

This gives using appropriate value of $\gamma$ and efficiencies

$$T_2 = T_1 (R)^{(\gamma-1)/(\gamma \eta_c)} = 586 \text{ K}$$

$$T_4 = T_3 (p_4/p_3)^{(\gamma-1)\eta_t/\gamma} = 755 \text{ K}$$

**Steam cycle.** Using steam table and the procedure, discussed before, for expansion in turbine, we get, $h_6 = 3272$, $h_7 = 2064.37$ and $h_8 = 192$ kJ/kg, $T_9 = 250.4$°C, $h_9 = 1087$ kJ/kg. Since pinch point temperature difference is given as 30°C, temperature of gas in WHRB at the point 9' corresponding to point 9 on the steam cycle will be

$$T_9' = 250.4 + 30 = 280.4°\text{C}$$

**Combustor.** Let us first find out the air fuel ratio so that the mass flow rate in the two cycles can be determined. Let 1 kg of gases be flowing over the gas turbine blades and $f$ kg of fuel per kg of gases formed in combustor is supplied. Then, mass balance of the combustor gives

$m_a$ = mass of air = $1 - f$ kg. Energy balance yields

$$f \times \text{C.V.} + (1-f)\, h_2 = 1 \times h_3 \qquad \text{(A)}$$

It is common to write for air, $h = c_p T$, where, $T$ is the temperature and $c_p$ is specific heat. However it should be remembered that two assumptions are hidden in this, which are (i) enthalpy is taken 0 at absolute zero of temperature and (ii) specific heat remains constant from 0 to $T$. Since we have to determine exergy, the dead state $T_0$ is to be taken reference state. Enthalpy can then be written as $h = c_p (T - T_0)$. Using this, we can write Eq. A as follows

$$f \times \text{C.V.} + (1-f)\, c_{pa} (T_2 - T_0) = 1 \times c_{pg} (T_3 - T_0)$$

where $T_0 = 298$ K. $\rightarrow f = 0.0.12 \rightarrow$ AF = 60.73

(i) **Percentage of excess air.** We may now find it out by writing chemical equation and determining stoichiometric air fuel ratio, as described in Chapter 6. The chemical reaction is

$$C_8H_{18} + 12.5\, O_2 \rightarrow 8\, CO_2 + 9\, H_2O$$

Stoichiometric air fuel ratio = (12.5 × 32 × 4.32)/(8 × 12 + 18 × 1)
= 15.12

$\therefore$ percentage excess air = (60.73 – 15.12)/15.12 = 302%

(ii) **Total power output and overall efficiency.** Energy balance of the recovery boiler gives

$$m_g c_{pg} (T_4 - T_9') = m_s (h_6 - h_9) \rightarrow m_g = 275.8$$

$$m_f = f \times m_g = 4.466 \text{ and}$$

$$m_a = (1-f) m_g = 271.3 \text{ kg/s, since } f = m_f / m_g$$

$\therefore$ $P$ = Power output

$$= m_g c_{pg} (T_3 - T_4) - m_a (T_2 - T_1) + m_s (h_6 - h_7) = 82607 \text{ kW.}$$

and overall efficiency = $P/(m_f \times \text{C.V.})$ = 41.63%

(iii) **Lost coefficient.** In Chapter 7, we discussed binary vapour cycle and combined cycles wherein we assumed that there is no loss or gain of heat between the two cycles. However, in actual implementation there is always some heat loss. (In combined cycle it is also possible to add heat by burning some fuel in the exhaust from gas turbine before the waste heat recovery boiler, as done in Example AP 8.9.) In such a case, overall efficiency of the cycle as given by Eq. 7.6.1 is modified to

$$\eta = \eta_1 + \eta_2 - \eta_1 \eta_2 - \eta_2 x_L$$

where, $\eta_1$ are $\eta_2$ are the thermal efficiencies of topping and bottoming cycle and $x_L$ is the fraction of heat supplied to the topping cycle that is lost and is called heat lost coefficient. Heat is lost through stack gases and therefore, we have to find out temperature of gases leaving the recovery boiler stack. Refer to points $f$ and $c$ in Fig. AP 8.10, which correspond to the point at which water is fed to the WHRB. We are given temperature of feed water at this point, which is $T_f$ = 170.4°C. Temperature of gases at this section is say $T_c$, the temperature of the gases leaving boiler. Energy balance of the economiser yields

$$m_g c_{pg} (T_9' - T_c) = m_s (h_9 - h_f), \text{ where}$$

$$h_f \text{ at } 170.4°\text{C} \approx 4.18 \times 170.4 = 713 \text{ kJ/kg}$$

$$\therefore T_c = 247°\text{C, heat lost} = Q_L = m_g c_{pg} (T_c - 25) \text{ and}$$

$$Q_s = \text{heat supplied to topping cycle} = m_f \times \text{C.V.}$$

$$\therefore x_L = Q_L/Q_s = 0.354; \text{ significant loss}$$

Power output, $P_{st}$ = output of steam turbine

$$= m_s (h_6 - h_7) = 28863 \text{ kW}$$

and $P_{gt}$ = output of gas turbine = $m_g (h_3 - h_4)$ = 53744 kW

(iv) **Exergetic efficiency.** We shall now determine exergy loss due to irreversibility in different units. Change in entropy of gases are determined using Eq. 4.25

$$s_2 - s_1 = c_p \ln (T_2/T_1) - R \ln (p_2/p_1) \qquad (4.25)$$

*Proper value of R is to be taken since in the problem it is different for air and the gas. Care should be observed while making substitution.*

Irreversibility of a process is obtained using equation

$$I = T_0 (s_2 - s_1)$$

**Compressor.** $I_c = m_a T_0 (s_2 - s_1) = 6613$ kW

**Combustor.** Using notation *p*, *f* and a for products, fuel and air respectively and 1, 2, etc. for the state point under consideration, we can write

$$I_{\text{comb}} = T_0 [S_{p3} - S_{a2} - S_f]$$
$$= T_0 [(S_{p3} - S_{p0}) - (S_{a2} - S_{a0}) + (S_{p0} - S_{a0} - S_f)] \quad \text{(A)}$$

The first two quantities in the [ ] can be determined for the products and air using Eq. 4.25. The last term is evaluated from the knowledge of Gibbs function of the fuel, which is given to us. For fuel,

$$h/c = (18 \times 1)/(8 \times 12)$$
$$\rightarrow \psi = 1.0401 + 0.1729\,(h/c) = 1.0725, \text{ and}$$
$$\Delta H_0 = m_f \times \text{C.V.} = 198424 \text{ kW}$$
$$\therefore \; \Delta G_0 = \psi \times \Delta H_0 = 212810 \text{ kW}$$
$$= \Delta H_0 + T_0 \Delta S_0, \text{ where}$$
$$\Delta S_0 = S_{p0} - S_{a0} - S_f$$
$$\rightarrow T_0\, \Delta S_0 = T_0 (S_{p0} - S_{a0} - S_f) = 14386 \text{ kW}$$
$$S_{p3} - S_{p0} = m_g\,[c_{pg} \ln (T_3/T_0) - R \ln (p_3/p_0) \text{ and similarly}$$
$$S_{a2} - S_{a0} = m_a\,[c_{pa} \ln (T_2/T_0) - R \ln (p_2/p_0)$$

Substituting these values in Eq. A, we get, $I_{\text{comb}} = 88661$ kW

**Gas turbine.** $I_{gt} = m_g c_{pg} (s_3 - s_4) = 5646$ kW

*WHRB.* $I_{\text{WHRB}} = T_0 [m_s (s_6 - s_f) + m_g (s_4 - s_c)]$

$$s_6 - s_f = 6.853 - 2.046 = 4.807 \text{ from steam table and}$$
$$s_4 - s_c = c_{pg} \ln (T_c/T_6) - R \ln (p_1/p_4)]$$
$$\therefore \; I_{\text{WHRB}} = 7853 \text{ kW}$$

**Exhaust loss.** For this case,

$$\int_{T_c}^{T_0} (1 - T_0/T_1)\, dQ = m_g\, c_{pg}\, [(T_c - T_0) - T_0 \ln (T_c/T_0]$$
$$= 17760 \text{ kW}$$

**Steam turbine.** $I_{st} = m_s T_0 (s_6 - s_7) = 6412$ kW

We have not determined entropy at point 7 that can be obtained either showing process 67 on Mollier chart or using process equation and isentropic efficiency of turbine.

Exergy Balance

| Exergy input, kW | Power output, kW | Exergy loss, kW |
|---|---|---|
| $-\Delta G_0 = 212810$ | $P_{gt} = 53744$<br>$P_{st} = 28863$ | Compressor: 6613<br>Combustor: 88661<br>Gas turbine: 5646<br>WHRB: 7853<br>Exhaust loss: 17760<br>Steam turbine: 6412 |
| Input = 212810 | Output = 82607 | Total losses: 132945 |

Exergetic efficiency = output/input = 82607/212810 = 38.8%.

**Example AP 8.11**

A turbo-jet aircraft is flying at a speed of 792 km/h at an altitude where ambient pressure is 0.799 bar and the temperature is 25°C. The compressor pressure ratio is 4 and the maximum temperature is limited to 860°C. Find the specific fuel thrust produced in a nozzle, using the following data: efficiency of ram compression = 90%, efficiency of compressor = 80%, efficiency of turbine = 85%, nozzle efficiency = 93%, transmission efficiency = 98%.

***Solution*** Given: $p_a = 0.799$ bar, $T_a = 298$ K, $C_a = 792$ km/h = 220 m/s, $\eta_r = 0.9$, $R = 4$, $\eta_c = 0.8$, $T_{03} = 1133$ K, $\eta_t = 0.85$, $\eta_n = 0.93$ and $\eta_{tr} = 0.98$.

Refer Fig. AP 8.7, which shows different zones of a jet aircraft.

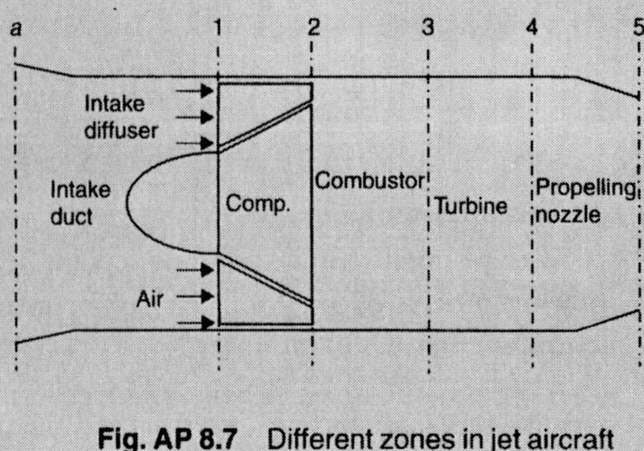


**Fig. AP 8.7** Different zones in jet aircraft

**Diffuser.** In this section there is no change in the stagnation temperature. This means, using a prime to represent isentropic process in the diffuser,

$$T'_{01} = T_{oa} = T_a + C_a^2/(2c_p) = 346.5 \text{ K, and}$$

$$p_a/p_{01} = (T_a/T'_{01})^{\gamma/(\gamma-1)}$$

$$\rightarrow p_{01} = 1.35 \text{ bar}, p_{02} = R\, p_{01} = 5.4 \text{ bar}$$

and $\eta_r = (T'_{01} - T_a)/(T_{01} - T_a) \rightarrow T_{01} = 351.5 \text{ K}$

**Compressor.** $T'_{02} = T_{01}\,(R)^{(\gamma-1)/\gamma} = 522.3$ K and

$$\eta_c = (T_{02} - T_{01})/(T_{02} - T_{01}) \rightarrow T_{02} = 565\text{K}$$

**Turbine.** In jets, turbine is used to run the compressor. Therefore, it produces power required for the purpose. This gives

$$w_c/\eta_{tr} = w_t, \text{ or } (T_{02} - T_{01})/0.85$$

$$= (T_{03} - T_{04}) \rightarrow T_{04} = 915.1 \text{ K}$$

and $\eta_t = (T_{03} - T_{04})/(T'_{03} - T'_{04})$

$$\rightarrow T'_{04} = 876.6 \text{ K and}$$

$$T_{03}/T'_{04} = (p_{03}/p_{04})^{(\gamma-1)/\gamma} / p_{04} = 2.2 \text{ bar}$$

**Nozzle.** Assuming isentropic expansion in the nozzle throughout its length so that $p_s = p_a$, we may write

$$T'_{05} = T_{04}\,(p_a/p_{04})^{(\gamma-1)/\gamma} \rightarrow T'_{05} = 685.4 \text{ K and}$$

$$\eta_n = (T_{04} - T_{05})/(T_{04} - T'_{05}) \rightarrow T_{05} = 701.5 \text{ K}$$

$\therefore$ $Cr_5$ = Relative velocity of jet w.r.t. jet itself

$$= \sqrt{\{2c_p(T_{04} - T_{05})\}} = 655.2 \text{ m/s}$$

But $Cr_5 + (-Ca) = C_5 \rightarrow C_5 = 435.2$ m/s

(–ve sign with Ca is used since the turbo-jet moves in forward direction whereas we are measuring velocities positive in the direction of flow of air through the jet, whihc is opposite to that of motioin of jet).

$\therefore$ specific thrust = $m \times C_5 = 435.2$ N/kg

**Mach number.** $T_5 = T_{05} - C_5^2/(2c_p)$

$= 607.\ 27$ and

$a$ = sound velocity = $\sqrt{(\gamma\ RT_5)} = 494$ m/s

$\therefore$ $M$ = Mach number

$= C_{r5}/a = 1.326$

This shows that the nozzle is convergent–divergent type

**Example AP 8.12**

A jet plane having 2 jets works on turbo-jet system. It flies at a speed of 800 km/h at an altitude where density of air is 0.15 kg/m$^3$. The propulsive efficiency is 55%. The drag on the plane is 6500 N. Calculate (i) absolute velocity of jet, (ii) quantity of air compressed and (iii) diameter of jet.

***Solution*** Given: $C_a$ = 800 km/h = 222.2 m/s, $\eta_p$ = 0.55,

$\rho$ = 0.15 kg/m$^3$, $F$ = 6500 N.

Refer Fig. 8.21 (a) for the notations used.

$$\eta_p = 2C_a/(C_j + C_a) \rightarrow C_j = 585.8 \text{ m/s}$$

$\therefore$ absolute velocity of jet = $C_j - C_a$ = 363.6 m/s

Thrust = $m \times C_j$ = 6500 (given)

$\rightarrow m$ = mass of air = 17.88 kg/s

$Q$ = discharge from nozzle = 17.88/0.15

= 119.2 m$^3$/s = $2 \times A \times C_j$

$\rightarrow A$ = 0.1018 m$^2$ and

$d$ = nozzle diameter = 0.36 m

**Example AP 8.13**

A turbo-jet engine takes 50 kg/s of air and propels an aircraft with a uniform flight of 800 km/h. Isentropic enthalpy change for nozzle is 210 kJ/kg and its velocity coefficient is 0.95. The fuel air ratio is 1.2%, combustion efficiency is 95% and LCV of fuel is 42000 kJ/kg. Determine: (i) propulsive power, (ii) fuel flow in kg/h and thrust specific fuel consumption, (iii) thrust power, propulsive efficiency and overall efficiency. Neglect mass of fuel wherever necessary.

***Solution*** Given: $C_a$ = 800 km/h = 222.2 m/s, $C_v$ = 0.95,

$\Delta h_{i,n}$ = 210 kJ/kg, $m$ = 50 kg/s, f/a = 0.012

Velocity of jet = $C_j = C_v \times \sqrt{(2000 \times \Delta h_{i,n})}$ = 615.7 m/s

$\therefore$ Propulsive power = $m\,(C_j^2 - C_a^2)/2000$

= 8242.8 kW

$Fs$ = specific thrust = $C_j - C_a$ = 393.5 N/kg and

$m_f$ = fuel consumed per hour = $(f/a) \times m \times 3600$ = 2160 kg/h

$\therefore$ Thrust specific fuel consumption = $m_f/(Fs \times m)$ = 0.1098 kg/h/N

Thrust power = $m\,C_a\,(C_j - C_a)$ = 4371.8 kW

$\therefore$ Propulsive efficiency = Thrust power/Propulsive power = 53%

$\eta_0$ = overall efficiency = Thrust power/heat supplied

= 4371.8/(50 × 0.012 × 42000 × 0.95) = 18.3%

## Chapter 9

### Example AP 9.1

Calculate the rms speed of $O_2$ molecules at 300 K. What is the mean translational kinetic energy?

***Solution*** Given: $T$ = 300 K, $No = 6.023 \times 10^{23}$, $M$ = mole,

Weight of $O_2$ = 32. Also $k = 1.38 \times 10^{-23}$ J/molecule K

$m$ = mass of $O_2$ molecule = $M/No = 5.32 \times 10^{-26}$ kg

Root mean square velocity is given by Eq. 9.27

$$C_{rms} = \sqrt{(3kT/m)} = \sqrt{\{3 \times 1.38 \times 10^{-23} \times 300/(5.32 \times 10^{-26})\}}$$

= 483 m/s

Mean translational kinetic energy is (by Eq. 9.33) = 3/2 kT

= $6.21 \times 10^{-21}$ J/molecule

### Example AP 9.2

A cylinder containing $O_2$ at 300 K, 1 atm is placed in an evacuated chamber. If a hole of area 0.01 mm² is made in the cylinder, calculate the number of molecules leaking through the hole per second. Take the average velocity of molecules as 0.925 of the rms speed.

***Solution*** Given: $T$ = 300 K, $No = 6.023 \times 10^{23}$, $\bar{C} = 0.925 \times C_{rms}$, $p$ =1 bar, $M$ = mole. Weight of $O_2$ = 32, k = $1.38 \times 10^{-23}$ J/molecule K

$m$ = mass of $O_2$ molecule = $M/No = 5.32 \times 10^{-26}$ kg.

Root mean square velocity is given by Eq. 9.27.

$$C_{rms} = \sqrt{(3KT/m)} = \sqrt{\{3 \times 1.38 \times 10^{-23} \times 300/(5.32 \times 10^{-26})\}}$$

= 483 m/s

$\therefore \bar{C} = 0.925 \times C_{rms}$ = 447 m/s

Number of molecules per m³ is obtained from equation of state,

$n = pV/(kT) = 2.46 \times 10^{25}$ molecules/m³

Rate of flux of molecules per unit area (Eq. 9.44)

$= 2.75 \times 10^{27}$ molecules/m$^3$-s

$\therefore$ rate of leakage of $O_2$ molecules through hole,

$= 2.75 \times 10^{27} \times 0.01 \times 10^{-6} = 2.75 \times 10^{19}$ molecules/s

**Example AP 9.3**

Determine the fraction of molecules of a gas having velocities for which $f$ lies between 29.5° and 30.5°, while $\theta$ lies between 44.5° and 45.5°.

***Solution*** Given: $\Delta\theta = \Delta\phi = 1^0 = \pi/180$ rad, $\theta = 45°$.

Eq. 9.3 gives population of all those molecules having any velocity but moving in the direction given by the two limits on $\theta$ and $\phi$ given in the question.

$$d^2 n_{\theta\phi} = (n/4\pi) \sin\theta \cdot d\theta \cdot d\phi$$

Since $d\theta$ and $d\phi$ are small, there is no need to integrate the equation; rather, these can be replaced by $\Delta\theta$ and $\Delta\phi$, respectively. Substitution gives

$$d^2 n_{\theta\phi}/n = 1.714 \times 10^{-5}$$

**Example AP 9.4**

Determine fraction of molecules having $x$ component velocities in the range $C_{mp}$ and 1.6 $C_{mp}$. Also, determine the fraction of molecules having positive $x$-component of velocities greater than 1.2 $C_{mp}$.

***Solution*** Eq. 9.14, 9.19 and 9.22, when combined, the Maxwell–Boltzman distribution of the $x$ component of velocities is given by

$$dN_{Cx}/N = f(C_x)\, dC_x, \text{ where } f(C_x)$$

$$= \alpha \exp(-\beta C_x^2),\ \alpha = \sqrt{(\beta/\pi)} \text{ and } \beta = m/(2\text{kT})$$

The required fraction is given by

$$f = \frac{1}{\sqrt{\pi}} \int_{C_{mp}}^{1.6C_{mp}} \frac{f(C_x)\, dC_x}{N}$$

$$= \frac{1}{\sqrt{\pi}} \int_{C_{mp}}^{1.6C_{mp}} \left(\frac{m}{2KT}\right)^{1/2} \exp\left(-\frac{mC_x^2}{2kT}\right) dC_x$$

$= 1/2$ (erf (1.6) – erf (1)} = 0.0668

where, erf ($x$) is the error function, whose values are given in tabular form for different values of $x$, defined as follows:

$$\text{erf}(x) = \sqrt{\pi}/2 \int_0^x x^2 \exp\left(-x^2\right) dx$$

*Fraction of molecules having positive x-component of velocities greater than 1.2 $C_{mp}$*

We may determine fraction of molecules having $x$-component of velocities from 0 to 1.2 $C_{mp}$ as above by carrying out the integration from 0 to 1.2 $C_{mp}$. This comes to 0.4552. However, the fraction of molecules having positive $x$ component of velocities is ½ that varies from 0 to $\infty$.

$\therefore$ fraction of molecules having positive $x$-component greater than 1.2 $C_{mp}$ = 0.5 – 0.4552 = 0.0448.

**Example AP 9.5**

Determine the fraction of molecules of a gas having speed between $C_{mp}$ and 1.6 $C_{mp}$.

***Solution*** This problem is similar to above problem except that now we have to consider molecules in any direction, since speed is a scalar quantity. For this, the Maxwell–Boltzman distribution function is given by Eq. 9.21,

$$dn_c/dC = 4\pi\ N\ \alpha^3 C^2 \exp\left(-\beta\ C^2\right)\ dC = f(C)\ dC$$

The required fraction is given by

$$f = \frac{4}{\pi} \int_{C_{mp}}^{1.6C_{mp}} \left(\frac{m}{2kT}\right)^{3/2} C^2 \exp\left(-\frac{mC^2}{2kT}\right) dC \qquad \text{(A)}$$

To evaluate the integral, let us assume $x = C/C_{mp}$ or $dC = C_{mp}\ dx$. Substituting it in (A) and using $C_{mp} = \sqrt{(2kT/m)}$ from Eq. 9.28, we get after simplification

$$f = \frac{4}{\sqrt{\pi}} \int_1^{1.6} x^2 \exp\left(-x^2\right) dx = \frac{4}{\sqrt{\pi}} \int_0^{1.6} x^2 \exp\left(-x^2\right) dx$$

$$-\frac{4}{\sqrt{\pi}} \int_0^{1} x^2 \exp\left(-x^2\right) dx$$

$$\therefore f = 2\left[\text{erf}(1.6) - \text{erf}(1)\right] = 0.2673$$

**Example AP 9.6**

The radius of oxygen molecule is 0.18 nm. Calculate the collision cross-section, the collision frequency and the mean free path for oxygen at 0.01 N/m$^2$ and 500 K based on simple kinetic theory of gases. Also, find (a) fraction of the free paths longer than 1.5 m and (b) fraction of the free paths longer than 1.5m but shorter than 2m?

***Solution*** $d = 0.18$ nm $\rightarrow r = 0.09$ nm $= 0.09 \times 10^{-9}$ m, $p = 0.01$ N/m$^2$, $No$ = Avagadro's number = $6.023 \times 10^{23}$, $R = 8.314$ kJ/kmol K, $T = 500$ K, $y$ = distance travelled by molecules without collision = 1.5m

**Collision cross-section** = $\sigma = 4\pi r^2 = 0.4072 \times 10^{-18}$

**Collision frequency.**

$$n = \text{No. of molecules/m}^3 = p\,No/(RT)$$
$$= 0.01 \times 6.023 \times 10^{23}/(8.314 \times 500)$$
$$= 1.489 \times 10^{18} \text{ molecules/m}^3$$

From Eq. 9.26,

$$\bar{C} = \sqrt{[8kT/(\pi m)]}, \text{ where } k = R/No \text{ and } m = M/No$$
$$\therefore \bar{C} = \sqrt{[8RT/(\pi M)]} = 575.2 \text{ m/s}$$

From Eq. 9.38,

$$Z = \text{collision frequency} = \sigma = 4\pi r^2 = 339 \text{ collisions/s}$$

*Mean free path* = $\lambda$ (from Eq. 9.38) = $1/(\sigma n) = 1.695$ m

(a) **Fraction of the free paths longer than 1.5 m.** Fraction of molecules after travelling distance 1.5 m is given by Eq. 9.42.

$$N/No = \exp(-y/\lambda)$$
$$= \exp(-1.5/1.695) = 0.4127$$

(b) **Fraction of the free paths longer than 1.5m but shorter than 2m.** we may similarly find the fraction of molecules, which survive after travelling distance 2m; this comes out to 0.3073.

$\therefore$ fraction that collides with each other during 1.5 to 2 m is

$0.4127 - 0.3073 = 0.1054$

**Example AP 9.7**

The mean free path in a certain gas at a temperature of 25°C is $2.63 \times 10^{-5}$ m. the radius of the molecule is $2.56 \times 10^{-10}$ m. (i) Find the pressure of the gas. (ii) Calculate the number of collisions made by a molecule per meter of path.

***Solution*** Given $r = 2.56 \times 10^{-10}$ m, $T = 298$ K, $\lambda = 2.63 \times 10^{-5}$ m

Collision cross-section, $\sigma = 4\pi r^2 = 82.36 \times 10^{-20}$ m$^2$.

From Eq. 9.3.6, assuming Maxwell–Boltzman velocity distribution,

$$n = 1/\sqrt{(\sigma\lambda)} = 0.326 \times 10^{23} \text{ molecules/m}^3.$$

(i) **Pressure of the gas.** From equation of state, we get

$$p = n\,k\,T = 0.326 \times 10^{23} \times 1.38 \times 10^{-23} \times 298$$
$$= 134.06\ Pa.$$

(ii) **Number of collisions made by a molecule per meter of path is**

$$1/\lambda = 3.8 \times 10^4.$$

**Example AP 9.8**

Assuming a triatomic molecule to be a triangular structure with each atom at one apex, determine the specific heat ratio of a gas consisting of triatomic molecules, with the help of principle of equipartition of energy.

***Solution*** Such a molecule will have three translational and three rotational degrees of freedom. In addition, each atom will have freedom of vibration along the line joining it to the other atom. Thus, in all there will be six degrees of freedom of vibration and 3 + 3 + 6 = 12 degrees in total. By the principle of equipartition of energy, internal energy of 1 mol of gas is

$$U = 12 \times 1/2 \times No\,kT$$
$$= 6\,n\,\overline{R}\,T \rightarrow u = 6\,\overline{R}\,T$$
$$C_v = 6\,\overline{R},\ C_p = 7\,\overline{R} \text{ and } \gamma = C_p/C_v = 7/6 = 1.167$$

**Example AP 9.9**

Calculate the coefficient of viscosity and thermal conductivity of $O_2$ at 1 atm pressure and 300 K.

***Solution*** Given $p$ = 1 atm, T = 300 K, $O_2$ is diatomic gas for which $f$ = degree of freedom = 5.

For oxygen, $m$ = mass of molecule = 5.31 × 10.26 kg/molecule, $r$ = 0.15 nm.

We may determine average velocity using Eq. 9.26 to give

$\overline{C}$ = 445 m/s, and

collision cross-section = $4\pi r^2 = 3.84 \times 10^{-19}$

From Eq. 9.48 and 9.38, $\mu$ = 1/3 mn, $\overline{C}\,\lambda$ = 1/3 m

$\overline{C}\,\sigma = 2.05 \times 10^{-5}$ kg/m-s = $2.05 \times 10^{-5}$ Ns/m$^2$

Similarly from Eq. 9.49, $k = \overline{C}\ kl/(6\sigma)$

$= 445 \times 5 \times 1.38 \times 10^{-23}/(6 \times 3.84 \times 10^{-19})$

$= 0.0133$ W/m K

**Example AP 9.10**

Viscosity of $O_2$ at 273 K is 19.2 $\mu$ N-s/m$^2$. Estimate the hard sphere radius of $O_2$ molecule.

***Solution*** We may assume Maxwellian distribution of velocities in which case the mean free path is given by Eq. 9.39 whereas viscosity is given by Eq. 9.48. Combining these two equations, we get

$$\mu = m\ \overline{C},\ \left(3\sqrt{2 \times \sigma}\right) \text{ where } m = M/No \qquad \text{(A)}$$

But $\overline{C} = \sqrt{[8RT/(\pi M)]} = 425$ m/s. Substituting values in Eq. A, we get

$$\sigma = 32 \times 425/[6.023 \times 10^{23} \times 3\sqrt{2} \times 19.2 \times 10^{-6}]$$

$$= 0.2772 \times 10^{-19}\ \text{m}^2.\ \text{But } \sigma = 4\pi r^2$$

$r$ = radius of $O_2$ molecule = 0.15 nm

**Example AP 9.11**

A spherical satellite is moving in the outer fringes of the earth's atmosphere where the molecular free paths are very much large compared to the satellite radius $r$. By treating the satellite as moving with velocity $u$ through a space with particle density $n$ of stationary molecules of mass $m$, derive an expression for the drag due to elastic collision with these molecules of the rarefied gas.

***Solution*** We may consider collision of particles with a wall with velocity $u$. this can be brought to rest by attaching the reference frame with the satellite. Due to elastic collision, the relative velocity before and after collision will be same. The loss of kinetic energy in one collision is (see Fig. AP 9.7 and Section 9.3.1)

$$\tfrac{1}{2}\, m\, (C\cos\theta)^2 - \tfrac{1}{2}\, (C\cos\theta - 2u)^2 = 2\, mCu\cos\theta - 2\, m\, u^2$$

$$\iiint_{\phi\ \theta\ C} \frac{1}{4\pi}\left[C dn_C \sin\theta\cos\theta\, d\theta\, d\phi\right]\left[2mCu\cos\theta - 2mu^2\right]$$

$$= \frac{mu}{2\pi}\left[\int_0^\infty C^2 dn_C \int_0^{\pi/2} \sin\theta \cos^2\theta \, d\theta \int_0^{2\pi} d\phi\right]$$

$$= 1/3mn\overline{C}^2u - 1/2mn\overline{C}u^2$$

∴ The drag in the satellite =

$$\frac{\left[1/3mn\overline{C}^2u - 1/2mn\overline{C}u^2\right]4\pi r^2}{u} = 2\pi mnr^2\left[\frac{2}{3}\overline{C}^2 - u\overline{C}\right]$$

**Example AP 9.12**

Consider a system consisting of a large number of particles at 300 K. The permissible energy levels of the particles is $3\varepsilon$, $6\varepsilon$, $9\varepsilon$ and $9\varepsilon$, where $\varepsilon$ = 300k. The degeneracies of these energy levels are 1, 3, and 3, respectively. Determine the fractional distribution of the particles among the energy levels at 300 K.

***Solution*** $T$ = 300 K

Eq. 9.70 gives partition function

$$Z = g_i \exp-(\beta\varepsilon_i), \text{ where } \beta = 1/(\text{kT}) \qquad \text{(Eq. 9.67)}$$

$$= \exp\{-3 \times 300\text{k}/(300\text{k})\} + 3\exp\{-6 \times 300\text{k}/(300\text{k})\}$$

$$+ 3\exp\{-9 \times 300\text{k}/(300\text{k})\}$$

$$= 4.9787 \times 10^{-2} + 7.4363 \times 10^{-3} + 3.7023 \times 10^{-4}$$

$$= 5.7594 \times 10^{-2}$$

Assuming MB distribution, Eq. 9.67 gives the distribution of particles in different energy levels. This gives

$n_1/N = 4.9787 \times 10^{-2}/(5.7594 \times 10^{-2}) = 0.8644$

$n_2/N = 7.4363 \times 10^{-3}/(5.7594 \times 10^{-2}) = 0.1291$, and

$n_3/N = 3.7023 \times 10–^{4}/(5.7594 \times 10^{-2}) = 0.0065$

# APPENDIX THREE

# List of Formulae

## Chapter 2

$$\oint \delta q \propto \oint \delta w \text{ or } \oint \delta w = J \oint \delta q \text{; } J = 1 \text{ in SI units} \quad (2.1)$$

$$\int_{2B}^{1} (\delta q - \delta w) = \int_{2C}^{1} (\delta q - \delta w) \quad (2.2)$$

$$\delta Q - \delta W = dE \quad (2.3)$$

$$\delta Q - \delta W = dU \quad (2.4)$$

$$m_1 e_1 + Q = m_2 e_2 + W + dE_\sigma \text{ or } Q - W = m_2 e_2 - m_1 e_1 + dE_\sigma \quad (2.5)$$

$$(Q - W)/m = e_2 - e_1 = de \text{ or, } q - w = d(h + ke + pe) \cong dh \quad (2.6)$$

## Chapter 3

$$Q_2/Q_1 = T_2/T_1 \text{ or, } Q_1/T_1 = Q_2/T_2 \quad (3.1)$$

$$dS = (\delta Q/T)_{\text{rev.}} \quad (3.2)$$

$$(\Delta S)_{\text{universe}} = (\Delta S)_{\text{system}} + (\Delta S)_{\text{surrounding}} > 0 \quad (3.5)$$

$$dS_\sigma = m_1 s_1 - m_2 s_2 + \sum \delta Q/T \quad (3.6)$$

$$\sum \delta Q/T = m_2 s_2 - m_1 s_1 \text{ for steady flow} \quad (3.7)$$

$$\delta Q/T = \delta Q_0/T_0 \quad \text{or,} \quad \delta Q = T_0\, \delta Q_0/T = T_0\, dS, \quad (3.8)$$

$$\delta W_{\text{rev}} = \delta W - p_0\, dV = -d(E + p_0 V - T_0 S), \quad (3.9)$$

$$\phi = U + p_0 V - T_0 S \quad (3.10)$$

and $$A_c = (\phi - \phi_0) + KE + PE \cong (\phi - \phi_0)$$

$$b = h - T_0 s \quad (3.12)$$

$$W_{rev} = b_1 - b_2 \quad (3.13)$$

$$\text{Exergy} = h - T_0 s + C \quad (3.15)$$

$C$ = 438.338 kJ/kg for air and 5.02416 kJ/kg for water

## Chapter 4

$p V = m R T$ (mass form) or $p V = n R_u T$, (mole form) (4.1)

$c_p - c_v = R$ (Mayer's relation) (4.2)

$$dU = m c_v dT \quad (4.3)$$

$$(p + a/v^2)(v - b) = \bar{R} T \quad (4.4)$$

$v_c = 3b$, $p_c = a/(27 b^2)$ $T_c = 8 a/(27 b \bar{R})$ and $p_c v_c/(\bar{R} T_c) = 3/8$ (4.5)

$$pv = Z \bar{R} T \quad (4.6)$$

$$\mu = (\beta T - 1) v/c_p \quad (4.24)$$

$$s_2 - s_1 = c_v \ln (T_2/T_1) + R \ln(v_2/v_1), \quad s_2 - s_1 = c_p \ln (T_2/T_1) - R \ln(p_2/p_1)$$

or $$s_2 - s_1 = c_p \ln (v_2/v_1) + c_v \ln(p_2/p_1) \quad (4.25)$$

## Chapter 5

$m_{fi}$ = mass fraction = $m_i/\sum m_i$ and $x_i$ = mole fraction = $n_i/\sum n_i$ (5.1)

$$p_i/p = n_i/n = x_i \quad (5.3)$$

$$p = \sum p_i \quad (5.4)$$

## Chapter 7

Work ratio, $WR = (w_e - w_c)/w_e$ (7.1)

$P = m_s \times w_{net}$ and ssc = $m_s \times 3600/P = 3600/w_{net}$ (7.2)

$f = m_f/m_a$ and sfc = $m_f \times 3600/P = (m_f \times 3600)/(m_a \times w_{net}) = 3600 f/w_{net}$ (7. 3)

$\therefore$ $\eta$ = Cycle efficiency = $w_{net}/q$ = 3600/(sfc × CV) (7.4)

Since, $V_1 - V_2 = V_s$, $V_1/V_2 = r$ and $V_2/V_s = c$, $r = (c + 1)/c$ (7.8)

$\therefore$ Rankine efficiency = $(h_2 - h_3)/(h_2 - h_{4'})$ (7.12)

$$\eta = 1 - Q_3/Q_1 = 1 - \{Q_2(1 - \eta_1)\}/Q_1$$
$$= 1 - (1 - \eta_1)(1 - \eta_2)$$
$$= \eta_1 + \eta_2 - \eta_1\eta_2 \tag{7.13}$$

## Chapter 8

$$T_2/T_1 = T_3/T_4 = r^{(\gamma - 1)} \text{ or } T_4/T_1 = T_3/T_2$$

$$\text{and} \quad \eta_0 = 1 - T_1/T_2 = 1 - 1/r^{(\gamma - 1)} \tag{8.1}$$

$$\eta_d = 1 - \frac{1}{\gamma}\frac{\rho^\gamma - 1}{\rho - 1}\frac{1}{r^{\gamma - 1}} \tag{8.2}$$

$$\eta_{dc} = 1 - \frac{\alpha_1 \rho^\gamma - 1}{(\alpha_1 - 1) + \gamma\alpha_1(\rho - 1)} = \frac{1}{r^{\gamma - 1}} \tag{8.3}$$

$\rho$ = Cut-off ratio, $\alpha_1$ = pressure ratio = $p_3/p_2$

# APPENDIX FOUR

# Answers to Multiple-choice Questions

## Chapter 1

| | | | | |
|---|---|---|---|---|
| 1.1 (b) | 1.2 (b) | 1.3 (c) | 1.4 (c) | 1.5 (c) |
| 1.6 (a) | 1.7 (c) | 1.8 (c) | 1.9 (a) | |

## Chapter 2

| | | | | |
|---|---|---|---|---|
| 2.1 (b) | 2.2 (a) | 2.3 (d) | 2.4 (a) | 2.5 (d) |
| 2.6 (c) | | | | |

## Chapter 3

| | | | | |
|---|---|---|---|---|
| 3.1 (d) | 3.2 (b) | 3.3 (a) | 3.4 (b) | 3.5 (d) |
| 3.6 (a) | 3.7 (b) | 3.8 (c) | 3.9 (c) | 3.10 (i) [a] |
| 3.10 (ii) [c] | 3.10 (iii) [b] | 3.10 (iv) [a] | 3.10 (v) [a] | 3.10 (vi) [d] |
| 3.11 (b) | 3.12 (d) | 3.13 (a) | 3.14 (a) | 3.15 (a) |
| 3.16 (a) | | | | |

## Chapter 4

| | | | | |
|---|---|---|---|---|
| 4.1 (d) | 4.2 (d) | 4.3 (c) | 4.4 (a) | 4.5 (c) |
| 4.6 (c) | 4.7 (a) | 4.8 (a) | 4.9 (d) | 4.10 (b) |
| 4.11 (b) | 4.12 (b) | 4.13 (c) | 4.14 (b) | 4.15 (a) |
| 4.16 (b) | 4.17 (b) | 4.18 (b) | 4.19 (a) | 4.20 (b) |
| 4.21 (b) | | | | |

## Chapter 5

| | | | | |
|---|---|---|---|---|
| 5.1 (d) | 5.2 (a) | 5.3 (b) | 5.4 (c) | 5.5 (b) |
| 5.6 (c) | 5.7 (c) | 5.8 (b) | 5.9 (d) | 5.10 (a) |
| 5.11 (b) | 5.12 (b) | 5.13 (c) | 5.14 (c) | 5.15 (b) |
| 5.16 (c) | 5.17 (c) | 5.18 (a) | 5.19 (a) | 5.20 (d) |
| 5.21 (d) | 5.22 (a) | 5.23 (b) | 5.24 (b) | 5.25 (b) |

## Chapter 6

| | | | | |
|---|---|---|---|---|
| 6.1 (c) | 6.2 (d) | 6.3 (b) | 6.4 (a) | 6.5 (c) |
| 6.6 (a) | 6.7 (b) | 6.8 (c) | 6.9 (d) | 6.10 (b) |
| 6.11 (d) | 6.12 (c) | | | |

## Chapter 7

| | | | | |
|---|---|---|---|---|
| 7.1 (b) | 7.2 (b) | 7.3 (b) | 7.4 (b) | 7.5 (a) |
| 7.6 (b) | 7.7 (b) | 7.8 (c) | 7.9 (a) | 7.10 (a) |
| 7.11 (b) | 7.12 (b) | 7.13 (a) | 7.14 (a) | 7.15 (b) |
| 7.16 (b) | 7.17 (d) | | | |

## Chapter 8

| | | | | |
|---|---|---|---|---|
| 8.1 (a) | 8.2 (d) | 8.3 (d) | 8.4 (a) | 8.5 (a) |
| 8.6 (b) | 8.7 (a) | 8.8 (a) | 8.9 (c) | 8.10 (c) |
| 8.11 (c) | 8.12 (c) | 8.13 (a) | 8.14 (a) | 8.15 (b) |
| 8.16 (b) | 8.17 (b) | 8.18 (c) | 8.19 (b) | 8.20 (b) |
| 8.21 (a) | 8.22 (b) | 8.23 (b) | | |

# Bibliography


1. Balmer, R., *Thermodynamics,* West Publ. Co., St. Paul, MN, 1990.
2. Benson, R. S., *Advanced Engineering Thermodynamics,* Pergamon, Oxford, 1967.
3. Cengel, Y. and Boles, M.A., *Thermodynamics,* Tata McGraw-Hill, New Delhi, 2002.
4. Collins, *Advanced Gas Turbines,* Power, 1993.
5. Dugalas, J.F., *An Introduction to Dimensional Analysis for Engineers,* Pitman, Boston, 1969.
6. Dugdale, J.S., *Entropy and Low Temperature Physics,* Hutchinson University Library, London, 1966.
7. Eastopop, T.D. and McConkey, A., *Applied Thermodynamics for Engineering Technologists,* Pearson, UK, 2002.
8. Giedt, W.H., *Thermophysics,* Von Nostrand Reinhold, New York, 1964.
9. Granet and Bluestein, *Thermodynamics and Heat Power,* Pearson, UK, 2000.
10. Haywood, R.W., *Analysis of Engineering Cycles,* Pergamon, Oxford, 1991.
11. Joel, R., *Basic Engineering Thermodynamics,* Longman/ELBS, 1991.
12. Kestin, J., *A Course in Thermodynamics,* Tata McGraw-Hill, New Delhi, 1979.
13. Lay, J.E., *Thermodynamics,* Pitman, Boston, 1963.
14. Mathur and Mehta, *Thermal Engineering,* Jain Bros., New Delhi, 1992.
15. Mathur and Gupta, *A New Mnemonic Method for Maxwell Relations,* J. Inst. Engrs, India.
16. Nag, P.K., *Development of Combined Cycles,* J. Inst. Engrs, India, 1976.
17. Nag, P.K., *Power Plant Engineering,* Tata McGraw-Hill, New Delhi, 2001.
18. Nash, L.K., *Elements of Statistical Thermodynamics,* Addison-Wesley, Reading, MA, 1968.
19. Rogers, G. and Mayhew, Y., *Engineering Thermodynamics,* Addison-Wesley, Reading, MA, 1999.
20. Saha, M.N. and Shrivastava, B.N., *A Treatise of Heat,* The Indian Press, Allahabad, 1950.
21. Sear, Richards and Zimansky, *Modern University Physics,* Pt I, Addison-Wesley, Reading, MA, 1960

22. Skortzki, *Basic Thermodynamics*, Tata McGraw-Hill, New Delhi, 1963.
23. Somasunderam, *Thermal Engineering*, New Age International, 1996.
24. Somasundram, S.L., *Gas Dynamics and Heat Propulsion*, New Age International, New Delhi, 1996.
25. Sontag, Borgnakke and Wylen, *Fundamentals of Thermodynamics*, Wiley, 2003.

# Index